World-Systems Evolution and Global Futures

This series seeks to promote understanding of large-scale and long-term processes of social change, in particular the many facets and implications of globalization. It critically explores the factors that affect the historical formation and current evolution of social systems, on both the regional and global level. Processes and factors that are examined include economies, technologies, geopolitics, institutions, conflicts, demographic trends, climate change, global culture, social movements, global inequalities, etc. Building on world-systems analysis, the series addresses topics such as globalization from historical and comparative perspectives, trends in global inequalities, core-periphery relations and the rise and fall of hegemonic core states, transnational institutions, and the long-term energy transition. This ambitious interdisciplinary and international series presents cutting-edge research by social scientists who study whole human systems and is relevant for all readers interested in systems approaches to the emerging world society, especially historians, political scientists, economists, sociologists, geographers and anthropologists.

This book series is indexed in Scopus.

All titles in this series are peer-reviewed.

Matthew Chidozie Ogwu ·
Sylvester Chibueze Izah

Climate Frontiers and Social Transformation

Socio-Ecological Shifts in the Age of Global Inequality

Springer

Matthew Chidozie Ogwu
Sustainable Development Department
Appalachian State University
Boone, NC, USA

Sylvester Chibueze Izah
Department of Community Medicine
Bayelsa Medical University
Yenagoa, Nigeria

ISSN 2522-0985 ISSN 2522-0993 (electronic)
World-Systems Evolution and Global Futures
ISBN 978-3-032-17638-7 ISBN 978-3-032-17639-4 (eBook)
https://doi.org/10.1007/978-3-032-17639-4

This Springer imprint is published by the registered company Springer Nature Switzerland AG
The registered company address is: Gewerbestrasse 11, 6330 Cham, Switzerland

Preface

The climate crisis has rapidly become the defining challenge of our era. It is no longer an abstract scientific concern confined to laboratories and policy documents but a lived reality shaping economies, politics, cultures, and ecosystems across the planet. Rising seas encroach upon coastal cities, prolonged droughts threaten food systems, wildfires ravage landscapes, and climate-induced displacement reshapes human geographies. Yet, behind these visible signs of planetary stress lies a more profound and more complex story—one of historical trajectories, structural inequalities, and the systemic forces shaping how societies experience, respond to, and transform under conditions of ecological change.

Climate Frontiers and Social Transformation: Socio-Ecological Shifts in the Age of Global Inequality emerges from this recognition. We began this book with a simple yet profound question: how can we understand the climate crisis not only as an environmental phenomenon but also as a structural force that fundamentally alters the long-term dynamics of the global system? Too often, climate change is treated narrowly, thus reduced to rising carbon emissions, technological fixes, or isolated policy responses. Yet, as our research and teaching have consistently demonstrated, the crisis is deeply rooted in centuries of global capitalism, imperial expansion, urbanization, and geopolitical power struggles. It amplifies existing hierarchies between core and peripheral regions, disrupts social and ecological systems, and generates both risks and opportunities for transformative change.

This book represents our attempt to weave together insights from world-systems theory, ecology, ethnoscience, and sustainability science to illuminate these intersections. Across fourteen chapters, we explore how climate change interacts with hegemonic power, economic rationalities, grassroots mobilization, and institutional responses to produce complex socio-ecological futures. We trace the historical roots of climate disruption, examine its uneven geographies, and engage critically with the limits of existing governance frameworks while highlighting the possibilities emerging from Indigenous movements, youth activism, and just energy transitions. While this book draws upon an extensive and diverse body of existing scholarship, its primary aim is not to introduce a single new theory but to integrate multiple strands of research into a coherent, system-wide framework. Much of the existing literature remains siloed, focusing independently on climate change, global inequality, corporate power, or local adaptation efforts. By applying a world-system perspective and synthesizing these domains, the book seeks to offer an original and unified understanding of the structural forces that reproduce vulnerability as well as the fragmented yet emerging responses across scales. The contribution lies in the scope, integration, and reframing of these materials rather than in presenting novel theories in isolation.

Writing this book has been both an intellectual journey and a moral imperative. We have drawn inspiration from scholars, activists, students, and community leaders who refuse to accept ecological collapse or systemic injustice as inevitable. Their struggles remind us that the future is not predetermined; it is shaped by collective choices, informed action, and the courage to imagine alternatives beyond the confines of fossil capitalism and extractive economies.

We hope this book will serve as a resource for scholars, policymakers, and practitioners seeking to understand the structural dimensions of the climate crisis and to envision equitable, post-carbon futures. Above all, we hope it invites readers to see climate change not as a distant threat but as a catalyst for rethinking the very foundations of our global social order.

Boone, USA Matthew Chidozie Ogwu
Yenagoa, Nigeria Sylvester Chibueze Izah

About This Book

Climate change has emerged as the defining crisis of the twenty-first century, reshaping ecosystems, societies, and economies at a planetary scale. Yet its impacts are far from uniform. Across the Global South, marginalized communities face the harshest consequences (from climate-induced displacement and collapsing agricultural systems to urban heat crises and fragile infrastructures) while contributing the least to global emissions. Meanwhile, powerful nations and corporate actors continue to drive carbon-intensive growth models that deepen inequality and ecological vulnerability. This asymmetry has created what many scholars call a "climate apartheid," where the burdens of planetary disruption fall most heavily on those with the fewest resources to adapt.

Climate Frontiers and Social Transformation: Socio-Ecological Shifts in the Age of Global Inequality offers a timely and integrative examination of how climate change is reconfiguring the world-system. The aim is not to propose entirely new theoretical constructs, but rather to connect existing evidence, frameworks, and lived experiences into a coherent, accessible narrative. By doing so, this book reflects the current scholarly consensus while highlighting pathways for climate action, justice, and structural reform. The book situates the climate crisis within centuries of global economic, political, and environmental change, thus, from the fossil-fueled industrial revolutions and imperial expansions of the past to today's neoliberal globalization and contested energy transitions. Through carefully crafted chapters, the book examines the historical drivers of ecological crises, the reproduction of global inequalities under climate stress, the politics of energy transitions, the limits of corporate "green capitalism," and the emergence of grassroots and youth-led climate justice movements. Drawing on world-systems analysis, political ecology, and environmental sociology, it explores how core-periphery relations, hegemonic transitions, urbanization, and demographic shifts interact with accelerating ecological pressures to shape the future of the planet.

Special attention is given to the vulnerabilities of the Global South—from peripheral ecologies exploited for extractive industries to rapidly urbanizing regions confronting infrastructural fragility and water-food-energy insecurities. The book highlights how Indigenous knowledge systems, grassroots activism, and climate justice coalitions are contesting dominant paradigms of growth and offering alternative visions for sustainable, equitable futures. Distinctive features include its systems-thinking approach, historical depth, and forward-looking scenario analysis integrating climate modeling with socio-political dynamics. Chapters analyze possible trajectories ranging from systemic collapse to just energy transitions and democratic, post-carbon transformations. By linking long-term historical processes with urgent contemporary debates, the book provides a framework for understanding the climate crisis as both a challenge and an opportunity for systemic reimagination.

With its interdisciplinary scope and emphasis on equity, *Climate Frontiers and Social Transformation* speaks to scholars, policymakers, students, and practitioners across environmental studies, sociology, political science, development studies, and sustainability science. It offers not only a critical diagnosis of planetary crisis but also pathways for action rooted in justice, inclusivity, and historical understanding—making it an essential resource for navigating the turbulent frontiers of climate and social change.

Contents

About the Authors

Dr. Matthew Chidozie Ogwu is an interdisciplinary environmental scientist and Assistant Professor of Sustainable Development at Appalachian State University, USA. He holds a Ph.D. in Biological Sciences from Seoul National University, South Korea, and degrees in Plant Biology and Biotechnology from the University of Benin, Nigeria. Dr. Ogwu's expertise spans integrated ecology, environmental health, and sustainability science, focusing on One Health and EcoHealth frameworks. His research includes natural resource management, microbiome applications in health and agriculture, environmental monitoring, and food waste valorization. An award-winning mentor and educator, Dr. Ogwu has published widely, including journal articles, book chapters, and edited volumes with Springer, Wiley, Taylor and Francis, and Elsevier. He is committed to participatory and decolonial approaches in environmental and human health governance, working with communities in the Global South and the United States to advance sustainability and foster the next generation of environmental leaders. His works continue to advance convergence research approaches integrating One Health and EcoHealth perspectives to address complex sustainability and health challenges globally.

Dr. Sylvester Chibueze Izah is a distinguished faculty member at Bayelsa Medical University in Yenagoa, Nigeria. He holds dual Ph.D. in Public Health, and Applied Microbiology and Environmental Health, an M.Sc. in Applied Microbiology, and a B.Sc. in Biological Sciences from Niger Delta University, Nigeria. In addition to his teaching role, he holds the position of Assistant Director of Academic Planning, Research, and Innovations at the university. Dr. Izah is a licensed environmental health specialist in Nigeria. His expertise spans various disciplines, including the assessment and management of air, soil, and water quality, as well as toxicology, hygiene, sanitation, food science, waste management, biodiversity, and sustainability. Dr. Izah's academic prowess is highlighted by an impressive research and publication portfolio, which includes more than 300 peer-reviewed publications. His work extends to the publication of edited books with renowned publishers such as Springer and Elsevier, contributions to book chapters in publications by Elsevier, Springer, and Taylor and Francis, as well as numerous journal articles.

Part I

Theoretical and Historical Foundations

1 Global Environmental Change and the Evolution of World-Systems Thinking

1.1 Introduction

The purpose of this chapter is to articulate the thesis that the ongoing climate crisis emerges directly from the dynamics of global world-systems, which are inherently tied to patterns of socio-economic development, structural inequalities, and political discourse. This linkage is significant as it provides a robust framework for understanding not only the environmental changes being experienced but also the underlying socio-economic processes that exacerbate these issues. World-systems analysis offers critical insights into how historical and contemporary patterns of resource exploitation, economic growth, and environmental degradation are interconnected. By framing climate change within the context of world-systems dynamics, it becomes possible to better grasp how local actions have global repercussions and how entrenched inequalities inhibit effective responses to the climate crisis (Izah et al., 2025; Mkodzongi, 2023; Simpson & Pizarro Choy, 2024). According to IPCC (2022), 3.3–3.6 billion people live in contexts highly vulnerable to climate change, with the highest burdens concentrated in Africa, South Asia, and Small Island States.

The interrelationship between climate change and world-systems theory is manifest in several key areas, which warrant detailed exploration. First, the role of global capitalism as a driving force behind environmental degradation is evident. Neoliberal economic policies prioritize short-term profit maximization over long-term sustainability, leading to the exploitation of natural resources without regard for ecological consequences. This phenomenon is exacerbated by a political economy that disproportionately disadvantages the Global South, where nations contribute little to greenhouse gas emissions yet bear the brunt of climate-related impacts (de Paula Sieverding et al., 2024; Mkodzongi, 2023). The narrative of the climate crisis as fundamentally tied to global capitalism highlights the need for systemic change rather than mere technological or policy fixes.

M. C. Ogwu and S. Chibueze Izah, *Climate Frontiers and Social Transformation*, World-Systems Evolution and Global Futures,
https://doi.org/10.1007/978-3-032-17639-4_1

Evidence suggests that environmental changes are intricately linked to human health outcomes (Iyiola et al., 2024; Izah et al., 2024; Ogwu & Izah, 2024), stressing the importance of integrating health perspectives into climate discourse. As climate change intensifies, health repercussions are projected to be especially severe in low-income areas, where existing socio-economic factors amplify vulnerabilities. The United Nations and numerous health organizations have increasingly framed climate change as a global health emergency, highlighting the pressing nature of this crisis (Bellizzi et al., 2023; Cardwell & Elliott, 2013; Maibach et al., 2010). This integration of health into climate discussions is critical as it highlights the human dimension of environmental policy, making the stakes more relatable to policymakers and the public. Without decisive action, climate change is projected to reduce global GDP by 11–14% by 2050 under high-emission scenarios (Swiss Re Institute, 2021).

Additionally, the dialogue around climate change is often dominated by a discourse that fails to recognize the multifaceted nature of environmental crises. Scholars argue for a polycrisis perspective, where climate change is recognized alongside other interlinked crises, including economic instability, food insecurity, and social inequality (Krogmann, 2025; Stoddard et al., 2021). This interconnectedness calls for a more holistic approach to problem-solving—one that recognizes the contributions of disciplines such as environmental science, sociology, and public health (Kalogirou et al., 2020). For example, the relationship between food regimes characterized by industrial agriculture and climate change underscores the need to consider agricultural policies within the broader framework of climate change strategies (McMichael, 2013).

Furthermore, the climate crisis disproportionately affects marginalized communities, raising issues of environmental justice that cannot be overlooked in world-systems analysis. Historical patterns of colonialism and exploitation have created enduring inequalities in both social and ecological domains (Legagneux et al., 2018; Simpson & Pizarro Choy, 2024). The need for decolonizing climate justice movements has become increasingly apparent as communities advocate for policies that address historical injustices and promote equitable resource distribution (Kang et al., 2023; Simpson & Pizarro Choy, 2024). By framing climate action through a lens of justice, the ethical dimensions of the climate crisis are stressed, along with the moral imperative for equitable solutions that prioritize the needs and voices of the most vulnerable populations. To address the climate crisis effectively, it is essential to understand the mechanisms through which political and economic power operate globally. Global governance institutions tasked with managing climate change often reflect the interests of Western, industrialized nations, sidelining the concerns of developing countries (Kang et al., 2023; Mkodzongi, 2023). This asymmetry in power dynamics should be rectified to foster a genuinely participatory approach to climate governance that includes diverse stakeholder perspectives from around the world (Legagneux et al., 2018; Lori et al., 2025; Simpson & Pizarro Choy, 2024). It is through this inclusive governance framework that innovative solutions may emerge, grounded in local knowledge and aimed at sustainable development pathways. The integration of health narratives into climate discourse also highlights the need for community-driven initiatives that empower individuals and regional organizations. Research suggests that communities that engage actively in climate adaptation and mitigation strategies experience better health outcomes and greater resilience (Cardwell & Elliott, 2013; Maibach et al., 2010; Workman et al., 2018). Leveraging local capacities, knowledge, and resources can lead to innovative solutions and practices that align with both environmental sustainability and health enhancement, creating co-benefits that are easier to communicate and mobilize (Cardwell & Elliott, 2013; Maibach et al., 2010;) As such, fostering community engagement remains a crucial element of addressing the climate crisis from a world-systems perspective.

In addition to these socio-political dimensions, the technological landscape surrounding climate change interventions also warrants scrutiny. Technological solutions, if implemented alongside existing socio-economic structures, may inadvertently reinforce inequalities if not carefully designed (Kalogirou et al., 2020; Le Quéré et al., 2012; Stoddard et al., 2021). For instance, while renewable energy technologies are essential for reducing carbon emissions, their deployment should consider local needs, socio-cultural contexts, and the existing socio-economic fabric to avoid exacerbating disparities (Maibach et al., 2010; Zhang et al., 2011). This calls for a paradigm shift in how technological interventions are viewed, moving from a purely technical lens to one that considers socio-economic implications within a world-systems framework. The role of education and public awareness in addressing climate change cannot be understated either. An informed public is key to mobilizing action and demanding accountability from political and corporate leaders (Legagneux et al., 2018; Maibach et al., 2010; Workman et al., 2018). Campaigns that successfully translate complex scientific data into actionable insights enable more effective public engagement and policy influence. Furthermore, fostering a collective understanding of climate impacts can galvanize social movements that advocate for systemic change, thus aligning public sentiment with environmental goals (Barretto, 2024; de Paula Sieverding et al., 2024; Mkodzongi, 2023). Education thus serves as a catalyst for both individual and collective agency against climate change. As the climate crisis evolves, the discourse surrounding legacy and intergenerational ethics becomes particularly poignant. The desire to leave a habitable planet for future generations adds a moral impetus to climate action that transcends traditional economic rationales (de Paula Sieverding et al., 2024; Ogwu & Kosoe, 2025). This legacy motivation can drive significant shifts in personal and collective behaviors related to climate protection, fostering a commitment to sustainable practices that consider their long-term ramifications. The imperative to act for the sake of future generations reinforces the need for urgent, equitable, and proactive climate strategies that reflect an understanding of our interconnectedness in a world-systems context.

This chapter posits that the climate crisis is fundamentally rooted in the dynamics of global world-systems, requiring an integrated and interdisciplinary approach to understanding and addressing its multifaceted challenges. By linking world-systems analysis with environmental change scholarship, the socio-economic drivers of ecological degradation can be unveiled, the systems of inequality recognized, and nuanced strategies developed to prioritize human health and social equity. Such an approach not only illuminates the underpinnings of the climate crisis but also inspires rigorous action towards a more sustainable and equitable future. This chapter integrates world-systems analysis with environmental change scholarship. It argues that the climate crisis emerges from the structure of the capitalist world-economy, not from isolated shocks. The analysis spans the longue durée of capitalism, colonial expansion, and technological development from 1500 to the present. Comparative cases ground the claim, including Dutch peat and timber with maritime control, British coal and steam, and United States oil, automobility, and petrodollar finance. The chapter develops an analytical framework linking world-system position, energy and material regimes, Earth system pressures, and distributional outcomes. The framework specifies indicators, data sources, and methods, including material and energy flow analysis, multi-regional input-output tables, footprinting, and remote sensing. The contribution is a rigorous bridge between social theory and Earth system science, with inequality and accountability at the center. The chapter sets a baseline for subsequent chapters and defines testable hypotheses for empirical evaluation.

1.2 Definitions and Scope of Environmental Change and Evolution of World-Systems Thinking

World-systems thinking involves a comprehensive framework for understanding the interconnectedness of social, economic, and ecological processes across different spatial and temporal scales. At its core, a "world-system" is defined as a complex structure that integrates multiple layers of societies, economies, and environments that interact on a global scale, reflecting the intricate dynamics of power, production, and ecological impact (Steffen et al., 2011). It extends beyond national borders and local phenomena, emphasizing the interactions and consequences arising from global integration of various socio-ecological systems, including trade networks and environmental policies.

Socio-ecological systems represent the mutual interdependence between human societies and their environments, recognizing that social and ecological dynamics are inseparable. These systems illustrate how human activities can lead to significant environmental transformations, thereby impacting biodiversity and ecosystem services (Larson et al., 2016; Mensing et al., 2018). The spatial and temporal dimensions of these systems have been extensively studied, particularly from 1500 to the present, highlighting significant shifts during periods of colonization, industrialization, and the Anthropocene—an era defined by profound human influence on Earth's geology and ecosystems (Steffen et al., 2011).

The concept of "planetary boundaries" delineates a safe operating space for humanity concerning the Earth's biophysical systems. Proposed by a group of interdisciplinary researchers, it identifies nine critical Earth system processes that should remain within certain thresholds to prevent irreversible environmental changes. These boundaries involve climate change, biodiversity loss, and nitrogen and phosphorus cycles, among others (Steffen et al., 2011). Crossing these boundaries poses risks of destabilizing the global environment, leading to consequences that can affect multiple generations.

The term "climate frontier" refers to the evolving boundary conditions under which climate change impacts human societies and ecosystems. This concept encapsulates the idea of thresholds beyond which communities may face increased vulnerability and diminished adaptability to environmental changes. Understanding the climate frontier allows for better strategic planning and adaptive responses to climate-induced challenges (Steffen et al., 2011).

1.2.1 Brief Historical Context

Historical responsibility is highly unequal: the Global North accounts for 92% of excess atmospheric CO_2 emissions since 1850, while many of the most vulnerable regions contributed little to the problem (Hickel, 2020). From the early 1500s to the present, the global environment has undergone transformative changes exacerbated by anthropogenic forces. Historical records, including archaeological and ecological datasets, provide essential insights into the evolution of world systems in response to environmental changes. The Renaissance marked the beginning of profound global interconnectedness, paving the way for the Age of Discovery and subsequent colonization efforts that significantly altered ecosystems through resource extraction and habitat modification (Mensing et al., 2018). Today, the richest 10% of the global population produce 48% of global emissions, whereas the poorest 50% contribute only 12% (Chancel, 2022). The interactions between human societies and the biosphere have been profound, with varying impacts across different regions, illustrating the concept of socio-ecological feedback loops. In the late twentieth century, the emergence of the Anthropocene brought new urgency to the study of world-systems, as scholars recognized the unprecedented scale of human impact on the planet. As outlined by Steffen et al., the Anthropocene highlights how human activity plays a deterministic role in shaping Earth's biophysical systems, emphasizing the urgency of sustainable practices (Steffen et al., 2011). The idea of socio-ecological resilience gained traction as researchers acknowledged the need for adaptable systems that could withstand environmental pressures without collapsing (Adger, 2000).

1.2.2 Methodological Approaches

To comprehend these complex interrelationships, scholars utilize various methodologies that integrate historical data, paleoecological studies, and socio-economic analyses. For instance, long-term ecological records serve as valuable resources for understanding biodiversity conservation strategies and ecosystem responses to past climate fluctuations (Willis et al., 2007). These records are crucial for making science-informed conservation decisions that consider species extinction risks and the dynamics of ecosystem stability (Barnosky et al., 2017). The study of historical ecology, particularly, reveals how human-induced changes have led to significant landscape transformations, thereby influencing current biodiversity patterns (Larson et al., 2016; Mensing et al., 2018). Data collected from paleolimnological studies provide insights into how ancient environments reacted to human interventions, helping to forecast future ecosystem shifts under changing climatic conditions (Kong et al., 2017). These interdisciplinary approaches emphasize the importance of integrating ecological, archaeological, and anthropological perspectives to form a comprehensive understanding of past, present, and future socio-ecological systems (Barnosky et al., 2017).

Numerous case studies underline the importance of integrating historical records with current environmental challenges. For instance, the transformation of the Italian landscape since the Roman era illustrates how agricultural practices and urbanization have influenced local biodiversity and ecosystem dynamics (Kosoe & Ogwu, 2025; Mensing et al., 2018). Similarly, the management of freshwater ecosystems emphasizes how historical consumption patterns and anthropogenic stressors have led to significant biodiversity loss, necessitating a shift toward more sustainable management practices (Dudgeon et al., 2006). Another illustrative case is the recognition of coastal anthromes, where human activity has reshaped ecosystems along coastlines, leading to unique management challenges that require a nuanced understanding of their vulnerabilities (Lazarus, 2017). The case of Lake Chaohu in China exemplifies how hydrological changes driven by human activity can lead to regime shifts in aquatic ecosystems, demonstrating the importance of comprehensive ecological and socio-economic data in informing conservation efforts (Kong et al., 2017).

1.3 Evolution of World-Systems Thoughts

The evolution of world-systems thought is a multifaceted discourse that emerged from the works of historians and sociologists such as Fernand Braudel and Immanuel Wallerstein, who laid the groundwork for understanding the interrelations of economies, societies, and the environment on a global scale. Braudel's concept of "longue durée" emphasizes the importance of long-term historical structures over short-term events, suggesting that the socio-economic dynamics of regions cannot be adequately understood without considering these enduring factors. Wallerstein expanded upon this by integrating the notion of a "world-system," characterizing it as a complex social system of interdependent countries divided broadly into core, semi-periphery, and periphery states. The core comprises highly industrialized, economically dominant nations, while the periphery consists of less developed states that rely on the export of raw materials. The semi-periphery includes countries that fall between these two extremes, serving as both suppliers of resources and consumers of manufactured goods (Rice, 2007; Wang et al., 2019). Figure 1.1 links time scales to system positions.

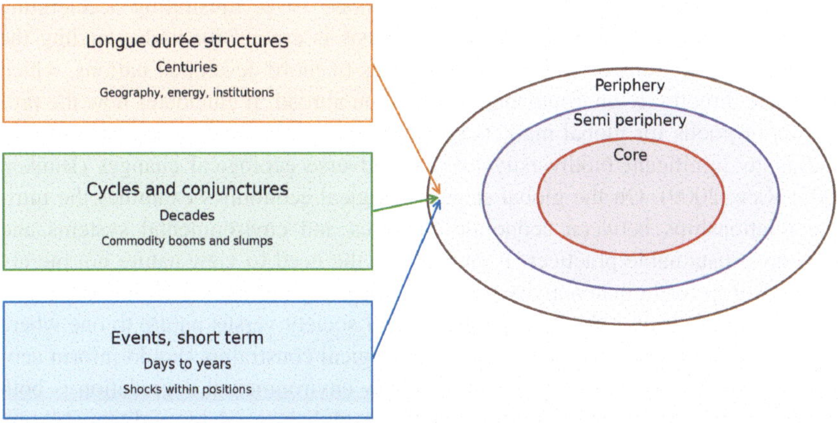

Fig. 1.1 Conceptual overview of Braudel's Temporal Layers and Wallerstein's World System

This three-tier structure reflects a palpable hierarchy within the global economy, one that is not only economic but also ecological. For instance, ecological unequal exchange involves industrialized nations externalizing environmental costs onto peripheral countries. More developed nations disproportionately exploit ecological resources from the periphery, resulting in environmental degradation in those less developed regions (Jorgenson, 2016a; Jorgenson, 2006). This systematic extraction and externalization of ecological burdens can be encapsulated in the concept of "metabolic rift," a term that highlights the disturbances in natural processes caused by capitalist production methods (Schneider & McMichael, 2010). The adverse consequences of these ecological practices often manifest in ecological disparities, demonstrating how resource-dependent economies in the periphery suffer the most (Sanderson & Frey, 2014).

The integration of frameworks such as political ecology and ecological economics enhances our understanding of these dynamics, in which the political economy of globalization interfaces directly with environmental sustainability. Political ecology critiques the structures that perpetuate inequalities and examines how power relationships shape environmental governance and resource allocation. For example, focusing on state relations within the context of ecological governance provides insight into how environmental policies can lead to disparities, with core nations benefiting from environmental laxity in peripheral nations (Henderson & Shorette, 2017; Jorgenson, 2016b). Through this lens, understanding how ecological inequalities are produced and maintained is essential for formulating effective sustainability strategies.

Moreover, research on well-documented cases, such as deforestation linked to soybean exports, has shown dire consequences for ecosystems in peripheral nations, illustrating ecologically unequal exchange (Shandra et al., 2009). The world has lost 420 million hectares of forest since 1990—an area larger than India (FAO, 2020).

Wildlife populations have declined 69% since 1970, illustrating accelerating ecological collapse (WWF, 2022). This analysis is crucial for understanding the profound implications of consumption patterns of more developed nations, which contribute directly to environmental degradation abroad. It elucidates how the production of goods for global markets typically disregards local ecological systems, leading to significant biodiversity loss and adverse ecological changes (Bunker, 2005; Rice, 2009). On the global stage, ecological economics examines the intricate relationships between economic activities and environmental systems and advocates sustainable practices. It emphasizes the need to view nature not merely as a backdrop for human activity but as an integral co-producer within social systems. This shift from a dualistic perspective of society versus nature to one where they are co-produced acknowledges that ecological constraints should inform economic decisions. These changes highlight how environmental degradation is both a consequence of and a contributing factor to global social inequalities (Moore, 2011; Jorgenson, 2016b). The transformation of nature from a mere backdrop to an active participant in socio-economic processes represents a significant conceptual shift vital to understanding contemporary environmental issues. As these discussions unfold, it is critical to recognize that the interplay between capital accumulation and environmental exploitation is continuous, fostering a cycle in which core nations, dependent on their economic success, exacerbate ecological conditions in peripheral contexts. This dynamic maintains an inequitable system whereby industrialized nations continue to thrive while peripheral countries face increasing ecological burdens.

Figure 1.2 provides a conceptual bridge between Braudel's historical temporality and Wallerstein's world-system hierarchy, clarifying how energy and material regimes unfold across different temporal and spatial scales. On the left, Braudel's three temporal layers—événementielle (short-term events), conjonctures (medium-term cycles), and longue durée (deep structural patterns)—capture the nested rhythms through which socio-ecological change occurs. On the right, Wallerstein's spatial hierarchy of core, semiperiphery, and periphery illustrates how global economic power is unevenly organized. By linking these temporal and spatial dimensions, Figure 1.2 highlights how long-term transformations in energy and material systems shape the relative positions of regions within the world-economy.

1.4 Link to Earth System Science

The examination of social drivers in the context of Earth system science is critical for understanding how human behavior interacts with vital Earth variables, particularly in the realm of climate change, biodiversity loss, and land use. This intersection has become increasingly significant in light of the Great Acceleration, a term describing the dramatic increase in human impact on the Earth system since the mid-twentieth century. This period marks substantial changes in various anthropogenic variables, including significant increases in the emissions of carbon dioxide (CO_2) and methane (CH_4), widespread conversion of natural land for

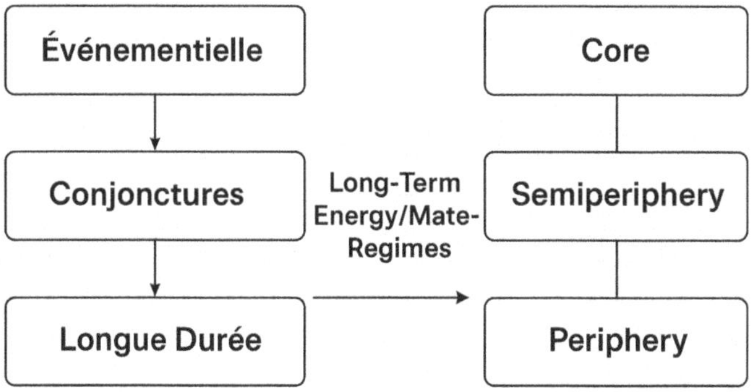

Fig. 1.2 Temporal Layers (Longue Durée) and World-System Positions

agricultural use, and notable declines in biodiversity. Land use change is a primary social driver impacting these Earth system variables. Research indicates that agriculture is a leading cause of biodiversity loss, significantly driven by rising demand for food and energy resources. This trend threatens the delicate balance of ecosystems, leading to profound implications for carbon cycling and habitat availability (Boakes et al., 2023; Rounsevell et al., 2013). For example, Fritz et al. emphasize the role of land cover as a climate variable and its biophysical significance for Earth system models, asserting that land use change can have cascading effects on biodiversity and climate systems (Fritz et al., 2017). As agricultural land expands, often at the expense of natural habitats, this necessitates scrutinizing agricultural practices and their intersection with biodiversity conservation (de Baan et al., 2015; Wilting et al., 2017).

The temporal and spatial scales of social drivers should align with social theories to truly grasp their impacts. Social theory often highlights how socio-economic factors drive land use decisions, impacting biodiversity and climate dynamics. Titeux et al. emphasize that habitat destruction via land use changes poses immediate threats to biodiversity, which are further exacerbated by climate change (Titeux et al., 2016). These socio-environmental dynamics emphasize the need for models that account for both current practices and future scenarios in projections of biodiversity change. The need for interdisciplinary approaches—including social theories and socio-economic variables—is emphasized by Winberg et al., who discuss the interplay between policy decisions and biodiversity outcomes in agricultural landscapes (Winberg et al., 2023).

The Great Acceleration itself provides a compelling reference point for understanding how sharply human activities have escalated since 1950. With advancements in technology and population growth, emissions of greenhouse gases like CO_2 and CH_4 have surged, resulting in accelerated warming and ecological stress (Estes et al., 2014; Schipper et al., 2020). The synergistic effects of land-use changes and climate-induced transformations drastically reshape ecosystems, leading to potential tipping points. A tipping point is a threshold beyond which ecosystems can transition to significantly different states, often irreversibly. For instance, as biodiversity declines and ecosystems become less resilient to climate change, we may reach a point where the ability of these systems to provide essential services is radically diminished, leading to cascading ecological failures (Schipper et al., 2020; Wilting et al., 2017).

In this context, climate change and biodiversity loss are interconnected, with land use changes serving as both causes and consequences in a self-reinforcing cycle. Jeswani et al. (2018) highlight the importance of considering these interactions in life cycle assessments (LCA), noting that decision-makers often overlook the ecological consequences of land use transformations, especially concerning biodiversity. The urgency for more sophisticated assessments that account for these feedbacks is highlightd by Geyer et al. (2010), who discuss the importance of integrating various impact models for a comprehensive understanding of biodiversity shifts caused by land use changes.

Another layer of complexity emerges when considering the temporal and spatial dimensions associated with these social drivers. Tzanopoulos et al. (2013) argue that social dynamics operate across scales; local land-use decisions are often influenced by national agricultural policies or global market pressures. Such scales of interaction should be aligned when estimating ecological impacts. This spatial-temporal alignment is crucial, particularly when modeling biodiversity loss resulting from land use changes, which are not uniform and often reflect broader trends driven by socio-economic forces towards agricultural products (Boakes et al., 2023; Schipper et al., 2020). Research also shows that the impacts of land use intensification are unevenly distributed globally. Regions with rich biodiversity, such as tropical and Mediterranean areas, are disproportionately threatened by these changes, as indicated by Newbold et al. in their study on biodiversity sensitivity to land-use and climate change (Newbold, Bentley, et al., 2020a). Consequently, it is necessary to tailor conservation strategies to specific regions and contexts while considering how social drivers might differ across these landscapes.

Furthermore, the concept of thresholds can be expanded to include human systems and the socio-economic contexts in which they operate. As human activities push ecosystems toward these ecological tipping points, social resilience becomes a critical factor in biodiversity conservation. Policies that incorporate social equity and environmental justice can facilitate better conservation outcomes, aligning social drivers with Earth system sustainability (Boakes et al., 2023; Wilting et al., 2017). In light of these considerations, strategies aimed at climate mitigation and biodiversity conservation should incorporate interdisciplinary frameworks that understand the complex interactions among land use, ecological systems, and social drivers. Approaches such as integrative land use planning and ecosystem-based management can effectively address these multifaceted relationships. By recognizing and acting upon the interconnectedness of ecological systems and human decision-making processes, policymakers can better navigate the precursory signals of tipping points and work to avert catastrophic ecological outcomes.

1.5 Longue Durée of Global Capitalism

The longue durée of global capitalism is marked by various phases that involve mercantile expansion, industrialization, the Great Acceleration, and the contemporary neoliberal era. This expansive trajectory has been shaped significantly by evolving energy sources, key sectors, and geopolitical frontier zones. Understanding these phases requires a nuanced exploration of how each has contributed to material throughput, the accompanying waste sinks, and the continuities and breaks that inform current global risks. Table 1.1 condenses the four phases into one view.

1.5.1 Mercantile Expansion (16th–18th Centuries)

The initial phase of global capitalism was characterized by mercantile expansion, where the foundation of capitalist practices was laid. During this period, energy sources were predominantly biomass and early forms of wind and water, facilitating trade across oceans and rivers. The economic policy was heavily state-driven, as mercantilism emphasized the role of the state in controlling trade and maximizing national wealth through the accumulation of precious metals and trade surpluses (Asmi et al., 2023). The expanding networks of trade routes allowed for the emergence of significant economic powers, notably Spain and Portugal, which dominated global trade through colonization and the exploitation of resources (Milne, 2021). Key sectors during mercantile expansion typically included agriculture, textiles, and commodities, such as sugar and tobacco, integral to the mercantile economies and fueling trade between colonies and European markets. Frontier zones during this period often manifested in colonial territories where resource extraction and the introduction of cash crops were prevalent, aligning with the mercantilist focus on accumulating wealth (Williams & May, 2025). The material throughput associated with this era predominantly consisted of raw materials being exported from the colonies to Europe, while the corresponding waste consisted of environmental degradation and social dislocation among indigenous populations (Edziah et al., 2021).

1.5.2 Industrialization (18th–19th Centuries)

The transition to industrialization marked a significant break from the prior mercantile systems. Energy sources shifted primarily to coal, initiating profound transformations in industry and urban life. The Industrial Revolution catalyzed the transition from agrarian economies to industrial powerhouses, with coal serving as a foundational energy source powering steam engines and electricity generation. Industrialization fostered key sectors such as metallurgy, textiles, and manufacturing, reshaping not only the economy but also societal structures. Frontier zones in this era became urban industrial centers, leading to massive migrations from rural areas to cities. The material throughput increased substantially, as did the

Table 1.1 Longue Durée of Global Capitalism, Phase Summary

Phase	Time	Energy base	Key sectors	Frontier zones	Throughput pattern	Waste sinks	Signature risks
Mercantile expansion	16th to 18th centuries	Biomass, wind, water	Agriculture, textiles, sugar, tobacco	Colonies in the Americas, Africa, and Asia	Raw materials flow to Europe	Deforestation, soil loss, and coastal pollution	Dispossession, famine, and disease spread
Industrialization	18th to 19th centuries	Coal, steam	Metallurgy, factories, railways	Industrial cities, mining districts	Rapid output growth and urbanization	Air and river pollution, slag, and coal ash	Labor exploitation, urban health crises
Great acceleration	Mid twentieth century	Oil, gas, electricity	Autos, chemicals, aerospace, electronics	Global South industrial zones, highways, ports	Mass consumption and global trade surge	CO_2 and CH_4 rise, plastics, nitrates	Climate forcing, biodiversity loss, waste mountains
Neoliberal era	Late twentieth century to present	Fossils plus rising renewables	Finance, digital services, global supply chains, clean tech	Export platforms, data hubs, mineral belts	Just-in-time trade, offshored production	E-waste, carbon lock-in, land and water stress	Inequality, supply chain shocks, and slow decarbonization

generation of waste, notably in the form of air and water pollution from industrial activities (Li et al., 2023). The introduction of transportation infrastructure, such as railroads, allowed for unprecedented speeds in the movement of goods, further integrating global markets but also exacerbating disparities in labor and capital (Wang et al., 2022). This period is characterized by relentless resource extraction, which laid the groundwork for ongoing environmental crises (Adepoju et al., 2022). Climate change may displace 216 million people by 2050 across Sub-Saharan Africa, South Asia, and Latin America (World Bank, 2021). Wider ecosystem and economic stressors could put 1.2 billion people at risk of displacement by mid-century (IEP, 2021).

1.5.3 Great Acceleration (Mid-Twentieth Century)

Following World War II, the Great Acceleration marked a vital growth phase fueled by technological innovations and intense globalization. This era was propelled by diverse energy sources, including oil and natural gas, enabling mass production and consumption. The post-war economic boom led to the expansion of consumer culture, underpinned by key sectors such as automotive, aerospace, and electronics (Mueller & Farhat, 2022). The establishment of institutions like the International Monetary Fund and the World Bank illustrated this period's focus on facilitating global trade and investment flows. The Great Acceleration also saw the emergence of new frontier zones, particularly in the Global South, where industrial capabilities developed alongside global supply chains. The material throughput reached unprecedented levels, resulting not only in economic growth but also in significant ecological impacts, given the heightened levels of waste generation due to increased consumerism (Munshi, 2007). The reliance on fossil fuels contributed to carbon emissions, setting the stage for climate-related challenges that dominate contemporary discourse.

1.5.4 Neoliberal Era (Late Twentieth Century–Present)

The late twentieth century ushered in the neoliberal era, characterized by deregulation, privatization, and the increased influence of market forces over state power. The energy landscape evolved further with a strong push toward renewable energy sources, coinciding with concerns regarding climate change and energy sustainability (Feldman et al., 2018). Neoliberalism has significantly impacted economic policies in various nations, reflecting a broader trend toward economic liberalization and globalization (Harris & Lauderdale, 2002). In this current phase, key sectors have shifted towards technology-focused industries such as digital services and renewable energy. Frontier zones have likewise transformed, with emerging markets playing crucial roles in global supply chains. Continuous reshaping of material throughput now necessitates sustainable practices that emphasize the circular economy, which seeks to minimize waste and enhance resource efficiency

(Edziah et al., 2021). However, despite these advancements, existing inequities and environmental injustices have become pronounced, revealing ongoing challenges posed by neoliberal doctrines, particularly in their impact on social and economic disparities (Zagelmeyer, 2024).

1.5.5 Continuities and Breaks Impacting Present Risks

Throughout these phases, there are notable continuities in capital accumulation, labor exploitation, and environmental degradation. Capitalism's tendency to perpetuate inequalities—both social and economic—remains evident, with persistent risks arising from resource depletion and climate change (Mueller & Farhat, 2022). The historical reliance on fossil fuels and the political economies built around them have established a trajectory of environmental crises affecting global stability today. Conversely, breaks can be identified at each transition, such as the industrial shift from manual labor to mechanization or from coal to more diverse energy sources. These transitions have opened up new opportunities for technological advancements and economic models, but have also introduced risks associated with socio-economic disruption and environmental impact (Kerslake & Krishnamurti, 2024). The Longue durée of global capitalism illustrates a complex interplay of phases shaped by energy sources, economic sectors, and geopolitical dynamics, creating a rich tapestry of material throughput and waste production that continues to evolve. The linkages between historical practices and contemporary challenges reveal critical insights into how these narratives inform future trajectories of capitalism and its associated risks.

Shifts in dominant energy regimes have fundamentally shaped the evolution of the capitalist world-system. Across the longue durée, transitions from biomass and wind to coal and steam, and later to oil, petrochemicals, and contemporary fossil-renewable hybrids, have enabled successive hegemonic powers to expand resource frontiers, deepen extractive infrastructures, and generate escalating waste and emissions profiles (Fig. 1.3). These energy regimes underpin the economic, political, and ecological structures that define each historical phase of global capitalism. Figure 1.3 highlights how energy systems interact with world-system dynamics and how each era constructs new patterns of environmental pressure, inequality, and geopolitical influence.

1.6 Colonial Expansion and Ecological Change

Colonial expansion has been intimately connected to significant ecological changes, primarily through practices such as the establishment of plantation complexes, the exploitation of mineral resources, and the seizure of land. Table 1.2 links core colonial practices to their ecological and socio-economic impacts. It condenses mechanisms, direct effects, and lasting legacies across plantation agriculture, mining, and commodity chains. It supports Sect. 6 on colonial expansion

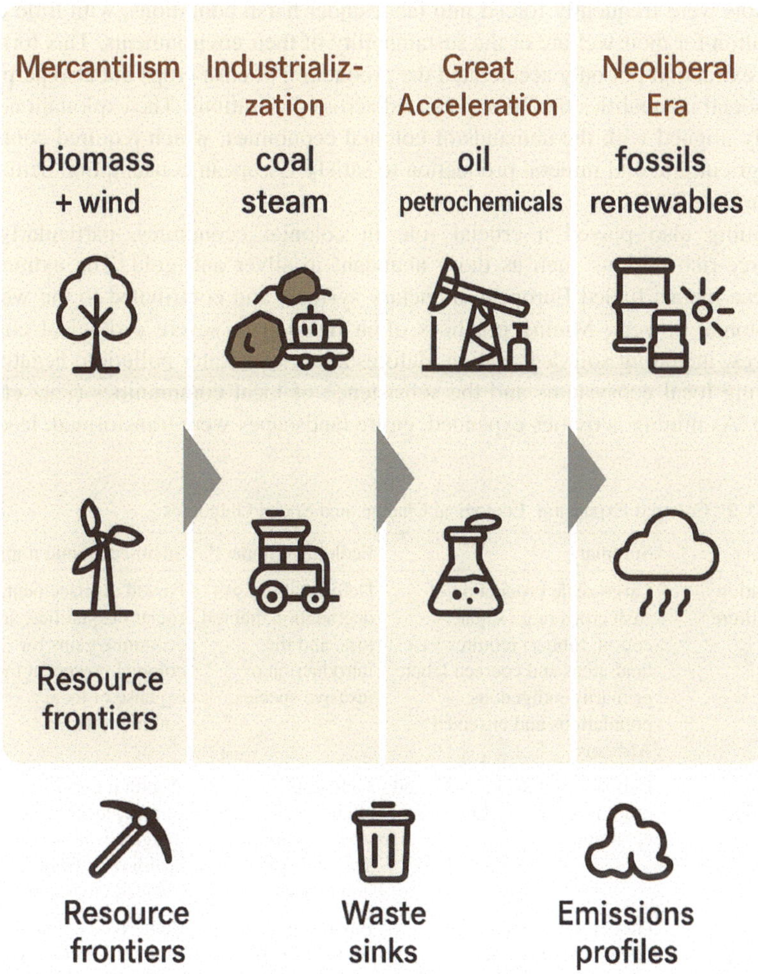

Fig. 1.3 Longue Durée Energy Regimes and World-System Hegemony

and ecological change. The plantation system evolved as a key economic model in colonial territories, particularly for cash crops like sugar, cotton, and rubber. These plantations required large tracts of land, which often resulted in systematic land seizures accompanied by the displacement of indigenous populations. In many regions, the establishment of plantations altered the landscape, as extensive areas of forest were cleared for agriculture, fundamentally transforming local ecosystems and biodiversity (Crabb et al., 2024). Coerced labor on these plantations was essential for their economic viability. Indigenous populations and enslaved Africans were frequently forced into labor under harsh conditions, with little consideration for their welfare or the sustainability of their environments. This form of labor extraction not only accelerated the production of cash crops but also perpetuated social inequalities that have persisted across generations. The exploitation was closely aligned with the demands of colonial economies, which required continuous agricultural and mineral production to satisfy European consumption demands (Mohamed, 2004).

Mining also played a crucial role in colonial economies, particularly in resource-rich regions such as those abundant in silver and gold. The extraction of these metals fueled European monetary systems and contributed to the wealth of colonial powers. Mining practices often resulted in severe ecological consequences, including soil degradation, deforestation, and water pollution, negatively affecting local ecosystems and the subsistence of local communities (Patz et al., 2004). As mining activities expanded, entire landscapes were transformed, leading

Table 1.2 Colonial Expansion, Ecological Change, and Social Outcomes

Aspect	Summary	Ecological impact	Socio-economic impact
Plantation agriculture	Large-scale production of cash crops (e.g., sugar, cotton, rubber) requires vast land areas and coerced labor, primarily indigenous populations and enslaved Africans	Deforestation, soil degradation, habitat loss, and the introduction of invasive species	Forced displacement, social inequalities, and economic gains for colonial powers at the expense of local communities
Mining activities	Extraction of silver, gold, and other minerals for European economies	Landscape alteration, soil and water pollution, and ecological imbalance	Wealth transfer to Europe, local economic marginalization, and health risks from polluted environments
Commodity chains	Global networks linking resource extraction, processing, and European markets (e.g., sugar, silver, cotton, palm oil)	Environmental degradation occurs across multiple stages of production and transport	Reinforced colonial economic hierarchies, exploitation of labor, and economic dependency of colonies
Ecological consequences	Deforestation, soil erosion, loss of biodiversity, and altered disease ecologies due to human encroachment on natural habitats	Habitat fragmentation, species decline, and increased zoonotic disease risks	Long-term vulnerability of communities to environmental crises
Legacy of colonialism	Persistent socio-economic disparities, food insecurity, and environmental unsustainability in formerly colonized regions	Ongoing ecological degradation from colonial-era land-use practices	Structural economic vulnerabilities, underdevelopment, and inequality

to ecological imbalances that exacerbated socio-economic disparities among local populations, as wealth derived from these resources was predominantly siphoned off to benefit colonial empires (Yasuoka & Levins, 2007). Commodity chains are vital for understanding the interconnectedness of colonial practices and their ecological impacts. The sugar commodity chain, for example, which began with the cultivation of sugarcane on plantations, involved a vast labor network that extended from fields to markets in Europe and beyond. This chain illustrated the transformation of raw materials into valuable goods, reinforcing economic and social hierarchies entrenched within colonial frameworks. The movement of commodities such as sugar, silver, cotton, and palm oil exemplify how colonial economies functioned at the expense of local environments and communities, exacerbating inequalities that persist to this day (Mohamed, 2004).

The ecological consequences of colonial expansion are profound. Deforestation occurred on a large scale as forests were cleared for plantations and mines. This widespread tree removal not only reduced biodiversity but also led to soil erosion, which had lasting impacts on agricultural productivity. Disruption of natural habitats facilitated the introduction of invasive species, which could outcompete native flora and fauna, further destabilizing local ecosystems (Chan et al., 2023). Moreover, landscape alteration often initiated new disease ecologies; as humans encroached on previously undisturbed environments, the risk of zoonotic diseases increased, with ecological changes creating conditions favorable for pathogen emergence (Patz et al., 2004). The legacy of colonial ecologies remains significant. The socio-economic disparities established during the colonial era continue to have implications for current global dynamics. Regions formerly exploited for their resources often struggle with economic underdevelopment, compounded by ongoing ecological degradation. For instance, countries deeply affected by plantation agriculture face challenges related to food security and environmental sustainability, issues rooted in their colonial pasts (Yasuoka & Levins, 2007). The structural vulnerabilities established in the colonial period frequently resurface in modern contexts, where environmental challenges intertwine with economic inequities (Mohamed, 2004).

1.7 Technology and Energy Regimes

The transition to sustainable energy systems involves multiple factors, including technological advancements, environmental impacts, energy densities, and the socio-political landscape surrounding infrastructure development. Historical contrasts, such as timber and peat, coal and steam, oil and automobility, gas and grids, as well as critical minerals for renewables, delineate the evolution of energy regimes. Countries like the Netherlands, Britain, and the United States are increasingly focusing on renewable technologies that promise sustainability while addressing the legacies of their existing infrastructures and the associated challenges.

1.7.1 Historical Context of Energy Regimes

Timber and peat were among the earliest energy sources, distinguishing themselves from fossil fuels like coal and oil. Timber was a sustainable resource during its early use in dwellings and heating, representing a much less environmentally damaging option than fossil fuels. Current assessments have reignited interest in sustainable biomass as a modern solution to fossil fuel dependence, advocating for a circular economy that highlights timber's potential through enhanced forest management and innovative processing techniques (Garcia-Lillo et al., 2023). Coal and steam were crucial during the Industrial Revolution, catalyzing profound transformations in energy consumption patterns characterized by high energy densities. This era saw the establishment of coal-centric infrastructure that, in turn, reinforced systemic power dynamics linked to industrial interests, culminating in significant environmental consequences, including greenhouse gas emissions from coal use (Bogdanov et al., 2019; He et al., 2016). Thus, these entrenched systems exhibit path lock-in—resistance to change due to substantial pre-existing investments and established networks.

1.7.2 Transport Costs and Energy Density

Understanding energy regimes requires a grasp of energy density, defined as the amount of energy per unit volume or mass, which significantly influences transportation costs. Coal's high energy density made it an advantageous choice for steam engines, driving productivity but also incurring severe environmental costs (Sgouridis et al., 2016). This dynamic extends to oil, whose liquid form supported the emergence of automobility. The extensive development of road infrastructure and the automotive sector embedded an oil dependency into economic frameworks (Jacobson et al., 2017). Natural gas has also played a significant role in urban electricity generation via gas grids, offering greater efficiency but also creating complex regulatory and supply chain issues. This interplay of energy density and transport costs is further complicated by the rising demand for critical minerals needed in renewable technologies, such as lithium for batteries, which introduces geopolitical considerations regarding resource availability and technological adaptation (Greim et al., 2020; Zhang et al., 2021).

1.7.3 Systemic Power and State Capacity

State capacity is crucial to discussions of energy transitions. Governments that invest in renewable energy infrastructure can exercise greater authority in shaping sustainable energy policy. In the United States, federal incentives, such as tax credits for solar energy and electric vehicles, exemplify initiatives to combat climate change (Clack et al., 2017; Rosenow et al., 2017). However, these efforts often contend with entrenched fossil fuel interests benefiting from prior infrastructure.

The acquisition of critical minerals necessary for renewable technologies is closely tied to state capacity. Countries rich in resources such as lithium or cobalt emerge as central players in the transition to renewable energy, given their vital role in battery production essential for energy storage (Greim et al., 2020; Zhang et al., 2021). Hence, the intersection of state capacity and systemic power is integral to understanding how resource management and regulatory frameworks influence broader energy transitions.

In the Netherlands, the shift towards a circular economy has highlighted timber as a renewable energy alternative, necessitating innovative forest management strategies. Dutch energy policies aim to incorporate renewable sources while managing the legacy of historically dominant gas infrastructure. The focus on wind and solar energy represents a strategic pivot from a gas-dependent history, underlining challenges inherent in overcoming path lock-in (Schlachtberger et al., 2018). Britain serves as a noteworthy example of effective policy frameworks driving systematic change. Through proactive government policies and investments in renewable technologies, the UK has successfully increased its share of renewable energy while phasing out reliance on coal. The UK's energy transition illustrates how political will can reshape systemic dynamics, fostering an infrastructure conducive to achieving net-zero emissions and promoting smart grid development (Bulatov et al., 2018; Nakano et al., 2018). In contrast, the U.S. presents a diverse narrative marked by regional variations in energy policies and resource availability. Fossil fuel reliance remains both a legacy and a challenge, with states like Texas prioritizing oil and gas, while others push towards renewables. This complex energy landscape reflects a balance of state capacity and systemic changes prompted by local governance, market conditions, and community needs (Clack et al., 2017; Rosenow et al., 2017).

1.7.4 Rebound Effects and Infrastructure Choices

Rebound effects can substantially influence the efficacy of energy transitions, as efficiency improvements may lead to increased consumption due to lower costs. This phenomenon has been historically documented (Sgouridis et al., 2016; Zhu et al., 2024). These implications necessitate careful strategic planning at both macro and micro levels to ensure that efficiency measures do not inadvertently escalate consumption, negating their environmental benefits (Shu et al., 2021). Infrastructure investments are vital in shaping the trajectory of energy transitions, particularly regarding energy storage and smart grid technologies. The demand for flexibility in energy systems is increasing as variable renewable energy sources like solar and wind become dominant. The integration of smart grids allows for improved energy management and enhances consumer participation, a critical factor as communities adopt renewable technologies.

1.8 Climate Crisis as System Outcome

The complexities of the climate crisis necessitate a multifaceted approach to understand and rectify the intertwined issues of greenhouse gas emissions, resource depletion, and socio-economic inequities. Central to this discourse is the relationship between accumulation, emissions growth, sink depletion, and mechanisms of risk transfer. Accumulation of capital, especially in industrialized nations, often leads to increased emissions as consumption rises without adequate regulatory frameworks to address environmental impacts. For instance, as suggested by Liu et al. (2016) and Hoekstra and Wiedmann (2014), consumption-driven demand fosters emissions growth, particularly in global supply chains serving affluent consumers. These emissions not only heighten the urgency for action but also highlight the depletion of natural sinks, such as forests and oceans, which are critical for carbon sequestration. Ocean acidity has increased 30% since the industrial era, and is now rising faster than at any time in 26,000 years (IPCC, 2021). Global sea level is rising at 3.7 mm per year, nearly double the twentieth-century average (IPCC, 2021).

Moreover, risk transfer occurs when the impacts of climate change disproportionately affect vulnerable populations, while wealthier nations and entities escape the immediate burdens of those effects. This divergence is analyzed by Shaikh et al. Shaikh et al. (2020) explored unequal vulnerability connected to consumption patterns and resource accessibility. This structural mismatch exacerbates inequity, as the poorest regions bear the brunt of climate disasters despite their minimal contributions to global emissions. The disparities between production-based and consumption-based footprints further illuminate this conversation. Production-based (PB) emissions account for the greenhouse gases emitted during the manufacture of goods within a country, while consumption-based (CB) emissions account for the GHGs associated with the consumption of goods, regardless of where they are produced. For example, MRIO (Multi-Regional Input-Output) models have been vital for illustrating these distinctions, in which economic activities in one country can significantly affect emissions in another (Giljum et al., 2015; Sodersten et al., 2018). This is significant for global policy discussions, as consumption in wealthier countries often leads to environmental degradation in less developed regions, as articulated by Lin et al. (2015). Climate-related disasters caused 45,000 deaths and over USD 260 billion in economic losses in 2022 alone (UNDRR, 2023). Air pollution linked to fossil fuels causes 8.7 million premature deaths annually (Vohra et al., 2021).

Sustainability should address not just emissions metrics but also the profound inequalities tied to class and regional disparities in exposure and capacity for climate adaptation. Class-based analyses, as explored by Lettenmeier et al. (2012), reveal that lower-income households typically bear a heavier burden of environmental impacts, owing to their limited capacity to implement adaptive measures in the face of climate change. This situation is exacerbated in regions highly dependent on agriculture, where soil degradation and resource depletion occur, further trapping these populations in cycles of poverty and vulnerability. Incorporating solutions such as border carbon adjustments and the formation of climate clubs emerges as a potential structural response to address both emissions and the equity challenges posed by climate change. Border carbon adjustments, which impose tariffs on imported goods based on their carbon content, can mitigate the outsourcing of emissions and encourage domestic industries to adopt greener practices. This concept is presented by Lin et al. (2015) and Hu et al. (2016), who advocate for such measures in trade agreements to promote sustainable development across borders.

Furthermore, climate clubs, consortia of countries that commit to rigorous climate policies, create forums for collaboration and mutual benefits to support collective climate action. This notion is highlighted by the argument for shared responsibility among nations, particularly in addressing transnational environmental challenges. Some research highlights the importance of integrating sustainability into dietary guidelines as a systemic response to emissions growth and health outcomes (Kovacs et al., 2021). The disparities evident in dietary carbon footprints reinforce the argument for policy interventions that target not only production methods but also consumption habits, particularly among affluent populations whose dietary choices impose significant environmental burdens (Sáez-Almendros et al., 2013). Key studies thus illustrate that a comprehensive approach to addressing the climate crisis hinges on understanding the interplay among various footprint measures, the mechanisms of emissions generation, and socio-economic disparities across different classes and regions (Heinonen & Junnila, 2011). Effective structural changes and policy initiatives are paramount in shifting the systemic roots of the climate crisis toward a sustainable and equitable global future.

Figure 1.4 shows the difference between production-based (territorial) emissions and consumption-based (import-adjusted) emissions in Multi-Regional Input–Output (MRIO) modeling. While production-based accounting assigns emissions to where goods are made, consumption-based accounting reassigns those emissions to the final consumers through global supply chains. Figure 1.4 demonstrates how this approach uncovers carbon transfer from the Global North to the Global South, where resource-intensive manufacturing often happens.

1.9 Hegemony and Ecology: Case of the Dutch, British, and the United States

The evolution of hegemony in civilizations such as the Dutch, British, and United States has been significantly influenced by their access to resources, military capabilities, energy advantages, and financial architectures, which all shape their geopolitical orientations and strengths. Historically, the Dutch Republic emerged as a formidable maritime power in the seventeenth century, dominating global trade through its innovative methods of resource extraction and aggressive trade policies. The establishment of a vast colonial empire enabled it to exploit natural resources from various territories, thereby reinforcing its wealth and military capabilities.

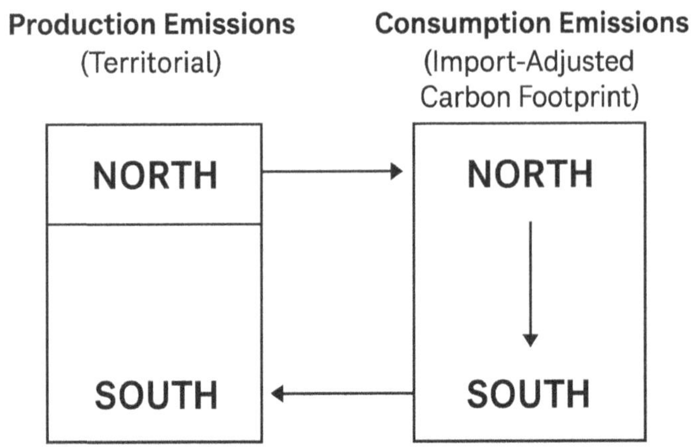

Fig. 1.4 Production-Based vs. Consumption-Based Emissions in MRIO Frameworks

However, this hegemony was undermined by changing dynamics of international power, leading to its decline as Britain ascended to become the new hegemon. The British Empire expanded its reach through industrialization, characterized by significant advances in military logistics and energy acquisition, particularly through coal and later oil, which were vital for maintaining its naval dominance and colonial interests (Costigan et al., 2017; Vlados, 2020). The United States rose to hegemony in the twentieth century, particularly after World War II, supported by the establishment of the US dollar as the global reserve currency. This status afforded the US significant political leverage and control over international trade (Costigan et al., 2017). The confluence of military capabilities, unmatched technological innovation, and control over vital energy resources further solidified US dominance, especially during the Cold War era (Vlados, 2020).

As these hegemonic powers rose and fell, they invariably impacted ecological systems. The extraction of resources has not only altered economies but has led to significant environmental degradation, highlighting ecological limits that challenge existing structures. Restructuring is triggered when these limits are reached, as evidenced by the increasing frequency of environmental crises and the recognition that unchecked resource exploitation yields dire consequences for biodiversity (Silver, 2017). Empirical studies indicate that human activities, particularly land-use transformation, significantly diminish species richness and disrupt ecological functions, signaling that ecological realities are increasingly imposing constraints on hegemonic pursuits (Newbold, Oppenheimer, et al., 2020b).

The present landscape of global hegemony includes rising contenders such as China and the European Union, both of which assert their influence amid ongoing ecological challenges. China, through its Belt and Road Initiative and resource acquisition strategies, aims to shape global trade patterns while facing ecological constraints tied to pollution and resource depletion (Vlados, 2020). The EU positions itself as a counterweight to US hegemony, advocating for ecological governance and climate action as integral to its political and economic frameworks (Bulkeley & Castán Broto, 2013). The competition among these powers raises additional questions about the future of global governance in the face of climate change. Urban political ecologies illustrate how local governance systems adapt to global pressures, though often replete with inequities that exacerbate social strife. The reconfiguration of governance structures through public-private partnerships signifies innovative responses to climate change, yet they risk reinforcing existing inequities (Bulkeley & Castán Broto, 2013; Heynen et al., 2006). As nations pursue their hegemonic aspirations, the inevitable restructuring initiated by ecological limits emerges as a dominant theme in political discourse.

As the climate crisis intensifies, financial architectures built on resource extraction are under scrutiny, compelling nations to reconsider their strategies. The historical precedent of hegemonic transitions shows that reliance on specific resources is tenuous, as exemplified by the British Empire's dependence on coal, which contributed to industrial pollution and global civil discontent. Similarly, the US's reliance on fossil fuels underscores the need to reevaluate energy strategies that prioritize sustainability (Costigan et al., 2017; Dixon, 2014). The dynamics of hegemonic power appear increasingly fragile as ecological realities necessitate comprehensive restructuring. Agricultural transformations driven by resource maximization have contributed to critical biodiversity losses, exposing the limitations of current practices in sustaining long-term ecological integrity and human livelihoods (Newbold, Oppenheimer, et al., 2020b). As nations grapple with climate-induced disruptions, the resulting policy adaptations signal a shift toward more sustainable and equitable governance frameworks that prioritize ecological health alongside economic imperatives.

1.10 Comparative World Systems Vignettes

Examining world systems through the lens of environmental impacts and socioeconomic adaptations reveals extensive challenges and intricate interrelations. Table 1.3 compares five regions based on their drivers, risks, impacts, levers, and indicators. Apply the indicators for fast screening and tracking. In the Arctic, climate change has catalyzed a transformative shift in shipping routes, significantly altering global maritime transit patterns. As polar ice melts, new shipping lanes emerge, dramatically reducing travel distances between major economic centers, thereby posing various geopolitical security risks. The thawing Arctic not only reveals access to previously unreachable natural resources, such as oil and minerals, but simultaneously heightens the potential for territorial disputes and environmental degradation (Kundzewicz et al., 2014; Rocha et al., 2022). The implications of increased maritime traffic also extend to ecological consequences, such as heightened risks of spills and increased greenhouse gas emissions as shipping activities intensify in these fragile ecosystems (Kundzewicz et al., 2014).

Table 1.3 Comparative World Systems Vignettes

Region	Focal Area	Key drivers	Ecological risk	Socio-economic impact	Governance lever	Indicator examples
Arctic	New shipping and extraction	Warming, ice loss, and resource rush	Oil spills, habitat loss, and black carbon	Route competition, security tensions, and Indigenous rights pressure	Polar codes, no drilling zones, and black carbon limits	Summer sea ice extent, shipping days per season, and spill incidents
Amazon	Commodity frontier and tipping risk	Beef, soy, mining, and roads	Forest dieback, fire, and rainfall shift	Land conflict, health costs, and loss of ecosystem work	Deforestation moratoria, supply chain due diligence, and protected areas	Annual forest loss percent, fire alerts, and embargoed farms count
Sahel	Mobility and agro-pastoral adaptation	Rainfall shocks, pasture decline, and border controls	Desertification and water stress	Food insecurity, displacement, and conflict events	Pasture corridors, water points, and climate services	Rainfall coefficient of variation, NDVI trend, and conflict events per year
Small island states	Loss, damage, and relocation	Sea level rise, cyclones, and coral loss	Coastal erosion, saltwater intrusion	GDP hits, housing loss, and cultural loss	Loss and damage finance, resilient housing codes, and planned relocation	Sea level rise mm per year, disaster loss as a percent of GDP, and households relocated
Megacities	Heat, flood, and infrastructure triage	Urban sprawl, paving, and river encroachment	Urban heat islands, pluvial and fluvial floods	Mortality spikes, service outages, and inequality	Heat action plans, green infrastructure, and risk-based zoning	Days above 35 C WBGT, people in floodplains, and tree canopy percent

Transitioning to the Amazon, the concept of tipping points becomes central regarding environmental degradation and socio-economic exploitation. The Amazon serves as a critical agricultural frontier, where agribusiness expansion has led to rampant deforestation and biodiversity loss. Research indicates that the interplay between local smallholder farmers and multinational agribusinesses creates a complex land-use dynamic that often exacerbates socio-economic inequalities and environmental vulnerabilities (Coca et al., 2023; Ioris, 2019). Tipping risks are exacerbated by agribusiness strategies that not only marginalize local communities but also destabilize the regional ecological balance, leading to potential irreversible changes (Coca et al., 2023). This transition fuels conflict over land rights and resource access, further complicating agrarian reform efforts Nguyen et al., 2021).

In the Sahel region, movement patterns among agro-pastoral communities illustrate adaptive strategies in response to environmental changes. The Sahel faces significant climate variability, where erratic rainfall and droughts challenge traditional farming and pastoral systems. These changes not only threaten food security but also heighten conflict risks as pastoralists and agriculturalists compete over dwindling resources (Coca et al., 2023; Kundzewicz et al., 2014). The movement of people and livestock in response to fluctuating environmental conditions showcases resilience through mobility. However, the persistence of conflict and competition for land renders these adaptive strategies precarious (Nguyen et al., 2021). Effective solutions require cross-border cooperation and investments in infrastructural resilience to manage these dynamic interdependencies (Coca et al., 2023; Nguyen et al., 2021).

Small island nations epitomize vulnerability in the context of climate change, grappling with loss and damage finance as they face existential threats from rising sea levels and extreme weather events. The debate regarding relocation becomes increasingly prominent as communities consider the feasibility of moving populations from increasingly uninhabitable areas (Kapetas & Fenner, 2020). The economic implications of such decisions are profound, as loss of land entails not merely physical displacement but also losses in cultural identity and connectivity (O'Donnell et al., 2021). Funding mechanisms for climate-induced damages remain inadequate and contentious, posing obstacles to effective policy responses and adaptation strategies (Chang et al., 2014.

Cities, as central nodes in socio-economic networks, face acute environmental challenges, including urban heat and flooding. Rapid urbanization exacerbates these vulnerabilities due to inadequate infrastructure and poor planning (Green et al., 2021; Takin et al., 2023). Approaches such as green infrastructure and sustainable drainage systems (SuDS) are gaining traction as part of holistic urban flood management strategies designed to enhance resilience (Green et al., 2021; Vercruysse et al., 2019). These adaptive frameworks not only mitigate flooding risks but also foster social and ecological benefits, suggesting a shift from conventional engineering solutions to integrated approaches that consider community needs and environmental health (Takin et al., 2023). The interdependence of urban infrastructures requires a reconceptualization of risk management, emphasizing the

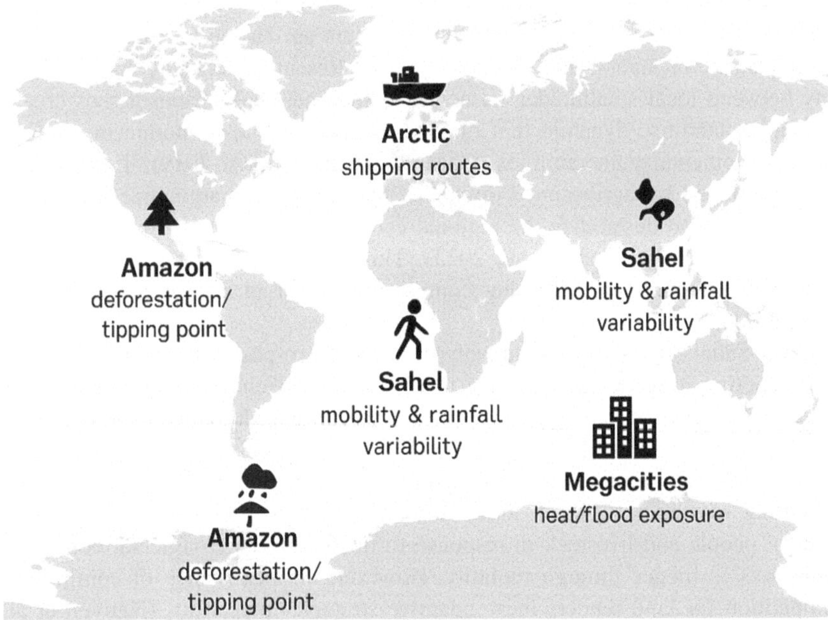

Fig. 1.5 Global Risk Hotspots Across World-System Zones

need for interconnected systems capable of responding to compounded hazards (Ogwu et al., 2025; Vercruysse et al., 2019).

Figure 1.5 attempts to synthesize the five world-system vignettes, highlighting major geospatial hotspots where climate and socio-ecological risks intersect. Figure 1.5 identifies the Arctic (rapid sea-ice decline and emerging shipping routes), the Amazon (accelerating deforestation and approaching biome tipping thresholds), the Sahel (rainfall variability, mobility systems, and livelihood precarity), Small Island Developing States (extreme vulnerability to sea-level rise and coastal hazards), and global megacities (concentrated exposure to heat extremes, flooding, and infrastructural stress). Together, these hotspots illustrate the differentiated yet interconnected pressures shaping contemporary planetary risk landscapes and underscore the need for region-specific and justice-centered adaptation strategies.

1.11 Integrated Framework for World Systems

In the domain of systems thinking sustainability research, various methods and analytical frameworks are vital for understanding material and energy flows, as well as the environmental impacts associated with human activities. Tools such as material and energy flow analyses, MRIO analysis, ecological footprinting, remote sensing, and network analysis have been extensively utilized. Each of these methods offers unique insights into the interplay between human consumption and environmental sustainability. The concept of ecological footprinting, particularly in studies examining carbon and water footprints, is essential for assessing the sustainability of agricultural practices (Cruz-Pérez et al., 2022; Mohan et al., 2022; Scheepers & Jordaan, 2016). This approach enables quantification of various environmental impacts associated with production systems, facilitating the evaluation of resource use over time and across regions. For instance, scholars have effectively quantified carbon footprints in various agricultural settings using LCA, enriching our understanding of sustainability (Mohan et al., 2022). Furthermore, studies employing the water footprint methodology emphasize the importance of climate data in assessing water use and its implications for sustainable practices (Azhar et al., 2023; Zhang et al., 2019).

Data sources critical to these analyses include institutions such as the International Energy Agency (IEA), FAOSTAT, the World Bank, and EXIOBASE. These organizations provide comprehensive datasets that allow researchers to conduct rigorous analyses of energy consumption, agricultural outputs, and environmental impacts. For instance, the IEA provides vital energy statistics that inform assessments of carbon footprints associated with different energy consumption patterns (Lee, 2019a). FAOSTAT offers data on agriculture and food production, which are crucial for understanding the relationships between land use, resource consumption, and environmental sustainability (Asube & Sinadjan, 2021). To ensure a reproducible workflow in research, the integration of code notebooks and versioned datasets is indispensable. Tools such as Jupyter Notebooks can be utilized to document the analytical process, making it easier for researchers to share methodologies and results. This practice not only enhances transparency but also allows for continuous improvement and adaptation of methodologies based on new findings or shifting paradigms in the field (Fortun et al., 2021).

Moreover, pairing quantitative analyses with community-based fieldwork can significantly enhance the relevance and applicability of research findings. Conducting fieldwork allows researchers to engage with local communities and incorporate their insights and values into sustainability assessments. This collaborative approach can reveal local variations in resource use and environmental impacts, leading to more nuanced understandings of sustainability challenges and solutions. For instance, collaborative studies in Taiwan and Kenya have demonstrated the importance of local data and community perspectives in informing sustainable agricultural practices (Lee, 2019b; Bosire et al., 2015).

To create a comprehensive integrated framework that presents a matrix involving system position, energy regime, planetary pressure, and distributional outcomes, it is essential to define specific indicators for each category and outline data sources for replication. Simultaneously, step-by-step guidance for applying this matrix and developing testable hypotheses tied to measurable thresholds will be provided.

1. **System Position**: This refers to the current state of the system under evaluation, involving operational, environmental, and sociocultural dimensions. Indicators may include:
   - **Operational Efficiency**: Measured by energy consumption per unit of output.
   - **Environmental Impact**: Evaluated through emissions data from relevant regulatory frameworks (Grunwald & Rösch, 2011).
   - **Social Inclusion**: Assessed via community engagement and impact metrics, as outlined in sustainability assessment research (Pope et al., 2017).
   - **Data Sources**: This could include sustainability reports from organizations, governmental databases, and non-governmental organizations focused on sustainability metrics.
2. **Energy Regime**: This aspect evaluates the source and use of energy within the system. Indicators might consist of:
   - **Type of Energy Utilized**: Categorized by renewable versus non-renewable sources.
   - **Efficiency of Energy Use**: Tracked via the ratio of energy input to effective output (Persdotter Isaksson et al., 2019).
   - **Greenhouse Gas Emissions**: Quantified to assess the carbon footprint relative to operations (Grunwald & Rösch, 2011).
   - **Data Sources**: This could be national energy statistics or reports from energy regulatory bodies.
3. **Planetary Pressure**: This component focuses on the ecological footprint and environmental degradation linked to the system's operations. Indicators may include:
   - **Biodiversity Impact**: Measured through species preservation rates in operational areas (Hussain et al., 2017).
   - **Resource Depletion Metrics**: Assessed with frameworks for tracking sustainable resource use (Liu et al., 2018).
   - **Pollution Levels**: Documented via environmental monitoring agencies and studies (Slätmo et al., 2017).
   - **Data Sources**: this may include environmental impact assessments and databases from environmental protection agencies.
4. **Distributional Outcome**: This area evaluates the equity of resource distribution and social benefits derived from the system's operations. Indicators might be:
   - **Economic Equality Index**: A measure of wealth distribution within affected communities (Whitelock, 2019).
   - **Access to Resources**: Quantified by demographic outreach metrics and resource availability studies (Shelton et al., 2020).
   - **Health Impacts**: Analyzed through public health surveys reflecting community health outcomes (Ramani et al., 2011).
   - **Data Sources**: This can be derived from socio-economic studies, public health records, and community welfare assessments.

Step-by-Step Guidance for Application

1. **Define Scope**: Clearly outline the objectives and boundaries of the matrix application. Identify stakeholders involved and their roles.
2. **Select Indicators**: Choose relevant indicators from the above categories that align with specific goals and existing frameworks for sustainability assessment tailored to the specific scenario under investigation (Pope et al., 2017).
3. **Collect Data**: Utilize the identified data sources to gather quantitative and qualitative data pertinent to each indicator (Bastas & Liyanage, 2018). Following established protocols ensures data integrity and comparability.
4. **Analyze Data**: Apply statistical and modeling methods to assess the relationships between indicators within each cell of the matrix. Emphasize integrative approaches that reveal underlying patterns (Asif et al., 2011).
5. **Develop Testable Hypotheses**: Formulate hypotheses based on findings such as:
   - "Increasing renewable energy usage by 20% will lead to a 10% reduction in greenhouse gas emissions and improve community well-being metrics by 15%."
   - Assign measurable thresholds to these hypotheses to facilitate future evaluations (Aghajani et al., 2023).
6. **Implement Monitoring Mechanisms**: Establish a feedback loop for continuous monitoring and refinement of strategies based on real-time data and stakeholder feedback (Shiroyama et al., 2012).

1.12 Conclusion

Climate crisis arises from the structure of the capitalist world economy. Longue durée history shows energy and material regimes drive hegemonic rise and press decline. Colonial expansion organized extraction, coerced labor, and ecological burdens across peripheries. Earth system pressures couple with accumulation, trade, and finance to produce systemic risk. This chapter advances a framework linking system position, energy regime, planetary pressure, and distributional outcome. The framework sets hypotheses, indicators, and methods grounded in material and energy flow accounts, multiregional input-output models, footprinting, and remote sensing. Comparative cases on Dutch, British, and United States orders support the argument and clarify limits. Analysis also flags feedbacks that shape present constraints, including carbon lock-in and infrastructural path dependence. Future research should test the framework across present contenders and sectoral frontiers, including critical minerals, agroecosystems, and coastal zones. Policy design should target unequal exchange, supply chain governance, border carbon adjustments, and loss and damage finance. Measurement should include distributional audits, exposure and capacity metrics, and transparent data pipelines.

Governance should align industrial policy, labor protections, and community rights with decarbonization. The next part develops metrics and data strategies for replication and cross-scale comparison.

References

Adepoju, O. O., David, L. O., & Nwulu, N. I. (2022). Analysing the impact of human capital on renewable energy penetration: A bibliometric reviews. *Sustainability, 14*(14), 8852.

Adger, W. N. (2000). Social and ecological resilience: are they related? *Progress in Human Geography, 24*(3), 347–364.

Aghajani, M., Ruge, G., & Jugdev, K. (2023). An integrative review of project portfolio management literature: Thematic findings on sustainability mindset, assessment, and integration. *Project Management Journal, 54*(6), 629–650.

Asif, M., Searcy, C., Zutshi, A., & Ahmad, N. (2011). An integrated management systems approach to corporate sustainability. *European Business Review, 23*(4), 353–367.

Asmi, N., Andriani, R., & Siradjuddin, S. (2023). Early economic thought and the dynamics of economic thought. *APLIKATIF: Journal of Research Trends in Social Sciences and Humanities, 2*(2), 123–134.

Asube, L. C. S., & Sinadjan, R. L. (2021). Estimations of carbon footprint from electricity consumption during Covid-19 lockdown and pre-lockdown in Butuan city. *The International Archives of the Photogrammetry, Remote Sensing and Spatial Information Sciences, 46*, 49–56.

Azhar, N., Ahmad, W. A. A. W., Mohammed, S. A., Ahmad, A. G., Santiagoo, R., Vijean, V., & Ahmad, R. (2023, July). Assessment on water footprint of paddy cultivation in Kedah. In IOP Conference Series: Earth and Environmental Science (Vol. 1216 1, p. 012038). IOP Publishing.

de Baan, L., Curran, M., Rondinini, C., Visconti, P., Hellweg, S., & Koellner, T. (2015). High-resolution assessment of land use impacts on biodiversity in life cycle assessment using species habitat suitability models. *Environmental Science & Technology, 49*(4), 2237–2244.

Barnosky, A. D., Hadly, E. A., Gonzalez, P., Head, J., Polly, P. D., Lawing, A. M., Eronen, J. T., Ackerly, D. D., Alex, K., Biber, E., & Blois, J. (2017). Merging paleobiology with conservation biology to guide the future of terrestrial ecosystems. *Science, 355*(6325), eaah4787.

Barretto, R. (2024). Psyche, body, and environment: Integrations of the self in times of climate change. https://doi.org/10.56238/sevened2024.031-060.

Bastas, A., & Liyanage, K. (2018). ISO 9001 and supply chain integration principles based sustainable development: A Delphi study. *Sustainability, 10*(12), 4569.

Bellizzi, S., Popescu, C., Napodano, C. M. P., Fiamma, M., & Cegolon, L. (2023). Global health, climate change and migration: The need for recognition of "climate refugees". *Journal of Global Health, 13*, 03011.

Boakes, E. H., Dalin, C., Etard, A., & Newbold, T. (2023). Impacts of global food supply on biodiversity via land use and climate change. *bioRxiv*, 2023-05.

Bogdanov, D., Farfan, J., Sadovskaia, K., Aghahosseini, A., Child, M., Gulagi, A., Oyewo, A. S., & de Souza Noel Simas Barbosa, L. and Breyer, C. (2019). Radical transformation pathway towards sustainable electricity via evolutionary steps. *Nature Communications, 10*(1), 1–16.

Larson, G., Boivin, N., Zeder, M., Fuller, D., Alison, C., Erlandson, J., Tim, D., & Petraglia, M. (2016). Ecological consequences of human niche construction: Examining long-term anthropogenic shaping of global species distributions. *Proceedings of the National Academy of Sciences of USA, 113*(23), 6388–6396.

Bosire, C. K., Ogutu, J. O., Said, M. Y., Krol, M. S., de Leeuw, J., & Hoekstra, A. Y. (2015). Trends and spatial variation in water and land footprints of meat and milk production systems in Kenya. *Agriculture, Ecosystems & Environment, 205*, 36–47.

Tobias, J. A., Durant, S. M., & Pettorelli, N. (2021). Improving predictions of climate change–land use change interactions. *Trends in Ecology & Evolution, 36*(1), 29–38.

Bulatov, Y., Kryukov, A., & Arsentyev, G. (2018). Application of energy routers in railway power supply systems. In *MATEC Web of Conferences* (Vol. 239, p. 01047). EDP Sciences.

Bulkeley, H., & Castán Broto, V. (2013). Government by experiment? Global cities and the governing of climate change. *Transactions of the Institute of British Geographers, 38*(3), 361–375.

Bunker, S. G. (2005). How ecologically uneven developments put the spin on the treadmill of production. *Organization & Environment, 18*(1), 38–54.

Cardwell, F. S., & Elliott, S. J. (2013). Making the links: Do we connect climate change with health? A qualitative case study from Canada. *BMC Public Health, 13*(1), 208.

Chan, S. F., Rubenstein, D. R., Chen, I. C., Fan, Y. M., Tsai, H. Y., Zheng, Y. W., & Shen, S. F. (2023). Higher temperature variability in deforested mountain regions impacts the competitive advantage of nocturnal species. *Proceedings of the Royal Society B, 290*(1999), 20230529.

Chang, S. E., McDaniels, T., Fox, J., Dhariwal, R., & Longstaff, H. (2014). Toward disaster-resilient cities: Characterizing resilience of infrastructure systems with expert judgments. *Risk Analysis, 34*(3), 416–434.

Clack, C. T., Qvist, S. A., Apt, J., Bazilian, M., Brandt, A. R., Caldeira, K., Davis, S. J., Diakov, V., Handschy, M. A., Hines, P. D., & Jaramillo, P. (2017). Evaluation of a proposal for reliable low-cost grid power with 100% wind, water, and solar. *Proceedings of the National Academy of Sciences, 114*(26), 6722–6727.

Coca, E., Soyer, G., & Barbosa Jr, R. (2023). Matopiba's disputed agricultural frontier: Between commodity crops and agrarian reform.

Costigan, T., Cottle, D., & Keys, A. (2017). The US dollar as the global reserve currency: Implications for US hegemony. *World Review of Political Economy, 8*(1), 104–122.

Crabb, L., Agar, C. C., & Böhm, S. (2024). Internal colonialism as socio-ecological fix: The case of New Clark City in the Philippines. *Antipode, 56*(4), 1233–1263.

Cruz-Pérez, N., Santamarta, J. C., García-Gil, A., Rodríguez-Martín, J., Miralles-Wilhelm, F., Hernández-Alemán, A., & Aldaya, M. M. (2022). Water footprint of the water cycle of Gran Canaria and Tenerife (Canary Islands, Spain). *Water, 14*(6), 934.

Dixon, M. (2014). The land grab, finance capital, and food regime restructuring: the case of Egypt. *Review of African Political Economy, 41*(140), 232–248.

Dudgeon, D., Arthington, A. H., Gessner, M. O., Kawabata, Z. I., Knowler, D. J., Lévêque, C., Naiman, R. J., Prieur-Richard, A. H., Soto, D., Stiassny, M. L., & Sullivan, C. A. (2006). Freshwater biodiversity: Importance, threats, status and conservation challenges. *Biological Reviews, 81*(2), 163–182.

Edziah, B. K., Sun, H., Anyigbah, E., & Li, L. (2021). Human capital and energy efficiency: Evidence from developing countries. *American Journal of Industrial and Business Management, 11*(6), 599–610.

Estes, L. D., Paroz, L. L., Bradley, B. A., Green, J. M., Hole, D. G., Holness, S., Ziv, G., Oppenheimer, M. G., & Wilcove, D. S. (2014). Using changes in agricultural utility to quantify future climate-induced risk to conservation. *Conservation Biology, 28*(2), 427–437.

Feldman, D. J., Margolis, R. M., & Jones-Albertus, R. (2018). *Impact of research and development, analysis, and standardization on PV project financing costs* (No. NREL/TP-6A20-70939). National Renewable Energy Lab. (NREL), Golden, CO (United States).

Fortun, K., Adams, J., Schütz, T., & Knowles, S. G. (2021). Knowledge infrastructure and research agendas for quotidian Anthropocenes: Critical localism with planetary scope. *The Anthropocene Review, 8*(2), 169–182.

Fritz, S., See, L., Perger, C., McCallum, I., Schill, C., Schepaschenko, D., Duerauer, M., Karner, M., Dresel, C., Laso-Bayas, J. C., & Lesiv, M. (2017). A global dataset of crowdsourced land cover and land use reference data. *Scientific Data, 4*(1), 1–8.

Geyer, R., Lindner, J. P., Stoms, D. M., Davis, F. W., & Wittstock, B. (2010). Coupling GIS and LCA for biodiversity assessments of land use: Part 2: Impact assessment. *The International Journal of Life Cycle Assessment, 15*(7), 692–703.

Giljum, S., Bruckner, M., & Martinez, A. (2015). Material footprint assessment in a global input-output framework. *Journal of Industrial Ecology, 19*(5), 792–804.

Green, D., O'Donnell, E., Johnson, M., Slater, L., Thorne, C., Zheng, S., Stirling, R., Chan, F. K., Li, L., & Boothroyd, R. J. (2021). Green infrastructure: The future of urban flood risk management? *Wiley Interdisciplinary Reviews: Water, 8*(6), e1560.
Greim, P., Solomon, A. A., & Breyer, C. (2020). Assessment of lithium criticality in the global energy transition and addressing policy gaps in transportation. *Nature Communications, 11*(1), 4570.
Grunwald, A., & Rösch, C. (2011). Sustainability assessment of energy technologies: Towards an integrative framework. *Energy, Sustainability and Society, 1*(1), 3.
Harris, R., & Lauderdale, P. (2002). Globalization, neoliberalism, and post-apartheid South Africa. *Journal of Asian and African Studies, 37*(3–5), 422–435.
He, G., Avrin, A. P., Nelson, J. H., Johnston, J., Mileva, A., Tian, J., & Kammen, D. M. (2016). SWITCH-China: A systems approach to decarbonizing China's power system. *Environmental Science & Technology, 50*(11), 5467–5473.
Heinonen, J., & Junnila, S. (2011). A carbon consumption comparison of rural and urban lifestyles. *Sustainability, 3*(8), 1234–1249.
Henderson, K., & Shorette, K. (2017). Environmentalism in the periphery: Institutional embeddedness and deforestation among fifteen palm oil producers, 1990–2012. *Journal of World-Systems Research, 23*(2), 269–297.
Heynen, N., Kaika, M., & Swyngedouw, E. (2006). Urban political ecology: Politicizing the production of urban natures. In *In the nature of cities* (pp. 16–35). Routledge.
Hoekstra, A. Y., & Wiedmann, T. O. (2014). Humanity's unsustainable environmental footprint. *Science, 344*(6188), 1114–1117.
Hussain, M., Alameeri, A., & Ajmal, M. M. (2017). Prioritizing sustainable practices of service organizations: An empirical evidence from automobile dealers in UAE. *International Journal of Information Systems in the Service Sector (IJISSS), 9*(1), 22–36.
Ioris, A. A. (2019). Peasant farming in the southern tracts of the Amazon: The reluctant alterity of agribusiness. *Perspectives on Global Development and Technology, 18*(4), 375–400.
Persdotter Isaksson, M., Hulthén, H., & Forslund, H. (2019). Environmentally sustainable logistics performance management process integration between buyers and 3PLs. *Sustainability, 11*(11), 3061.
Iyiola, A. O., Afolabi, O. A., Alimi, S. K., Akingba, O. O., Izah, S. C., & Ogwu, M. C. (2024). Climate change and water crisis in the Global South. In S. C. Izah, M. C. Ogwu, A. Loukas, & H. Hamidifar (Eds.), *Water crises and sustainable management in the Global South* (pp. 111–140). Springer.
Izah, S. C., Ogwu, M. C., & Shahsavani, A. (2024). Air Pollutants in the context of one health: Fundamentals, sources and impacts. In *Part of the book series: The handbook of Environmental Chemistry*. Springer. https://link.springer.com/book/9783031741647
Jacobson, M. Z., Delucchi, M. A., Bauer, Z. A., Goodman, S. C., Chapman, W. E., Cameron, M. A., Bozonnat, C., Chobadi, L., Clonts, H. A., Enevoldsen, P., & Erwin, J. R. (2017). 100% clean and renewable wind, water, and sunlight all-sector energy roadmaps for 139 countries of the world. *Joule, 1*(1), 108–121.
Jeswani, H. K., Hellweg, S., & Azapagic, A. (2018). Accounting for land use, biodiversity and ecosystem services in life cycle assessment: Impacts of breakfast cereals. *Science of the Total Environment, 645*, 51–59.
Jorgenson, A. K. (2006). Unequal ecological exchange and environmental degradation: A theoretical proposition and cross-national study of deforestation, 1990–2000. *Rural Sociology, 71*(4), 685–712.
Jorgenson, A. K. (2016a). Environment, development, and ecologically unequal exchange. *Sustainability, 8*(3), 227.
Jorgenson, A. K. (2016b). The sociology of ecologically unequal exchange, foreign investment dependence and environmental load displacement: Summary of the literature and implications for sustainability. *Journal of Political Ecology, 23*(1), 334–349.
Kalogirou, M. R., Olson, J., & Davidson, S. (2020). Nursing's metaparadigm, climate change and planetary health. *Nursing Inquiry, 27*(3), e12356.

Kang, S., Havercroft, J., Eisler, J., Wiener, A., & Shaw, J. (2023). Climate change and the challenge to liberalism. *Global Constitutionalism, 12*(1), 1–10.

Kapetas, L., & Fenner, R. (2020). Integrating blue-green and grey infrastructure through an adaptation pathways approach to surface water flooding. *Philosophical Transactions of the Royal Society A, 378*(2168), 20190204.

Kerslake, R. M., & Krishnamurti, C. (2024). New world, or out of this world? Columbus–an exploratory study of HASS and STEM success factors in the first "space" race. *Accounting, Auditing & Accountability Journal, 37*(5), 1457–1486.

Koellner, T., De Baan, L., Beck, T., Brandão, M., Civit, B., Margni, M., L.M., i. C., Saad, R., De Souza, D. M., & Müller-Wenk, R. (2013). UNEP-SETAC guideline on global land use impact assessment on biodiversity and ecosystem services in LCA. *The International Journal of Life Cycle Assessment, 18*(6), 1188–1202.

Kong, X., He, Q., Yang, B., He, W., Xu, F., Janssen, A. B., Kuiper, J. J., van Gerven, L. P., Qin, N., Jiang, Y., & Liu, W. (2017). Hydrological regulation drives regime shifts: Evidence from paleolimnology and ecosystem modeling of a large shallow Chinese lake. *Global Change Biology, 23*(2), 737–754.

Kovacs, B., Miller, L., Heller, M. C., & Rose, D. (2021). The carbon footprint of dietary guidelines around the world: A seven country modeling study. *Nutrition Journal, 20*(1), 15.

Krogmann, D. (2025). Liberal environmentalism and climate change in the polycrisis. *Global Sustainability, 8*, e8.

Kundzewicz, W., Kanae, S., Seneviratne, S. I., Handmer, J., Nicholls, N., Peduzzi, P., Mechler, R., Bouwer, L. M., Arnell, N., Mach, K., & Muir-Wood, R. (2014). Flood risk and climate change: Global and regional perspectives. *Hydrological Sciences Journal, 59*(1), 1–28.

Kosoe, E. A., & Ogwu, M. C. (2025). Sustainable urban planning and environmental management. In M. C. Ogwu & S. Chibueze Izah (Eds.), *Evaluating environmental processes and technologies*. Environmental Science and Engineering Series. Springer. https://doi.org/10.1007/978-3-031-85327-2_14

Lazarus, E. D. (2017). Toward a global classification of coastal anthromes. *Land, 6*(1), 13.

Lee, Y. J. (2019a). Carbon footprint and built-up land footprint: Taipei case study. In *E3S web of conferences* (Vol. 118, p. 03052). EDP Sciences.

Lee, Y. J. (2019b). Ecological footprint and water footprint of Taipei. *Sustainability, 11*(20), 5714.

Legagneux, P., Casajus, N., Cazelles, K., Chevallier, C., Chevrinais, M., Guéry, L., Jacquet, C., Jaffré, M., Naud, M. J., Noisette, F., & Ropars, P. (2018). Our house is burning: Discrepancy in climate change vs. biodiversity coverage in the media as compared to scientific literature. *Frontiers in Ecology and Evolution, 5*, 297310.

Lettenmeier, M., Hirvilammi, T., Laakso, S., Lähteenoja, S., & Aalto, K. (2012). Material footprint of low-income households in Finland—Consequences for the sustainability debate. *Sustainability, 4*(7), 1426–1447.

Li, L., Xiong, W., Duan, W., & Xiong, Y. (2023). Evaluation on substitution of energy transition—An empirical analysis based on factor elasticity. *Frontiers in Energy Research, 10*, 1068936.

Garcia-Lillo, F., Sanchez-Garcia, E., Marco-Lajara, B., & Seva-Larrosa, P. (2023). Renewable energies and sustainable development: A bibliometric overview. *Energies, 16*(3), 1211.

Lin, J., Hu, Y., Cui, S., Kang, J., & Ramaswami, A. (2015). Tracking urban carbon footprints from production and consumption perspectives. *Environmental Research Letters, 10*(5), 054001.

Liu, Y., van Nederveen, S., Wu, C., & Hertogh, M. (2018). Sustainable infrastructure design framework through integration of rating systems and building information modeling. *Advances in Civil Engineering, 2018*(1), 8183536.

Liu, Z., Davis, S. J., Feng, K., Hubacek, K., Liang, S., Anadon, L. D., Chen, B., Liu, J., Yan, J., & Guan, D. (2016). Targeted opportunities to address the climate–trade dilemma in China. *Nature Climate Change, 6*(2), 201–206.

Maibach, E. W., Nisbet, M., Baldwin, P., Akerlof, K., & Diao, G. (2010). Reframing climate change as a public health issue: An exploratory study of public reactions. *BMC Public Health, 10*(1), 299.

McMichael, P. (2013). A food regime genealogy. In *Critical perspectives in rural development studies* (pp. 129–158). Routledge.
Mensing, S. A., Schoolman, E. M., Tunno, I., Noble, P. J., Sagnotti, L., Florindo, F., & Piovesan, G. (2018). Historical ecology reveals landscape transformation coincident with cultural development in central Italy since the Roman Period. *Scientific Reports, 8*(1), 2138.
Milne, S. A. (2021). The Erber: Tracing global trade through a London building. *Urban History, 48*(3), 435–460.
Mkodzongi, G. (2023). The political economy of the climate crisis in Southern Africa. *Review of African Political Economy, 50*(177–178), 374–387.
Mohamed, J. (2004). The political ecology of colonial Somaliland. *Africa, 74*(4), 534–566.
Mohan, S. M., Vidhyavathi, A., Padmarani, S., & Balaji, P. (2022). Carbon footprints and conventional rice cultivation; a case study in Thanjavur District. *Asian Journal of Agricultural Extension, Economics & Sociology, 40*(10), 233–243.
Moore, J. W. (2011). Ecology, capital, and the nature of our times: Accumulation and crisis in the capitalist world-ecology. *Journal of World-Systems Research, 17*(1), 108–147.
Mueller, M. L., & Farhat, K. (2022). Regulation of platform market access by the United States and China: Neo-mercantilism in digital services. *Policy & Internet, 14*(2), 348–367.
Munshi, K. (2007). From farming to international business: The social auspices of entrepreneurship in a growing economy.
Nakano, R., Miwa, T., & Morikawa, T. (2018). Comparative analysis on citizen's subjective responses related to their willingness to pay for renewable energy in Japan using latent variables. *Sustainability, 10*(7), 2423.
Newbold, T., Bentley, L. F., Hill, S. L., Edgar, M. J., Horton, M., Su, G., Şekercioğlu, Ç. H., Collen, B., & Purvis, A. (2020a). Global effects of land use on biodiversity differ among functional groups. *Functional Ecology, 34*(3), 684–693.
Newbold, T., Oppenheimer, P., Etard, A., & Williams, J. J. (2020b). Tropical and Mediterranean biodiversity is disproportionately sensitive to land-use and climate change. *Nature Ecology & Evolution, 4*(12), 1630–1638.
Nguyen, H. D., Fox, D., Dang, D. K., Pham, L. T., Viet Du, Q. V., Nguyen, T. H. T., Dang, T. N., Tran, V. T., Vu, P. L., Nguyen, Q. H., & Nguyen, T. G. (2021). Predicting future urban flood risk using land change and hydraulic modeling in a river watershed in the central Province of Vietnam. *Remote Sensing, 13*(2), 262.
Ogwu, M. C., & Izah, S. C. (2024). Sustainable strategies for air pollution mitigation: Developments economics and technologies. In *Part of the book series: The handbook of Environmental Chemistry* (p. 572). Springer. https://link.springer.com/book/9783031741562
Ogwu, M. C., Kosoe, E. A., Alaka, A. C., & Xu, J. (2025). Green infrastructure and environmental health security: The role of nature-based solutions in improving urban health. In S. C. Izah & M. C. Ogwu (Eds.), *Innovative approaches in environmental health management: Processes, technologies, and strategies for a sustainable future*. Environmental Science and Engineering Series. Springer Nature. https://doi.org/10.1007/978-3-031-81966-7_9
O'Donnell, E. C., Gosling, S. N., Netusil, N. R., Chan, F. S., & Dolman, N. J. (2021). Perceptions of blue-green and grey infrastructure as climate change adaptation strategies for urban water resilience. *Journal of the British Academy, 9*(s9), 143–182.
Patz, J. A., Daszak, P., Tabor, G. M., Aguirre, A. A., Pearl, M., Epstein, J., Wolfe, N. D., Kilpatrick, A. M., Foufopoulos, J., Molyneux, D., & Bradley, D. J. (2004). Unhealthy landscapes: Policy recommendations on land use change and infectious disease emergence. *Environmental Health Perspectives, 112*(10), 1092–1098.
Pope, J., Bond, A., Hugé, J., & Morrison-Saunders, A. (2017). Reconceptualising sustainability assessment. *Environmental Impact Assessment Review, 62*, 205–215.
Le Quéré, C., Andres, R. J., Boden, T., Conway, T., Houghton, R. A., House, J. I., Marland, G., Peters, G. P., Van der Werf, G., Ahlström, A., & Andrew, R. M. (2012). The global carbon budget 1959-2011. *Earth System Science Data Discussions, 5*(2), 1107–1157.

Ramani, T. L., Zietsman, J., Gudmundsson, H., Hall, R. P., & Marsden, G. (2011). Framework for sustainability assessment by transportation agencies. *Transportation Research Record, 2242*(1), 9–18.

Rice, J. (2007). Ecological unequal exchange: Consumption, equity, and unsustainable structural relationships within the global economy. *International Journal of Comparative Sociology, 48*(1), 43–72.

Rice, J. (2009). The transnational organization of production and uneven environmental degradation and change in the world economy. *International Journal of Comparative Sociology, 50*(3–4), 215–236.

Rocha, C. B., de Majo, C., & Dutra, S. (2022). A geo-historical analysis of expanding soybean frontiers in the Brazilian Cerrado. *Historia Ambiental Latinoamericana y Caribeña (HALAC) revista de la Solcha, 12*(2), 217–252.

Rosenow, J., Kern, F., & Rogge, K. (2017). The need for comprehensive and well targeted instrument mixes to stimulate energy transitions: The case of energy efficiency policy. *Energy Research & Social Science, 33*, 95–104.

Rounsevell, M. D. A., Arneth, A., Alexander, P., Brown, D. G., de Noblet-Ducoudré, N., Ellis, E., Finnigan, J., Galvin, K., Grigg, N., Harman, I., & Lennox, J. (2013). Towards decision-based global land use models for improved understanding of the Earth system. *Earth System Dynamics Discussions, 4*(2), 875–925.

Sáez-Almendros, S., Obrador, B., Bach-Faig, A., & Serra-Majem, L. (2013). Environmental footprints of Mediterranean versus Western dietary patterns: Beyond the health benefits of the Mediterranean diet. *Environmental Health, 12*(1), 118.

Sanderson, M. R., & Frey, R. S. (2014). From desert to breadbasket… to desert again? A metabolic rift in the High Plains Aquifer. *Journal of Political Ecology, 21*(1), 516–532.

Sanderson, M. R., & Frey, R. S. (2015). Structural impediments to sustainable groundwater management in the High Plains Aquifer of western Kansas. *Agriculture and Human Values, 32*(3), 401–417.

Scheepers, M. E., & Jordaan, H. (2016). Assessing the blue and green water footprint of lucerne for milk production in South Africa. *Sustainability, 8*(1), 49.

Schipper, A. M., Hilbers, J. P., Meijer, J. R., Antão, L. H., Benítez-López, A., de Jonge, M. M., Leemans, L. H., Scheper, E., Alkemade, R., Doelman, J. C., & Mylius, S. (2020). Projecting terrestrial biodiversity intactness with GLOBIO 4. *Global Change Biology, 26*(2), 760–771.

Schlachtberger, D. P., Brown, T., Schäfer, M., Schramm, S., & Greiner, M. (2018). Cost optimal scenarios of a future highly renewable European electricity system: Exploring the influence of weather data, cost parameters and policy constraints. *Energy, 163*, 100–114.

Schneider, M., & McMichael, P. (2010). Deepening, and repairing, the metabolic rift. *The Journal of Peasant Studies, 37*(3), 461–484.

Sgouridis, S., Csala, D., & Bardi, U. (2016). The sower's way: Quantifying the narrowing net-energy pathways to a global energy transition. *Environmental Research Letters, 11*(9), 094009.

Shaikh, M. A., Hadjikakou, M., & Bryan, B. A. (2020). Patterns and trends in national-level consumption and production-based utilisation of the land-system change planetary boundary. *agriRxiv*, (2020), 20203233479.

Shandra, J. M., Leckband, C., McKinney, L. A., & London, B. (2009). Ecologically unequal exchange, world polity, and biodiversity loss: A cross-national analysis of threatened mammals. *International Journal of Comparative Sociology, 50*(3–4), 285–310.

Shelton, R. C., Chambers, D. A., & Glasgow, R. E. (2020). An extension of RE-AIM to enhance sustainability: Addressing dynamic context and promoting health equity over time. *Frontiers in Public Health, 8*, 134.

Shiroyama, H., Yarime, M., Matsuo, M., Schroeder, H., Scholz, R., & Ulrich, A. E. (2012). Governance for sustainability: Knowledge integration and multi-actor dimensions in risk management. *Sustainability Science, 7*(Suppl 1), 45–55.

Shu, Y., Zhang, L., Zhang, Y., Wang, Y., Lu, G., Yuan, B., & Xia, P. (2021). Carbon peak and carbon neutrality path for China's power industry. *Strategic Study of Chinese Academy of Engineering, 23*(6), 1–14.

de Paula Sieverding, T., Kulcar, V., & Schmidt, K. (2024). Act like there is a tomorrow—Contact and affinity with younger people and legacy motivation as predictors of climate protection among older people. *Sustainability, 16*(4), 1477.

Silver, J. (2017). The climate crisis, carbon capital and urbanisation: An urban political ecology of low-carbon restructuring in Mbale. *Environment and Planning A, 49*(7), 1477–1499.

Simpson, M., & Pizarro Choy, A. (2024). Building decolonial climate justice movements: Four tensions. *Dialogues in Human Geography, 14*(2), 269–272.

Slätmo, E., Fischer, K., & Röös, E. (2017). The framing of sustainability in sustainability assessment frameworks for agriculture. *Sociologia Ruralis, 57*(3), 378–395.

Sodersten, C. J. H., Wood, R., & Hertwich, E. G. (2018). Endogenizing capital in MRIO models: The implications for consumption-based accounting. *Environmental Science & Technology, 52*(22), 13250–13259.

Steffen, W., Grinevald, J., Crutzen, P., & McNeill, J. (2011). The Anthropocene: Conceptual and historical perspectives. *Philosophical Transactions of the Royal Society A: Mathematical, Physical and Engineering Sciences, 369*(1938), 842–867.

Stoddard, I., Anderson, K., Capstick, S., Carton, W., Depledge, J., Facer, K., Gough, C., Hache, F., Hoolohan, C., Hultman, M., & Hällström, N. (2021). Three decades of climate mitigation: Why haven't we bent the global emissions curve? *Annual Review of Environment and Resources, 46*(1), 653–689.

Takin, M., Cilliers, E. J., & Ghosh, S. (2023). Advancing flood resilience: The nexus between flood risk management, green infrastructure, and resilience. *Frontiers in Sustainable Cities, 5*, 1186885.

Titeux, N., Henle, K., Mihoub, J. B., Regos, A., Geijzendorffer, I. R., Cramer, W., Verburg, P. H., & Brotons, L. (2016). Biodiversity scenarios neglect future land-use changes. *Global Change Biology, 22*(7), 2505–2515.

Tzanopoulos, J., Mouttet, R., Letourneau, A., Vogiatzakis, I. N., Potts, S. G., Henle, K., Mathevet, R., & Marty, P. (2013). Scale sensitivity of drivers of environmental change across Europe. *Global Environmental Change, 23*(1), 167–178.

Vercruysse, K., Dawson, D. A., & Wright, N. (2019). Interoperability: A conceptual framework to bridge the gap between multifunctional and multisystem urban flood management. *Journal of Flood Risk Management, 12*(S2), e12535.

Vlados, C. (2020). The dynamics of the current global restructuring and contemporary framework of the US–China trade war. *Global Journal of Emerging Market Economies, 12*(1), 4–23.

Wang, F., Jin, M., Li, J., Zhang, Y., & Chen, J. (2022). Profound impact of economic openness and digital economy towards a sustainable development: A new look at RCEP economies. *Sustainability, 14*(21), 13922.

Wang, Y., Zhou, T., Chen, H., & Rong, Z. (2019). Environmental homogenization or heterogenization? The effects of globalization on carbon dioxide emissions, 1970–2014. *Sustainability, 11*(10), 2752.

Whitelock, V. G. (2019). Multidimensional environmental social governance sustainability framework: Integration, using a purchasing, operations, and supply chain management context. *Sustainable Development, 27*(5), 923–931.

Williams, A., & May, J. (2025). Racial capitalism and the workhouse–Plantation Nexus in the Atlantic World. Antipode 57 5 2015–2044.

Willis, K. J., Araujo, M. B., Bennett, K. D., Figueroa-Rangel, B., Froyd, C. A., & Myers, N. (2007). How can a knowledge of the past help to conserve the future? Biodiversity conservation and the relevance of long-term ecological studies. *Philosophical Transactions of the Royal Society B: Biological Sciences, 362*(1478), 175–187.

Wilting, H. C., Schipper, A. M., Bakkenes, M., Meijer, J. R., & Huijbregts, M. A. (2017). Quantifying biodiversity losses due to human consumption: A global-scale footprint analysis. *Environmental Science & Technology, 51*(6), 3298–3306.

Winberg, J., Smith, H. G., & Ekroos, J. (2023). Bioenergy crops, biodiversity and ecosystem services in temperate agricultural landscapes—A review of synergies and trade-offs. *GCB Bioenergy, 15*(10), 1204–1220.

Workman, A., Blashki, G., Bowen, K. J., Karoly, D. J., & Wiseman, J. (2018). The political economy of health co-benefits: embedding health in the climate change agenda. *International Journal of Environmental Research and Public Health, 15*(4), 674.

Yasuoka, J., & Levins, R. (2007). Impact of deforestation and agricultural development on anopheline ecology and malaria epidemiology. *The American Journal of Tropical Medicine and Hygiene, 76*(3), 450–460.

Zagelmeyer, S. (2024). Moving beyond delinking, decoloniality and the pluriverse: Reflections on the "decolonizing international business" debate. *Critical Perspectives on International Business, 20*(1), 71–93.

Zhang, D. D., Lee, H. F., Wang, C., Li, B., Zhang, J., Pei, Q., & Chen, J. (2011). Climate change and large-scale human population collapses in the pre-industrial era. *Global Ecology and Biogeography, 20*(4), 520–531.

Zhang, N., Dai, H., Wang, Y., Zhang, Y., & Yang, Y. (2021). Power system transition in China under the coordinated development of power sources, network, demand response, and energy storage. *Wiley Interdisciplinary Reviews: Energy and Environment, 10*(2), e392.

Zhang, Y., Tian, Q., Hu, H., & Yu, M. (2019). Water footprint of food consumption by Chinese residents. *International Journal of Environmental Research and Public Health, 16*(20), 3979.

Zhu, Z., Wang, W., Lin, Z., Dong, H., Gao, Y., Hu, L., ... & Hao, W. (2024, February). Multi-scenario research on the low-carbon transition in energy-deficient cities-view from the power supply side. In Journal of Physics: Conference Series (Vol. 2703, No. 1, p.012007). IOP Publishing.

Lori, T., Osei-Kwarteng, M., Ojija, F., Ojo, A. O., & Ogwu, M. C. (2025). Participatory approaches to environmental health management. In S. C. Izah & M. C. Ogwu (Eds.), *Innovative approaches in environmental health management: Processes, technologies, and strategies for a sustainable future*. Environmental Science and Engineering Series. Springer Nature. https://doi.org/10.1007/978-3-031-81966-7_11

Ogwu, M. C., & Kosoe, E. A. (2025). Environmental economics: Valuing ecosystem services and green technologies. In M. C. Ogwu & S. Chibueze Izah (Eds.), *Evaluating environmental processes and technologies*. Environmental Science and Engineering Series. Springer. https://doi.org/10.1007/978-3-031-85327-2_15

Izah, S. C., Ogwu, M. C., & Salimon, S. S. (2025). Principles of sustainable development and environmental management. In M. C. Ogwu & S. Chibueze Izah (Eds.), *Evaluating environmental processes and technologies*. Environmental Science and Engineering Series. Springer. https://doi.org/10.1007/978-3-031-85327-2_2

Chancel, L. (2022). *Climate inequality report 2022: Unequal contributions to climate change*. World Inequality Lab. https://wid.world.

Chenoweth, E., & Stephan, M. J. (2020). *Why civil resistance works: The strategic logic of nonviolent conflict* (Updated ed.). Columbia University Press.

FAO. (2020). *Global forest resources assessment 2020*. Food and agriculture organization of the United Nations. https://www.fao.org

FAO. (2023). *State of the world's forests 2023*. Food and Agriculture Organization of the United Nations.

Hickel, J. (2020). Quantifying national responsibility for climate breakdown: An equality-based attribution approach for carbon dioxide emissions in excess of the planetary boundary. *The Lancet Planetary Health, 4*(9), e399–e404.

Hickel, J., Dorninger, C., Wieland, H., & Suwandi, I. (2022). Imperialist appropriation in the world economy: Drain from the global South through unequal exchange, 1990–2015. *Global Environmental Change, 73*, 102467.

IEA. (2023). *World Energy Outlook 2023*. International Energy Agency.

Institute for Economics & Peace. (2021). Ecological threat report 2021. https://visionofhumanity.org

IPCC. (2021). *AR6: The physical science basis*. Intergovernmental Panel on Climate Change.

IPCC. (2022). *AR6: Impacts, adaptation, and vulnerability*. Intergovernmental Panel on Climate Change.

Kulp, S. A., & Strauss, B. H. (2019). New elevation data triple estimates of global vulnerability to sea-level rise and coastal flooding. *Nature Communications, 10*, 4844.
NSIDC. (2022). *Arctic sea ice news and analysis*. National Snow and Ice Data Center. https://nsidc.org
Robine, J.-M., Cheung, S. L. K., Le Roy, S., Van Oyen, H., Griffiths, C., Michel, J.-P., & Herrmann, F. R. (2008). Death toll exceeded 70,000 in Europe during the summer of 2003. *Comptes Rendus Biologies, 331*(2), 171–178.
Science Panel for the Amazon. (2022). Amazon assessment report 2022. https://www.theamazonwewant.org
Swiss Re Institute. (2021). *The economics of climate change: No action not an option*. Swiss Re.
UNDRR. (2020). *Human cost of disasters 2000–2019*. United Nations Office for Disaster Risk Reduction.
UNDRR. (2023). *Global disaster loss report 2023*. United Nations Office for Disaster Risk Reduction.
Vohra, K., Vodonos, A., Schwartz, J., Marais, E. A., Sulprizio, M. P., & Mickley, L. J. (2021). Global mortality from outdoor fine particle pollution generated by fossil fuel combustion: Results from GEOS-Chem. *Environmental Research, 195*, 110754.
World Bank. (2021). *Groundswell report: Preparing for internal climate migration*. World Bank.
World Bank. (2020). Indigenous peoples overview. https://www.worldbank.org.
World Resources Institute. (2021). Global forest review. https://research.wri.org.
WWF. (2022). *Living planet report 2022: Building a nature-positive society*. World Wildlife Fund.

Capital, Carbon, and the Climate System: Historical Drivers of Ecological Crisis

2

2.1 Introduction

The industrial era represents a vital turning point in human-environment interactions (Izah et al., 2024a,b; Ogwu et al., 2024a,b), fundamentally altering the dynamics between society and the ecological systems upon which it depends. The 19th and 20th centuries, characterized by rapid industrialization, heralded not only unprecedented economic growth but also significant shifts in energy consumption patterns, particularly through the adoption of fossil fuels (Chen et al., 2011). This transition has been marked by an escalation in carbon emissions, which can now be viewed as a systemic outcome of fossil-fueled capitalism rather than mere coincidental lapses in policy (Chen et al., 2011; Zhang, 2017). Historically, industrial activities have led to profound changes in land use, urbanization, and resource extraction, creating complex feedback loops that have driven ecological crises (Dillehay & Kolata, 2004). Pre-industrial societies exhibited more sustainable human-environment interactions due to their localized production systems and a generally lower environmental footprint (Dillehay & Kolata, 2004). In contrast, the advent of industrial capitalism introduced a model in which economic growth became inextricably linked to carbon-intensive energy sources, leading to rampant environmental degradation and climate change across ecosystems (Chen et al., 2011).

Furthermore, the proliferation of consumer culture during the industrial era has exacerbated the ecological crisis. As societies became more entrenched in the capitalist framework, the demand for energy-intensive goods surged, leading to significant spikes in carbon emissions (Mas' ud, A. A.,, & Tenriyola, A. P., 2023). The introduction of low-carbon technologies, particularly in the energy sector, illustrates efforts to mitigate these emissions; however, the systemic nature of the issue indicates that such technologies alone cannot dislodge the deep-rooted practices of fossil-fueled capitalism (Chen et al., 2011; Zhang, 2017). This demonstrates the need for broader systemic changes that prioritize environmental sustainability

M. C. Ogwu and S. Chibueze Izah, *Climate Frontiers and Social Transformation*, World-Systems Evolution and Global Futures,
https://doi.org/10.1007/978-3-032-17639-4_2

over traditional economic imperatives (Chen et al., 2015). The post-World War II period exemplifies the shift toward a more mechanized approach to labor and technology, increasingly favoring mechanization over human labor. The technological advances symbolized by Industry 4.0 showcase this transition, in which machines began to substitute for human labor at a significant scale, affecting labor dynamics and ecological impacts simultaneously (Ejsmont, 2021). For instance, while Industry 4.0 opened pathways for improved efficiency in production processes, it also contributed to a growing misalignment between humanity's ecological footprint and the planet's regenerative capacities (Ejsmont, 2021).

In light of these historical drivers, it becomes apparent that carbon emissions should be viewed through the lens of systemic outcomes rather than isolated events. The correlation between economic growth and environmental degradation is evident—the more factories produce for consumption without regard for ecological sustainability, the higher the carbon emissions released into the atmosphere (Yuan et al., 2011). Solutions, therefore, need to transcend technological fixes and integrate sustainability into the core economic models that drive capitalism, necessitating a fundamental reevaluation of human-environment relations rooted in the tenets of social and climate justice (Dillehay & Kolata, 2004). By understanding these interrelationships as systemic problems, clarity emerges on the necessity of coordinated approaches to address climate change that involve stakeholders across political and economic landscapes. Policymaking should pivot from viewing carbon emissions as a problem of compliance to recognizing them as a critical dimension of economic viability—a vital aspect to rethink in the face of climate change (Zhang, 2017). Engaging communities, businesses, and governments collectively repositions this discussion toward sustainability, encouraging accountability within systemic governance frameworks centered on long-term ecological stewardship (Dillehay & Kolata, 2004).

Adopting strategies that align economic incentives with environmental outcomes can catalyze required changes. For example, carbon trading systems have emerged as a significant policy instrument to control emissions (Chen et al., 2015). Such mechanisms aim at internalizing the external costs of carbon emissions, thereby reshaping market dynamics to favor low-carbon technologies and ecological integrity (Zhang, 2017). However, success in these endeavors remains contingent on a collective acknowledgment of the systemic drivers of the ecological crisis rather than mere technical solutions sanctioned by policy frameworks. Understanding carbon emissions as a systemic outcome of these dynamics emphasizes the need for transformative strategies that simultaneously promote environmental sustainability and economic resilience. Addressing these challenges requires a shift from reactive policies to proactive, systemic approaches that incorporate insights gained from historical trends and contemporary practices, facilitating more harmonious human-environment interactions that can sustain both current and future generations.

Contemporary evidence makes clear that the climate crisis is unfolding within—and accelerating—the structural dynamics described in world-systems analysis. Human activities have already warmed the planet by approximately 1.1 °C above pre-industrial levels, with global atmospheric CO_2 concentrations reaching about 419 ppm in 2023, compared to roughly 280 ppm before industrialization (Global Carbon Project, 2023; IPCC, 2022). Annual fossil CO_2 emissions reached 37.7 $GtCO_2$ in 2023, with total anthropogenic emissions (including land use change) rising to 40.9 $GtCO_2$, underscoring the persistence of carbon-intensive development pathways even amid expanding renewable capacity (Global Carbon Project, 2023). These patterns reflect deep infrastructural and economic lock-ins: emissions from already existing and planned fossil-fuel infrastructure alone are projected to exceed the remaining 1.5 °C carbon budget of 510 $GtCO_2$ by roughly 340 $GtCO_2$, illustrating the structural inertia of the fossil regime (IPCC, 2022). Inequality further shapes global emissions trajectories. The wealthiest 10% of the world's population are responsible for nearly half of global emissions. In comparison, the poorest 50% generate only about one-tenth, revealing a profound asymmetry between those who benefit from carbon-intensive growth and those who bear its harms (Oxfam., 2023). Historical responsibility follows a similar pattern: the United States alone accounts for roughly 20% of all cumulative CO_2 emissions since 1850, disproportionately contributing to the planetary pressures now constraining development prospects in the Global South (Carbon Brief, 2021). Together, these statistics demonstrate that the climate crisis is not merely an environmental disruption but a deeply structured outcome of global political-economic systems, marked by unequal energy access, uneven development, and historically accumulated carbon debts.

This chapter aims to uncover the historical roots of today's climate crisis by tracing how the industrial era's fossil-fueled capitalism embedded carbon-intensive practices into the global economy. By situating carbon emissions within the long trajectory of industrialization, capital accumulation, and energy dependence, the chapter moves beyond interpretations of climate change as the outcome of discrete policy failures. Instead, it frames emissions as systemic consequences of economic, technological, and political path dependencies that have locked societies into carbon-intensive development pathways. The chapter contributes to understanding socio-ecological shifts in the age of global inequality *by* revealing how historical processes of industrialization and global economic integration not only generated profound ecological disruptions but also entrenched social and spatial inequalities that shape contemporary vulnerabilities. By integrating perspectives from political economy, environmental history, and climate science, the chapter provides a foundation for understanding why equitable and transformative climate transitions require confronting the structural forces that have historically driven both ecological degradation and global inequality.

2.2 The Rise of Fossil-Fueled Capitalism

The rise of fossil fuel capitalism marks a vital moment in human history, marked by an energy transition that began with the Industrial Revolution and advanced through technological developments favoring coal, oil, and gas (Fig. 2.1). This

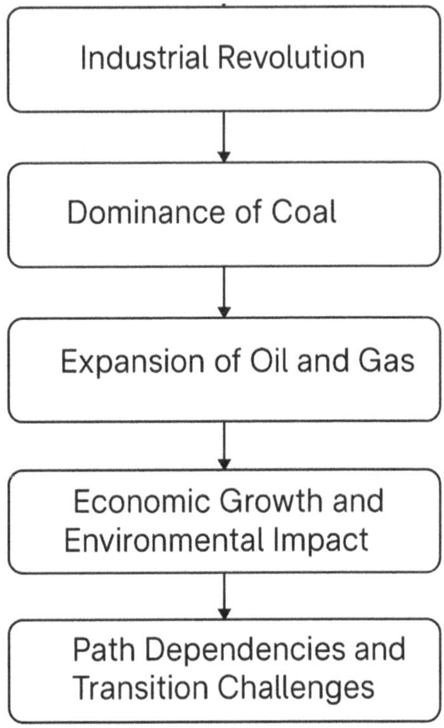

Fig. 2.1 From Coal to Carbon Capitalism: Tracing the Energy Roots of the Modern Economy

transition transformed production methods, economic structures, and societal dynamics. Central to this shift from a biomass-based economy to one dominated by fossil fuels was the intricate interplay between energy consumption, industrial growth, and environmental consequences.

The Industrial Revolution, which began in the late eighteenth century, saw coal become the dominant energy source, catering to the needs of expanding industries such as textiles, iron, and glassmaking. This marked a significant departure from earlier reliance on organic sources of energy, such as firewood and animal power. The geological abundance of coal provided a dense, accessible energy source, allowing industries to develop in proximity to coalfields due to logistical difficulties in transporting heavy materials over land. Consequently, areas rich in coal experienced unparalleled industrial concentration and urbanization, significantly changing the socio-economic landscape. Additionally, the growing dependence on coal enabled the adoption of enhanced mechanization and mass-production methods typical of this period.

The fusion of technological innovation with coal utilization intensified production capabilities and resource extraction methods. This development spread across Europe, leading to a remarkable increase in output and labor demand, further solidifying coal's status as a crucial economic factor. In this context, critical factors shaping the energy landscape included economic principles related to demand elasticity, suggesting that energy consumption would continue to rise, particularly in developing areas.

As industrialization advanced in the 19th and 20th centuries, coal's dominance began to be challenged by the emergence of oil and gas, which offered higher energy densities and greater efficiencies, thereby expanding capitalist growth aspirations. This change diversified energy markets and created new economic motivations for extraction and consumption, illustrating a trend of increasing carbon dependency and atmospheric emissions. Indeed, the search for inexpensive energy has driven geopolitical initiatives, establishing a global energy market dominated by fossil fuels. The rise of capitalism necessitated the ongoing extraction of energy resources to sustain economic growth. However, this relentless pursuit has severe implications for environmental stability. Current evidence reveals a consistent increase in greenhouse gas emissions linked to fossil fuel consumption, largely driven by the ongoing global energy demand propelled by population growth and consumption behaviors. This scenario highlights the critical issues of energy security and climate change, stressing the urgent need for a transition to renewable energy sources (Liu et al., 2023).

This transition is complicated by the established infrastructure and societal behaviors developed during the fossil fuel era, which resist swift changes toward sustainability. The ideas surrounding energy determinism often emerge, suggesting that coal and other fossil fuels have historically dictated economic and social outcomes, creating a path dependency that hinders the shift to renewable resources. Therefore, addressing the historical narrative of energy transitions requires an acknowledgment of both the advantages that fossil fuels have provided to industrial societies and the significant environmental challenges they engendered. Recognizing the health and ecological repercussions of fossil fuel dependency, contemporary discussions advocate for redirecting investments from fossil fuels toward renewable energy sources. Biomass has emerged as a viable alternative, leveraging existing technologies such as co-firing to reduce reliance on fossil fuels while transitioning to more sustainable energy systems. The feasibility of integrating biomass with coal in current power generation infrastructure underscores the need for hybrid energy solutions capable of mitigating emissions without compromising current energy demands. In summary, the intricate relationship between energy demand, economic growth, and environmental factors necessitates a critical reevaluation of our energy frameworks.

Historical patterns indicate that technological advancements and capital investments in energy sectors are consistently tied to both economic growth and ecological degradation. The current imperative is to learn from this trajectory and redefine the frameworks governing energy policy, capital investment, and sustainable development practices. The past century has demonstrated that while fossil fuels have driven remarkable growth, they have also introduced significant challenges that cannot be overlooked in the pursuit of future sustainability. As the world transitions into a new energy era, it is crucial to leverage lessons from historical energy shifts and the socio-economic dynamics that facilitated fossil fuel capitalism. Understanding the history, implications, and potential futures of energy production and consumption is vital for informing policies that minimize environmental impacts while meeting human energy demands. The importance of a collaborative global approach in transforming energy markets, fostering technological innovation, and promoting sustainable practices cannot be overstated in navigating the complexities of a climate-uncertain world.

2.3 Post–World War II Carbon Acceleration and Globalization

The post-World War II era marked a significant period of transformation, characterized by the acceleration of carbon emissions and the intensification of globalization processes. This acceleration, often referred to as the "Great Acceleration," encapsulated an unprecedented surge in anthropogenic influences on the environment, driven primarily by industrial production and technological advancements. Such developments not only exacerbated climate change but also fundamentally reshaped the global economic landscape, demonstrating the intricate interdependencies between environmental and economic systems (O'Sullivan et al., 2023). Table 2.1 summarizes the relationship between carbon emissions and globalization in the decades following World War II. It highlights how industrial reconstruction, technological innovation, economic expansion, and global interconnectedness accelerated carbon output while shaping modern trade, production, and consumption patterns. Each period reflects the interplay between economic growth and environmental impact, illustrating the rising tension between development and sustainability in the global era.

The evidence suggests that, following World War II, the introduction of machine-based production increased energy demand, further contributing to resource depletion and environmental degradation. This transition was not merely a statistical anomaly but a profound shift in the operational frameworks of economies worldwide. The relationship between production methodologies and carbon emissions became increasingly pronounced, as societies transitioned from agrarian economies to industrial powerhouses—often at the expense of ecological integrity (O'Sullivan et al., 2023). Furthermore, the geopolitical landscape of the post-war era facilitated the establishment of liberal international economic policies that reinforced these production paradigms, thus deepening the global reliance on fossil fuels (Tariq et al., 2020).

As nations endeavored to recover from the devastation of World War II, a concurrent rise in globalization began to take shape. The establishment of institutions aimed at fostering international cooperation, such as the United Nations, reflected an emerging consensus around the need for solidarity in the face of everyday adversities—chief among them being the environmental fallout from industrialization (Vlados, 2019). These frameworks have been instrumental in promoting multilayered dialogues that emphasized economic growth, often sidelining the urgent need for sustainable practices (Fernández-Villaverde & Ohanian, 2018). As the

Table 2.1 Post–World War II Carbon Acceleration and Globalization: A Chronological Overview

Period	Key drivers of carbon acceleration	Globalization trends	Environmental implications
1945–1960	Industrial reconstruction in Europe and Asia; rise in energy use	Early trade networks, colonial transitions	Moderate rise in carbon emissions
1960–1980	Green Revolution, mass production, and oil boom	Expansion of global trade; multinational corporations	Significant increase in fossil fuel consumption
1980–2000	Neoliberal economic policies, consumerism, and technological growth	Financial globalization, global supply chains	Rapid growth in emissions; emerging climate concerns
2000–2020	Industrialization in China, India, and the digital economy	Hyperglobalization; e-commerce; international agreements	Record-high CO_2 emissions; climate change awareness rises
2020–Present	Energy transitions; renewable technologies; climate negotiations	Digital interconnectivity; regional trade blocs	Efforts toward decarbonization; climate adaptation focus

U.S. emerged as a global leader, its economic policies and military interventions abroad can be seen as both a product of and a catalyst for globalization processes (Tariq et al., 2020).

Moreover, the relationship between economic policies and migration patterns during this period is noteworthy. Post-World War II migration policies tended to liberalize, despite the tumultuous political rhetoric around immigration, thus facilitating the movement of labor critical for industrial growth (Shintaro, 2020). These dynamics are closely intertwined with the environmental challenges posed by increased migratory pressures, which were often the result of economic disparities exacerbated by emerging neoliberal policies. The global economic system that developed in the ensuing decades necessitated a level of resource extraction and energy consumption that directly correlated with rising carbon emissions (O'Sullivan et al., 2023). The structural transformations observed across sectors, particularly manufacturing, underscore the profound shifts in labor and productivity dynamics during this period. Industries in the U.S. South, for instance, experienced substantial diversification since World War II, with a notable shift toward higher technology and higher-paying roles as a response to global competitiveness (Fisher, 1981). Such transformations facilitated economic growth but also increased carbon footprints as demand for energy-intensive production processes escalated. These economic phenomena should be viewed through the lens of environmental sustainability to understand the trajectory of post-war development fully.

Simulation of scenarios based on historical data reveals that volatility in macroeconomic indicators diminished post-World War II, suggesting a shift toward more stable economic frameworks that, while effective in reducing recession frequency, often gloss over the underlying ecological costs associated with that stability. This newfound stability did not come without its downsides; it inadvertently perpetuated a model of development that prioritized economic expansion over ecological health, culminating in an existential crisis due to climate change (Fernández-Villaverde & Ohanian, 2018). At a sociocultural level, the post-war period fostered an ethos surrounding consumerism that fueled both economic and environmental consequences. Citizens, buoyed by post-war optimism and expansion, engaged in unsustainable consumption patterns, emphasizing short-term gains over long-term ecological consequences. Public health policies evolved during this time, reflecting shifting social norms influenced by economic prosperity; however, they failed to address the all-involving ramifications of the intertwined economic and environmental crises.

As global economic interconnectedness deepened, reliance on cross-border trade grew. While economically beneficial, this reliance continued to exacerbate global carbon emissions (Tariq et al., 2020). The liberalization of trade policies after World War II not only opened new markets but also intensified competition, prompting nations to engage in environmentally damaging practices in pursuit of economic advantage. These developments necessitated a reevaluation of globalization, as emerging critiques highlighted the environmental injustices perpetuated by global economic structures. For instance, lower-income nations often bear the brunt of industrial pollution driven by higher production demands, highlighting a glaring disparity in the global economic framework (O'Sullivan et al., 2023). This has initiated discourse surrounding the need for equitable global systems that reconcile economic growth with environmental stewardship, particularly in light of the substantial inequalities brought to the fore by globalization (Tang & Long, 2012).

2.4 Carbon-Intensive Production and Consumption Systems

2.4.1 Industrial Manufacturing and Emissions

The industrial sector is a significant contributor to global carbon emissions, particularly in the manufacturing of iron, steel, textiles, and chemicals. The production and rolling of steel generate substantial carbon emissions, underscoring the need to optimize these processes to reduce emissions (Stavropoulos & Panagiotopoulou, 2022). Advanced methods, such as recycling, can significantly reduce emissions; specifically, producing secondary steel can cut energy intensity by approximately 50% compared to primary production (Stavropoulos & Panagiotopoulou, 2022). Moreover, the incorporation of innovative materials designed for lower energy intensity further promotes sustainability in manufacturing practices (Stavropoulos & Panagiotopoulou, 2022). Aligning industrial processes with sustainable practices is urgent as current systems significantly exacerbate global carbon footprints, necessitating a transition toward greener alternatives (Panagiotopoulou et al., 2022).

Additionally, the chemical industry, heavily reliant on fossil fuels, contributes significantly to greenhouse gas emissions. Understanding the carbon footprint of each industrial product is essential for assessing overall contributions to climate change (Panagiotopoulou et al., 2022). Manufacturers are encouraged to adopt comprehensive carbon accounting methods, such as the GHG Protocol, to quantify emissions accurately and implement reductions effectively (Panagiotopoulou et al., 2022). Consequently, industries should evolve to meet both economic and environmental demands by introducing eco-efficient technologies that streamline production processes while minimizing emissions.

2.4.2 Agriculture and Land-Use Change

Agriculture plays an essential role in emissions linked to carbon-intensive practices, notably through mechanized farming, fertilizer usage, and land-use changes such as deforestation. Research emphasizes the crucial link between agricultural production efficiency and carbon emissions, highlighting that inefficiencies in farming practices can exacerbate carbon output (Zhu & Huo, 2022). By improving agricultural efficiency, sectors could significantly mitigate emissions while still harnessing the productivity needed to meet global food demands. Furthermore, differences in livestock management practices, such as grain-finished versus pasture-finished beef production, impact carbon footprints, with grain-finished operations generally demonstrating lower emissions (Blaustein-Rejto et al., 2023).

Deforestation driven by agricultural expansion, particularly in regions such as the Brazilian Amazon, poses serious threats to global carbon dynamics. Research shows that land-use changes significantly affect carbon stocks in critical ecosystems, underscoring the need for policy strategies to halt deforestation and mitigate climate impacts. Therefore, comprehensive land management policies that balance agricultural needs with conservation priorities are vital. Thus, the intersection of modern agricultural practices and conservation is crucial for maintaining ecological stability while addressing carbon emissions.

2.4.3 Consumerism and Mass Logistics

The expansion of consumer markets alongside the development of extensive logistics networks, such as railways, shipping routes, and automobile transportation, compounds the global carbon footprint. The shift towards carbon-neutral economies emphasizes the importance of logistics operations in achieving broader sustainability goals. In this context, the logistics industry is heavily reliant on fossil fuels, underscoring the urgency of adopting renewable energy sources to power transportation systems. The integration of green logistics initiatives can significantly reduce emissions produced from the movement of goods and services across global supply chains (Yu et al., 2021).

Simultaneously, consumer behavior plays a crucial role in driving demand for advanced and comprehensive logistics systems. As consumerism escalates, energy demand rises proportionately, thereby compounding atmospheric greenhouse gas concentrations (Githiru et al., 2017). This phenomenon necessitates transforming logistical frameworks to incorporate energy-efficient practices and sustainable resource management approaches. For instance, initiatives that promote low-carbon warehousing demonstrate how adopting environmentally friendly designs can lead to substantial reductions in energy consumption within logistics operations (Freis et al., 2016). Moreover, integrating technology and automation into warehousing and logistics operations enhances operational efficiencies, further lowering carbon footprints (Liu et al., 2023). Highlighting the intricate connections between consumerism and logistics, along with their associated carbon costs, indicates a pressing need for sustainable consumption models. The interplay between stakeholder forces and corporate responsibility emphasizes that consumer preferences and demands for greener alternatives influence corporate practices (Ifekanandu et al., 2025). Companies may be incentivized to adopt environmentally friendly logistics and distribution methods, thereby improving their market positioning in increasingly eco-conscious consumer markets.

2.5 Path Dependencies and Systemic Lock-Ins in Energy Systems

The notions of path dependency and systemic lock-in are particularly salient when examining transitions within carbon-intensive infrastructures, transport networks, and industrial frameworks. Infrastructure built upon fossil fuels entails substantial investments in physical assets, which create significant inertia against transitions towards low-carbon alternatives. These infrastructures are not merely technical configurations; they embody social and political complexities that reinforce existing energy systems. Historical dependencies often perpetuate the status quo at the expense of innovative solutions like direct air capture technologies, which could alter this trajectory (Scott-Buechler et al., 2024). The cultural dimensions of energy systems further complicate this dynamic, as the embedded norms surrounding energy consumption and growth significantly influence, and often stymie, the willingness to adopt more sustainable alternatives (Atchison et al., 2024). Table 2.2 examines the multiple dimensions through which historical energy choices have created self-reinforcing dependencies in industrial, economic, and cultural systems. From long-lived infrastructure and entrenched fossil fuel interests to deeply rooted cultural norms of progress and growth, these lock-ins constrain efforts to transition toward low-carbon futures.

The historical emphasis on fossil fuel resource extraction has created economic incentives that entrench communities into carbon-intensive practices. For

Table 2.2 Fossil Fuel Path Dependencies and Barriers to a Sustainable Energy Transition

Dimension	Key drivers	Examples / Manifestations	Consequences
Infrastructure and technology	Long lifespans of energy infrastructure, capital-intensive investments, and technological specialization	Coal-fired power plants, oil refineries, pipelines, and centralized power grids	Locked-in fossil fuel use, slow adoption of renewables due to sunk costs and technical inertia
Economic and political systems	Dependence on fossil fuel revenues, industrial lobbying, fossil fuel subsidies, and global trade regimes	OPEC influence, fossil fuel subsidies, and carbon-intensive trade routes	Policy resistance to energy transitions, protection of fossil fuel interests, and uneven climate policies
Cultural and ideological norms	Beliefs in endless economic growth, modernity linked to industrial energy use, and consumer culture	"Progress" narratives are tied to high energy consumption, car-centric lifestyles, and expectations of energy abundance	Public resistance to degrowth ideas, preference for convenience over sustainability, and cultural inertia
Path dependency feedback loops	Interactions between technological, economic, and cultural factors reinforce fossil fuel dominance	Fossil fuel infrastructure is shaping transport, urban design, and industrialization patterns	Self-reinforcing carbon-intensive systems complex to disrupt without systemic interventions

example, the fossil fuel industry's historical provision of local economic benefits, including job creation and infrastructure support, has fostered a sense of reliability among communities regarding these industries (Billon et al., 2021). Schimpf et al. argue that public confidence in fossil fuel companies significantly reduces support for clean energy transitions, highlighting that constituents often prefer to maintain existing economic structures even in the face of environmental urgencies (Schimpf et al., 2021). This ingrained infrastructure and the associated socio-economic dependencies exemplify how path dependencies lock in carbon-intensive systems, resisting the policy changes necessary for a more sustainable future.

Economic and political path dependencies are inextricably linked with fossil fuel interests, which shape policy frameworks, financial instruments, and international trade regimes. The entrenched economic interests in fossil fuels often translate into legislative agendas that prioritize industry welfare over environmental considerations, leading to a landscape in which regulatory risks against fossil fuel companies are politically marginalized. Studies highlight that policy support for climate measures is influenced by individuals' trust in both their governing systems and the fossil fuel industry (Harring et al. 2019; Ogwu et al. 2025b), indicating how a socio-political environment can impede progressive climate initiatives.

The political framework frequently aligns with sustaining fossil fuel dominance, presenting another layer of systemic lock-in.

Moreover, the ideological dimensions of economic and political path dependencies further entrench carbon dependencies. The norms of continuous growth and progress, often guided by neoliberal ideologies, foster a societal valuation of fossil fuel-based innovations. This cultural backdrop suggests that economic performance is often measured against carbon-intensive outputs, relegating alternative energy sources to secondary status. Luo and Tang elucidate how national culture influences corporate behaviors regarding carbon management, suggesting that cultural ideologies can prompt corporations to sustain practices that undermine climate objectives (Luo & Tang, 2021). These cultural and ideological constructs reflect the systemic dependencies that prolong carbon use, reinforcing the vital need for systemic cultural shifts to facilitate a transition towards sustainable alternatives.

Notably, path dependencies manifest in individual and collective behaviors towards climate action, where cognitive and emotional mindsets shaped by community identity and cultural values influence perceptions of energy transitions. Public support for climate initiatives is closely bound to how individuals perceive the political and economic climate surrounding them, a phenomenon highlighted by several studies examining public attitudes toward climate policies (Muttakin et al., 2022). The prevailing narrative constructed around fossil fuels suggests that a lack of immediate catastrophic effects of climate change often leads to public complacency regarding energy transition policies, indicating that changing the framing of the climate discourse is integral to unveiling latent public support for stronger action (Le Billon et al., 2021).

Despite clear signs of the detrimental impacts of climate dependence, public acceptance remains tied to entrenched socio-political legacies. The reluctance to abandon the familiar benefits associated with fossil fuel infrastructures, coupled with cultural narratives regarding modernity, results in resistance to change that extends beyond mere policy frameworks (Drews & Van den Bergh, 2016). Organizations should navigate these contested terrains, acknowledge existing structures while advocating for alternative energy pathways that resonate with established cultural ideologies (Lori et al., 2025).

The implications of systemic lock-in extend to decision-makers tasked with crafting meaningful interventions. Political culture markedly shapes climate policy preferences, as demonstrated by Harring et al., who argue that the interaction of political and economic contexts with cultural dimensions critically dictates public attitudes towards climate policies (Harring et al., 2019). The influence of fossil fuel interests extends beyond lobbying; it embeds itself within governance frameworks, complicating trajectories toward recognized sustainability goals. This systemic entrenchment illustrates how adaptive governance strategies should confront the material realities of existing infrastructures and the deeply held cultural narratives that support them.

Transition pathways necessitate a comprehensive understanding of existing infrastructures not as static elements but as dynamic social constructs shaped by intricate interplays between economic viability, cultural values, and political frameworks. An effective transition towards cleaner energy systems should address these systemic dependencies, leverage support for innovative technologies, and involve transformational leadership that can reframe cultural narratives towards sustainable practices. It is important to highlight the need to build adaptive leadership frameworks capable of managing the multidimensional nature of energy transitions by fostering stakeholder engagement and promoting inclusive policy discussions (Mohamed et al., 2016; Izah et al., 2025).

2.6 Global Inequalities and the Carbon Economy

The analysis of global inequalities and their connections to the carbon economy reveals intricate relationships among colonialism, trade dynamics, and carbon footprints. The historical injustices rooted in colonialism are perpetuated in the structure of global trade, which has largely benefited developed nations at the expense of developing countries. This colonial legacy has not only created uneven carbon emissions across nations but has also intensified the vulnerability of less developed regions to climate change impacts. Semieniuk and Yakovenko (2020) argue that the historical evolution of carbon emissions has contributed to global inequalities, positing that emissions and footprints converge towards a disproportionately unequal distribution, with their results indicating significant inequality in carbon emissions over time. The researchers highlight that this disparity can largely be traced back to the colonial practices that established exploitative economic systems in colonized nations, leading to their current high levels of vulnerability in the carbon economy. The carbon economy has evolved unevenly across the world, reflecting deep historical, economic, and technological inequalities. While industrialized nations have long benefited from carbon-intensive development, emerging economies have accelerated emissions in recent decades to fuel growth, and low-income countries remain marginal contributors yet bear a disproportionate share of climate impacts. Table 2.3 below highlights these contrasts, illustrating differences in historical responsibility, current emissions patterns, economic benefits, technological access, and climate vulnerability across income groups (Ogwu & Hait, 2025).

Continuing this historical narrative, Chen and Işıkara (2022) investigate cross-country variations in carbon consumption patterns, emphasizing that prevailing inequalities within and between nations exacerbate the vulnerability of specific populations. Their study highlights the need to address these disparities not just within national boundaries but also through an international lens. By extending the analysis beyond nation-state narratives, the work illuminates how systemic inequities in resource allocation and infrastructure contribute to higher carbon footprints in developing economies than in their developed counterparts.

Furthermore, the notion of climate debt, which arises from historical emissions and inequitable responsibility for climate change, emphasizes the ethical ramifications of these inequalities. In this context, concepts of reparation emerge as

Table 2.3 Global Inequalities and the Carbon Economy: Emissions, Benefits, and Vulnerabilities

Aspect	High-income countries	Middle-income countries	Low-income countries
Historical emissions	Most significant contributors since the Industrial Revolution	Growing emissions since the mid-twentieth century	Minimal contributions historically
Current emissions share	High but stabilizing or declining due to technology and regulation	Rapidly increasing with industrialization and urbanization	Very low but rising with population growth
Economic benefits	Industrial wealth, advanced infrastructure, and high energy access	Export-oriented economies, moderate infrastructure growth	Limited benefits, weak industrial base
Climate vulnerability	Higher capacity for adaptation, lower exposure in many regions	Mixed capacity, some areas highly exposed to climate risks	Highest vulnerability, limited resources for adaptation

vital in discussions surrounding climate justice. Li et al. (2021) provide insight into how emissions contributions have evolved, suggesting that historical accountability for carbon emissions remains disproportionately concentrated in developed nations, while developing countries, burdened by the consequences, bear minimal responsibility for the crisis despite their heightened vulnerability. This highlights the moral imperative for affluent nations to take responsibility for their historical and ongoing impact on climate change. Examining the dynamics of unequal vulnerabilities and responsibilities reveals that various socio-economic structures play a critical role. Matarira et al. (2013) explore how subsistence communities are impacted by climate change and the differentiated adaptation strategies they employ. Their findings suggest that social hierarchies dictate access to resources and adaptive capacities, further entrenching existing inequalities. Similar sentiments are echoed by Ahmed and Eklund (2024), who apply an intersectional lens to assess how various social identities, such as religion, gender, and ethnicity, influence access to adaptive resources, thereby shaping differential vulnerability to climate change. This intersectional perspective reinforces the notion that vulnerability is not homogeneously distributed across populations and should be approached through a nuanced understanding of social inequalities.

As the climate crisis advances, the relationship between income inequality and environmental degradation becomes increasingly relevant. Research suggests that such inequalities lead to long-term ecological deterioration, indicating that economic policies should transition towards sustainability to mitigate against these inequities (Ali, 2022). This assertion is critical for evaluating the connections among economic development, carbon emissions, and climate vulnerability across various global regions. Thus, discussions surrounding reparative measures are not merely ethical considerations; they are essential for future-proofing vulnerable populations against the adverse effects of climate change. Building understanding of carbon debt and climate justice, Bolea et al. aim to frame policy discussions to address environmental inequalities (Bolea et al., 2020). They argue that prevailing frameworks in international climate negotiations fail to adequately account for the historical context of emissions and their impact on today's Global South, thereby calling for a restructuring of dialogues to include reparative justice. These arguments should be met with policy actions that reflect an understanding of historical emissions disparities, current responsibilities, and nations' respective vulnerabilities.

Addressing the multifaceted challenges posed by climate change necessitates a reevaluation of power relations and inequalities established through colonial histories and contemporary economic practices. The interplay between these historical legacies and current socio-economic conditions aims to inform a more equitable approach to climate resilience and adaptation, ensuring that the burden of reparations is rightly placed upon those nations historically responsible for the majority of carbon emissions. Indeed, as Ahmed notes, addressing these inequalities requires robust policy frameworks that recognize the intertwined nature of power, vulnerability, and environmental degradation (Ahmed, 2024).

2.7 Carbon Emissions as Systemic Outcomes

The discourse surrounding carbon emissions often shifts focus toward the actions of individual countries or specific policies, leading to narratives that overlook the underlying systemic forces shaping these emissions. A deeper analysis reveals that attributing the stasis in carbon emissions solely to national policies or the shortcomings of particular countries simplifies the complex interplay of global economic dynamics, historical precedents, and embedded structures that govern production and consumption. High-income countries often exhibit higher consumption-based carbon emissions than production-based emissions, reflecting their status as consumers of goods produced in countries with lower emissions standards. Knight et al. articulate that this discrepancy is essential to understanding the broader economic implications of carbon emissions, as it highlights the role of wealth inequality in shaping both production processes and emissions outcomes in affluent nations (Knight et al., 2017). Figure 2.2 illustrates how carbon emissions emerge not as isolated policy failures but as the cumulative result of interconnected economic, technological, political, and cultural systems. It highlights how industrialization, fossil-fueled capitalism, consumerism, and global trade have created feedback loops that lock societies into carbon-intensive development pathways. By visualizing these structural linkages, Figure 2.2 highlights the need for systemic transformation rather than piecemeal reforms to address the climate crisis effectively.

Governments frequently grapple with policy uncertainty, which can directly affect economic activity and, consequently, carbon emissions. Ashena and Shahpari (2022) indicate that policy fluctuations can lead to non-linear relationships between economic development and emissions, complicating the formulation of

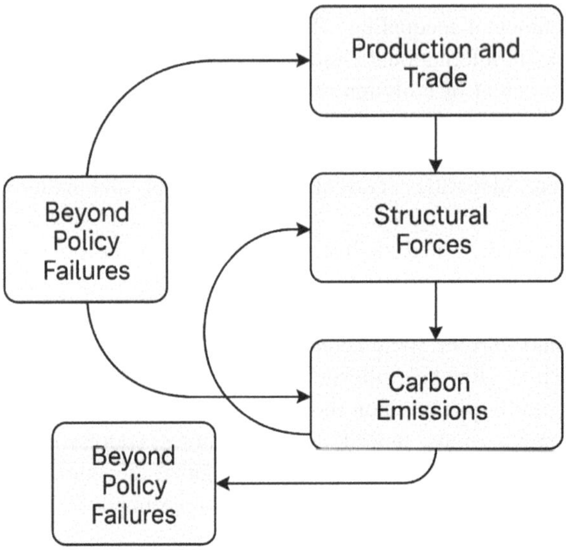

Fig. 2.2 Carbon Emissions as Systemic Outcomes in the Global Economy

effective climate strategies. This uncertainty is exacerbated within a capitalist framework, where short-term economic imperatives often overshadow long-term sustainability goals. The entrenched architecture of capitalist growth promotes increased production and consumption, inherently tied to energy usage and carbon emissions, making it imperative to understand these dynamics when seeking pathways for mitigation.

Moreover, the systemic nature of carbon emissions cannot be disentangled from the capitalist growth imperative driving global economic systems. As cities expand and develop, so do their emissions, often influenced by energy consumption patterns. Wang et al. (2019) show that economic growth is strongly positively correlated with carbon emissions in urban areas of China, revealing that this structural linkage exists globally. This phenomenon suggests that decoupling economic growth from carbon emissions without systemic changes in production and consumption patterns is increasingly challenging. Rethinking climate mitigation strategies should therefore transcend the piecemeal reforms typically advocated. The literature is consensus that promotes comprehensive systemic transitions rather than isolated policy measures that fail to account for the interconnectedness between economic growth and carbon emissions. Huang et al. (2017) highlight the necessity for structural reforms that incentivize low-carbon technologies and practices across sectors to achieve sustainable development goals. Investments in renewable energy, innovations in energy efficiency, and urban planning are essential components for fostering a holistic approach to mitigating carbon emissions.

Furthermore, transitioning toward sustainable practices requires acknowledging the drivers of carbon emissions across different industrial contexts. For instance, rapid urbanization in countries like China intensifies emissions but also presents opportunities to implement green building practices that can mitigate such impacts. Jiang et al. (2022) provide empirical evidence demonstrating that urban development strategies should integrate sustainability as a core objective for effective emissions management by focusing on urban carbon management through integrated design that considers local environmental contexts, thereby establishing a more sustainable trajectory. The intricate relationships between economic development and emissions also warrant a broader understanding of energy consumption patterns. Khan et al. (2020) emphasize that fossil fuel usage remains dominant in many developing economies, where carbon dioxide emissions significantly contribute to environmental degradation. This reliance on traditional energy sources serves as a barrier to effective carbon mitigation, necessitating a focus on developing and integrating alternative energy sources into national economic plans.

Additionally, disparities in regional development can obscure the effects of regional inequalities, leading to uneven contributions to carbon outputs. Studies by Su et al. (2018) have identified that regional inequalities significantly impact city-level emissions patterns, with some urban areas experiencing higher per capita emissions due to industrial concentration and lower efficiency standards. These disparities underscore the importance of adopting equitable policy solutions to address both regional and global emissions challenges. The necessity of systemic transitions is further underscored by the need for strategies that decouple growth from increases in carbon emissions. Scholarly explorations, such as those by Yang et al. (2022), suggest that a dual approach (combining economic growth with energy transition initiatives) is essential to effective global climate commitments. Implementing carbon capture technologies, enhancing energy efficiency, and promoting low-carbon modalities in transportation and industry can facilitate this decoupling. Pervasive narratives that criticize individual countries or policies overlook the deeper structural forces and capitalist imperatives that continuously embed carbon emissions into production and trade. The resolution to the escalating climate crisis requires not merely policy adjustments to absolve individual responsibility but a fundamental rethinking of our global economic systems. Extensive systemic reforms are essential to meet the ambitious targets needed to mitigate climate change and ensure sustainable future growth across nations.

2.8 Lessons for Contemporary Climate Transitions

The examination of historical path dependencies in the context of climate transitions reveals that past decisions profoundly shape present and future responses to climate change. Across sectors, the legacy of earlier policies, cultural norms, and institutional commitments often shapes the viability of current alternatives. These historical configurations can endorse not only the persistence of certain practices but also the limitations on adaptability essential for effective climate resilience. By examining how these dependencies manifest, it becomes possible to identify both obstacles and opportunities in the ongoing struggle against climate change. Path dependence, as a conceptual framework, elucidates how historical decisions create constraints that hinder transformative change. As described by Munck af Rosenschöld, J., Rozema, J. G.,, and Frye-Levine, L. A. (2014), the notion of institutional inertia is critical in understanding climate governance; historical precedents can create self-reinforcing cycles that favor the status quo, making it challenging to pivot towards sustainable trajectories. Specifically, the authors argue that proactive policymakers may nurture countervailing policies that can potentially establish path-dependent low-carbon trajectories, which subsequently influence future policy decisions (Munck af Rosenschöld, J., Rozema, J. G.,, & Frye-Levine, L. A., 2014). This notion emphasizes the necessity for a keen awareness of historical context in current policy formation.

Highlighting the limitations imposed by historical decisions, Rosen (2015) discusses the failure of the Kyoto Protocol, in which a flawed institutional framework confined ideological and operational paths, resulting in magnified challenges for climate action. The fixity associated with such institutions suggests that policymakers and advocates often miss opportunities to explore alternative approaches, reinforcing the idea that present dilemmas are usually embedded in past choices. The analysis notes that alternative methods for climate action could have performed better, particularly highlighting missed opportunities for 'soft commitments' that could have opened multifaceted avenues for engagement in international climate agreements.

In addition to policy frameworks, path dependencies shape adaptive responses across sectors such as agriculture, urban development, and coastal management. In their investigation of adaptation strategies within Quebec's coastal territories, Noblet and Brisson (2017) emphasize that existing institutional frameworks lead to stalled actions due to bureaucratic dependencies. This highlights the danger of allowing historical processes to dictate the pace and nature of adaptation efforts, potentially locking communities into maladaptive practices. The global shift toward a low-carbon future requires learning from the historical trajectory of industrialization, energy use, and climate governance. Past patterns reveal how delayed action, uneven technological adoption, and global inequalities have shaped today's climate challenges. Drawing on these experiences, Table 2.4 outlines key lessons for guiding contemporary climate transitions, emphasizing technological innovation, equitable policies, public engagement, and resilience-building to ensure sustainable and just outcomes.

The significance of the agricultural sector cannot be overlooked, as demonstrated by Chhetri et al. (2010), who argue that agricultural adaptations are often formulated based on past agrarian practices, potentially leading to the persistent use of ineffective strategies amid changing climatic conditions. Their study reveals that reliance on traditional agrarian pathways can exacerbate vulnerabilities when climate variability intensifies (Chhetri et al., 2010). This exemplifies how path dependence can inhibit the adoption of innovative adaptations necessary

Table 2.4 Key Takeaways for Sustainable Climate Transition

Lesson area	Key insights	Implications for climate transitions
Historical industrialization	Past growth relied heavily on fossil fuels, ignoring long-term environmental costs	Shift toward low-carbon energy sources to avoid repeating historical patterns of emissions
Global inequalities	Industrialized nations emitted the most, while poorer nations bear the brunt of climate change	Ensure climate justice through technology transfer, climate finance, and equitable policies
Technological innovation	Delayed adoption of clean technologies worsened environmental impacts	Accelerate R&D and deployment of renewable energy, storage, and carbon capture technologies
Policy and governance	Weak early climate agreements slowed coordinated global action	Strengthen international treaties, carbon pricing, and enforcement mechanisms

for sustainable agricultural practices. Historical legacies similarly influence urban transitions. Emelianoff (2014) analyzes local climate action in cities like Hanover and Växjö (2013), illustrating that while historical investments in particular pathways can catalyze transitions, these same investments can also perpetuate divergent political and ecological outcomes resulting from entrenched interests (Emelianoff, 2014). The political dimensions of path dependence complicate these urban transitions, as the initial impetus for change often encounters resistance from established frameworks and institutions, ultimately shaping whether cities can adapt effectively to climate challenges.

Furthermore, as explored by Klenert et al. (2020), lessons from large-scale shocks such as the COVID-19 pandemic can reveal vital insights into the resilience required in climate governance. They emphasize learning from such events to dismantle existing path dependencies that may limit proactive climate mitigation strategies. This suggests that effective climate governance should reflect a continuity of practices while remaining sufficiently dynamic to adapt to new global realities and emerging threats (Klenert et al., 2020). It is essential to recognize the ecological dimensions of climate dependencies. Seto et al. (2016) highlight the concept of "carbon lock-in," which posits that existing carbon-intensive systems hinder the transition towards less carbon-intensive alternatives. The persistence of entrenched fossil fuel infrastructures not only perpetuates greenhouse gas emissions but also shapes the economic and social landscapes in which climate policies are formulated, effectively locking societies into specific pathways that inhibit flexibility in climate response adaptation.

In the discourse around climate adaptation, measures have been proposed to mitigate the risks presented by historical dependencies. The analysis by Matthews indicates that successful climate adaptation strategies should address the barriers posed by path dependence, suggesting that institutional changes should facilitate innovative practices in the built environment (Matthews, 2013a). By integrating green infrastructure into planning frameworks, municipalities can break free from historical reliance on conventional infrastructure, enabling more sustainable urban development. The interplay of historical path dependencies significantly impacts the capacity for successful climate transitions. Acknowledging the constraints imposed by past decisions and institutional frameworks can help policymakers develop more effective strategies that anticipate contemporary challenges and future uncertainties. As historical legacies continue to shape present opportunities and obstacles, it becomes imperative to devise innovative approaches to climate governance that are responsive to these dynamics, ultimately fostering adaptive pathways capable of addressing the exigencies of climate change.

2.9 Conclusion

The historical trajectory of fossil-fueled capitalism demonstrates that the contemporary climate crisis is not a coincidental byproduct of industrial progress or the result of fragmented policy lapses. Instead, it is the cumulative outcome of centuries of systemic economic, technological, and political transformations that embedded carbon-intensive practices into the very fabric of the global economy. The industrial era catalyzed a profound shift toward energy-intensive production, consumption, and transportation systems, creating enduring path dependencies that have locked societies into high-carbon development trajectories. These infrastructures, technologies, and economic logics—built around the imperatives of capital accumulation and continuous growth—continue to constrain present-day efforts to decarbonize. By tracing these historical linkages, this chapter reveals how industrialization and global economic integration produced not only ecological degradation but also entrenched socio-spatial inequalities. The wealth and power generated through fossil-fueled expansion were unevenly distributed, often concentrating benefits in industrialized centers while externalizing environmental harms and economic vulnerabilities to peripheral regions. These asymmetries remain embedded in current global climate politics, shaping both the capacity for and the terms of climate action.

Viewing carbon emissions as structural outcomes rather than isolated policy failures highlights the need for systemic solutions. Addressing the climate crisis requires moving beyond incremental technological fixes and market-based reforms toward transformative changes that confront the deep-seated historical forces underpinning the carbon economy. This includes rethinking growth-centered development models, challenging the dominance of fossil fuel interests, and acknowledging the historical responsibilities of industrialized nations. Only by confronting these intertwined legacies of capitalism, colonialism, and ecological exploitation can climate transitions become truly equitable, sustainable, and capable of reshaping the socio-ecological relations that have defined the modern era.

References

Ahmed, S. (2024). Unequal power relations at the center of social vulnerability to climate change: Empirical insights from coastal Bangladesh. *PS Political Science & Politics, 57*(4), 540–544.

Ahmed, S., & Eklund, E. (2024). Intersectionality shapes the access to various adaptive resources in climate-vulnerable contexts. *Environmental Research: Climate, 3*(4), 045021.

Ali, I. M. A. (2022). Income inequality and environmental degradation in Egypt: Evidence from dynamic ARDL approach. *Environmental Science and Pollution Research, 29*(6), 8408–8422.

Ashena, M., & Shahpari, G. (2022). Policy uncertainty, economic activity, and carbon emissions: A nonlinear autoregressive distributed lag approach. *Environmental Science and Pollution Research, 29*(34), 52233–52247.

Atchison, J., Foster, R., & Bell-James, J. (2024). Blue carbon as just transition? A structured literature review. *Global Sustainability, 7*, e27.

Blaustein-Rejto, D., Soltis, N., & Blomqvist, L. (2023). Carbon opportunity cost increases carbon footprint advantage of grain-finished beef. *PLoS One, 18*(12), e0295035.

Bolea, L., Duarte, R., & Sánchez-Chóliz, J. (2020). Exploring carbon emissions and international inequality in a globalized world: A multiregional-multisectoral perspective. *Resources, Conservation and Recycling, 152*, 104516.

Carbon Brief. (2021). Analysis: Which countries are historically responsible for climate change? Carbon Brief.. https://www.carbonbrief.org.

Chen, Q., Kang, C., Xia, Q., & Guan, D. (2011). Preliminary exploration on low-carbon technology roadmap of China's power sector. *Energy, 36*(3), 1500–1512.

Chen, S., Sun, B., He, Z., & Wu, Z. (2015, December). Calculation method of carbon emission flow in power system based on the theory of power flow calculation. In *2015 6th international conference on manufacturing science and engineering* (pp. 443–450). Atlantis Press.

Chen, Y., & Işıkara, G. (2022). Beyond the nation-state narrative: An empirical inquiry into the cross-country and cross-income-group carbon consumption patterns. *Environment and Development Economics, 27*(1), 67–85.

Chhetri, N. B., Easterling, W. E., Terando, A., & Mearns, L. (2010). Modeling path dependence in agricultural adaptation to climate variability and change. *Annals of the Association of American Geographers, 100*(4), 894–907.

Crivelari-Costa, P. M., Lima, M., La Scala Jr, N., Rossi, F. S., Della-Silva, J. L., Dalagnol, R., Teodoro, P. E., Teodoro, L. P. R., Oliveira, G. D., Junior, J. F. D. O., & Silva Junior, C. A. D. (2023). Changes in carbon dioxide balance associated with land use and land cover in Brazilian Legal Amazon based on remotely sensed imagery. *Remote Sensing, 15*(11), 2780.

Dillehay, T., & Kolata, A. L. (2004). Pre-industrial human and environment interactions in Northern Peru during the late Holocene. *The Holocene, 14*(2), 272–281.

Drews, S., & Van den Bergh, J. C. (2016). What explains public support for climate policies? A review of empirical and experimental studies. *Climate Policy, 16*(7), 855–876.

Ejsmont, K. (2021). The impact of industry 4.0 on employees—insights from Australia. *Sustainability, 13*(6), 3095.

Emelianoff, C. (2014). Local energy transition and multilevel climate governance: The contrasted experiences of two pioneer cities (Hanover, Germany, and Växjö, Sweden). *Urban Studies, 51*(7), 1378–1393.

Fernández-Villaverde, J., & Ohanian, L. E. (2018). The lack of European productivity growth: Causes and lessons for the US.

Fisher, J. S. (1981). Structural adjustments in the Southern manufacturing sector. *The Professional Geographer, 33*(4), 466–474.

Freis, J., Vohlidka, P., & Günthner, W. A. (2016). Low-carbon warehousing: Examining impacts of building and intra-logistics design options on energy demand and the CO_2 emissions of logistics centers. *Sustainability, 8*(5), 448.

Githiru, M., Mutwiwa, U., Kasaine, S., & Schulte, B. (2017). A spanner in the works: Human–elephant conflict complicates the food–water–energy nexus in drylands of Africa. *Frontiers in Environmental Science, 5*, 69.
Global Carbon Project. (2023). Global carbon budget 2023. Global Carbon Project.. https://www.globalcarbonproject.org.
Harring, N., Jagers, S. C., & Matti, S. (2019). The significance of political culture, economic context and instrument type for climate policy support: A cross-national study. *Climate Policy, 19*(5), 636–650.
Huang, J., He, J., Xie, Z., Liang, L., Feng, C., Xiao, Z., & Zhong, C. (2017, December). Study on factor decomposition of per capita carbon emissions from household energy consumption in guangdong province based on LMDI method. In *2017 International Conference on Economic Development and Education Management (ICEDEM 2017)* (pp. 380–384). Atlantis Press.
Ifekanandu, C. C., Iloka, C. B., Anukwe, G. I., Eze, J. O., Uwakwe, I. V., & Asagba, S. (2025). Adoption of sustainable fuels and renewable energy in logistics: Examining the role of stakeholder theory in supply chain strategy. *PLoS One, 20*(3), e0320507.
IPCC. (2022). Climate Change 2022: Synthesis Report. Contribution of Working Groups I, II and *III to the Sixth Assessment Report of the Intergovernmental Panel on Climate Change*. Intergovernmental Panel on Climate Change.
Izah, S.C, Ogwu, M.C., and Hait, M. (2025). Clean energy solutions and public health: Innovations in renewable energy for a healthier environment. In: Izah, S.C., and Ogwu, M.C. (eds) Innovative approaches in environmental health management: Processes, technologies, and strategies for a sustainable future. *Environmental Science and Engineering Series*. Springer Nature, https://doi.org/10.1007/978-3-031-81966-7_7
Izah, S. C., Ogwu, M. C., & Hossein, H. (2024b). *Biomonitoring of pollutants in the Global South* (p. 891). Springer. https://link.springer.com/book/10.1007/978-981-97-1658-6
Izah, S. C., Ogwu, M. C., Loukas, A., & Hossein, H. (2024a). *Water crises and sustainable management in the Global South* (p. 667). Springer. https://link.springer.com/book/10.1007/978-981-97-4966-9
Jiang, J., Zhu, S., & Wang, W. (2022). Carbon emissions, economic growth, urbanization, and foreign trade in China: Empirical evidence from ARDL models. *Sustainability, 14*(15), 9396.
Khan, M. K., Khan, M. I., & Rehan, M. (2020). The relationship between energy consumption, economic growth and carbon dioxide emissions in Pakistan. *Financial Innovation, 6*(1), 1.
Klenert, D., Funke, F., Mattauch, L., & O'Callaghan, B. (2020). Five lessons from COVID-19 for advancing climate change mitigation. *Environmental and Resource Economics, 76*(4), 751–778.
Knight, K. W., Schor, J. B., & Jorgenson, A. K. (2017). Wealth inequality and carbon emissions in high-income countries. *Social Currents, 4*(5), 403–412.
Le Billon, P., Lujala, P., Singh, D., Culbert, V., & Kristoffersen, B. (2021). Fossil fuels, climate change, and the COVID-19 crisis: Pathways for a just and green post-pandemic recovery. *Climate Policy, 21*(10), 1347–1356.
Li, R., Wang, Q., Liu, Y., & Jiang, R. (2021). Is global carbon inequality getting better or worse? A decomposition analysis of carbon inequality in intraincome and interincome groups. *Management of Environmental Quality: An International Journal, 32*(6), 1255–1283.
Liu, K., He, Z., Luo, Y., Lin, P., & Chen, Y. (2023). Massive fabrication of flexible, form-stable, and self-repairing brine phase change material gels toward smart cold chain logistics. *ACS Applied Materials & Interfaces, 15*(13), 17091–17102.
Lori, T., Osei-Kwarteng, M., Ojija, F., Ojo, A. O., & Ogwu, M. C. (2025). Participatory approaches to environmental health management. In S. C. Izah & M. C. Ogwu (Eds.), *Innovative approaches in environmental health management: processes, technologies, and strategies for a sustainable future*. Environmental Science and Engineering Series. Springer Nature. https://doi.org/10.1007/978-3-031-81966-7_11
Luo, L., & Tang, Q. (2022). National culture and corporate carbon performance. *Australian Journal of Management, 47*(3), 503–538.
Mas' ud, A. A., & Tenriyola, A. P. (2023). HR competency analysis on increasing MSMEs performance in supporting industrial era 4.0. *Jambura Science of Management, 5*(2), 86–96.

Matarira, C. H., Pullanikkatil, D., Kaseke, T., Shava, E., & Manatsa, D. (2013). Socio-economic impacts of climate change on subsistence communities: Some observations from Lesotho. *International Journal of Climate Change Strategies and Management, 5*(4), 404–417.

Matthews, T. (2013). Institutional perspectives on operationalising climate adaptation through planning. *Planning Theory & Practice, 14*(2), 198–210.

Munck af Rosenschöld, J., Rozema, J. G., & Frye-Levine, L. A. (2014). Institutional inertia and climate change: A review of the new institutionalist literature. *Wiley Interdisciplinary Reviews: Climate Change, 5*(5), 639–648.

Muttakin, M. B., Rana, T., & Mihret, D. G. (2022). Democracy, national culture and greenhouse gas emissions: An international study. *Business Strategy and the Environment, 31*(7), 2978–2991.

Noblet, M., & Brisson, G. (2017). Adaptation to climate change in Quebec's coastal zone: A difficult transformation of public action. *International Journal of Climate Change Strategies and Management, 9*(03), 282–298.

Ogwu, M. C., El Malahi, S., & Izah, S. C. (2025b). Policy and regulatory frameworks for sustainable environmental practices. In M. C. Ogwu & S. Chibueze Izah (Eds.), *Evaluating environmental processes and technologies*. Environmental Science and Engineering Series. Springer. https://doi.org/10.1007/978-3-031-85327-2_16

Ogwu, M. C., Izah, S. C., Dessureault-Rompre, J., & Gasparatos, D. (2024a). *Sustainable soil systems in Global South* (p. 768). Springer. https://link.springer.com/book/10.1007/978-981-97-5276-8

Ogwu, M. C., Izah, S. C., & Ntuli, N. R. (2024b). *Food safety and quality in the Global South* (p. 760). Springer. https://link.springer.com/book/10.1007/978-981-97-2428-4

Ogwu, M. C., Kosoe, E. A., Alaka, A. C., & Xu, J. (2025a). Green infrastructure and environmental health security: The role of nature-based solutions in improving urban health. In S. C. Izah & M. C. Ogwu (Eds.), *Innovative approaches in environmental health management: Processes, technologies, and strategies for a sustainable future*. Environmental Science and Engineering Series. Springer Nature. https://doi.org/10.1007/978-3-031-81966-7_9

O'Sullivan, A., Omukuti, J., & Ryder, S. S. (2023). Global surpluses of extraction and slow climate violence: A sociological framework. *Sociological Inquiry, 93*(2), 320–340.

Oxfam. (2023). *Carbon inequality report: The global inequality of carbon emissions*. Oxfam International. https://www.oxfam.org

Panagiotopoulou, V. C., Stavropoulos, P., & Chryssolouris, G. (2022). A critical review on the environmental impact of manufacturing: A holistic perspective. *The International Journal of Advanced Manufacturing Technology, 118*(1), 603–625.

Rosen, A. M. (2015). The wrong solution at the right time: The failure of the kyoto protocol on climate change. *Politics & Policy, 43*(1), 30–58.

Schimpf, C., DeCillia, B., Sleptcov, N., Thomas, M., & Thorlakson, L. (2022). If it ain't broke, don't fix it: How the public's economic confidence in the fossil fuel industry reduces support for a clean energy transition. *Environmental Politics, 31*(6), 1081–1101.

Scott-Buechler, C., Wang, K., Fraser, C., & Scott, C. (2024). Complex socio-technical transitions in fossil fuel country: Considerations for direct air capture deployment in the US Gulf Coast region.

Semieniuk, G., & Yakovenko, V. M. (2020). Historical evolution of global inequality in carbon emissions and footprints versus redistributive scenarios. *Journal of Cleaner Production, 264*, 121420.

Seto, K. C., Davis, S. J., Mitchell, R. B., Stokes, E. C., Unruh, G., & Ürge-Vorsatz, D. (2016). Carbon lock-in: Types, causes, and policy implications. *Annual Review of Environment and Resources, 41*(1), 425–452.

Shaojian, W., Yongyuan, H., & Yuquan, Z. (2019). Spatial spillover effect and driving forces of carbon emission intensity at the city level in China. *Journal of Geographical Sciences, 29*(2), 231–252.

Shintaro, A. (2020). Effectiveness of international migration since early stage. *Available at SSRN 3796693.*

Stavropoulos, P., & Panagiotopoulou, V. C. (2022). Carbon footprint of manufacturing processes: Conventional vs. non-conventional. *Processes, 10*(9), 1858.
Su, W., Liu, Y., Wang, S., Zhao, Y., Su, Y., & Li, S. (2018). Regional inequality, spatial spillover effects, and the factors influencing city-level energy-related carbon emissions in China. *Journal of Geographical Sciences, 28*(4), 495–513.
Tang, S., & Long, S. J. (2012). America's military interventionism: A social evolutionary interpretation. *European Journal of International Relations, 18*(3), 509–538.
Tariq, M., Khan, A. A., & Khan, E. (2020). A critical analysis of the US role in the World. *Research Journal of Social Sciences and Economics Review, 1*(4), 355–360.
Vlados, C. (2019). The phases of the postwar evolution of capitalism: The transition from the current crisis into a new worldwide developmental trajectory. *Perspectives on Global Development and Technology, 18*(4), 457–488.
Yang, Z., Cai, J., Lu, Y., & Zhang, B. (2022). The impact of economic growth, industrial transition, and energy intensity on carbon dioxide emissions in China. *Sustainability, 14*(9), 4884.
Yu, Y., Zhu, W., & Tian, Y. (2021). Green supply chain management, environmental degradation, and energy: Evidence from Asian countries. *Discrete Dynamics in Nature and Society, 2021*(1), 5179964.
Yuan, C., Li, F., & Kuri, B. (2011, July). Optimal power generation mix towards an emission target. In *2011 IEEE Power and Energy Society General Meeting* (pp. 1–7). IEEE.
Zhang, Q. (2017, January). Research on Carbon Emissions Trading System in Electricity Market. In *2016 2nd International Conference on Economics, Management Engineering and Education Technology (ICEMEET 2016)* (pp. 119–123). Atlantis Press.
Zhu, Y., & Huo, C. (2022). The impact of agricultural production efficiency on agricultural carbon emissions in China. *Energies, 15*(12), 4464.

Climate as Structure and Process: A Systems Perspective on Ecological Transformation

3

3.1 Introduction

In examining climate as both a structural condition and an active driving force within socio-ecological systems, it is critical to adopt a systems perspective that accounts for the complexity and dynamism of climate processes. This approach positions climate not merely as a backdrop of environmental conditions, but as an active agent that shapes and reshapes socio-ecological interactions. As Ballew et al. (2019) point out, understanding the systemic nature of climate change is vital for effectively communicating its complexities and the multifaceted processes involved in natural resource usage. Public awareness of these invisible systems remains limited, often leading to misconceptions about the causes of climate change, such as greenhouse gas emissions and energy consumption, which are frequently oversimplified in public discourse (Ballew et al., 2019; Izah et al., 2025a, 2025b). Atmospheric carbon dioxide levels reached about 420 ppm in 2023, the highest in at least two million years, while global mean sea levels rose at an accelerated 4.7 mm per year between 2013 and 2022, demonstrating the nonlinear, feedback-driven nature of contemporary climate dynamics (Global Carbon Project, 2023; IPCC, 2022).

The framing of climate within a world-system perspective necessitates acknowledging that historical, cultural, and economic contexts shape socio-ecological interactions. Garmestani (2013) emphasizes the limitations of traditional management approaches that favor stability and control, proposing that adaptive governance, which involves nonlinear dynamics, is essential for navigating the ever-evolving landscape of social-ecological systems. Recognizing climate as a catalyst for change, rather than a static condition, underscores the need for responsive, adaptive frameworks to address the complexities of climate-driven transformations. In situating climate as a dynamic force, it becomes pertinent to consider how various levels of governance and social structures influence ecological outcomes. Theories of complexity highlight that climate-induced changes often

M. C. Ogwu and S. Chibueze Izah, *Climate Frontiers and Social Transformation*, World-Systems Evolution and Global Futures,
https://doi.org/10.1007/978-3-032-17639-4_3

manifest nonlinearly, meaning small alterations in climate variables can produce disproportionately large effects on ecosystems (Ballew et al., 2019). This perspective aligns with the research conducted by Ratinen et al. (2023), which posits that fostering systems thinking within organizations is crucial for addressing carbon emissions and promoting sustainable practices. Without a comprehensive understanding of interconnections within ecological systems, efforts to mitigate climate change may fall short.

Moreover, the emergent properties of social-ecological systems significantly complicate the interplay between climate change and governance structures. As Garmestani (2013) points out, institutional rigidity can lead to disastrous long-term outcomes, as these systems are inherently dynamic and influenced by both environmental changes and human activity. Thus, effective governance should involve flexible, adaptable policy frameworks to accommodate the complex interactions arising from climate variability (Ogwu et al., 2025a, 2025b). The insights from Hunsicker et al. (2016) reveal how marine ecosystems illustrate the nonlinear relationships between climate metrics and ecological responses. By understanding these dynamics, better ocean management practices can be developed that are responsive to the immediate feedback from climate change. This synergy of climate dynamics and ecological management exemplifies the need for an integrated framework that considers the full spectrum of socio-ecological relationships.

Complexity theory and systems thinking provide essential tools for analyzing climate-driven change. These theories encourage a holistic view of systems, wherein interdependent components interact dynamically and influence one another. Burato et al. (2023) highlight that these cognitive capacities, particularly systems thinking, are crucial for understanding the challenges posed by climate change and for supporting policies aimed at mitigating it. The intricate feedback mechanisms within ecological systems imply that traditional linear models of cause and effect may be insufficient to address the realities of climate change. Sterman and Sweeney (2007) argue that public complacency regarding climate change is partly due to a failure to acknowledge the complex interrelations within ecological systems, which often leads to misinterpretations of how activities affect climate dynamics. The challenge thus lies in enhancing individuals' capacity to engage with these complex systems through education and innovative modeling approaches.

Systems thinking offers a pathway to bridge the knowledge gap between scientific understanding and public perception by fostering skills that deepen comprehension of climate systems and their dynamics. According to Nagarajan and Overton (2019), incorporating project- and problem-based learning in educational settings can significantly promote systems thinking among students and render them more adept at addressing contemporary climate challenges. This pedagogical shift embeds a more nuanced understanding of complexity within the education system, equipping future generations with the tools to engage actively with socio-ecological resilience. Integrating these theories into practical frameworks is crucial. Sharma and Mathur (2021) propose utilizing systems thinking and modeling to engage communities in understanding and addressing climate impacts experimentally. The feedback loops inherent in these models allow users to observe the repercussions of their decisions on the climate system, thereby fostering a more responsible attitude toward environmental stewardship. This approach embodies a transformative adaptation strategy that not only acknowledges the limitations of incremental changes but promotes systemic shifts in behavior and policy. It is worth noting that the nonlinear dynamics characteristic of ecological systems necessitates an understanding of thresholds and potential abrupt shifts in state—notions that should be central to climate governance and policymaking (Litzow & Hunsicker, 2016). As Lezak and Thibodeau (2016) acknowledge, public understanding of climate risks and dynamic systems is often hampered by cognitive biases and oversimplified perceptions of climate change consequences, suggesting that initiatives aimed at enhancing systems thinking should be prioritized in educational curricula.

This chapter aims to conceptualize climate change as both a structural condition and a dynamic process shaping socio-ecological systems through the lens of complexity theory and systems thinking. By modeling the interlinked feedback loops among climate variability, resource availability, population displacement, and socio-political instability, the chapter offers a framework for understanding how climate shocks propagate across ecological and social domains, producing cascading crises. The contribution lies in integrating multiple scales of analysis—biophysical, economic, political, and cultural—into a unified systems perspective that transcends linear cause–and–effect models. In doing so, the chapter advances scholarship on climate–society interactions by highlighting thresholds, tipping points, and emergent properties that drive nonlinear transformations. It also bridges theoretical modeling with practical implications for resilience-building, adaptive governance, and policy interventions, thus providing both a conceptual foundation and actionable insights for researchers, policymakers, and practitioners working at the intersection of climate change, sustainability, and socio-ecological resilience.

3.2 Climate as a Structuring Condition

Climate serves as a critical structuring condition that shapes ecosystems, human settlements, and socio-economic developments (Iyiola et al., 2024; Izah et al., 2023). The biophysical constraints imposed by climate, primarily temperature, precipitation, and atmospheric patterns, affect both ecological productivity and the viability of human habitation in various regions. Elevated temperatures and altered precipitation patterns limit agricultural output and alter the availability of natural resources crucial to human settlements (Wang, 2012). Climate change, as evidenced by the increased frequency of severe weather events, poses challenges to food security and threatens livelihoods, especially for vulnerable populations (Sultan et al., 2014). Understanding these constraints is essential for effective climate risk management strategies. For instance, city administrations often struggle to grasp local hazard exposures and the spatial distribution of vulnerabilities stemming from changing climatic conditions (Fünfgeld, 2010). Figure 3.1 illustrates the

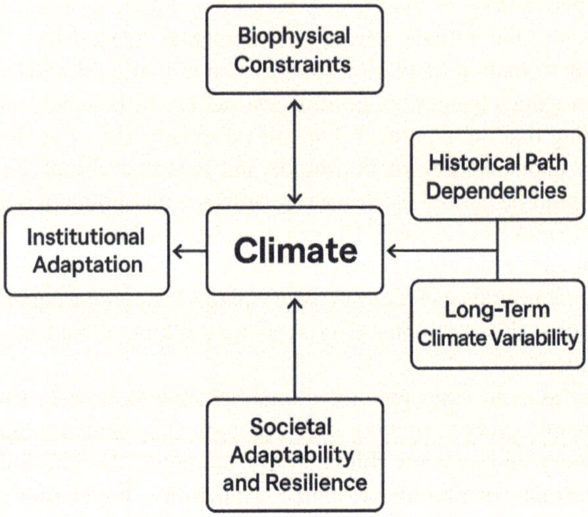

Fig. 3.1 Interconnected dimensions of climate as a structuring condition in socio-ecological systems

conceptualization of climate as a structuring condition shaping socio-ecological systems. The central role of climate is depicted through its interactions with five interconnected dimensions: biophysical constraints, historical path dependencies, long-term climate variability, institutional adaptation, and societal adaptability and resilience. Together, these dimensions highlight how climate acts as both a limiting factor and an enabling force, influencing ecological productivity, human settlement patterns, governance systems, and the capacity for societal adaptation over time. There is a need for integrated, cross-scalar analyses that account for environmental, historical, and institutional contexts when addressing climate change and its impacts on human and ecological systems.

Variations in climate not only set immediate biophysical limits but also create historical path dependencies that dictate the long-term development of economies, political contexts, and cultural practices. Historical climate regimes have significantly influenced settlement patterns, with many societies evolving under the constraints of their climatic contexts (Faust & Ashkenazy, 2007). For example, in the Near East, research suggests that periods of increased humidity were conducive to denser settlement, whereas prolonged aridity led to population declines and migration (Faust & Ashkenazy, 2007). These historical precedents echo throughout societal development, mirroring how current regions engage with their climatic challenges. Moreover, the provisioning of social and economic structures often adapts to these climatic realities, with institutions developing as responses to historical climate challenges (Dovers & Hezri, 2010). Institutions are thus shaped not only by human actions but also by climate factors that necessitate adaptive governance (Bouwer & Aerts, 2006). Sustained long-term climate patterns have highlighted fundamental changes in human adaptation mechanisms over extensive periods, notably during glacial cycles and critical climate episodes such as the Little Ice Age. Historical ecological dynamics highlight that climatic variabilities have consistently modified human societies. For example, Quaternary paleoecological research indicates three biological responses to climate change: persistence, migration, and extinction (Davis et al., 2005). This triad of potential outcomes indicates that climate pressures necessitate substantial evolutionary processes. The adaptability of societies often hinges on both historical climatic conditions and emergent climate anomalies, resulting in varying degrees of economic resilience and capability to trigger adaptive cultural practices (Mazer et al., 2020).

The interplay between climate variability and human adaptation is further illustrated by studies documenting how societies transitioned in response to climatic shifts. During the Little Ice Age, changing climates prompted significant alterations in agricultural practices, leading to increased food insecurity in places poorly prepared to adapt to temperature drops and changes in precipitation (Kabir et al., 2016). For instance, subsistence farming patterns that thrived during warmer periods often collapsed under harsher climes, leading to socio-political unrest and significant transformations within governance structures (Sultan et al., 2014). Thus, climate serves as a constraint against which societies measure their adaptability and resilience over time. Global greenhouse gas emissions climbed to approximately 57.1 $GtCO_2$-equivalent in 2023, reaffirming the deep structural forces intensifying climate pressures on ecosystems and human societies worldwide (Global Carbon Project, 2023).

Furthermore, the long-term observability of climate elements plays a central role in informing effective adaptation strategies. By analyzing long-term climatic data, regions can better forecast agricultural yield potentials and calibrate policies accordingly (Taylor & Hakim, 2023). Additionally, specific sectors, such as aquaculture, have begun recognizing the need for innovative, responsive strategies to address climate-induced challenges (Soto et al., 2019). For instance, species selection and cultivation practices are adapted in real-time as farmers confront the biophysical exigencies of their environments. This suggestive adaptability relies on historical context and continuous learning from climate engagements, underscoring the pivotal need for integrated assessments and sustainable frameworks in policymaking (Schultz & Adler, 2017).

The conceptualization of climate as a structuring condition highlights the dynamic interaction between ecological and human systems, as they adapt to both long-term climatic trends and immediate climatic anomalies. This notion aligns with the understanding that effective climate resilience requires not only reactive adaptations but also anticipatory strategies that minimize long-term societal costs. As historical and contemporary case studies show, societies that successfully integrate environmental considerations into their adaptation strategies possess the foundational knowledge to thrive amid climatic pressures, thereby contributing to robust resilience to inevitable climatic shifts (Saba et al., 2013). Overall, integrating climate knowledge across various facets of governance and societal design remains paramount for ensuring sustainable human settlements capable of enduring the pressures of both temporal climate variability and long-term change.

3.3 Climate as a Dynamic Process

3.3.1 Variability, Extremes, and Gradual Shifts

Climate variability involves both short-term anomalies, such as droughts and floods, and long-term directional changes, including global warming. Recent studies indicate that ecological responses to climate change, such as altered phenology and species distributions, are increasingly noticeable due to rising temperatures and shifting precipitation patterns (Hua et al., 2021; Huang et al., 2014). Specifically, Huang et al. (2014) note that ecosystems are beginning to exhibit changes in their physiological and compositional dynamics in response to climate shifts. Furthermore, Foden et al. (2013) stress that these shifts also affect species vulnerability, illuminating the broader implications of climate-induced variability beyond mere fluctuations in weather patterns.

Droughts and floods, as short-term climate anomalies, are becoming more frequent and intense in various regions worldwide, challenging ecological balances and agricultural sustainability (Runting et al., 2017). In 2023 alone, climate-related hazards triggered over 26 million new internal displacements, illustrating how short-term climatic shocks rapidly reshape demographic patterns and exacerbate social vulnerability (IDMC, 2024). These extreme weather events are not only symptoms of climate change but are also intertwined with socio-economic factors that amplify their impacts (Reidsma et al., 2015). Long-term directional changes, such as those observed in average global temperatures, indicate a significant and continuous warming trend that leads to gradual shifts in biomes and ecosystems, particularly in sensitive areas like the Arctic and alpine environments (Li et al., 2018a, 2018b). As indicated by Li et al. (2018a, 2018b), ecosystems such as tundra's are undergoing substantial changes that pose further risks, and they rely heavily on their intrinsic ability to resist or adapt to climatic disturbances (Kosoe & Ogwu, 2025). Historical climate data substantiates the presence of both variability and gradual shifts, with evidence suggesting a clear trajectory of warming over the past century, underscoring the urgency of adaptive measures tailored to both acute and chronic stressors induced by climate change (Zhang et al., 2022). Indeed, gradual shifts are often difficult to discern amidst short-term anomalies, but contemporary research employs advanced models that facilitate the differentiation and understanding of these distinct but interconnected phenomena (Mina et al., 2017; Zhang et al., 2022).

Moreover, ecosystem responses are often nonlinear; small fluctuations in climate can trigger abrupt ecological shifts, referred to as tipping points that may lead to dramatic alterations in habitat and species composition (Seidl et al., 2017). This dynamic interaction highlights the importance of integrating temporal scales into climate assessments, ensuring that both immediate and long-term impacts are considered (Li et al., 2018a, 2018b; Seidl et al., 2017).

3.3.2 Nonlinear Responses and Emergent Properties

Understanding nonlinear responses and emergent properties is critical when evaluating the impacts of climate change on socio-ecological systems. These systems often exhibit complex interrelationships that do not respond in simple, linear ways to stressors. As highlighted by Seidl et al. (2017), disturbances within ecosystems, such as fires or pest outbreaks, can have cascading effects that disrupt equilibria within these biological communities. The interactions between climatic variables and ecosystem dynamics can also result in feedback loops that amplify change and lead to regime shifts, where the fundamental structure and function of an ecosystem change, often irreversibly (Rowland et al., 2025). Feedback amplification is particularly prevalent in regions vulnerable to climate extremes, as emphasized by multiple studies documenting how rising temperatures contribute to increased frequency and severity of floods and droughts, ultimately exacerbating existing vulnerabilities within both natural and human systems (Foden et al., 2013). For instance, Runting et al. (2017) demonstrate how climate change can negatively affect vital ecosystem services for human welfare, illustrating the complex ramifications of ecological shifts on economic and social structures.

Figure 3.2 illustrates the complex, nonlinear dynamics through which climate change may affect socio-ecological systems. It depicts how disturbances such as rising temperatures, extreme weather events, or pest outbreaks can trigger feedback loops that amplify ecosystem stress, leading to tipping points and regime shifts. These thresholds represent critical moments when small changes in climatic variables can trigger disproportionate, and often irreversible, transformations in ecosystem structure and function. There is also a role for anthropogenic pressures, such as urbanization and land-use changes, in intensifying vulnerabilities and generating emergent properties that challenge ecological stability. By visualizing the interconnectedness between ecological, social, and economic components, Figure 3.2 highlights the need for interdisciplinary approaches and adaptive management strategies to enhance resilience in the face of climatic uncertainties.

The concept of tipping points further illustrates the potential for sudden, irreversible changes in ecosystems. Studies indicate that as ecosystems approach these critical thresholds, small climatic shifts can trigger disproportionate responses, leading to ecosystem collapse or drastic alterations in function and services (Estaque et al., 2023; Rowland et al., 2025). Moreover, the challenges of predicting these nonlinear responses highlight the need for interdisciplinary approaches that integrate ecological, social, and economic knowledge into adaptive management frameworks (Li et al., 2018a, 2018b; Rowland et al., 2025). This integration is crucial for developing resilience strategies that can withstand variations and shocks in climate patterns. Regime shifts, evident in socio-ecological interactions, highlight how complexity arises not solely from ecological components but also

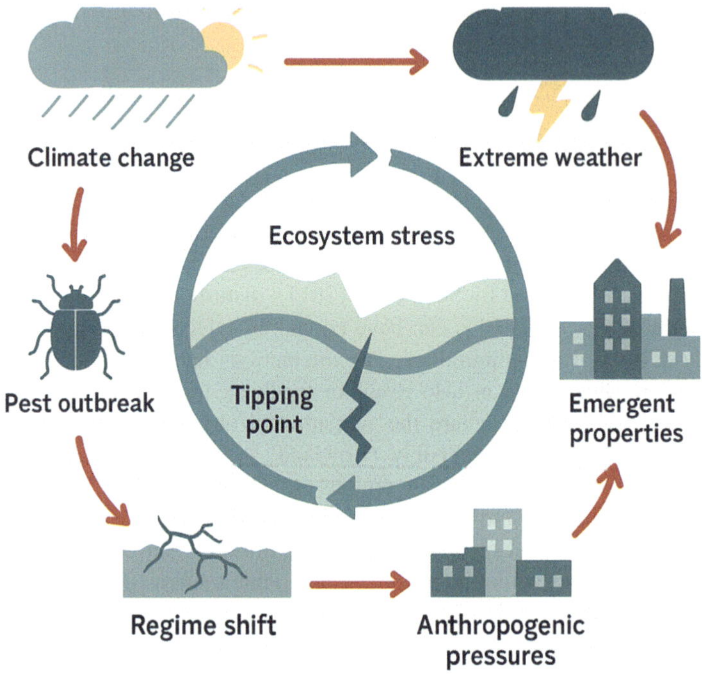

Fig. 3.2 Nonlinear responses, feedback loops, and emergent properties in socio-ecological systems under climate stress

from human activities that alter natural feedback mechanisms (Li et al., 2018a, 2018b; Seidl et al., 2017). As noted by Meiguirong (2024), the interplay between urbanization, climate, and ecological responses demonstrates how anthropogenic influences exacerbate natural vulnerabilities, often leading to emergent properties that challenge ecosystem health and sustainability. Ultimately, understanding these complex dynamics enables more effective conservation strategies and enhances our capacity to respond to unpredictable climate impacts, thereby fostering adaptive capacity among both ecosystems and human communities (Rowland et al., 2025).

3.3.3 Climate Change as an Active Agent of Transformation

Climate change is a powerful agent of transformation, shaping the resilience and sustainability of both natural ecosystems and human communities. As ecosystems are subjected to continuous climatic pressures, they experience disruptions that generate new vulnerabilities and affect their ability to provide essential services (Chiu et al., 2017; Hua et al., 2021). For instance, Kang et al. (2020) elucidate how climate change reconfigures the service values of grassland ecosystems in China, underscoring the adaptive necessity for communities that depend on them.

The disruptions caused by climate change have the potential to shift established equilibria, moving ecosystems into novel states that may no longer be conducive to their former functions. As documented by Shi et al. (2015), dynamic responses of ecosystems to climate change can alter service provision, impacting human livelihoods and ecological health. Here, the challenge lies in human societies' ability to anticipate and adapt to these shifts, requiring an acknowledgment of climate change not merely as an external threat but as a transformative force that necessitates proactive measures (Kang et al., 2020; Zhang et al., 2022).

Opportunities for adaptation may arise alongside the risks posed by climate change. For example, adaptive management strategies that emphasize ecosystem-based approaches, as suggested by Roberts et al. (2012), have gained traction as communities seek to develop resilience amid these rapid transformations. By harnessing ecosystem services in conservation plans, communities can leverage both ecological and social capital to navigate the uncertainties posed by climate variations (Chiu et al., 2017; Roberts et. al., 2012). However, the risk of marginalization of affected communities is significant, particularly in regions where adaptive capacity is low due to socio-economic constraints (Li et al., 2018a, 2018b; Rowland et al., 2025). Thus, the interplay between climate change and social factors becomes pivotal in shaping adaptive responses and the effectiveness of interventions aimed at maintaining ecosystem integrity (Zhang et al., 2022).

Furthermore, as communities are urged to consider climate change through a multidimensional lens, it is crucial to foster conditions that enable knowledge sharing and learning from transformative experiences, thus enhancing their capacity to adapt to ongoing changes (Mina et al., 2017; P, 2015). Long-term perspectives rooted in ecosystem health and resilience not only facilitate the management of climate-induced transformations but also generate pathways to sustainable futures that embrace change as an integral part of ecological and societal evolution (Deressa et al., 2009; Rowland et al., 2025). Climate change should be recognized as an active agent of transformation, presenting both challenges and opportunities across ecological and human networks, urging an integrated, forward-thinking approach that embraces adaptive governance, community engagement, and sustainable practices to confront the uncertainties bred from climatic shifts.

3.4 Feedback Loops in Socio-ecological Systems

Table 3.1 summarizes key reinforcing and balancing feedback loops linking climate variability, ecosystem functions, resource availability, and socio-political responses. Each loop captures the chain of interactions through which climate extremes, such as droughts, floods, and heatwaves, trigger ecological and economic disruptions, shape human decision-making, and influence long-term system trajectories. Reinforcing (R) loops highlight self-amplifying processes that can intensify vulnerabilities, leading to regime shifts, ecosystem degradation, or socio-political instability. In contrast, balancing (B) loops emphasizes adaptive mechanisms such as early warning systems, ecosystem restoration, and social safety nets that can dampen shocks and foster resilience. Table 3.1 provides a multi-scalar lens for understanding how local climate events cascade into broader systemic transformations and identifies leverage points for adaptive governance and sustainable resource management. By 2024, approximately 295 million people across 53 countries were experiencing crisis-level acute food insecurity, with climate extremes serving as a major compounding driver alongside conflict and economic shocks (FAO et al., 2024).

The interconnected nature of socio-ecological systems (SES) underscores the multifaceted feedback loops that shape climate variability and resource availability, thereby significantly affecting agriculture, water security, biodiversity, and ecosystem services. The impacts of climate change have been documented to alter biodiversity patterns, which in turn affect ecosystem functionality and service delivery essential for agricultural productivity and water resource management. In particular, changes in climate systems can disrupt ecological balance in fragile environments, as evidenced in pastoral systems that rely heavily on stable climatic conditions for livestock and agriculture (Kakinuma et al., 2019).

Research indicates that operationalizing the SES framework is vital for developing sustainable agricultural practices that can adapt to changing climatic conditions. Climatic variables, such as water availability and soil quality, are crucial for agriculture, thereby necessitating an integrative approach that captures the interactions between human activities and ecological systems (Leslie et al., 2015). Global climate change not only directly affects physical environments, such as precipitation patterns, but also triggers socio-political dynamics that complicate resource management across different scales (Li et al., 2022). Integrated assessments show that shifts in resource availability can undermine food security and exacerbate socio-political tensions, leading to migration as communities seek more favorable living conditions (Angeler et al., 2016).

The concept of feedback loops illustrates how changes in agricultural practices driven by climate variability can cascade into effects on socio-ecological resilience. For example, disturbances caused by drought or flood can lead to shifts in agricultural output, which may further exacerbate socio-political instability as competition for diminishing resources increases (Herrero-Jáuregui et al., 2018). These dynamics are particularly evident in regions with strained governance systems, where resource scarcity may escalate into unrest and conflict, linking environmental factors directly with socio-political challenges (Swart et al., 2018). Such interactions demonstrate significant cross-scale effects; agricultural decline in one area can create migration pressures that strain urban resources, thereby intensifying challenges associated with urbanization (York et al., 2019). The reconfiguration of rural-urban linkages can create complex urban ecosystems struggling to accommodate the influx of populations seeking better access to essential resources. Computational models of socio-ecological properties have been used to analyze how local disturbances cascade through broader economic networks, ultimately impacting larger socio-political systems (Kasthurirathna & Piraveenan, 2015).

Understanding population displacement requires a nuanced approach since various factors, such as climate-induced resource scarcity, urbanization pressures, and

Table 3.1 Feedback loops linking climate variability, ecological dynamics, and socio-political systems

Loop ID	Example causal pathway	Primary state variables	Near-term effects	Systemic trajectory/ risk	Key monitoring indicators	High-leverage interventions
R1: Climate–resource–GHG	Climate variability → reduced water/soil moisture → lower yields → land-use change & intensification → higher GHG emissions → further climate variability	Precipitation, soil moisture, crop yield, land cover, emissions	Food/water stress; shortfalls in farm income	Entrenchment of maladaptation; emissions lock-in	SPEI/Palmer drought index, NDVI, fertilizer use, deforestation alerts, sectoral GHG	Drought-smart agronomy, water harvesting, peat/ forest protection, low-emission inputs
R2: Drought–price–unrest–capacity	Drought → crop failure → food-price spikes → unrest/ political instability → weakened institutions → poorer risk management → larger drought impacts	Crop output, prices, governance capacity	Market volatility; emergency aid needs	Governance erosion; conflict risk	FAO price indices, protest/ event data, disaster response lead-time	Social safety nets, targeted subsidies, early warning→early action, depoliticized relief
R3: Flood–infrastructure–urbanization	Extreme rainfall → infrastructure damage → disrupted markets/ livelihoods → peri-urban expansion & impervious surfaces → heightened flood hazard	Rainfall extremes, road/bridge status, urban land cover	Service outages; displacement	Hazard amplification; repeated losses	Gauge/river levels, SAR flood maps, imperviousness trend	Nature-based flood management, risk-sensitive urban planning, green/blue networks

(continued)

Table 3.1 (continued)

Loop ID	Example causal pathway	Primary state variables	Near-term effects	Systemic trajectory/risk	Key monitoring indicators	High-leverage interventions
R4: Heat–energy–emissions	Heatwaves → peak cooling demand → fossil power dispatch → higher emissions → more heat extremes	Temperature anomalies, grid load, fuel mix	Rolling blackouts; health stress	Warming feedback; energy poverty	Cooling-degree days, grid carbon intensity, mortality	Demand response, efficiency retrofits, district cooling, renewable baseload + storage
R5: Biodiversity–services–intensification	Habitat loss/heat/drought → biodiversity decline → reduced pollination & pest control → yield risk → chemical inputs & land clearing → further biodiversity loss	Species richness, pollinator abundance, pesticide load	Short-term yield gains with externalities	Ecosystem service collapse	eDNA/biotic indices, IPM adoption, pesticide residues	Habitat corridors, IPM, agroforestry, set-asides, payments for ecosystem services
R6: Pastoral systems degradation	Rainfall deficit → pasture decline → herd mobility/overgrazing → land degradation → lower carrying capacity → livelihood stress	Forage biomass, grazing pressure, herd size	Conflict over water/pasture	Desertification, chronic displacement	Forage NDVI, borehole levels, conflict reports	Rotational grazing, fodder banks, negotiated resource access, drought insurance

(continued)

Table 3.1 (continued)

Loop ID	Example causal pathway	Primary state variables	Near-term effects	Systemic trajectory/ risk	Key monitoring indicators	High-leverage interventions
B1: Anticipatory risk management	Forecast skill ↑ → pre-positioned aid & buffer stocks → stabilized prices & livelihoods → reduced exposure	Forecast accuracy, contingency funds, stocks	Softer shock impacts	Dampened hazard impacts across cycles	Forecast Brier score, lead-time, stock-to-use	Impact-based forecasting, parametric triggers, local warehouses
B2: Ecosystem-based adaptation	Wetlands/forests restoration → hazard attenuation & carbon sequestration → fewer losses → lower pressure for risky coping	Wetland extent, canopy cover, carbon stocks	Reduced flood peaks; microclimate cooling	Risk reduction + mitigation co-benefits	Canopy/peat extent, flood peak ratios	Riparian buffers, mangrove rehab, peat rewetting, urban trees
B3: Social learning & risk communication	Trusted communication → protective behaviors & coordination → lower mortality/ damage → higher trust for next event	Trust indices, warning uptake	Faster, safer responses	Virtuous preparedness cycle	Warning open-rate, evacuation compliance	Community drills, co-production of advisories, and multilingual messaging

(continued)

Table 3.1 (continued)

Loop ID	Example causal pathway	Primary state variables	Near-term effects	Systemic trajectory/risk	Key monitoring indicators	High-leverage interventions
B4: Safety nets vs. erosion	Shock → safety nets enable coping **without** resource liquidation → ecosystems are maintained → faster recovery.	Cash transfers, insurance coverage	Avoidance of distress sales	Prevents the poverty-environment trap	Coverage rates, payout timeliness	Scalable cash/insurance, livelihood grants, anti-charcoal alternatives

Key (1) Arrows indicate dominant directions; real systems include cross-links and delays. (2) Indicators prioritize operational monitoring (remote sensing + administrative data). (3) Interventions are framed as leverage points suited to policy, planning, and community practice.
GHG – Greenhouse Gas Emissions, SPEI – Standardized Precipitation Evapotranspiration Index, NDVI – Normalized Difference Vegetation Index, FAO – Food and Agriculture Organization of the United Nations, SAR – Synthetic Aperture Radar, IPM – Integrated Pest Management, eDNA – Environmental DNA, Brier Score – A measure of the accuracy of probabilistic forecasts, B – Balancing Feedback Loop (negative feedback loop), R – Reinforcing Feedback Loop (positive feedback loop).

socio-political dynamics, drive migration across scales. Studies emphasize that socio-ecological resilience should be a fundamental consideration in managing urbanization, as emergent population dynamics impact service delivery and food security(Dentoni et al., 2021). Analyses at multiple scales highlight the necessity for adaptable governance frameworks that can respond dynamically to changing population pressures while incorporating principles of ecological sustainability (Çiftçioğlu, 2019). The interplay between food insecurity and political unrest reflects another critical feedback loop within the SES framework. Climate shocks that undermine agricultural viability can precipitate food insecurity, potentially leading to socio-political conflict. The interdependencies among socio-economic vitality and food systems mean that disruptions in one area can lead to cascading vulnerabilities in others (Paolisso et al., 2019). For regions facing chronic food insecurity, economic shocks can lead to increased dissent and instability, underscoring the need for economic frameworks that account for environmental impacts and human behavior (Collins et al., 2011).

Moreover, interactions across scales reveal how local disturbances can influence both regional and global systems. The panarchy concept emphasizes understanding these interactions, as disturbances at the micro-level can significantly affect macro-level socio-ecological outcomes. Grasping these dynamics requires an understanding of nested adaptive cycles, where smaller-scale interactions inform broader ecological and social systems (Jan et al., 2023). As societies face the realities of climate change, a critical aspect of building resilience lies in institutional structures and governance. Effective management of socio-ecological systems necessitates robust institutions that promote cooperation, adaptive management, and localized decision-making (Pétursdóttir et al., 2013). Participatory governance approaches illustrate how stakeholder engagement enhances socio-ecological resilience by ensuring diverse perspectives guide resource management, thereby mitigating the impacts of climate variability on vulnerable populations (Herrera-Franco et al., 2018; Lori et al., 2025). Feedback loops within socio-ecological systems are essential for understanding and addressing challenges stemming from climate variability, resource scarcity, and socio-political instability. The interdependencies among ecological health, resource availability, and socio-political dynamics necessitate integrated management frameworks that are resilient, adaptable, and inclusive of diverse stakeholder perspectives. This comprehensive approach can help stabilize communities facing climate challenges, ensuring sustainable livelihoods and environmental stewardship.

3.5 Modeling Cascading Crises

Modeling cascading crises, particularly in the context of climate-society interactions, requires employing complexity-based approaches that account for the dynamic interrelationships and threshold effects in socio-ecological systems. Figure 3.3 illustrates the interlinked processes and feedback mechanisms underlying the modeling of cascading crises arising from climate shocks. It integrates climate drivers, systemic vulnerabilities, and socio-ecological responses using approaches such as system dynamics, agent-based modeling, and network analysis. Arrows represent both direct pathways (e.g., climate variability → infrastructure disruption) and amplifying feedback loops (e.g., economic stress → governance instability → reduced adaptive capacity). Threshold points and tipping cascades are marked to highlight potential nonlinearities and regime shifts.

Complexity-based modeling utilizes methodologies such as system dynamics, agent-based modeling (ABMs), and network analysis to simulate the interactions between climate variables and societal responses. These models establish frameworks for understanding how systemic vulnerabilities propagate through networks, whether in financial or environmental systems. For instance, agent-based models have emerged as a powerful tool in environmental governance, enabling the integration of multiple, often divergent disciplinary perspectives to develop climate policies with comprehensive implications (Savin et al., 2023). These ABMs facilitate explorations of how various agents respond to climatic changes. This kind of integrated modeling allows for the examination of individual behaviors and their macro-level implications, thereby enhancing our understanding of climate-induced transformations. Complex network structures also play a crucial role; Lee et al. (2011) found that the topology of global macroeconomic networks significantly influences the spread and propagation of economic crises, stressing the interconnectedness of financial and ecological systems within the broader socio-economic landscape (Ogwu & Kosoe, 2025). The accelerating rise in global temperatures, CO_2 concentrations, and sea-level trends amplifies the probability that climate shocks—such as concurrent heatwaves, droughts, and food-price spikes—will trigger cascading failures across socio-ecological and economic systems (Global Carbon Project, 2023; IPCC, 2022).

Identifying thresholds and tipping points is another critical aspect of modeling cascading crises, as minor perturbations in a system can led to disproportionately massive shifts, resulting in large-scale transformations. The notion of thresholds in climate and socio-ecological systems has gained traction, establishing that small shocks can catalyze dramatic changes across varied domains (Naqvi et al., 2020).

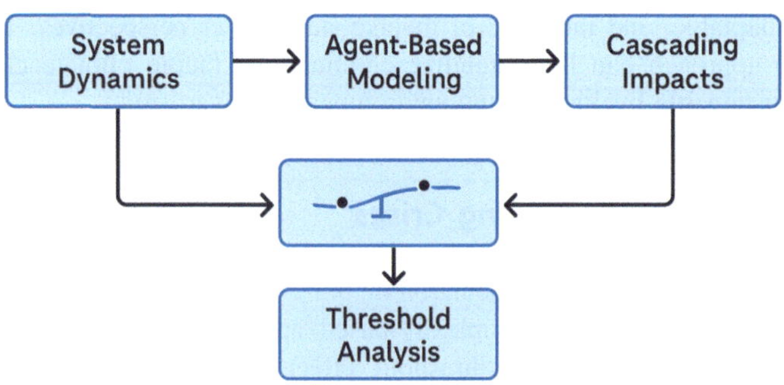

Fig. 3.3 Interconnected dynamics in modeling cascading crises across socio-ecological systems

For example, as articulated by Challinor et al. (2018), the interplay between climate variability and systemic vulnerabilities highlights the substantial risk of cascading failures across sectors, with ramifications that extend beyond immediate environmental impacts to include socio-economic repercussions. The concept of cascading failures is vividly illustrated through scenarios demonstrating how isolated climate shocks can precipitate sequential disruptions across ecological, economic, and political domains. An increasing body of literature emphasizes that crises in one area often reverberate through interconnected systems. Monasterolo et al. (2017) highlighted how vulnerabilities in financial sectors, especially due to climate-related risks, can lead to acute economic stress, thereby provoking further socio-ecological disarray. These cascading pathways can also manifest in agriculture, where a climate-induced crop failure in one region can lead to food insecurity and inflated prices, triggering migration or political unrest in neighboring areas (Naqvi et al., 2020).

Uncertainty and nonlinearity are intrinsic characteristics of climate systems that exacerbate the challenges in modeling cascading crises. The role of stochastic events and unexpected feedback mechanisms is especially prominent in socio-ecological frameworks, wherein nonlinear interactions may produce outcomes that defy intuitive predictions. For instance, the systemic risk emerging from the interconnectedness of financial institutions can amplify losses in unforeseen ways, as noted by Elliott et al. (2014). The inherent unpredictability of ecological dynamics also poses challenges; Gao et al. (2018) demonstrated that cascading failures can stem from seemingly unrelated financial behaviors, illustrating the complexity of the interactions that can ensue.

Furthermore, the implications of these cascading failures extend into urban planning and infrastructure resilience (Ogwu et al., 2025a, 2025b). As described by Derrible (2017), integrating urban infrastructure systems should account for the cascading effects of climate-related disruptions, which can lead to extensive service outages and failures. The interdependencies of critical infrastructures necessitate robust modeling frameworks capable of anticipating the multifaceted responses to climate-induced stressors. Navigating these interconnected systems requires an iterative approach to model development. By leveraging methodologies that account for the dynamic interplay between diverse variables, researchers can discern vital patterns that contribute to systemic resilience. In this context, the work of Grêt-Regamey et al. (2019) elucidates how actor diversity within social-ecological systems enhances flexibility and adaptation in the face of global changes, demonstrating that collective capability can mitigate the adverse effects of cascading failures.

The implications of these integrated modeling approaches extend beyond academic inquiry; they shape actionable insights for policymakers and practitioners aiming to counter the adverse effects of climate change. For example, understanding the thresholds that precipitate disruptions can inform better risk assessment methodologies, vital for anticipating crises and mitigating negative outcomes. As scholars continue to refine these modeling frameworks, achieving a nuanced understanding of climatic interactions requires a commitment to interdisciplinary methodologies, the acknowledgment of uncertainties, and the recognition of the systemic interdependencies that characterize our world today. By elucidating the interconnections between climate shocks and socio-ecological systems, more resilient infrastructures capable of absorbing and adapting to shocks can be put in place. Thereby enhancing the overall sustainability of our societies.

3.6 Implications for Resilience and Adaptation

Adaptive governance is increasingly recognized as a vital element in addressing the complex challenges posed by climate change and evolving socio-ecological dynamics. Its design is essential for institutions and policies that effectively respond to these challenges. The ability to adapt governance frameworks to fluctuating environmental conditions involves multiple dimensions, including integrating social learning, fostering institutional flexibility, and ensuring inclusivity in policy processes. Research has highlighted that awareness of scientific uncertainties, as discussed by Anderegg et al. (2014), is crucial in shaping adaptive governance structures. Understanding the implications of both type 1 and type 2 errors in climate science empowers policymakers to develop robust decision-making frameworks that reflect the nuanced realities of climate change impacts (Izah et al., 2025a, 2025b).

Moreover, a comparative analysis of climate adaptation policies across diverse regions, such as Australia and the United Kingdom, indicates that conflicts arising from norms and values associated with scientific understanding and political realities pose significant obstacles to effective governance (Tangney & Howes, 2016). As such, reforming existing governance structures to foster collaborative partnerships between scientists, policymakers, and local communities is essential for enhancing adaptive capacity. The case of climate service integration for coastal adaptation in the United Arab Emirates illustrates how structured collaborations can improve policy outcomes, thereby fostering better-informed decision-making and community engagement (Melville-Rea et al., 2021).

Within this context, resilience thinking emerges as a critical paradigm, shifting from merely reducing vulnerabilities to actively promoting transformational pathways within socio-ecological systems. The notion of resilience emphasizes the ability of systems to bounce back from disturbances while simultaneously adapting to changing conditions. The concept is grounded in recognizing how natural and human systems interconnect and influence each other, necessitating a holistic approach to resilience-building that incorporates environmental, social, and economic dimensions (Pilato et al., 2018). For instance, integrating climate change adaptation principles into development strategies has been shown to yield substantial benefits, as evidenced by findings in Tanzania (Pilato et al., 2018). By encouraging proactive transformation, such strategies can not only mitigate immediate climate impacts but also establish foundational resilience for future challenges. This transformation is also facilitated by reconsidering the traditional roles of institutions and infrastructures, which should evolve to meet the demands of a changing climate (Helmrich et al., 2021).

Additionally, transformative pathways require integrating local knowledge and practices, particularly in vulnerable communities that are often the most affected by climate change. The inclusion of indigenous knowledge in adaptation policymaking has been shown to foster resilience by creating avenues for meaningful dialogue between scientific and traditional knowledge systems (Pham, 2020). Such integrative approaches ensure that policies address the unique challenges faced by local communities while also leveraging their inherent adaptive capacities. There is also a pressing need to bridge the gap between climate science and policy. This requires accurately incorporating scientific modeling into local adaptation strategies, which has proven effective in multiple contexts, including the realms of infrastructure planning and community development (Hoff et al., 2023). Research highlights how adaptive capacity can be enhanced through integrating climate considerations into existing governance frameworks and community-led initiatives (De Groot et al., 2020). This integration helps create flexible infrastructures that can withstand potential climate-induced shocks, thereby fostering long-term resilience in socio-economic systems.

Furthermore, the design of systemic resilience in infrastructure requires a thorough understanding of dynamic capabilities that prioritize adaptive responses to uncertainties in environmental conditions. Recent studies emphasize that the capacity of infrastructure to respond to complex and volatile environments is integral to overall resilience (Hoff et al., 2023). Building adaptive capacities at both the organizational and individual levels ensures timely, effective responses to crises, thereby reducing the potential for long-term damage to communities and ecosystems. The promotion of collaborative frameworks among diverse stakeholders enhances system resilience by facilitating knowledge and resource sharing. This is particularly evident in urban settings, where partnerships between academia, government, and the private sector have been instrumental in advancing climate adaptation initiatives (Solecki et al., 2021). Such collaborative approaches can lead to innovative solutions tailored to specific urban challenges while promoting inclusive governance that reflects the voices of affected communities.

The establishment of comprehensive climate services is a vital strategy in this regard, allowing for the dissemination of crucial climate information to decision-makers and local actors (Hewitt et al., 2020). Effective climate services can significantly improve resilience and adaptation outcomes by providing relevant data that informs risk assessments and planning processes. This is especially important in the context of the Global Framework for Climate Services, which aims to transform weather and climate data into actionable knowledge for communities (Hewitt et al., 2020). Fostering adaptive governance and resilience thinking necessitates an integrative approach that combines scientific insight, policy innovation, and local wisdom. By addressing the multifaceted challenges posed by climate change through adaptive governance, we can not only build capacities for resilience but also empower communities to transform vulnerabilities into

Table 3.2 Key strategies and mechanisms for building climate resilience and adaptation

Strategic approach	Core mechanisms	Key stakeholders	Monitoring indicators	Expected outcomes
Adaptive governance	Flexible policies, multi-level coordination, and inclusive decision-making	Governments, local councils, and NGOs	Policy reforms and stakeholder participation levels	Improved adaptive capacity and policy integration
Ecosystem-based adaptation	Wetland restoration, agroforestry, urban green spaces	Environmental agencies, farmers, and urban planners	Canopy cover, carbon storage, and biodiversity indices	Risk reduction and co-benefits for mitigation
Infrastructure resilience	Climate-proof construction, risk-sensitive urban planning, nature-based defenses	Engineers, urban planners, and municipalities	Infrastructure loss/damage rates and service continuity	Reduced exposure and safer urban systems
Social safety nets & risk finance	Insurance, cash transfers, contingency funds	Governments, banks, and humanitarian agencies	Payout timeliness and coverage rates	Faster recovery and poverty trap prevention
Knowledge integration & learning	Local knowledge, climate services, risk communication, and education	Research institutions and local communities	Forecast skill, warning uptake, and education coverage	Informed communities and behavioral adaptation
Transformational adaptation pathways	Proactive adaptation, livelihood diversification, and community-led planning	NGOs, community groups, and policymakers	Livelihood diversification index, and migration trends	Long-term resilience and reduced vulnerability

strengths. This holistic approach is instrumental for navigating the complexities of socio-ecological systems in an era defined by uncertainty and change (Table 2).

3.7 Conclusion

This chapter has presented climate change as both a structural backdrop and a dynamic, evolving process that shapes the trajectory of socio-ecological systems. By applying complexity theory and systems modeling, single-cause explanations have revealed the interconnected feedback loops linking climate variability, resource availability, population movements, and socio-political instability. The analysis highlights how these interdependencies create the potential for cascading crises, in which localized climate shocks can reverberate across ecological, economic, and political scales, generating systemic disruptions. Importantly, understanding climate as both structure and process demands attention to thresholds, tipping points, and emergent properties that characterize complex adaptive systems. These insights shift the focus from merely predicting climate impacts to anticipating nonlinear transformations and building capacities for adaptation and resilience. The framework suggested here thus contributes to ongoing debates about climate governance by emphasizing cross-scale interactions, socio-ecological interdependencies, and the importance of integrating scientific, policy, and local knowledge systems. This systems perspective lays the groundwork for subsequent chapters to explore how resilience, adaptive governance, and transformative responses can be designed in the face of uncertainty and complexity. It calls for an integrated approach that combines theoretical rigor with practical action, enabling societies to navigate the intertwined ecological and social dimensions of the climate crisis with greater foresight and adaptability.

References

Anderegg, W. R., Callaway, E. S., Boykoff, M. T., Yohe, G., & Root, T. Y. L. (2014). Awareness of both type 1 and 2 errors in climate science and assessment. *Bulletin of the American Meteorological Society, 95*(9), 1445–1451.

Angeler, D. G., Allen, C. R., Garmestani, A. S., Gunderson, L. H., & Linkov, I. (2016). Panarchy use in environmental science for risk and resilience planning. *Environment Systems and Decisions, 36*(3), 225–228.

Ballew, M. T., Goldberg, M. H., Rosenthal, S. A., Gustafson, A., & Leiserowitz, A. (2019). Systems thinking as a pathway to global warming beliefs and attitudes through an ecological worldview. *Proceedings of the National Academy of Sciences, 116*(17), 8214–8219.

Bouwer, L. M., & Aerts, J. C. (2006). Financing climate change adaptation. *Disasters, 30*(1), 49–63.

Burato, M., Tang, S., Vastola, V., & Cenci, S. (2023). Organizational system thinking as a cognitive framework to meet climate targets. *Proceedings of the National Academy of Sciences, 120*(41), e2309510120.

Challinor, A. J., Adger, W. N., Benton, T. G., Conway, D., Joshi, M., & Frame, D. (2018). Transmission of climate risks across sectors and borders. *Philosophical Transactions of the Royal Society A: Mathematical, Physical and Engineering Sciences, 376*(2121), 20170301.

Chiu, M. C., Pan, C. W., & Lin, H. J. (2017). A framework for assessing risk to coastal ecosystems in Taiwan due to climate change. *Terrestrial, Atmospheric & Oceanic Sciences, 28*(1), 57–66.

Çiftçioğlu, G. Ç. (2019). Exploring the function of home gardens in strengthening the resilience of social-ecological landscapes through cross-scale interactions: A case study from Lefke City of the Northern Cyprus. *Resilience, 3*(2), 327–347.

Collins, S. L., Carpenter, S. R., Swinton, S. M., Orenstein, D. E., Childers, D. L., Gragson, T. L., Grimm, N. B., Grove, J. M., Harlan, S. L., Kaye, J. P., & Knapp, A. K. (2011). An integrated conceptual framework for long-term social–ecological research. *Frontiers in Ecology and the Environment, 9*(6), 351–357.

Davis, M. B., Shaw, R. G., & Etterson, J. R. (2005). Evolutionary responses to changing climate. *Ecology, 86*(7), 1704–1714.

De Groot, B., Leendertse, W., & Arts, J. (2020). Building adaptive capacity through learning in project-oriented organisations in infrastructure planning. *Urban Planning, 5*(1), 33–45.

Dentoni, D., Pinkse, J., & Lubberink, R. (2021). Linking sustainable business models to socio-ecological resilience through cross-sector partnerships: A complex adaptive systems view. *Business & Society, 60*(5), 1216–1252.
Deressa, T. T., Hassan, R. M., Ringler, C., Alemu, T., & Yesuf, M. (2009). Determinants of farmers' choice of adaptation methods to climate change in the Nile Basin of Ethiopia. *Global Environmental Change, 19*(2), 248–255.
Derrible, S. (2017). Urban infrastructure is not a tree: Integrating and decentralizing urban infrastructure systems. *Environment and Planning B: Urban Analytics and City Science, 44*(3), 553–569.
Dovers, S. R., & Hezri, A. A. (2010). Institutions and policy processes: The means to the ends of adaptation. *Wiley Interdisciplinary Reviews: Climate Change, 1*(2), 212–231.
Elliott, M., Golub, B., & Jackson, M. O. (2014). Financial networks and contagion. *American Economic Review, 104*(10), 3115–3153.
Estaque, T., Richaume, J., Bianchimani, O., Schull, Q., Mérigot, B., Bensoussan, N., Bonhomme, P., Vouriot, P., Sartoretto, S., Monfort, T., & Basthard-Bogain, S. (2023). Marine heatwaves on the rise: One of the strongest ever observed mass mortality event in temperate gorgonians. *Global Change Biology, 29*, 6159–6162.
FAO, IFAD, UNICEF, WFP, & WHO. (2024). *The state of food security and nutrition in the world 2024*. Food and Agriculture Organization of the United Nations.
Faust, A., & Ashkenazy, Y. (2007). Excess in precipitation as a cause for settlement decline along the Israeli coastal plain during the third millennium BC. *Quaternary Research, 68*(1), 37–44.
Foden, W. B., Butchart, S. H., Stuart, S. N., Vié, J. C., Akçakaya, H. R., Angulo, A., DeVantier, L. M., Gutsche, A., Turak, E., Cao, L., & Donner, S. D. (2013). Identifying the world's most climate change vulnerable species: A systematic trait-based assessment of all birds, amphibians and corals. *PLoS One, 8*(6), e65427.
Fünfgeld, H. (2010). Institutional challenges to climate risk management in cities. *Current Opinion in Environmental Sustainability, 2*(3), 156–160.
Gao, Y. C., Tang, H. L., Cai, S. M., Gao, J. J., & Stanley, H. E. (2018). The impact of margin trading on share price evolution: A cascading failure model investigation. *Physica A: Statistical Mechanics and Its Applications, 505*, 69–76.
Global Carbon Project. (2023). *Global carbon budget 2023*. Global Carbon Project. https://www.globalcarbonproject.org
Helmrich, A., Markolf, S., Li, R., Carvalhaes, T., Kim, Y., Bondank, E., Natarajan, M., Ahmad, N., & Chester, M. (2021). Centralization and decentralization for resilient infrastructure and complexity. *Environmental Research: Infrastructure and Sustainability, 1*(2), 021001.
Herrera-Franco, G., Alvarado-Macancela, N., Gavín-Quinchuela, T., & Carrión-Mero, P. (2018). Participatory socio-ecological system: Manglaralto-santa elena, Ecuador. *Geology, Ecology, and Landscapes, 2*(4), 303–310.
Herrero-Jáuregui, C., Arnaiz-Schmitz, C., Reyes, M. F., Telesnicki, M., Agramonte, I., Easdale, M. H., Schmitz, M. F., Aguiar, M., Gómez-Sal, A., & Montes, C. (2018). What do we talk about when we talk about social-ecological systems? A literature review. *Sustainability, 10*(8), 2950.
Hewitt, C., Allis, E., Mason, S., Muth, M., Pulwarty, R., Shumake-Guillemot, J., Bucher, A. E., Brunet, M., Fischer, A. M., Hama, A. M., & Kolli, R. K. (2020). Toward climate-resilient development. *Bulletin of the American Meteorological Society, 101*(3), 227–232.
Hoff, R., Helmrich, A., Dirks, A., Kim, Y., Li, R., & Chester, M. (2023). Dynamic criticality for infrastructure prioritization in complex environments. *Environmental Research: Infrastructure and Sustainability, 3*(1), 015011.
Hua, T., Zhao, W., Cherubini, F., Hu, X., & Pereira, P. (2021). Sensitivity and future exposure of ecosystem services to climate change on the Tibetan Plateau of China. *Landscape Ecology, 36*(12), 3451–3471.
Huang, L., Xu, X., Shao, Q., & Liu, J. (2014). Improving carbon mitigation potential through grassland ecosystem restoration under climatic change in Northeastern Tibetan Plateau. *Advances in Meteorology, 2014*(1), 379306.

Hunsicker, M. E., Kappel, C. V., Selkoe, K. A., Halpern, B. S., Scarborough, C., Mease, L., & Amrhein, A. (2016). Characterizing driver–response relationships in marine pelagic ecosystems for improved ocean management. *Ecological Applications, 26*(3), 651–663.

IDMC. (2024). *Global report on internal displacement 2024*. Internal Displacement Monitoring Centre. https://www.internal-displacement.org

IPCC. (2022). *Climate change 2022: Synthesis report. Contribution of working groups I, II, and III to the sixth assessment report of the intergovernmental panel on climate change*. Intergovernmental Panel on Climate Change.

Iyiola, A. O., Afolabi, O. A., Alimi, S. K., Akingba, O. O., Izah, S. C., & Ogwu, M. C. (2024). Climate change and water crisis in the Global South. In S. C. Izah, M. C. Ogwu, A. Loukas, & H. Hamidifar (Eds.), *Water crises and sustainable management in the global south* (pp. 111–140). Springer.

Izah, S. C., Iyiola, A. O., Yarkwan, B., & Richard, G. (2023). Impact of air quality as a component of climate change on biodiversity-based ecosystem services. In A. L. Srivastav, A. K. Dubey, A. Kumar, S. K. Narang, & M. A. Khan (Eds.), *Visualization techniques for climate change with machine learning and artificial intelligence* (pp. 123–148). Elsevier. isbn:978-0-323-99714-0.

Izah, S. C., Ogwu, M. C., & Hait, M. (2025a). Clean energy solutions and public health: Innovations in renewable energy for a healthier environment. In S. C. Izah & M. C. Ogwu (Eds.), *Innovative approaches in environmental health management: Processes, technologies, and strategies for a sustainable future. Environmental science and engineering series*. Springer Nature. https://doi.org/10.1007/978-3-031-81966-7_7

Izah, S. C., Ogwu, M. C., & Salimon, S. S. (2025b). Principles of sustainable development and environmental management. In M. C. Ogwu & S. Chibueze Izah (Eds.), *Evaluating environmental processes and technologies. Environmental science and engineering series*. Springer. https://doi.org/10.1007/978-3-031-85327-2_2

Jan, M. A., Zubair, M., & Aman, S. (2023). Panarchies of transformation and resilience: Delineating systemic change in the society, economy, and ecology of Malakand. *Asian Journal of Comparative Politics, 8*(2), 576–593.

Kabir, M. I., Rahman, M. B., Smith, W., Lusha, M. A. F., Azim, S., & Milton, A. H. (2016). Knowledge and perception about climate change and human health: Findings from a baseline survey among vulnerable communities in Bangladesh. *BMC Public Health, 16*(1), 266.

Kakinuma, K., Yanagawa, A., Sasaki, T., Rao, M. P., & Kanae, S. (2019). Socio-ecological interactions in a changing climate: A review of the Mongolian pastoral system. *Sustainability, 11*(21), 5883.

Kang, B., Shao, Q., Xu, H., Jiang, F., Wei, X., & Shao, X. (2020). Research on grassland ecosystem service value in China under climate change based on meta-analysis: A case study of Qinghai province. *International Journal of Climate Change Strategies and Management, 12*(5), 617–637.

Kasthurirathna, D., & Piraveenan, M. (2015). Emergence of scale-free characteristics in socio-ecological systems with bounded rationality. *Scientific Reports, 5*(1), 10448.

Kosoe, E. A., & Ogwu, M. C. (2025). Sustainable urban planning and environmental management. In M. C. Ogwu & S. Chibueze Izah (Eds.), *Evaluating environmental processes and technologies. Environmental science and engineering series*. Springer. https://doi.org/10.1007/978-3-031-85327-2_14

Lee, K. M., Yang, J. S., Kim, G., Lee, J., Goh, K. I., & Kim, I. M. (2011). Impact of the topology of global macroeconomic network on the spreading of economic crises. *PLoS One, 6*(3), e18443.

Leslie, H. M., Basurto, X., Nenadovic, M., Sievanen, L., Cavanaugh, K. C., Cota-Nieto, J. J., Erisman, B. E., Finkbeiner, E., Hinojosa-Arango, G., Moreno-Báez, M., Nagavarapu, S., Reddy, S. M. W., Sánchez-Rodríguez, A., Siegel, K., Ulibarria-Valenzuela, J. J., Weaver, A. H., & Aburto-Oropeza, O. (2015). Operationalizing the social-ecological systems framework to assess sustainability. *Proceedings of the National Academy of Sciences, 112*(19), 5979–5984.

Lezak, S. B., & Thibodeau, P. H. (2016). Systems thinking and environmental concern. *Journal of Environmental Psychology, 46*, 143–153.

Li, D., Wu, S., Liu, L., Zhang, Y., & Li, S. (2018a). Vulnerability of the global terrestrial ecosystems to climate change. *Global Change Biology, 24*(9), 4095–4106.

Li, L., Zhang, Y., Liu, L., Wu, J., Li, S., Zhang, H., Zhang, B., Ding, M., Wang, Z., & Paudel, B. (2018b). Current challenges in distinguishing climatic and anthropogenic contributions to alpine grassland variation on the Tibetan Plateau. *Ecology and Evolution, 8*(11), 5949–5963.

Li, Y., Ke, Q., & Zhang, Z. (2022). Millennial evolution of a karst socio-ecological system: A case study of Guizhou Province, Southwest China. *International Journal of Environmental Research and Public Health, 19*(22), 15151.

Litzow, M. A., & Hunsicker, M. E. (2016). Early warning signals, nonlinearity, and signs of hysteresis in real ecosystems. *Ecosphere, 7*(12), e01614.

Lori, T., Osei-Kwarteng, M., Ojija, F., Ojo, A. O., & Ogwu, M. C. (2025). Participatory approaches to environmental health management. In S. C. Izah & M. C. Ogwu (Eds.), *Innovative approaches in environmental health management: Processes, technologies, and strategies for a sustainable future. Environmental science and engineering series.* Springer Nature. https://doi.org/10.1007/978-3-031-81966-7_11

Mazer, S. J., Park, I. M., Kimura, M., Maul, E. M., Yim, A. M., & Peach, K. (2020). Mating system and historical climate conditions affect population mean seed mass: Evidence for adaptation and a new component of the selfing syndrome in Clarkia. *Journal of Ecology, 108*(4), 1523–1539.

Melville-Rea, H., Eayrs, C., Anwahi, N., Burt, J. A., Holland, D., Samara, F., Paparella, F., Al Murshidi, A. H., Al-Shehhi, M. R., & Holland, D. M. (2021). A roadmap for policy-relevant sea-level rise research in the United Arab Emirates. *Frontiers in Marine Science, 8*, 670089.

Mina, M., Bugmann, H., Cordonnier, T., Irauschek, F., Klopcic, M., Pardos, M., & Cailleret, M. (2017). Future ecosystem services from European mountain forests under climate change. *Journal of Applied Ecology, 54*(2), 389–401.

Monasterolo, I., Battiston, S., Janetos, A. C., & Zheng, Z. (2017). Vulnerable yet relevant: The two dimensions of climate-related financial disclosure. *Climatic Change, 145*(3), 495–507.

Nagarajan, S., & Overton, T. (2019). Promoting systems thinking using project-and problem-based learning. *Journal of Chemical Education, 96*(12), 2901–2909.

Naqvi, A., Gaupp, F., & Hochrainer-Stigler, S. (2020). The risk and consequences of multiple breadbasket failures: An integrated copula and multilayer agent-based modeling approach. *OR Spectrum, 42*(3), 727–754.

Ogwu, M. C., El Malahi, S., & Izah, S. C. (2025a). Policy and regulatory frameworks for sustainable environmental practices. In M. C. Ogwu & S. Chibueze Izah (Eds.), *Evaluating environmental processes and technologies. Environmental science and engineering series.* Springer. https://doi.org/10.1007/978-3-031-85327-2_16

Ogwu, M. C., & Kosoe, E. A. (2025). Environmental economics: Valuing ecosystem services and green technologies. In M. C. Ogwu & S. Chibueze Izah (Eds.), *Evaluating environmental processes and technologies. Environmental science and engineering series.* Springer. https://doi.org/10.1007/978-3-031-85327-2_15

Ogwu, M. C., Kosoe, E. A., Alaka, A. C., & Xu, J. (2025b). Green infrastructure and environmental health security: The role of nature-based solutions in improving urban health. In S. C. Izah & M. C. Ogwu (Eds.), *Innovative approaches in environmental health management: Processes, technologies, and strategies for a sustainable future. Environmental science and engineering series.* Springer Nature. https://doi.org/10.1007/978-3-031-81966-7_9

Paolisso, M., Prell, C., Johnson, K. J., Needelman, B., Khan, I. M., & Hubacek, K. (2019). Enhancing socio-ecological resilience in coastal regions through collaborative science, knowledge exchange and social networks: A case study of the Deal Island Peninsula, USA. *Socio-Ecological Practice Research, 1*(2), 109–123.

Pétursdóttir, T., Arnalds, O., Baker, S., Montanarella, L., & Aradóttir, Á. L. (2013). A social–ecological system approach to analyze stakeholders' interactions within a large-scale rangeland restoration program. *Ecology and Society, 18*(2).

Pham, H. (2020). The questions of who, what, and how in the science-policy dialogue: Experiences from the Pan-Canadian framework on clean growth and climate change.

Pilato, G., Sallu, S. M., & Gaworek-Michalczenia, M. (2018). Assessing the integration of climate change and development strategies at local levels: Insights from Muheza District, Tanzania. *Sustainability, 10*(1), 174.

Ratinen, I., Linnanen, L., Claudelin, A., & Halonen, V. (2023). Toward sustainable development: Connecting systems thinking competency and carbon footprint knowledge. *Sustainable Development, 31*(3), 1593–1605.

Reidsma, P., Wolf, J., Kanellopoulos, A., Schaap, B. F., Mandryk, M., Verhagen, J., & van Ittersum, M. K. (2015). Climate change impact and adaptation research requires integrated assessment and farming systems analysis: A case study in the Netherlands. *Environmental Research Letters, 10*(4), 045004.

Roberts, D., Boon, R., Diederichs, N., Douwes, E., Govender, N., Mcinnes, A., Mclean, C., O'Donoghue, S., & Spires, M. (2012). Exploring ecosystem-based adaptation in Durban, South Africa: "Learning-by-doing" at the local government coal face. *Environment and Urbanization, 24*(1), 167–195.

Rowland, J. A., Nicholson, E., Ferrer-Paris, J. R., Keith, D. A., Murray, N. J., Sato, C. F., Tóth, A. B., Tolsma, A., Venn, S., Asmüssen, M. V., & Pliscoff, P. (2025). Assessing risk of ecosystem collapse in a changing climate. *Nature Climate Change*, 1–13.

Runting, R. K., Bryan, B. A., Dee, L. E., Maseyk, F. J., Mandle, L., Hamel, P., Wilson, K. A., Yetka, K., Possingham, H. P., & Rhodes, J. R. (2017). Incorporating climate change into ecosystem service assessments and decisions: A review. *Global Change Biology, 23*(1), 28–41.

Saba, A., Biasutti, M., Gerrard, M. B., & Lobell, D. B. (2013). Getting ahead of the curve: Supporting adaptation to long-term climate change and short-term climate variability alike. *CCLR, 7*, 3.

Savin, I., Creutzig, F., Filatova, T., Foramitti, J., Konc, T., Niamir, L., Safarzynska, K., & van den Bergh, J. (2023). Agent-based modeling to integrate elements from different disciplines for ambitious climate policy. *Wiley Interdisciplinary Reviews: Climate Change, 14*(2), e811.

Schultz, K., & Adler, L. (2017). Addressing climate change impacts in the Sahel using vulnerability reduction credits. In *Renewing local planning to face climate change in the tropics* (pp. 343–363). Springer International Publishing.

Seidl, R., Thom, D., Kautz, M., Martin-Benito, D., Peltoniemi, M., Vacchiano, G., Wild, J., Ascoli, D., Petr, M., Honkaniemi, J., & Lexer, M. J. (2017). Forest disturbances under climate change. *Nature Climate Change, 7*(6), 395–402.

Sharma, K., & Mathur, M. (2021). Identifying climate adjacency for enhancing climate action using systems thinking and modelling. *Systems, 9*(4), 83.

Shi, C., Zhan, J., Yuan, Y., Wu, F., & Li, Z. (2015). Land use zoning for conserving ecosystem services under the impact of climate change: A case study in the middle reaches of the Heihe River Basin. *Advances in Meteorology, 2015*(1), 496942.

Solecki, W., Delgado Ramos, G. C., Roberts, D., Rosenzweig, C., & Walsh, B. (2021). Accelerating climate research and action in cities through advanced science-policy-practice partnerships. *Npj Urban Sustainability, 1*(1), 3.

Soto, D., León-Muñoz, J., Dresdner, J., Luengo, C., Tapia, F. J., & Garreaud, R. (2019). Salmon farming vulnerability to climate change in southern Chile: Understanding the biophysical, socioeconomic and governance links. *Reviews in Aquaculture, 11*(2), 354–374.

Sterman, J. D., & Sweeney, L. B. (2007). Understanding public complacency about climate change: Adults' mental models of climate change violate conservation of matter. *Climatic Change, 80*(3), 213–238.

Sultan, B., Guan, K., Kouressy, M., Biasutti, M., Piani, C., Hammer, G. L., McLean, G., & Lobell, D. B. (2014). Robust features of future climate change impacts on sorghum yields in West Africa. *Environmental Research Letters, 9*(10), 104006.

Swart, J. A., Zevenberg, J., Ho, P., Cortina, J., Reed, M., Derak, M., Vella, S., Zhao, H., & van der Windt, H. J. (2018). Involving society in restoration and conservation. *Restoration Ecology, 26*, S3–S6.

Tangney, P., & Howes, M. (2016). The politics of evidence-based policy: A comparative analysis of climate adaptation in Australia and the UK. *Environment and Planning C: Government and Policy, 34*(6), 1115–1134.
Taylor, L. M., & Hakim, G. J. (2023). Skillful coupled atmosphere-ocean forecasts on interannual to decadal timescales using a linear inverse model. *Earth and Space Science, 10*(4), e2022EA002679.
Wang, L. J. (2012). Sustainable human settlement and climate change adaptation in China. *Advanced Materials Research, 524*, 3674–3677.
York, A. M., Sullivan, A., & Bausch, J. C. (2019). Cross-scale interactions of socio-hydrological subsystems: Examining the frontier of common pool resource governance in Arizona. *Environmental Research Letters, 14*(12), 125019.
Zhang, F., Zeng, B., Yang, T., Zheng, Y., & Cao, Y. (2022). A multi-perspective assessment method with a dynamic benchmark for human activity impacts on alpine ecosystem under climate change. *Remote Sensing, 14*(1), 208.

The Energy Transition in Historical Perspective: From Biomass to Fossil Fuels and Beyond

4

4.1 Introduction: Energy as the Engine of Civilization

Historically, energy has been a fundamental driver of human civilization, shaping societal structures, technological advancements, and economic development. The term "energy transition" refers to the shift from one dominant energy regime to another, reflecting changes in the sources and methods of energy consumption throughout history. Conceptualizing energy transitions in world history involves understanding the systemic shifts from traditional to modern energy sources. Vaclav Smil identifies the current period as the fourth energy transition, characterized by structural changes in energy consumption patterns and the introduction of innovative technologies (Tumin et al., 2022). This shift can be contrasted with historical transitions, such as the nineteenth-century movement from biomass to coal, which facilitated industrialization and fundamentally changed labor dynamics and social organization (Malanima, 2012; Mulvaney, 2020). The interplay between energy systems and technological innovation reveals a dynamic relationship that has often been characterized by both conflict and cooperation. For instance, the Industrial Revolution marked a pivotal era in which the adoption of coal transformed energy availability and enabled monumental technological advancements in production processes (Barca, 2011; Izah et al. 2025a, 2025b). This relationship has continued into the modern era, in which fossil fuels have powered economic growth across continents but also led to complex socio-environmental challenges. The present discourse surrounding energy transitions increasingly emphasizes not only the technological dimensions but also the socio-political implications of these changes, especially as nations confront pressing climate challenges (Healy & Barry, 2017; Sovacool et al., 2019).

In discussing key concepts such as energy regimes, transitions, and energy justice, it is essential to recognize how power dynamics influence energy policies and practices. Energy regimes refer to the specific configurations of energy supply, production, and consumption in a society, which are often entrenched and

M. C. Ogwu and S. Chibueze Izah, *Climate Frontiers and Social Transformation*, World-Systems Evolution and Global Futures,
https://doi.org/10.1007/978-3-032-17639-4_4

resistant to change. Furthermore, the notion of energy justice holds that transitions should not only aim for sustainability but also prioritize fairness and equity in the distribution of benefits and burdens associated with energy production (Healy & Barry, 2017; Sovacool et al., 2019). This aligns with the contemporary recognition that disadvantaged groups often bear the brunt of environmental degradation and energy inequities resulting from traditional energy practices (Chukwuemeka et al., 2023; Finley-Brook & Holloman, 2016).

4.1.1 The Historical Context of Energy Transitions

Energy transitions are not novel phenomena; they have recurred throughout history, closely correlating with developments in energy access, resource exploitation, and related technological advancements. Figure 4.1 illustrates the major energy transitions timeline in human history, from the reliance on biomass in pre-industrial societies through the coal-powered Industrial Revolution, the petroleum era of the twentieth century, and the current shift toward renewable energy systems. Each transition reflects significant technological, economic, and ecological transformations shaping global development trajectories. Biomass provided the backbone of energy for early human societies, leading to the establishment of sedentary communities, agricultural practices, and metallurgy (Mulvaney, 2020). The ascendance of coal in the nineteenth century catalyzed the Industrial Revolution, transforming economies and social structures as energy consumption soared (Mulvaney, 2020; Tumin et al., 2022).

During this period, coal not only replaced biomass as the primary energy source but also enabled unprecedented economic growth and technological advancement. The shift also brought widespread social restructuring; urbanization surged as workers migrated to industrial cities, driven by new manufacturing job opportunities (Barca, 2011). Thus, the transition from biomass to coal marked a significant turning point in human history, with profound implications for energy consumption and societal organization.

In the twentieth century, the discovery and exploitation of fossil fuels, particularly oil and natural gas, propelled further energy transitions, leading to a complex interplay of economic growth, geopolitical tensions, and environmental ramifications. Johnstone and McLeish (2020) argue that global conflicts, such as the World Wars, significantly influenced energy consumption patterns and directed resources toward oil, further entrenching its role in modern economies. The socio-technical systems established around oil and gas, however, have engendered challenges related to climate change and sustainability, necessitating a re-examination of our energy dependencies and practices (Malanima, 2012; Newell & Simms, 2020). The persistent reliance on fossil fuels has prompted urgent calls for a transition to renewable energy sources. The contemporary energy landscape is thus marked by profound uncertainty regarding the future of energy systems amidst climate change discourse. As Norouzi and Fani (2021) note, the global endeavor to achieve 100% renewable energy underscores significant potential benefits, such as

Fig. 4.1 Energy transitions across centuries

improved energy security and environmental sustainability, yet it also underscores the need for comprehensive strategies to navigate this transition effectively.

4.1.2 Intersections of Technology, Policy, and Justice in the Global Energy Transition

Given the complexity surrounding energy transitions, a multidisciplinary approach is essential to understanding the implications for policy and society. The integration of digital technologies into energy systems has emerged as a facilitator of the transition, as evidenced by the push toward smart grids that enable better management and distribution of energy resources. This technological transformation highlights the critical role that innovation plays in reshaping energy landscapes. However, as Herranz-Surrallés (2024) highlights, the geopolitical context also profoundly influences energy transition efforts, raising concerns over technological leadership and resource accessibility. Nations increasingly face dilemmas that pit economic interests against sustainable practices, prompting calls for more substantive policies that reflect energy justice principles. Moreover, the recognition that historic injustices should be addressed in new energy systems informs discussions on energy equity and public participation in transition processes (Healy & Barry, 2017; Sovacool et al., 2019).

Collectively, these findings reveal a pressing need for coherent frameworks that address the multifaceted dynamics of energy transitions. A just transition aligned with the Sustainable Development Goals requires a comprehensive understanding of power relations, resource allocation, and the socio-ecological implications of energy practices (Healy & Barry, 2017; Miller & Richter, 2014). As such, the historical perspective elucidates the patterns and lessons that can inform present and future transitions toward sustainable energy systems. Examining energy transitions through a historical lens not only enriches our understanding of past dynamics but also informs current strategies for a sustainable future. The interplay among energy consumption, societal transformation, and environmental stewardship will be crucial as we navigate the complexities of contemporary energy systems and pursue equity and sustainability during these transitions.

This chapter provides a historical analysis of global energy transitions, tracing the evolution from biomass-based economies reliant on wood, water, and wind to the fossil fuel era of coal, oil, and nuclear power, and ultimately to the contemporary push for renewable energy. It examines the socio-political forces driving these transitions—industrialization, imperialism, warfare, and technological innovation—while critically assessing the ecological consequences of energy extraction and consumption across centuries. By foregrounding issues of global inequality, particularly the divergent experiences of the Global North and South, this chapter situates current debates on decarbonization and sustainability within a long-term historical framework. Its contribution lies in bridging historical, ecological, and political perspectives to illuminate energy transitions as multidimensional phenomena shaped by power relations, technological change, environmental limits, and justice concerns. This chapter offers a deeper understanding of how past energy choices have shaped development trajectories and provides conceptual and empirical tools for interpreting the ongoing renewable energy transition as a technological imperative and a profoundly social and ethical project.

4.2 Early Energy Economies: Biomass and Pre-industrial Power

The early energy economies relied heavily on biomass materials, primarily wood, dung, and charcoal, for domestic needs such as heating and cooking. Before the advent of fossil fuels, these materials were critical in supporting the lifestyles and industrial activities of agrarian societies. Table 4.1 summarizes the primary biomass materials used in pre-industrial energy systems, highlighting their functions, advantages, limitations, and socio-economic implications. Wood and dung served as ubiquitous sources of household fuel, thereby forming a crucial component of the energy ladder—a theoretical framework that describes how societies transition from using primary biomass to cleaner energy sources as they develop economically and technologically (Ndwiga et al., 2014; Woolley et al., 2020). Charcoal, recognized for its higher heating capacity and cleaner burning properties compared to raw wood fuel, was less common but significant due to its efficiency (Fall et al., 2015; Taylor & Nakai, 2012). The deliberate processing of wood and dung into charcoal was a significant innovation, enhancing fuel efficiency and reducing indoor air pollution—a substantial health risk for families reliant on traditional biomass fuels (Sana et al., 2019).

In agrarian societies, wood was often favored due to its availability; however, as deforestation increased, households increasingly turned to dung as a reliable fuel source (Fall et al., 2015). This shift altered the dynamics of fuel availability and influenced social and economic structures within rural communities. The transition to dung illustrated how resource availability informed community behavior and ultimately shaped early trade networks, as surplus agricultural yields could be exchanged to support households lacking adequate fuel supplies (Baame et al., 2021; Po et al., 2011). Charcoal emerged as a versatile product, better suited for cooking and heating due to its reduced emissions and higher energy output compared to wood and dung (Sana et al., 2019; Taylor & Nakai, 2012). However, charcoal's production was not without environmental concerns, including deforestation and soil erosion, highlighting the complex ecological footprint of early energy extraction strategies (Kadir et al., 2010; Woolley et al., 2020). Watermills and windmills epitomized the harnessing of mechanical energy in agrarian economies, facilitating the processing of agricultural goods and minimizing labor intensity in tasks such as grain milling and textile production (Braadbaart et al., 2017; Hassan et al., 2017. The strategic placement of these mechanisms near water sources or wind-prone areas revealed the ingenuity of early societies in harnessing natural resources to meet their economic needs (McCarrick et al., 2024). Moreover, the adoption of water and wind power granted rural communities greater autonomy

Table. 4.1 Biomass materials in early energy economies

Biomass Material	Primary uses	Advantages	Limitations/ Concerns	Socio-economic implications
Wood	Cooking, heating, metallurgy	Readily available, easy to process	Deforestation, soil erosion, and habitat loss	Supported rural households; triggered trade in wood-scarce areas
Dung	Household cooking and heating	Alternative to wood in deforested regions; low-cost fuel	Low energy content, smoke emissions, and indoor air pollution	Shaped agricultural waste use; influenced rural fuel trade patterns
Charcoal	Cooking, metallurgy, and small-scale industries	Higher energy density, cleaner burning than wood	Deforestation for wood production and soil degradation	Catalyzed local markets; favored for urban and trade-oriented economies
Crop residues	Cooking, heating, and fodder after harvest	Waste reuse, and a supplement to wood or dung	Air pollution, limited seasonal availability	Encouraged recycling practices in agrarian households
Peat (in some regions)	Heating, small-scale industrial processes	Accessible where deposits exist; higher energy content than dung	Ecosystem disruption, carbon emissions	Regional energy trade; early local industrial applications

and economic stability, shifting labor dynamics by enabling fewer individuals to manage larger volumes of work, ultimately increasing agricultural surpluses and supporting the expansion of trade networks (Aremu & Aremu, 2025; Braadbaart et al., 2017).

The advent of surplus agricultural production catalyzed the development of more complex community structures. Societies, previously characterized by subsistence living, began to establish market systems and trade relations, which were both a response to and contributor to increased interaction among diverse cultures—embedding the socio-economic significance of energy resources within societal frameworks (Juntarawijit & Juntarawijit, 2020; McCarrick et al., 2024). The reliance on biomass fuels dictated energy consumption patterns and regulated population health, as exposure to indoor air pollution became a significant public health concern, particularly for vulnerable groups such as women and children (Rajper et al., 2020; Torres-Duque et al., 2008). Transitioning to cleaner fuel options became a matter of economic development and public health advocacy, as chronic respiratory issues emerged from sustained biomass combustion exposure (Saini et al., 2020; Taylor & Nakai, 2012).

The ecological impacts of early energy extraction methods were significant. Deforestation, mainly driven by demand for wood fuel, led to soil erosion and disrupted local ecosystems, as forests are crucial for maintaining biodiversity (Kadir et al., 2010; Woolley et al., 2020). Additionally, biomass extraction contributed to water scarcity in many regions; deforestation altered rainfall patterns and hindered soil moisture retention (McCarrick et al., 2024; Oguanobi & Joel, 2024). The adoption of watermills, while beneficial for mechanical power generation, also affected waterway ecosystems, highlighting the delicate balance between human enterprise and ecological health (McCarrick et al., 2024; Shillito et al., 2021). These developments foretold new challenges in sustainable resource management that remain pertinent in contemporary discussions of energy economies (Po et al., 2011; Baame et al., 2021). Early energy economies were characterized by a complex interplay of fuel sources, technological innovations, social organization, and ecological consequences. Wood, dung, and charcoal were integral to the formation of rural community identities, trade networks, and health paradigms. Understanding these dynamics is essential for contextualizing the evolution of energy consumption practices and their lasting legacies in modern energy systems. The historical reliance on biomass fuels reflects broader themes of environmental stewardship, social health, and technological evolution that resonate in today's discussions about sustainable development and energy transition.

4.3 The Age of Coal: Fueling the Industrial Revolution

The Age of the Coal-Fueled Industrial Revolution is marked by profound changes in technology, urban development, and environmental impacts largely driven by coal. The role of coal as a catalyst for industrial transformation is significant; it enabled the development of steam engines and mechanized manufacturing, which were crucial in transitioning economies from agrarian to industrialized societies. Figure 4.2 illustrates how coal fueled the Industrial Revolution, driving technological innovation, urban growth, and global trade while simultaneously triggering profound environmental and public health challenges. Mapping the interconnected pathways from coal extraction to industrialization, urbanization, pollution, and the eventual push for cleaner energy transitions, highlights both the transformative power and the enduring consequences of coal reliance in shaping modern societies.

Hanlon's (2016) investigation indicates that the reliance on coal significantly influenced local economic growth and cannot be dismissed as "trivial" as previously thought. The steam engine evolved from a coal-powered variant, which vastly improved efficiency in manufacturing processes, leading to unprecedented levels of production and utility (Hanlon, 2016). Urban growth was simultaneously fueled by the industrial advancements driven by coal, creating densely populated industrial cities. The combined effects of mechanization and urbanization illustrate a phenomenon where areas with high industrial concentration often attract labor movements seeking better working conditions and wages, spurred by the harsh realities of coal-induced labor environments. The interplay between coal mining activities and industrialization shaped the socio-economic landscape, leading to recognizable patterns of urbanization that are still evident today. For instance, research indicates that urban agglomerations in industrialized regions often exhibit

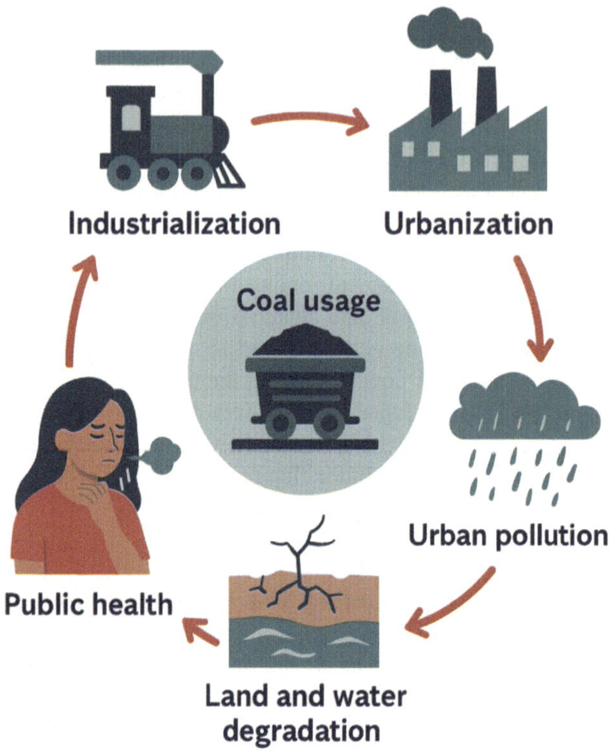

Fig. 4.2 Coal and the industrial revolution transformations impacts and legacies

pronounced air quality issues, exacerbated by coal consumption (Gao et al., 2023; Song et al., 2021). Energy reliance on coal extended beyond national borders, with the demand for it paving pathways for colonial expansion and the establishment of global trade networks. The demand for coal not only propelled industrialization but also supported imperial economies as they sought to control coal-rich territories. As industrial nations expanded their influence, they became increasingly reliant on coal, leading to profound social and economic interactions across continents (Beach & Hanlon, 2018; Ogwu et al., 2025a, 2025b). This connection is highlighted in the research conducted by Zhao et al. (2017), which describes significant environmental drawbacks of coal-dependent industries, including their considerable contribution to greenhouse gas emissions globally.

However, reliance on coal has led to serious environmental consequences. Urban pollution emerged as a critical issue, necessitating public health responses as coal combustion released substantial amounts of particulate matter and toxic pollutants into the atmosphere. Studies show that cities heavily reliant on coal revealed elevated concentrations of PM2.5 and other harmful air quality metrics, which were directly linked to severe public health risks, including respiratory diseases and other chronic conditions (Lin et al., 2014; Zhu, 2022). Urban areas such as those in Shanxi Province have been particularly affected, exposing how coal-centric economies can exacerbate public health crises through persistent air quality degradation, leading to public health emergencies and increased healthcare costs for affected populations Song et al., 2021; Xiao et al., 2022). The environmental degradation related to coal does not end with air quality; coal extraction processes have rendered extensive landscapes uninhabitable due to ecological disturbances and pollution. Regions noted for heavy coal mining practices suffer from significant soil and water contamination, leading to biodiversity loss and habitat destruction, which play a critical role in maintaining ecological balance (Hu et al., 2015, 2017). The urgency of environmental restoration approaches to reverse some detrimental impacts caused by coal mining practices has been echoed in many studies (Pinto et al., 1998).

In addition to immediate urban and environmental consequences, the role of coal as an energy source has shifted over time. The transition toward cleaner energy alternatives has gained momentum globally, especially considering the implications of climate change stemming from coal reliance. Research indicates that while the share of coal in energy consumption in developing countries remains substantial, efforts to reduce this dependency are critical for mitigating CO2 emissions and ensuring long-term sustainable development (Li & Wang, 2015; Tao et al., 2023). In essence, the Age of Coal was a multifaceted phase of transformation marked by significant technological advancements, urbanization, and environmental consequences. The benefits that arose from coal's exploitation were accompanied by long-term ramifications, shaping labor relations, public health realities, and urban landscapes. Continuing research into coal's past and implications helps illuminate current challenges and informs viable pathways toward sustainable practices in energy consumption and urban development.

4.4 The Petroleum Century: Oil, Modernity, and Global Power

The Petroleum Century signifies a vital transformation in global dynamics, emphasizing the connection between oil and modernity, global power structures, and ecological consequences. We will consider four primary dimensions of this phenomenon: the rise of oil economies, oil's role in warfare, the significance of petro-states and geopolitics, and the ecological ramifications associated with petroleum.

4.4.1 The Rise of Oil Economies

The early twentieth century witnessed the emergence of oil as a cornerstone of economic development. The discovery of oil prompted a refined understanding of energy production, leading to advancements in refining technology and the birth of the automobile industry (Babak & Deutsch, 2008). As energy demand increased, oil became the primary fuel source driving industrial growth, with the internal combustion engine revolutionizing transportation and subsequently transforming urban landscapes (Babak & Deutsch, 2008). The rise of oil led to the rapid development of oil markets, with countries rich in petroleum resources becoming significant players on the global economic stage. Nations capitalized on these resources, leading to the establishment of organizations such as the Organization of the Petroleum Exporting Countries (OPEC), which further solidified their economic dominance (Birdsey et al., 2006).

The refining process, a crucial aspect of oil economies, involved converting crude oil into usable products. This not only facilitated the mass production of fuels for heating and transportation but also spurred innovation in the chemical industry, where by-products of refining became essential for producing various goods and materials (Babak & Deutsch, 2008). The automobile age catalyzed consumer culture, fostering economic structures reliant on fossil fuels and creating both domestic and international dependencies on oil (Jones et al., 2005). As countries such as the United States and those in the Middle East capitalized on this newfound wealth, they also became key players in global geopolitics, with oil production rates directly influencing their standing and interactions (Birdsey et al., 2006).

4.4.2 Energy and Warfare

Oil's centrality in warfare cannot be overstated, particularly during the World Wars and the Cold War era. World War I marked the beginning of the significant militarization of fuel supplies, with oil being recognized as vital for naval and mechanized warfare (Hook, 2020) The reliance on oil-powered machinery changed the landscape of battle strategy, compelling nations to secure oil reserves and ensure continued access to this crucial energy source (Jones et al., 2005). During World War II, the United States' oil production outpaced that of any other nation, further establishing its status as a superpower capable of sustained military engagements (Barron, 2012). Control over oil reserves became a strategic objective for warring nations, as illustrated by campaigns targeting oil fields and refineries in North Africa and the Pacific (Barron, 2012). The Cold War continued this trend, with oil serving as a tool of power projection, economic manipulation, and ideological competition, such as when the United States sought to curb the expansion of oil-rich communist regimes (Legrand et al., 2018). Countries engaged in proxy wars often found oil wealth to be both a target and a means to fund military operations, thus entwining energy resources with geopolitical ambitions (Jones et al., 2005).

4.4.3 Petro-states and Geopolitics

The rise of petro-states fundamentally reshaped global political dynamics, emblematic of resource nationalism. Nations endowed with vast oil reserves wielded unprecedented influence, dictating terms in both economic and geopolitical realms (Birdsey et al., 2006). OPEC's formation in 1960 marked a concerted effort by oil-rich countries to control production, stabilize prices, and assert their influence on global markets, directly challenging Western dominance Hook, 2020. This coalition not only brought together major oil producers but also highlighted the growing political clout of these nations on the world stage. Additionally, the interdependence established between oil-rich nations and developed countries created complex relationships, characterized by both cooperation and conflict. The 1973 oil crisis illustrated the extent to which oil-producing nations could leverage their resources to affect global economies, prompting discussions on energy security and diversification (Pongratz et al., 2010). As some countries transitioned toward more diversified energy sources, the implications of resource nationalism became evident, underpinning contemporary debates on energy sovereignty and the role of fossil fuels in national security strategies (Okeke et al., 2024).

4.4.4 Ecological Consequences

The ecological ramifications of oil extraction, consumption, and its associated wars contribute significantly to environmental degradation. The twentieth century was marked by numerous oil spills, air pollution, and increased carbon emissions, leading to profound impacts on ecosystems and public health (Hook, 2020; Mamozai et al., 2024). Oil spills, such as the infamous Deepwater Horizon incident in 2010, not only resulted in immediate ecological disasters but also long-term consequences for marine and coastal habitats Barron, (2012). The overarching ecological crises sparked by oil have often highlighted the need for robust regulatory frameworks and innovative technologies to mitigate these environmental threats (Mamozai et al., 2024; Ogwu et al., 2025a, 2025b, 2025c, 2025d, 2025e).

Furthermore, emissions from fossil fuels have considerably contributed to climate change, exacerbating various ecological issues (Pu & Luo, 2021; Rahimi, 2021). The release of greenhouse gases, driven by industrial activities and vehicles, has altered atmospheric composition, ultimately affecting global climate patterns (Rahimi, 2021). Correspondingly, the increase in black carbon, primarily attributed to petroleum combustion, has been shown to impact air quality and contribute significantly to global warming (Skeie et al., 2011). To address the ecological consequences of oil dependency, a nuanced understanding of the interplay among energy production, environmental health, and climate science is essential. Striking a balance between energy needs and sustainable practices is imperative for addressing pressing ecological challenges, including biodiversity loss, habitat destruction, and pollution (Hook, 2020; Legrand et al., 2018).

4.5 The Nuclear Promise and Its Discontents

The perception of nuclear power as a catalyst for modern progress is multifaceted, deeply intertwined with historical, cultural, and political contexts. Table 4.2 summarizes the dual nature of nuclear energy, highlighting its potential to provide reliable, low-carbon power while stressing economic, environmental, and social challenges as existing and growing sources of discontent. Atomic energy has long been touted as a clean solution to the pressing challenge of climate change, especially given its lower greenhouse gas emissions compared to fossil fuels (Alwaeli & Mannheim, 2022). The evolving landscape of energy production, especially in the wake of heightened environmental awareness, highlights nuclear energy as a significant player capable of meeting baseload energy demands without directly contributing to atmospheric pollution (Alwaeli & Mannheim, 2022). Nevertheless, these advantages are often overshadowed by public apprehension, particularly fueled by past nuclear disasters such as Chernobyl (Ukraine) and Fukushima (Japan), which serve as stark reminders of the potential risks of nuclear technology (Zhang et al., 2019).

The geopolitical implications of nuclear power during the Cold War era further complicate its legacy as a symbol of progress. The adoption of nuclear energy was closely linked to deterrence politics, in which nations' nuclear capabilities were central to the security landscape (Ewing, 2008). As nations developed their nuclear arsenals, the distinction between civilian nuclear power and military applications became increasingly blurred. The dual-use nature of nuclear technology creates a precarious balance, raising ethical concerns about proliferation and global stability (Ahearne, 2000; Garwin, 2001). The specter of nuclear weapons, proliferating alongside civilian energy programs, adds layers of complexity to the debate over nuclear power's role in sustainable energy strategies. Public resistance to nuclear power is significantly shaped by fear of potential accidents, as demonstrated by

Table 4.2 The nuclear promise and some discontents linked to balancing energy potential and societal concerns

Aspect	Promise	Discontents
Energy security	Reliable, large-scale electricity generation	High costs, limited fuel availability
Climate mitigation	Low-carbon energy source	Radioactive waste, long-term storage issues
Technological progress	Advanced reactor designs, innovation potential	Risk of accidents (e.g., Chernobyl, Fukushima)
Economic development	Job creation and industrial growth	Decommissioning costs, economic risk of accidents
Global politics	Energy independence and reduced fossil fuel reliance	Nuclear weapons proliferation concerns
Public perception	Symbol of modernity and scientific advancement	Persistent fear, opposition from civil society

incidents such as Chernobyl and Fukushima. These disasters highlighted the dangers inherent in nuclear technology and have catalyzed widespread anti-nuclear movements globally, affecting policy decisions and public sentiment regarding energy production (Solomon et al., 2010). The memories of these events and their catastrophic implications have engendered a lingering mistrust in nuclear technology, suggesting that public perception is often more influenced by fear and risk perception than by rational evaluations of safety measures or technological advancements (Friederich & Boudry, 2022).

The environmental and ethical challenges posed by nuclear waste management are arguably one of the most contentious aspects of the nuclear debate. The long-lived nature of radioactive waste requires solutions that span generations, increasing the urgency for practical engineering, ethical considerations, and public policy (Cetina, 2024). Various approaches to manage high-level waste (HLW), such as deep geological repositories and advanced waste forms, have been proposed (Yanikomer et al., 2016). However, their implementation is encumbered not only by technical challenges but also by public opposition grounded in fears about the long-term integrity of storage solutions and potential environmental contamination (Tracy et al., 2021). The ethical implications are profound, raising questions about intergenerational equity and our responsibility toward future generations who will inherit the legacy of nuclear waste (Ahearne, 2000; Friederich & Boudry, 2022).

The transition toward a sustainable energy future amidst climate change pressures necessitates a reassessment of nuclear energy's role. Enduring public fear surrounding nuclear waste underscores the need for transparent communication strategies that effectively address safety concerns and environmental impacts (Yan et al., 2017). Countries continue to grapple with these dilemmas, and navigating between technological advancement and ethical responsibility becomes critical. The integration of innovative approaches to waste management and power generation could revitalize public support for nuclear energy, especially if perceived risks are mitigated through collaborative efforts among governments, communities, and scientific institutions (Kromp et al., 2023).

Moreover, the discourse surrounding nuclear waste management often reveals competing narratives shaped by socio-political dynamics. The alignment (or misalignment) of technological capabilities with public perceptions can create a stalemate in policy formulations, thus delaying necessary reforms in energy transitions (Solomon et al., 2010). Promoting a rigorous dialogue that involves diverse viewpoints is fundamental to overcoming obstacles to nuclear power's public acceptance (Schwenk-Ferrero & Andrianov, 2017). Furthermore, stakeholder engagement is necessary to ensure that local concerns and ethical considerations are adequately articulated in the formulation of nuclear energy policies. The complexities surrounding nuclear power encompass broader debates over energy sustainability, environmental ethics, and societal risk perception. As we contemplate the future of nuclear energy, it is vital to consider its implications not merely as a power source but as a multifaceted phenomenon that necessitates careful navigation through technical, ethical, and social landscapes. This integrated approach could lead to a more informed public discourse and potentially rejuvenate nuclear energy's reputation as not merely a technological option but a pragmatic solution for a sustainable future.

4.6 Transitions, Labor, and Ecological Transformations

The interconnections between energy transitions, labor practices, ecological transformations, and socio-economic consequences are complex and multifaceted. As the world transitions toward renewable energy systems to combat climate change and environmental degradation, the ramifications on industrial labor systems, urbanization, class structures, and global inequalities become evident. The effects of these intertwined phenomena can be analyzed through the lens of research studies that elucidate the challenges and opportunities they present.

Energy transitions significantly disrupt existing labor markets and industrial structures. In particular, the shift from fossil fuels to renewable energy sources creates substantial job displacement in traditional sectors such as coal mining. Research indicates that regions dependent on coal are experiencing high unemployment and may require targeted policy interventions to facilitate worker transitions into new industries. For instance, Michieka et al. (2022) highlight the need for regions to develop transportation systems linking displaced workers with educational institutions and new job sites, thereby bridging skill gaps in industries such as renewable energy. Moreover, Shutters (2021) notes that these transitions can exacerbate economic instability, potentially leading to increased migration and social unrest as workers grapple with job losses and economic uncertainty stemming from clean energy initiatives.

The socio-economic consequences of energy transitions are also evident in the changing dynamics of urbanization and class structures. As industries relocate or collapse, urban areas should adapt to changing workforce demographics. Union strategies play a crucial role in this adaptation process, particularly in politically sensitive regions where fossil fuel industries dominate. Brower Brown & Nelson, 2024 study of California's climate transitions demonstrate how unions can strategically navigate the shift from fossil fuel to renewable energy sectors, offering support to workers transitioning from "sunset" to "sunrise" industries. This transformation can yield new opportunities for workers, yet it also requires unions to redefine their roles in response to changing economic paradigms.

In examining the ecological costs of these transitions, the environmental consequences of resource extraction can no longer be overlooked. Global environmental policies increasingly focus on mitigating deforestation, air and water pollution, and biodiversity loss. For instance, while Butorina and Starodumova (2021) discuss the need for innovative financial forms to support transitions, their findings primarily relate to economic transformation rather than directly addressing environmental sustainability. Therefore, their relevance to ecological consequences requires caution. Additionally, the ecological footprint of resource extraction remains a concern, particularly in the Global South, where these activities often entail significant social and environmental costs (Myllylä & Takala, 2011). Indeed, global inequalities are stark, as resource-rich nations in the Global South often see their resources exploited for the benefit of industrialized countries in the Global North. Studies indicate that this pattern reinforces existing socio-economic disparities, as local communities often bear the brunt of environmental degradation resulting from industrial practices that prioritize profit over ecological sustainability (Mori, 2020). The role of foreign actors also emerges as crucial, as they can provide essential technological and financial resources to facilitate energy transitions but may also perpetuate dependence and the exploitation of local resources (Mori, 2020).

Furthermore, the interplay between labor markets and ecological concerns raises critical discussions about just transitions, which seek to ensure that workers displaced by environmental regulations are afforded equitable opportunities in emerging sectors. Abraham (2017) discusses how labor environmentalism in regions like Appalachia necessitates the establishment of just transition policies that safeguard workers' rights while championing environmental sustainability. This perspective is increasingly vital as workers in traditional energy sectors face significant barriers in accessing new careers in renewable industries without adequate support and customization of training programs.

Emerging technologies are also essential in shaping both labor markets and energy systems. The integration of innovations such as artificial intelligence and automation brings about transformative impacts on the labor force. According to a study by Frank et al. (2019), the interplay between these technologies and labor markets can either enhance productivity in some areas or displace jobs in others, thereby necessitating policy interventions to mitigate adverse outcomes for workers.

4.7 Renewable Energy and the Emerging Green Transition

The transition toward renewable energy is an essential component of addressing global environmental concerns, particularly climate change, and promoting sustainable development. Technological innovations across renewable energy sectors such as solar, wind, geothermal, hydro, and bioenergy are crucial to facilitating this transition. Renewable energy technologies have demonstrated significant potential to mitigate carbon dioxide emissions while fostering green economic growth. Research indicates that renewable energy transitions can enhance ecological sustainability and stimulate economic development by improving energy efficiency and reducing reliance on fossil fuels (Deshuai et al., 2022; Krzemień et al., 2023). Figure 4.3 illustrates the key elements driving the global shift toward renewable energy systems and highlights the value of technological innovations, policy frameworks, environmental considerations, and economic implications in shaping the emerging green transition.

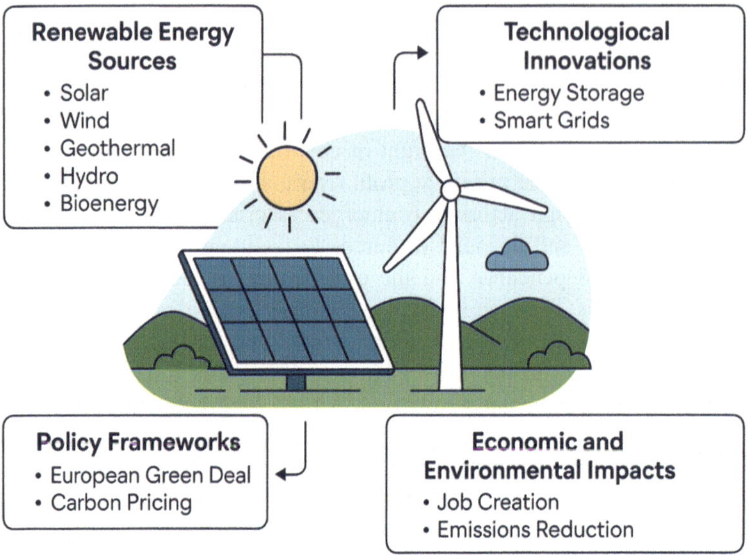

Fig.4.3 Renewable energy and the emerging green transition

In the realm of technological innovations, solar and wind energy systems are at the forefront. These sources not only offer cleaner alternatives to fossil fuels but also align closely with various international climate agreements aimed at reducing greenhouse gas emissions (Fratita et al., 2023). The convergence of technological advancements and public policy efforts is vital in this context as innovations have reduced costs and improved the efficiency of renewable energy systems, making them more viable for widespread adoption (Olczak et al., 2021). For instance, the dramatic decrease in solar panel costs over the past decade has made solar energy one of the most economical forms of energy generation globally (Olczak et al., 2021). Frameworks such as the European Green Deal and carbon pricing mechanisms significantly influence the policy landscape surrounding renewable energy. The European Union's Green Deal aims to transform Europe into the first climate-neutral continent by 2050, underlining the importance of renewable energy sources (Dyczko et al., 2021). The Green Deal promotes investments in renewable technologies and incentivizes reductions in greenhouse gas emissions through tailored policy measures and regulations (Simionescu et al., 2020). Furthermore, carbon pricing mechanisms internalize the environmental costs of fossil fuel consumption, thereby encouraging the transition to cleaner energy sources. Studies indicate that implementing these climate policies can lead to considerable growth in renewable energy investments and the creation of green jobs (Kayakuş et al., 2023; Pakulska, 2021).

Moreover, a significant aspect of energy transition is addressing the intermittency and storage challenges associated with renewable energy generation. The integration of energy storage solutions and smart grid technologies is essential for managing the variable nature of renewable energy sources (Zhang et al., 2016). Technological advancements in energy storage, including battery technologies and thermal storage solutions, contribute to the resilience and reliability of energy systems by providing backup power during periods of high demand or low generation (Dyczko et al., 2021). Smart grid technologies enhance the efficiency of energy distribution, allowing for better integration of decentralized renewable energy sources into the existing energy infrastructure (Ali et al., 2022). The intersections of energy transition with other critical sectors—such as food, water, and land use—present both challenges and opportunities for sustainable development. Energy production from renewable sources often competes with agricultural production for land, raising concerns about food security (Jackson, 2011). Additionally, water use in energy generation, particularly in hydropower and bioenergy, highlights the interconnectedness of these sectors. Policymakers must adopt an integrated approach that considers the implications of energy policies on water and land resources to mitigate adverse outcomes while maximizing sustainable benefits (Peña-Ramos et al., 2021).

Incorporating environmental considerations into renewable energy planning is also vital for resolving potential conflicts with biodiversity. Certain renewable technologies, such as large-scale bioenergy and hydropower installations, may have detrimental impacts on local ecosystems, thus necessitating careful environmental impact assessments and strategic planning (Geißler, 2013). Balancing the development of renewable energy infrastructure with the preservation of biodiversity can lead to more sustainable outcomes, which requires the involvement of multiple stakeholders, including local communities and conservation groups (Jackson, 2011). The economic implications of the renewable energy transition are profound. The shift to renewable sources can stimulate economic growth by creating new jobs in the energy sector, emerging technologies, and related fields (Simionescu et al., 2020). For instance, the establishment of local renewable energy projects can bolster regional economies by fostering job creation and modernization of infrastructure (Ali et al., 2022). However, the economic benefits of transitioning to renewable energy are not uniformly distributed, often requiring targeted support and incentives to ensure equitable participation, particularly in developing economies (Lori et al., 2025; Peña-Ramos et al., 2021).

Moreover, the role of international cooperation and agreements cannot be overstated in the pursuit of a global green transition. Collaborative efforts among nations promote knowledge sharing, technological exchange, and financial investment in renewable energy projects, thereby enhancing the overall impact of global climate goals (Kayakuş et al., 2023). Such international commitments not only facilitate the deployment of renewable energy technologies but also build robust mechanisms for accountability and progress monitoring toward achieving sustainability targets (Iyiola et al., 2025; Iyiola & Ogwu, 2025; Khorishko & Rudnieva, 2023). As nations continue to navigate this complex landscape, a cohesive strategy that clearly outlines technological, policy, and economic pathways toward a green future is essential to achieving long-term sustainability goals.

4.8 Energy, Inequality, and the Politics of Transition

The discourse surrounding energy, inequality, and the politics of transition is increasingly essential in contemporary academia, particularly amid growing environmental concerns and socio-economic disparities. Exploring frameworks of energy justice reveals critical facets such as access, affordability, and fairness, especially as the world pivots toward energy transitions necessitated by climate change. Scheier and Kittner (2022) emphasize the urgency of addressing residential energy burdens, arguing that targeted policies, including proactive tools and community participation, are vital to facilitating a just energy transition in the EU context. They argue that implementing strategies such as "One Stop Shops" (OSS) for energy efficiency upgrades can alleviate these burdens and serve as a model for inclusive policymaking.

The intersection of energy poverty and income inequality remains a pervasive issue. Nguyen and Nasir (2021) demonstrate that government policies focused on energy consumption levels tend to exacerbate existing inequalities, thus perpetuating a cycle of poverty. This sentiment is echoed by Khamjalas (2024), whose research indicates that in Sub-Saharan Africa, access to modern energy disproportionately benefits wealthier populations, thereby widening the income gap. Such findings underscore the importance of inclusive electricity programs that address access disparities and ensure the energy transition does not deepen socio-economic divides. Yu et al. (2023) highlight the critical issue of embodied renewable energy transfer and the inherent inequalities within global supply chains. They argue that underdeveloped renewable energy infrastructure in many regions further contributes to global energy welfare inequality, making it difficult for low-access communities to benefit from otherwise accessible technologies (Izah et al., 2025a, 2025b). This systemic imbalance highlights the urgent need for re-evaluated global trading systems and equity in energy transitions that empower all communities, particularly in the Global South.

Additionally, the implications of climate adaptation and environmental justice cannot be discounted, especially in vulnerable regions. Tornel (2023) asserts that energy justice frameworks should actively contest entrenched colonial power relations that have historically marginalized communities in the Global South. This decolonization perspective advocates for local agency and control over energy resources, which is a necessary shift to cultivate equitable energy policies. The interconnectedness of social inequality and climate impacts is further illustrated in Dias' (2024) analysis, which shows how climate change exacerbates vulnerabilities among the poorest, necessitating interventions that simultaneously address energy poverty and systemic inequalities. In addressing the politics of energy transition, it is crucial to consider the role of governance and agency. Sinha et al. (2022) advocate for a policy framework that appreciates the decomposition of inequalities in energy access. Similarly, Peddibhotla et al.'s (2024) examination of intermediaries in informal settlements highlights how grassroots movements can influence energy justice outcomes through localized political mobilization.

These perspectives illustrate the transformative potential of grassroots actions to counteract systemic inequalities in energy distribution and access.

Moreover, energy justice should proactively integrate considerations of gender, as Götzmann and Dicalou (2025) suggest a feminist approach to energy justice that emphasizes rectifying gendered inequalities within energy transition narratives. This lens is critical, as it recognizes the unique vulnerabilities faced by women, often exacerbated in the context of energy poverty and climate impacts. Ensuring that women's voices are incorporated into energy policymaking is essential for achieving holistic and equitable energy solutions. The significant role of technological innovation in bridging the energy access gap cannot be overlooked. Research by Li et al., 2025shows that while increased access to clean cooking fuels can yield marginal reductions in urban income inequality, disparities persist due to systemic barriers to access. As such, energy access programs should be designed not just to provide resources but also to enhance communities' capabilities and agency to utilize these resources effectively.

As nations strive toward sustainable energy futures, considerations of equity and justice should remain at the forefront of the dialogue. The analysis provided by McCauley and Heffron (2018) outlines a framework of justice—distributional, procedural, and restorative—that could underpin just transition efforts. This approach promotes an understanding of how benefits and burdens of energy systems are shared across different communities, presenting a pathway toward inclusive energy policies that consider past grievances while fostering future resilience. Achieving fairness in energy transitions cannot be an afterthought; it should be integrated from the outset. Chambers et al. (2022) highlight the importance of digital inclusion in energy systems, recognizing that even as technological advancements promise greater efficiency, they risk perpetuating existing inequalities if not carefully managed. Therefore, the dialogues surrounding energy justice, equity, and effective transition strategies should be sustained and dynamic, continually addressing the nuances and complexities inherent in such significant systemic shifts.

The transition to sustainable energy is critical for addressing climate change, reducing socio-economic inequalities, and ensuring energy justice (Ogwu et al., 2025a; 2025b). However, this transition faces multiple, interconnected challenges spanning policy, technology, economics, and social dimensions. These barriers hinder equitable access, slow technological adoption, and risk deepening existing inequalities if left unaddressed. Table 4.3 summarizes the key challenges affecting the sustainability and inclusivity of the energy transition.

4.9 Conclusion: Historical Lessons and Future Pathways

The history of energy transitions reveals that shifts in dominant energy regimes have never been purely technological; instead, they have been embedded in broader social, political, and ecological transformations. From the biomass-dependent

Table 4.3 Challenges in achieving a sustainable energy transition

Challenge area	Key issues	Implications
Energy access & affordability	Unequal access, high costs for low-income households	Deepens energy poverty and socio-economic inequalities
Policy & governance	Weak policies, lack of inclusive decision-making	Excludes marginalized groups from benefits and decision processes
Global Inequalities	Unequal renewable technology transfer, colonial legacies	Limits progress in developing economies, reinforces dependency
Technological gaps	Limited storage, smart grid infrastructure gaps	Reduces the reliability and scalability of renewable energy systems
Gender inequities	Gendered energy burdens, underrepresentation in policy processes	Disproportionate impacts on women and vulnerable populations
Economic barriers	Unequal distribution of green jobs, financing challenges	Concentrates economic gains in wealthier regions or groups
Environmental trade-offs	Land, water, and biodiversity conflict with renewable projects	Threatens ecosystems and food security
Social acceptance & participation	Lack of community engagement, resistance to change	Slows the adoption of energy innovations and inclusive solutions

economies of agrarian societies to the coal-driven industrial revolution, the oil-dominated twentieth century, and the emerging era of renewables, each transition has reconfigured patterns of production, urbanization, warfare, and imperialism while leaving lasting environmental footprints. Crucially, these transitions have been uneven, with the benefits and burdens of energy choices distributed along lines of class, empire, and geography, shaping a global landscape of energy inequality that persists today. As the world confronts the twin crises of climate change and resource depletion, the ongoing renewable energy transition offers opportunities to decarbonize economies and reimagine energy systems around principles of sustainability and justice; however, historical patterns caution against assuming that technological innovation alone can guarantee equitable outcomes. Without attention to power dynamics, financing mechanisms, and the needs of vulnerable populations, the green transition risks reproducing the same asymmetries that have long defined global energy politics. By situating contemporary debates within centuries of energy history, this chapter highlights the need for a just and inclusive energy transition—one that learns from past ecological degradation, prioritizes the rights of marginalized communities, and recognizes energy as a social as well as environmental good. Future pathways should therefore combine technological innovation, policy reforms, and ethical considerations to build energy systems capable of sustaining both people and planet in the long term.

References

Abraham, J. (2017). Just transitions for the miners: Labor environmentalism in the Ruhr and Appalachian coalfields. *New Political Science, 39*(2), 218–240.

Ahearne, J. F. (2000). Intergenerational issues regarding nuclear power, nuclear waste, and nuclear weapons. *Risk Analysis, 20*(6), 763–770.

Ali, S., Yan, Q., Irfan, M., Ameer, W., Atchike, D. W., & Acevedo-Duque, A. (2022). RETRACTED: Green investment for sustainable business development: The influence of policy instruments on solar technology adoption. *Frontiers in Energy Research, 10*, 874824.

Alwaeli, M., & Mannheim, V. (2022). Investigation into the current state of nuclear energy and nuclear waste management—A state-of-the-art review. *Energies, 15*(12), 4275.

Aremu, O., & Aremu, O. O. (2025). Effect of household air pollution and neighbourhood deprivation on the risk of acute respiratory infection among under-five children in Chad: A multilevel analysis. *International Journal of Environmental Research and Public Health, 22*(5), 710.

Baame, E. M., Goura, A. P., Mbatchou, B. H. N., Walage, B., Simo, H. S. Y., Medjou, R. M., Sonkoue, M. P., Djouda, C. D., Ngnewa, R. S. F., Guiagain, M. S. T., & Agokeng, B. D. K. (2021). Distribution of sources of household air pollution: a cross-sectional study in Cameroon. *BMC Public Health, 21*, 318.

Babak, O., & Deutsch, C. V. (2008). Simulating oil recovery during CO2 sequestration into a mature oil reservoir. *Journal of Canadian Petroleum Technology, 47*(08), 34–41.

Barca, S. (2011). Energy, property, and the industrial revolution narrative. *Ecological Economics, 70*(7), 1309–1315.

Barron, M. G. (2012). Ecological impacts of the deepwater horizon oil spill: Implications for immunotoxicity. *Toxicologic Pathology, 40*(2), 315–320.

Beach, B., & Hanlon, W. W. (2018). Coal smoke and mortality in an early industrial economy. *The Economic Journal, 128*(615), 2652–2675.

Birdsey, R., Pregitzer, K., & Lucier, A. (2006). Forest carbon management in the United States: 1600–2100. *Journal of Environmental Quality, 35*(4), 1461–1469.

Braadbaart, F., van Brussel, T., van Os, B., & Eijskoot, Y. (2017). Fuel remains in archaeological contexts: Experimental and archaeological evidence for recognizing remains in hearths used by Iron Age farmers who lived in peatlands. *The Holocene, 27*(11), 1682–1693.

Brower Brown, K., & Nelson, S. H. (2024). Working sunset to sunrise: Union strategies in three California climate transitions. *Environmental Politics, 33*(4), 657–677.

Butorina, O., & Starodumova, Y. (2021). Monetary and informational form of capital as the basis of investment support for the inter-phase transition to an innovative economy. In *SHS Web of Conferences* (Vol. 116, p. 00028). EDP Sciences.

Cetina, I. (2024). Nuclear waste management: Recycling and long-term safe storage. *Kemija u Industriji, 73*, 249–261.

Chambers, J., Robinson, C., & Scott, M. (2022). Digitalisation without detriment: A research agenda for digital inclusion in the future energy system. *People, Place and Policy Online, 16*(2), 177–192.

Chukwuemeka, N. S., Ugonna, A. P., Ugochukwu, O. B., Immaculata, E. N., Chiziterem, E. K., & Chukwubuikem, O. P. (2023). The challenges and opportunities of energy transition across Africa. *International Journal of Environment and Climate Change, 13*(10), 4312–4339.

Deshuai, M., Hui, L., & Ullah, S. (2022). RETRACTED: Pro-environmental behavior–Renewable energy transitions nexus: Exploring the role of higher education and information and communications technology diffusion. *Frontiers in Psychology, 13*, 1010627.

Dias, R. (2024). The complex interconnectedness of climate change and social inequality. In *Harmony of knowledge: Exploring interdisciplinary synergie*. Seven Editora. https://doi.org/10.56238/sevened2023.006-119

Dyczko, A., Kamiński, P., Stecuła, K., Prostański, D., Kopacz, M., & Kowol, D. (2021). Thermal and mechanical energy storage as a chance for energy transformation in Poland. *Polityka Energetyczna, 24*, 43–60.

Ewing, R. C. (2008). Nuclear fuel cycle: Environmental impact. *MRS Bulletin, 33*(4), 338–340.

Fall, P. L., Falconer, S. E., & Klinge, J. (2015). Bronze age fuel use and its implications for agrarian landscapes in the eastern Mediterranean. *Journal of Archaeological Science: Reports, 4*, 182–191.

Finley-Brook, M., & Holloman, E. L. (2016). Empowering energy justice. *International Journal of Environmental Research and Public Health, 13*(9), 926.

Frank, M. R., Autor, D., Bessen, J. E., Brynjolfsson, E., Cebrian, M., Deming, D. J., Feldman, M., Groh, M., Lobo, J., Moro, E., & Wang, D. (2019). Toward understanding the impact of artificial intelligence on labor. *Proceedings of the National Academy of Sciences, 116*(14), 6531–6539.

Fratita, M., Popescu, F., Rusu, E., Ion, I. V., & Mahu, R. (2023). Romanian energy system analysis (production, consumption, and distribution). *Energies, 16*(16), 6045.

Friederich, S., & Boudry, M. (2022). Ethics of nuclear energy in times of climate change: Escaping the collective action problem. *Philosophy & Technology, 35*(2), 30.

Gao, L., Zhang, J., Tian, Y., Liu, X., Guan, S., & Wu, Y. (2023). Study on the impact of collaborative agglomeration of manufacturing and producer services on PM2. 5 pollution: Evidence from urban agglomerations in the middle reaches of the Yangtze River in China. *International Journal of Environmental Research and Public Health, 20*(4), 3216.

Garwin, R. L. (2001). Can the world do without nuclear power? Can the world live with nuclear power? *Interdisciplinary Science Reviews, 26*(4), 265–271.

Geißler, G. (2013). Strategic environmental assessments for renewable energy development—Comparing the United States and Germany. *Journal of Environmental Assessment Policy and Management, 15*(02), 1340003.

Götzmann, N., & Dicalou, M. (2025). Towards a feminist energy justice framework. *Business and Human Rights Journal, 10*(1), 56–78.

Hanlon, W. W. 2016. *Coal smoke and the costs of the industrial revolution* (No. w22921). National Bureau of Economic Research.

Hassan, L. G., Sani, N. A., Sokoto, A. M., & Tukur, U. G. (2017). Comparative studies of burning rates and water boiling time of wood charcoal and briquettes produced from carbonized *Martynia annua* woody shells. *Nigerian Journal of Basic and Applied Sciences, 25*(2), 21–27.

Healy, N., & Barry, J. (2017). Politicizing energy justice and energy system transitions: Fossil fuel divestment and a "just transition". *Energy Policy, 108*, 451–459.

Hook, S. E. (2020). Beyond thresholds: A holistic approach to impact assessment is needed to enable accurate predictions of environmental risk from oil spills. *Integrated Environmental Assessment and Management, 16*(6), 813–830.

Hu, J., Wang, P., Ying, Q., Zhang, H., Chen, J., Ge, X., Li, X., Jiang, J., Wang, S., Zhang, J., Zhao, Y., & Zhang, Y. (2017). Modeling biogenic and anthropogenic secondary organic aerosol in China. *Atmospheric Chemistry and Physics, 17*(1), 77–92.

Hu, Z., Fu, Y., Xiao, W., Zhao, Y., & Wei, T. (2015). Ecological restoration plan for abandoned underground coal mine site in Eastern China. *International Journal of Mining, Reclamation and Environment, 29*(4), 316–330.

Iyiola, A. O., Akingba, O. O., Alimi, S. K., Akinsorotan, A. M., Osei, S. A., Idowu, C. F., Izah, S. C., & Ogwu, M. C. (2025). Impact assessment of biobased products in the global south. In M. C. Ogwu, S. C. Izah, J. G. Vazquez-Arenas, S. T. Feleke, & X. Wei (Eds.), *Sustainable bioeconomy development in the global south: Volume II bioeconomy techniques*. Springer. https://doi.org/10.1007/978-981-96-0305-3_16

Iyiola, A. O., & Ogwu, M. C. (2025). Environmental management using renewable biological resources in the global south. In M. C. Ogwu, S. C. Izah, J. G. Vazquez-Arenas, S. T. Feleke, & X. Wei (Eds.), *Sustainable bioeconomy development in the global south: Volume II bioeconomy techniques*. Springer. https://doi.org/10.1007/978-981-96-0305-3_11

Izah, S. C., Ogwu, M. C., & Hait, M. (2025a). Clean energy solutions and public health: Innovations in renewable energy for a healthier environment. In S. C. Izah & M. C. Ogwu (Eds.), *Innovative approaches in environmental health management: Processes, technologies, and strategies for a sustainable future. Environmental science and engineering series*. Springer Nature. https://doi.org/10.1007/978-3-031-81966-7_7

Izah, S. C., Ogwu, M. C., & Salimon, S. S. (2025b). Principles of sustainable development and environmental management. In M. C. Ogwu & S. Chibueze Izah (Eds.), *Evaluating environmental processes and technologies. Environmental science and engineering series.* Springer. https://doi.org/10.1007/978-3-031-85327-2_2

Izah, S. C., & Ohimain, E. I. (2013). The challenge of biodiesel production from oil palm feedstock in Nigeria. *Greener Journal of Biological Sciences, 3*(1), 1–12.

Jackson, A. L. (2011). Renewable energy vs. biodiversity: Policy conflicts and the future of nature conservation. *Global Environmental Change, 21*(4), 1195–1208.

Johnstone, P., & McLeish, C. (2020). World wars and the age of oil: Exploring directionality in deep energy transitions. *Energy Research & Social Science, 69*, 101732.

Jones, G. S., Jones, A., Roberts, D. L., Stott, P. A., & Williams, K. D. (2005). Sensitivity of global-scale climate change attribution results to inclusion of fossil fuel black carbon aerosol. *Geophysical Research Letters, 32*(14).

Juntarawijit, C., & Juntarawijit, Y. (2020). Cooking with biomass fuel and cardiovascular disease: A cross-sectional study among rural villagers in Phitsanulok, Thailand. *F1000Research, 9*, 307.

Kadir, M. M., McClure, E. M., Goudar, S. S., Garces, A. L., Moore, J., Onyamboko, M., Kaseba, C., Althabe, F., Castilla, E. E., Freire, S., & Parida, S. (2010). Exposure of pregnant women to indoor air pollution: A study from nine low and middle income countries. *Acta Obstetricia et Gynecologica Scandinavica, 89*(4), 540–548.

Kayakuş, M., Terzioğlu, M., Erdoğan, D., Zetter, S. A., Kabas, O., & Moiceanu, G. (2023). European Union 2030 carbon emission target: The case of Turkey. *Sustainability, 15*(17), 13025.

Khamjalas, K. (2024). Estimating rural premium and financial crisis effect on the nexuses between food, energy, and water consumption on urban-rural income gap in South–Eastern Asian countries using pooled regression analysis. *American Journal of Applied Statistics and Economics, 3*(1), 80–91.

Khorishko, L., & Rudnieva, A. (2023). Latvian renewable energy policy in the context of EU initiatives. *Baltic Journal of Economic Studies, 9*(2), 211–215.

Kromp, M. G., Zambetakis, F. P., & Wood, W. (2023). Innovative approaches to nuclear energy density optimization for enhanced power generation and waste minimization. *Vertex, 12*(2), 89–98.

Krzemień, A., Frejowski, A., Fidalgo Valverde, G., Riesgo Fernández, P., & Garcia-Cortes, S. (2023). Repurposing end-of-life coal mines with business models based on renewable energy and circular economy technologies. *Energies, 16*(22), 7617.

Legrand, M., McConnell, J. R., Preunkert, S., Arienzo, M., Chellman, N., Gleason, K., Sherwen, T., Evans, M. J., & Carpenter, L. J. (2018). Alpine ice evidence of a three-fold increase in atmospheric iodine deposition since 1950 in Europe due to increasing oceanic emissions. *Proceedings of the National Academy of Sciences, 115*(48), 12136–12141.

Li, C., Li, M., Zhang, L., Li, Q., Zheng, H., & Feldman, M. W. (2025). Energy–poverty–inequality SDGs: A large-scale household analysis and forecasting in China. *Proceedings of the National Academy of Sciences, 122*(1), e2408167121.

Li, L., & Wang, J. (2015). The effects of coal switching and improvements in electricity production efficiency and consumption on CO2 mitigation goals in China. *Sustainability, 7*(7), 9540–9559.

Lin, G., Fu, J., Jiang, D., Hu, W., Dong, D., Huang, Y., & Zhao, M. (2014). Spatio-temporal variation of PM2. 5 concentrations and their relationship with geographic and socioeconomic factors in China. *International Journal of Environmental Research and Public Health, 11*(1), 173–186.

Lori, T., Osei-Kwarteng, M., Ojija, F., Ojo, A. O., & Ogwu, M. C. (2025). Participatory approaches to environmental health management. In S. C. Izah & M. C. Ogwu (Eds.), *Innovative approaches in environmental health management: Processes, technologies, and strategies for a sustainable future. Environmental science and engineering series.* Springer Nature. https://doi.org/10.1007/978-3-031-81966-7_11

Malanima, P. (2012). The path towards the modern economy the role of energy. In *From Malthus' stagnation to sustained growth: Social, demographic and economic factors* (pp. 71–99). Palgrave Macmillan UK.

Mamozai, W., Hesam, A. M., & Hemma, W. H. (2024). Impacts of crude oils on water quality: A comprehensive review. *European Journal of Theoretical and Applied Sciences, 2*(1), 126–138.

McCarrick, S., Delaval, M. N., Dauter, U. M., Krais, A. M., Snigireva, A., Abera, A., Broberg, K., Eriksson, A. C., Isaxon, C., & Gliga, A. R. (2024). Toxicity of particles derived from combustion of Ethiopian traditional biomass fuels in human bronchial and macrophage-like cells. *Archives of Toxicology, 98*(5), 1515–1532.

McCauley, D., & Heffron, R. (2018). Just transition: Integrating climate, energy and environmental justice. *Energy Policy, 119*, 1–7.

Michieka, N. M., Graziano, M., Musso, M., & Fouquet, R. (2022). Energy transitions and labor market patterns in the US coal industry. *Structural Change and Economic Dynamics, 63*, 501–514.

Miller, C. A., & Richter, J. (2014). Social planning for energy transitions. *Current Sustainable/Renewable Energy Reports, 1*(3), 77–84.

Mori, A. (2020). Foreign actors, faster transitions? Co-evolution of complementarities, perspectives and sociotechnical systems in the case of Indonesia's electricity supply system. *Energy Research & Social Science, 69*, 101594.

Mulvaney, D. (2020). Energy transitions. In *Sustainable energy transitions: Socio-ecological dimensions of decarbonization* (pp. 1–32). Springer International Publishing.

Myllylä, S., & Takala, T. (2011). Leaking legitimacies: The Finnish forest sectors entanglement in the land conflicts of Atlantic coastal Brazil. *Social Responsibility Journal, 7*(1), 42–60.

Ndwiga, T., Kei, R. M., Jepngetich, H., & Korrir, K. (2014). Assessment of health effects related to the use of biomass fuel and indoor air pollution in Kapkokwon sub-location, Bomet country, Kenya. *Open Journal of Air Pollution, 3*(03), 61–69.

Nguyen, C. P., & Nasir, M. A. (2021). An inquiry into the nexus between energy poverty and income inequality in the light of global evidence. *Energy Economics, 99*, 105289.

Norouzi, N., & Fani, M. (2021). World on the road to 100% renewable energy. *Trends in Renewable Energy, 7*(1), 114–126.

Oguanobi, V. U., & Joel, O. T. (2024). Geoscientific research's influence on renewable energy policies and ecological balancing. *Open Access Research Journal of Multidisciplinary Studies, 7*(02), 073–085.

Ogwu, M. C., El Malahi, S., & Izah, S. C. (2025a). Policy and regulatory frameworks for sustainable environmental practices. In M. C. Ogwu & S. Chibueze Izah (Eds.), *Evaluating environmental processes and technologies. Environmental science and engineering series*. Springer. https://doi.org/10.1007/978-3-031-85327-2_16

Ogwu, M. C., Iyiola, A. O., Fawole, W. O., & Izah, S. C. (2025b). Biobased resource recovery techniques in the global south. In M. C. Ogwu, S. C. Izah, J. G. Vazquez-Arenas, S. T. Feleke, & X. Wei (Eds.), *Sustainable bioeconomy development in the global south*. Springer. https://doi.org/10.1007/978-981-96-0305-3_17

Ogwu, M. C., Izah, S. C., Merritt, H., Wise, R. D., & Jintrawet, A. (2025c). *Sustainable bioeconomy development in the global south: Volume I status and perspectives*. Springer. https://doi.org/10.1007/978-981-96-0640-5

Ogwu, M. C., Izah, S. C., Vazquez-Arenas, J. G., Feleke, S. T., & Wei, X. (2025d). *Sustainable bioeconomy development in the global south: Volume II bioeconomy techniques*. Springer. https://doi.org/10.1007/978-981-96-0305-3

Ogwu, M. C., Kosoe, E. A., & Osawaru, M. E. (2025e). Raw materials for sustainable bioeconomy development in the global south. In M. C. Ogwu, S. C. Izah, H. Merritt, R. D. Wise, & A. Jintrawet (Eds.), *Sustainable bioeconomy development in the global south: Volume II bioeconomy techniques*. Springer. https://doi.org/10.1007/978-981-96-0640-5_2

Okeke, T. E., Ewuim, S. C., Uhuo, C., Ononye, B. U., Akunne, C. E., & Okafor, K. P. (2024). Evaluating the ecological consequences of heavy metal contamination in soil induced by spent engine oil and palm oil mill effluents for sustainable development. *Sustainable Society Development, 2*(2), 2410.

Olczak, P., Żelazna, A., Matuszewska, D., & Olek, M. (2021). The "My Electricity" program as one of the ways to reduce CO2 emissions in Poland. *Energies, 14*(22), 7679.

Pakulska, T. (2021). Green energy in central and eastern European (CEE) countries: New challenges on the path to sustainable development. *Energies, 14*(4), 884.

Peddibhotla, A., Haldar, S., & Bazaz, A. (2024). Watts fair in Mumbai: Intermediaries impact on energy justice in informal settlements. *Environmental Research: Energy, 1*(4), 045006.

Peña-Ramos, J. A., del Pino-García, M., & Sánchez-Bayón, A. (2021). The Spanish energy transition into the EU green deal: Alignments and paradoxes. *Energies, 14*(9), 2535.

Pinto, J. P., Stevens, R. K., Willis, R. D., Kellogg, R., Mamane, Y., Novak, J., Šantroch, J., Beneš, I., Lenicek, J., & Bureš, V. (1998). Czech air quality monitoring and receptor modeling study. *Environmental Science & Technology, 32*(7), 843–854.

Po, J. Y., FitzGerald, J. M., & Carlsten, C. (2011). Respiratory disease associated with solid biomass fuel exposure in rural women and children: Systematic review and meta-analysis. *Thorax, 66*(3), 232–239.

Pongratz, J., Reick, C. H., Raddatz, T., & Claussen, M. (2010). Biogeophysical versus biogeochemical climate response to historical anthropogenic land cover change. *Geophysical Research Letters, 37*(8), L08702.

Pu, A., & Luo, X. (2021). Li-doped beryllonitrene for enhanced carbon dioxide capture. *RSC Advances, 11*(60), 37842–37850.

Rahimi, M. (2021). Lithium-ion batteries: Latest advances and prospects. *Batteries, 7*(1), 8.

Rajper, S. A., Nazir, A., & Ullah, S. (2020). Perceived health risks of exposure to indoor air pollution from cooking fuels in Sindh, Pakistan. *Polish Journal of Environmental Studies, 29*(4), 2833–2844.

Saini, J., Dutta, M., & Marques, G. (2020). A comprehensive review on indoor air quality monitoring systems for enhanced public health. *Sustainable Environment Research, 30*(1), 6.

Sana, A., Meda, N., Badoum, G., Kafando, B., & Bouland, C. (2019). Primary cooking fuel choice and respiratory health outcomes among women in charge of household cooking in Ouagadougou, Burkina Faso: Cross-sectional study. *International Journal of Environmental Research and Public Health, 16*(6), 1040.

Scheier, E., & Kittner, N. (2022). A measurement strategy to address disparities across household energy burdens. *Nature Communications, 13*(1), 288.

Schwenk-Ferrero, A., & Andrianov, A. (2017). Comparison and screening of NFC options in view of sustainable performance and waste management. *Sustainability, 9*(9), 1623.

Shillito, L., Namdeo, A., Bapat, A., Mackay, H., & Haddow, S. (2021). Analysis of fine particulates from fuel burning in a reconstructed building at Neolithic Çatalhöyük, Turkey, and implications for respiratory health in early settled communities.

Simionescu, M., Păuna, C. B., & Diaconescu, T. (2020). Renewable energy and economic performance in the context of the European green deal. *Energies, 13*(23), 6440.

Sinha, A., Balsalobre-Lorente, D., Zafar, M. W., & Saleem, M. M. (2022). Analyzing global inequality in access to energy: Developing policy framework by inequality decomposition. *Journal of Environmental Management, 304*, 114299.

Skeie, R. B., Berntsen, T., Myhre, G., Pedersen, C. A., Ström, J., Gerland, S., & Ogren, J. A. (2011). Black carbon in the atmosphere and snow, from pre-industrial times until present. *Atmospheric Chemistry and Physics, 11*(14), 6809–6836.

Solomon, B. D., Andrén, M., & Strandberg, U. (2010). Three decades of social science research on high-level nuclear waste: Achievements and future challenges. *Risk, Hazards & Crisis in Public Policy, 1*(4), 13–47.

Song, H., Zhuo, H., Fu, S., & Ren, L. (2021). Air pollution characteristics, health risks, and source analysis in Shanxi Province, China. *Environmental Geochemistry and Health, 43*(1), 391–405.

Sovacool, B. K., Martiskainen, M., Hook, A., & Baker, L. (2019). Decarbonization and its discontents: A critical energy justice perspective on four low-carbon transitions. *Climatic Change, 155*(4), 581–619.

Tao, M., Zheng, Y., Failler, P., Lau, W. Y., & Goh, L. T. (2023). The effect of energy structure on pollution emissions: The role of technological innovation. *Asian Economics Letters, 4.*

Taylor, E. T., & Nakai, S. (2012). Prevalence of acute respiratory infections in women and children in Western Sierra Leone due to smoke from wood and charcoal stoves. *International Journal of Environmental Research and Public Health, 9*(6), 2252–2265.

Tornel, C. (2023). Decolonizing energy justice from the ground up: Political ecology, ontology, and energy landscapes. *Progress in Human Geography, 47*(1), 43–65.

Torres-Duque, C., Maldonado, D., Pérez-Padilla, R., Ezzati, M., & Viegi, G. (2008). Biomass fuels and respiratory diseases: A review of the evidence. *Proceedings of the American Thoracic Society, 5*(5), 577–590.

Tracy, C. L., Park, S., Plevaka, M., & Bogdanova, E. (2021). Opportunities for US-Russian collaboration on the safe disposal of nuclear waste. *Bulletin of the Atomic Scientists, 77*(3), 146–152.

Tumin, V. M., Kharlanov, A. S., Zenkina, E. V., Kostromin, P. A., & Trifonov, V. A. (2022, February). Digital transformation of oil and gas sector during transition to renewable energy sources. In *International Scientific and Practical Conference Strategy of Development of Regional Ecosystems "Education-Science-Industry" (ISPCR 2021)* (pp. 484–489). Atlantis Press.

Woolley, K. E., Bagambe, T., Singh, A., Avis, W. R., Kabera, T., Weldetinsae, A., Mariga, S. T., Kirenga, B., Pope, F. D., Thomas, G. N., & Bartington, S. E. (2020). Investigating the association between wood and charcoal domestic cooking, respiratory symptoms and acute respiratory infections among children aged under 5 years in Uganda: A cross-sectional analysis of the 2016 demographic and health survey. *International Journal of Environmental Research and Public Health, 17*(11), 3974.

Xiao, Y., Zhang, G., Guo, J., Zhang, Z., Wang, H., Wang, Y., Wang, Z., Yuan, H., & Cui, D. (2022). Pollution characteristics and risk assessments of mercury in Jiutai, a county region thriving on coal mining in Northeastern China. *Sustainability, 14*(16), 10366.

Yan, X., Yang, L., Zhang, X., & Zhan, W. (2017). Concept of an accelerator-driven advanced nuclear energy system. *Energies, 10*(7), 944.

Yang, Y., Chen, Y., Poon, J., Qian, X., Zhou, Y., & Xia, S. 2023. Is embodied renewable energy transfer greening the global supply chain?

Yanikomer, N., Asal, S., Haciyakupoglu, S., & Erenturk, S. A. (2016). New solidification materials in nuclear waste management. *International Journal of Engineering Technologies IJET, 2*(2), 76–82.

Zhang, N., Sutanto, D., & Muttaqi, K. M. (2016). A review of topologies of three-port DC–DC converters for the integration of renewable energy and energy storage system. *Renewable and Sustainable Energy Reviews, 56*, 388–401.

Zhang, Q., Zheng, M., Huang, Y., Kunte, H. J., Wang, X., Liu, Y., & Zheng, C. (2019). Long term corrosion estimation of carbon steel, titanium and its alloy in backfill material of compacted bentonite for nuclear waste repository. *Scientific Reports, 9*(1), 3195.

Zhao, H., Ma, W., Dong, H., & Jiang, P. (2017). Analysis of co-effects on air pollutants and CO2 emissions generated by end-of-pipe measures of pollution control in China's coal-fired power plants. *Sustainability, 9*(4), 499.

Zhu, Y. (2022). Prediction modeling and research on the relationship between urban air pollutants and respiratory diseases. *Academic Journal of Science and Technology, 2*, 88–93.

Part II

Regional Dynamics and Systemic Inequalities

5 Climate Change and the Reproduction of Global Inequality

5.1 Introduction

Climate change has emerged as one of the most vital challenges of our era, transcending mere environmental degradation to manifest profound social, political, and economic implications that inevitably exacerbate global inequalities. The interconnection between climate change and socio-economic and ecological disparities is particularly salient in a world where more prosperous nations contribute disproportionately to carbon emissions but bear a lesser burden of climate impacts compared to poorer, more vulnerable countries. This phenomenon highlights the necessity of viewing climate change not only as an ecological crisis but also as a multifaceted socio-economic challenge, in which issues of injustice and inequality are intimately intertwined with environmental policies and climate responses. Recent estimates show that the richest 10% of the world's population are responsible for nearly 48% of global carbon emissions, while the poorest 50% contribute only 12%, highlighting the profound inequities underlying the climate crisis (Chancel, 2022). Historical analyses reveal that nations in the Global North have generated 92% of excess CO_2 emissions since the Industrial Revolution, underscoring their disproportionate responsibility for contemporary climate risks (Hickel, 2020).

The climate change–inequality nexus reveals a glaring discrepancy in the distribution of climate risks and resources for adaptation. Evidence from several studies demonstrates that poorer countries, despite contributing minimally to historical greenhouse gas emissions, bear the brunt of climate change, leading to a diminished capacity to pursue sustainable development. Wang et al. (2024) elucidate that region plagued by poverty, particularly in Africa, bear higher incidences of climate-related disasters such as compound droughts and heatwaves without the necessary socio-economic infrastructure to ameliorate their impacts. Hence,

M. C. Ogwu and S. Chibueze Izah, *Climate Frontiers and Social Transformation*, World-Systems Evolution and Global Futures,
https://doi.org/10.1007/978-3-032-17639-4_5

addressing climate change necessitates a comprehensive understanding of its socio-economic implications, especially regarding the widening developmental chasm between the global core, semi-periphery, and periphery regions.

5.1.1 Background and Rationale

Climate change fundamentally challenges existing global inequities by exacerbating the vulnerabilities of marginalized communities. The systemic disparities reinforced by both socio-economic structures and climatic phenomena are reflected in the exacerbation of poverty levels; unchecked climate change is projected to push as many as 130 million individuals into extreme poverty over the coming years (SenGupta & Atal, 2024). This projection underscores a grim reality for countries already facing economic hardships and limited adaptive capacity. As SenGupta and Atal (2024) point out, the unequal distribution of climate impacts poses severe threats to poverty eradication initiatives and risks entrenching pre-existing inequities across and within nations.

Further exploration of income inequality and climate change reveals a troubling correlation between uneven economic growth and inaction on climate change. Wang et al. (2025) highlight that perceptions of climate responsibility and social trust are diminished in contexts marked by pronounced economic inequality, which contributes to policy inaction and stymies collective efforts to combat climate change. Thus, addressing climate change effectively demands nuanced policies that not only mitigate environmental impacts but also intentionally rectify socio-economic disparities, fostering a more equitable global approach to climate governance.

5.1.2 Climate Change and Inequality Nexus

The nexus between climate change and inequality becomes glaringly apparent when examining the differential capacities of nations to cope with climatic stresses. Economic geography shows that more affluent countries, despite their substantial contributions to climate change, tend to have mechanisms in place to buffer against its impacts, thereby reducing vulnerabilities and protecting their economies from climate shocks (Diffenbaugh & Burke, 2019). Conversely, poorer nations lacking such resources and capacities suffer disproportionately, highlighting the structural imbalance entrenched within global systems. Indeed, as reported by Diffenbaugh and Burke (2019), the manifestations of global warming have been linked to increased economic inequality, further widening the gap between wealthy and impoverished nations. According to the IPCC, between 3.3 and 3.6 billion people currently live in contexts highly vulnerable to the impacts of climate change, with the greatest concentration of risk in low-income regions (IPCC, 2022). Although

African nations contribute only about 3% of global emissions, they experience 7–10 times higher mortality from climate-related disasters compared to high-income countries (UNEP, 2023).

Moreover, the complexities of climate change are coupled with underlying social injustices, as evidenced by the interaction between climate vulnerability and socio-economic hierarchies. Vulnerability to climate impacts is not only determined by geographical factors but is also significantly mediated by existing socio-economic inequalities. For instance, Castells-Quintana and McDermott (2024) indicate that the most adverse effects of climate change are projected to impact low-income countries, exacerbating existing inequalities and hindering their developmental pathways. The resulting feedback loops between climate vulnerabilities and social injustices necessitate policies that address both climate change mitigation and the underlying structural inequities perpetuating these vulnerabilities. The global health implications of climate change also reflect this interplay between climate and inequality. As explored by Paavola (2017), the health impacts of climate change disproportionately affect marginalized groups who often lack access to adequate healthcare and social services, compounding their vulnerabilities. The inequitable distribution of health impacts underscores that climate change is as much a health crisis as an environmental one, underscoring the importance of integrating health considerations into climate action strategies.

5.1.3 Compounding Vulnerabilities

The multidimensional nature of climate vulnerabilities in poorer regions often leads to compounding socio-economic effects. The myriad of climatic events, ranging from extreme weather phenomena to gradual environmental degradation, affect agricultural productivity, food security, and overall economic stability, thereby increasing income inequality in these regions (Palagi et al., 2022). Insufficient investment in climate-resilient infrastructure particularly contributes to this deterioration, as vulnerable regions may lack the necessary capital to implement protective measures, reinforcing the economic divide.

Moreover, armed conflicts in precarious regions serve as additional layers of vulnerability, where climate change acts as a threat multiplier. As Buhaug (2015) articulates, the vulnerabilities induced by armed conflict limit development and further compound the inequalities experienced by affected populations. This intertwined relationship between climate change and conflict underscores the need to address not only climate resilience but also the socio-political contexts in which these inequalities manifest. With climate-induced disruptions exacerbating pre-existing tensions, the interplay of climate change and conflict creates situations in which recovery and adaptation capabilities are severely compromised, thereby perpetuating cycles of disadvantage.

In addition to the socio-political variables at play, the economic implications of climate change reveal stark inequalities. For instance, regions heavily reliant on climate-sensitive sectors such as agriculture and natural resources are precariously affected by erratic climate patterns, whereas wealthier regions with diversified economic foundations exhibit greater resilience to such shocks (Castells-Quintana & McDermott, 2024). This disparity is fundamentally tied to the global economic structures that have historically disadvantaged poorer nations, resulting in uneven access to resources and differing adaptive capacities.

5.1.4 The Role of Policy and Governance

Effective climate governance should address these inherent inequities and their ramifications. Global climate agreements often fall short of addressing these disparities: wealthier nations may acknowledge their ecological footprints yet seldom commit to compensatory measures that benefit disproportionately affected nations. The inequitable contributions to climate change mean that those responsible for historical emissions frequently evade accountability, leaving poorer nations to grapple with the consequences (Chancel, 2022). As Guarino et al. (2024) elucidate, structural inequalities, coupled with international trading and economic policies, exacerbate global inequalities, particularly impacting developing economies that are often coerced into unfavorable positions within global markets. The importance of cooperation and equitable resource allocation cannot be overstated in the quest for effective climate solutions. A successful approach to mitigating the impacts of climate change on global inequality hinges on fostering a sense of collective responsibility among nations to address resource distribution and aid sustainably. Studies indicate that countries with higher economic inequalities exhibit lower levels of commitment to climate action measures, emphasizing the need for policies that incentivize equitable participation (Chancel, 2022; Wang et al., 2025). Furthermore, initiatives to enhance social safety nets and encourage income diversification are critical to bolstering resilience in vulnerable communities facing climate challenges (Eskander & Fankhauser, 2022).

5.1.5 Emerging Challenges and Future Directions

The intersectionality of climate change and inequality calls for innovative solutions that integrate social justice into the fabric of climate action. Policymakers should navigate the complexities of global inequalities while ensuring that initiatives prioritize the needs of the most marginalized populations. Approaches such as community-driven climate adaptation strategies and inclusive governance frameworks can help bridge disparities, enabling equitable access to resources and information (Soergel et al., 2021). Moreover, promoting local participation in climate policies not only fosters ownership among affected communities but also enhances the effectiveness of interventions suited to specific regional contexts. As the landscape of climate negotiations continues to evolve, it is imperative to prioritize equity-centered approaches that pragmatically assess climate impacts through an inequality lens. This entails integrating social, economic, and environmental considerations, ensuring that those most affected by climate change have a voice in decision-making processes that pertain to their futures (Tesselaar et al., 2020). The dynamic and often volatile nature of climate challenges necessitates responsive policies that adapt to the shifting realities faced by distinct communities worldwide.

This chapter aims to critically examine how climate change both reflects and deepens structural global inequalities, particularly across core, semi-periphery, and periphery regions. By mapping differential exposure to climate risks, analyzing disparities in adaptive capacities, and interrogating the inequitable distribution of climate finance, this chapter situates climate vulnerability within long-standing historical legacies of colonization, trade dependency, and uneven global governance structures. Its key contribution lies in synthesizing perspectives from world-systems theory, climate justice, and political ecology to reveal how contemporary climate crises reproduce centuries-old patterns of power and resource imbalances. This chapter illuminates the systemic roots of climate injustice. It highlights pathways for equitable climate governance and inclusive adaptation strategies, offering a critical framework for scholars and policymakers working toward sustainable and just climate futures.

5.2 Theoretical Overview of Climate Change and the Exacerbation of Global Inequality

The relationship between climate change and global inequality is multifaceted, drawing upon various theoretical frameworks to analyze how environmental issues exacerbate socio-economic disparities. Among these frameworks, World-Systems Theory, Climate Justice Frameworks, and Political Ecology Perspectives each present critical insights into the reproduction of inequality in the context of climate change.

5.2.1 World-Systems Theory and Economic-Environmental Disparities

World-Systems Theory, originating from the works of Immanuel Wallerstein, provides a lens through which to understand the global capitalist system comprising core, semi-peripheral, and peripheral nations. This framework illustrates how core nations exploit peripheral nations for resources and labor, leading to significant economic disparities and environmental degradation in the latter. The core–periphery model posits that economic benefits and ecological burdens are inequitably distributed, often leaving marginalized communities in peripheral regions to bear the brunt of adverse environmental impacts while core nations accrue the benefits (Boyce et al., 2016; Reich et al., 2022). Recent studies highlight this imbalance, noting that those living in the periphery, particularly indigenous peoples and rural agriculturalists, are disproportionately affected by climate mitigation efforts, such as large-scale renewable energy projects that may encroach on their land and traditional livelihoods (Reich et al., 2022; Whyte, 2020). Consequently, the adverse impacts of climate change are often felt more acutely in these nations, perpetuating a cycle of poverty and marginalization that aligns well with Wallerstein's analysis of global capitalism.

Furthermore, the negative environmental consequences faced by these communities often remain invisible within the broader economic narratives dominated by core nations (Boyce et al., 2016). Thus, World-Systems Theory serves as a foundational framework to understand the intersections of economic exploitation and environmental injustice, highlighting the systemic reproduction of global inequality.

5.2.2 Principles of Climate Justice Frameworks

The principles that underpin climate justice frameworks are essential for understanding the moral imperatives associated with climate adaptation and mitigation. Distributive justice emphasizes fair distribution of environmental burdens and benefits, while procedural justice insists that marginalized communities be included in decision-making processes that affect their environments and livelihoods (Baker, 2024; Schlosberg & Collins, 2014). Recognitional justice also plays a crucial role, advocating for the acknowledgment of the rights and identities of marginalized groups who often face a double burden of vulnerability to climate impacts and systemic injustices (Baker, 2024; Whyte, 2020).

Empirical studies illustrate that low-income and marginalized communities frequently experience environmental degradation due to deficient policies that fail to address their unique contexts (Martenies et al., 2017, 2018). For example, in cities like Lagos (Nigeria), New Delhi (India), and Detroit (United States), where segregated neighborhoods face poorer air quality and higher levels of pollutants, there are stark contrasts in environmental health based on socio-economic status (Martenies et al., 2017, 2018). As such, adherence to climate justice principles is necessary not only for ethical governance but also for practical outcomes that seek to redress these inequalities. Additionally, past policies often neglected procedural justice, thereby disenfranchising those most affected by climate change (Akbulut, 2012; Baker, 2024). Urban planning and resource management strategies should incorporate participatory approaches that empower local communities, ensuring that all voices are considered in climate resilience planning (Akbulut, 2012). Without these frameworks, the persistence of inequality will remain a fundamental challenge in addressing climate change globally.

5.2.3 Political Ecology Perspectives on Power and Resource Control

Political Ecology offers a critical perspective on how power dynamics shape environmental governance and resource control. This framework emphasizes that environmental change cannot be divorced from the socio-political contexts in which it occurs, highlighting the interplay between human activities and ecological change (Levy et al., 2006; Schlosberg & Collins, 2014). The uneven distribution of environmental benefits and burdens is critically influenced by power relations among societal actors, which include governments, corporations, and marginalized communities (Nagy, 2021; Schlosberg & Collins, 2014). Examining health disparities linked to environmental degradation reveals that the most vulnerable populations—often characterized by low socio-economic status—live in areas with the highest pollution levels (Bigman et al., 2021; Minas, 2022). The systemic inequities observed in response to environmental shifts reflect a broader pattern of exclusion where marginalized communities lack representation in policymaking processes (Minas, 2022).

Moreover, studies on labor mobility indicate that environmental quality can profoundly impact economic decisions, as individuals often seek to relocate to areas with better living conditions (Landrigan et al., 2010; Xing et al., 2024). This compels a reassessment of how environmental inequalities influence social and economic mobility, often reinforcing existing disparities (Tony, 2023; Xing et al., 2024). Political Ecology emphasizes that any effective response to climate injustice should account for these systemic inequities and advocate for equitable resource distribution and participatory governance (Nagy, 2021; Schlosberg & Collins, 2014).

5.2.4 Intersecting Frameworks and the Path Forward

These three frameworks—World-Systems Theory, Climate Justice Frameworks, and Political Ecology Perspectives—intersect to provide a comprehensive foundation for understanding the reproduction of global inequality in the context of climate change. The exploitation inherent in the world system reinforces environmental injustices that disproportionately affect marginalized communities. Concurrently, climate justice principles call for distributional fairness and procedural equity, while political ecology critiques the power dynamics that exacerbate these inequities. Collectively, they argue for integrated measures that consider historical injustices, recognize the rights of affected communities, and implement participatory governance practices. Addressing climate change effectively demands not only technological solutions but also a commitment to dismantling the structures perpetuating inequality. This includes fostering resilience among vulnerable communities, ensuring their voices are central to the formulation of adaptation and mitigation policies, and advocating for financial support to aid their transition to sustainable practices (Minas, 2022; Nagy, 2021).

As climate change continues to threaten global stability, the implications of ignoring these intersecting frameworks can be dire, further entrenching social inequities and exacerbating the impacts of environmental degradation on vulnerable populations. Therefore, the convergence of these theoretical foundations provides critical insights into addressing the interconnected challenges of climate change and global inequality. Through a nuanced understanding of power dynamics, justice principles, and systemic economic structures, researchers, policymakers, and activists can more effectively navigate the complexities of climate justice, paving the way toward a more equitable and sustainable future.

5.3 Mapping Differential Climate Risks

In addressing the complex interplay between climate change and the reproduction of global inequality, it is essential to map the differential climate risks that exacerbate existing vulnerabilities across geographical, social, and economic dimensions. Regions such as the Sahel, Borneo, Amazon, Andes, South Asia, and small island states are experiencing distinct climatic challenges that necessitate a nuanced understanding of their vulnerability patterns. For example, the Sahel is prone to severe droughts that jeopardize food security and economic stability, while flood-prone areas in South Asia are increasingly faced with the impacts of erratic monsoon patterns (Diffenbaugh & Burke, 2019). Furthermore, small island nations are facing dire threats from rising sea levels, which threaten their very existence and livelihoods, illustrating how geographical factors dictate the severity of climate risks (Sihotang et al., 2022). Figure 5.1 illustrates the actual and potential spatial distribution of major climate risks worldwide, highlighting how hazards such as severe drought, flooding, rising sea levels, wildfires, biodiversity loss, and heatwaves are unevenly distributed. These differential risks highlight the intersection between environmental stressors and pre-existing social and economic vulnerabilities, shaping diverse adaptation challenges across regions.

These geographical vulnerabilities are compounded by climate-related disasters such as hurricanes, wildfires, and heat waves, which uniquely impact various regions and socio-economic groups. For instance, hurricanes in the Caribbean and southern United States disproportionately affect low-income communities, often leading to a loss of homes and livelihoods, thereby widening the economic divide (Casson et al., 2023). Heat waves have become more frequent and intense due to climate change, disproportionately affecting vulnerable populations, including the elderly and those with pre-existing health conditions (Song et al., 2023). Recent studies stress how these disasters exacerbate existing inequalities and contribute to a cycle of disadvantage for marginalized communities (Khine & Langkulsen, 2023).

Moreover, the social and economic dimensions of climate vulnerability reveal critical inequities across gender, class, and racial lines. Gendered experiences of climate hazards are particularly pronounced, as women frequently encounter barriers to accessing resources, such as land and financial support in the aftermath

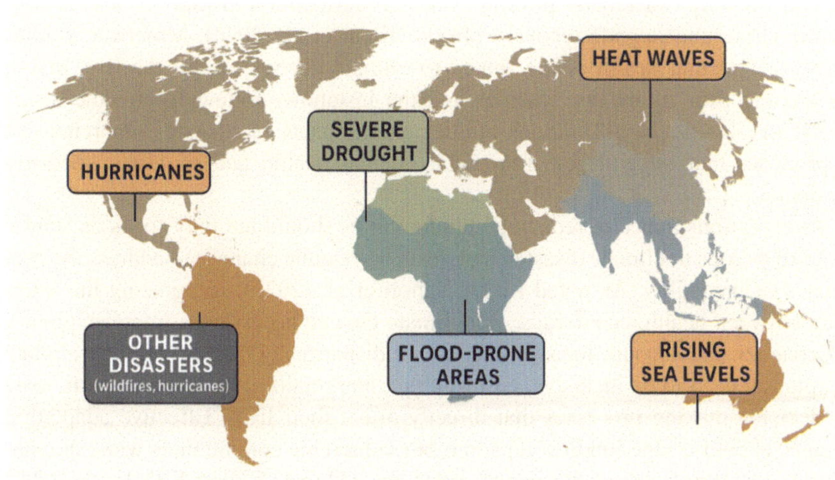

Fig. 5.1 Actual and potential global hotspots of differential climate risks and their socio-economic implications

of disasters (Awiti, 2022). Class disparities further fuel vulnerabilities, with economically disadvantaged groups having limited access to adaptive resources or emergency services (Pak et al., 2025). Racial inequities manifest in differing exposure to environmental hazards and disparities in health outcomes. This layering of vulnerabilities demonstrates a complex interaction where socio-economic status, gender roles, and racial identities influence how communities experience and adapt to climate change (da Silva & Requia, 2025).

A comparative analysis of case studies from diverse geographic contexts elucidates the varying impacts of climate-related disasters. In the United States, wealthier regions often recover more rapidly and effectively from disasters compared to economically challenged contexts like Brazil, where socio-economic inequalities hamper resilience efforts (Perry et al., 2021). For example, after a major hurricane, affluent neighborhoods may leverage resources and insurance mechanisms to rebuild quickly, while poorer neighborhoods face protracted recovery processes (Perry et al., 2021). Conversely, in countries like Mozambique, the lack of infrastructure and financial resources means that recovery from similar disasters is slower and leaves lasting scars on the community's socio-economic fabric (Diffenbaugh & Burke, 2019). These disparities highlight an urgent need for targeted interventions that consider the socio-economic and spatial dynamics of vulnerability and resilience.

Diving deeper into specific climate-related hazards, the historical context of each region sheds light on their unique vulnerabilities. For instance, droughts in the Sahel have been historically intertwined with cycles of poverty and food insecurity, which have worsened with changing climate patterns, highlighting the need for adaptive agricultural practices (Diffenbaugh & Burke, 2019). In South Asia, regular flooding exacerbates poverty cycles, as agricultural disruptions lead to economic strain and impede recovery efforts (Khan et al., 2024). Moreover, studies show that mental health issues linked to climate change are on the rise, driven by anxiety over future uncertainties and the instability following climatic disasters (Corvalan et al., 2022). Such multifaceted impacts necessitate comprehensive approaches that integrate environmental, mental health, and economic recovery strategies.

Interventions aimed at reducing vulnerabilities should not only focus on immediate responses to climate disasters but also on systemic changes to address the root causes of inequality. As noted by Di Napoli et al. (2022), recognizing the interplay between health, economics, and climate change necessitates interdisciplinary approaches that account for socio-economic disparities. Policies targeting climate adaptation need to be inclusive, ensuring that marginalized populations participate in decision-making processes that directly affect their lives. Effective adaptation finance should bridge funding gaps in more vulnerable communities while emphasizing community-based resilience strategies (Maina & Parádi-Dolgos, 2024). Such measures could enhance capacities to cope with climate impacts and facilitate robust recovery pathways following disasters. Since 1970, extreme weather events have increased fivefold, with 90% of associated deaths occurring in developing countries, demonstrating the unequal burden of climate disasters (WMO, 2023).

The integration of social determinants into climate vulnerability assessments offers a pathway for developing targeted adaptation strategies. Research indicates that geographic disparities in access to resources, educational opportunities, and healthcare contribute significantly to climate vulnerability and resilience (Kurani et al., 2020; Sihotang et al., 2022). By employing geospatial analysis tools, researchers can highlight areas with the most adverse socio-economic conditions and facilitate informed policy formulation that prioritizes these vulnerable communities (Piyasena & Bandara, 2025). Recognizing the interconnected nature of climate impacts encourages a holistic framework that links environmental sustainability with economic and social equity. The persistent link between climate change and social inequities highlights the importance of addressing structural inequalities to foster resilience against climatic threats. Simply implementing temporary fixes is insufficient for creating long-term adaptation strategies. Primary care settings and public health responses should integrate climate considerations to provide holistic care that acknowledges the health impacts of climate-related hazards (Badolo, 2024; Corvalan et al., 2022). This approach not only addresses immediate medical needs following disasters but also builds resilience against future climate impacts through preventative health measures. The adaptation finance gap is now estimated at USD 194–366 billion per year, far exceeding current international commitments (UNEP, 2023.

5.4 Adaptive Capacity and Technological Gaps

Infrastructure and resource endowments play a vital role in determining the adaptive capacity of communities worldwide. Adaptive capacity reflects the ability of societies to prepare for, respond to, and recover from the impacts of climate change. A combination of economic strength, technological innovation, institutional effectiveness, social infrastructure, and access to climate finance shapes it (Table 5.1). However, significant disparities exist between core, semi-periphery, and periphery regions, with marginalized areas facing technological gaps, limited funding, and weaker governance systems. Table 5.1 summarizes key dimensions of adaptive capacity and highlights regional differences, highlighting the structural inequities that exacerbate climate vulnerability and hinder effective adaptation. Access to resilient housing, healthcare systems, and emergency services can markedly improve a community's resilience to climate shocks. However, as highlighted in several empirical studies, weak governance and institutional barriers often frustrate efficient and effective adaptation efforts. For instance, in Vietnam, hierarchical governance systems have been observed to hinder managerial adaptability due to rigid bureaucratic structures, which can obstruct cross-agency collaboration essential for effective climate responses (Phuong et al., 2018). Similarly, in the Central African Republic, pervasive weak governance has limited the government's ability to address climate challenges effectively, illustrating the constraints faced in regions with deep-rooted institutional weaknesses (Brown et al., 2013). Furthermore, structural issues in countries like Nigeria, where political corruption and fragmented governance prevail, undermine efforts toward building adaptive systems in climate change contexts (Oramah & Olsen, 2021).

Technological innovation and transfer are equally crucial to enhancing adaptive capacity. Climate-smart agricultural practices, renewable energy technologies, and digital early warning systems are vital tools for adapting to climate-related risks. The underfunding of adaptation measures has been identified as a significant barrier, exacerbating disparities between developed and developing regions (Malik & Ford, 2024). Without equitable access to innovative solutions such as renewable energy technologies and climate-resilient agricultural practices, vulnerable populations remain further entrenched in cycles of poverty and neglect in the face of climate change (Malik & Ford, 2024). The pressing need for effective technology transfer mechanisms is underscored by the fact that many local communities possess limited knowledge of advanced scientific methods, which, when integrated with local wisdom, could enhance adaptive practices (Håkon Inderberg, 2011).

Knowledge production and accessibility are intertwined with the discourse on adaptive capacities. The predominance of Northern research institutions fosters a knowledge hierarchy where local insights are often undervalued or overlooked. This epistemic exclusion not only stifles the capacity for context-appropriate adaptation but also perpetuates inequities within global climate governance structures. The lack of integration between indigenous knowledge and scientific research hampers adaptation mechanisms that might be better suited to local environments

Table 5.1 Regional adaptive capacity and technological gaps in addressing climate change

Dimension	High-income (AKA Core) regions	Semi-periphery (AKA Middle Income) regions	Periphery (AKA Low-Income) regions	References
Infrastructure resilience	Advanced disaster-resilient infrastructure; widespread early warning systems; high urban planning capacity.	Moderate infrastructure; uneven urban planning; some disaster resilience initiatives.	Limited infrastructure; informal settlements are highly exposed to climate hazards.	Diffenbaugh and burke (2019) and Casson et al. (2023)
Technological innovation	Access to cutting-edge technology (e.g., precision agriculture, AI for climate modeling, renewable energy systems).	Emerging technology adoption, some renewable energy initiatives, and moderate R&D funding.	Low technology transfer; limited access to energy-efficient or climate-smart technologies.	Song et al. (2023) and da Silva and Requia (2025)
Financial resources	High per capita GDP; access to climate finance, insurance markets, and private investments.	Moderate GDP; reliance on multilateral loans; limited domestic climate financing.	Low GDP; minimal insurance coverage; heavy reliance on external aid.	Pak et al. (2025) and Di Napoli et al. (2022)
Institutional capacity	Strong governance, effective disaster management agencies, and research networks.	Mixed institutional capacity; varying governance effectiveness.	Weak institutions; political instability; fragmented disaster response systems.	Perry et al. (2021) and Kurani et al. (2020)
Knowledge & education	High research capacity; advanced climate science education; integration of indigenous knowledge.	Moderate research output; uneven knowledge transfer; rising higher education participation.	Low research funding, limited climate science literacy, and the undervaluation of indigenous knowledge.	Sihotang et al. (2022) and Maina and Parádi-Dolgos (2024)
Health systems & social safety nets	Advanced healthcare systems; well-funded public health adaptation programs.	Expanding healthcare access; uneven distribution of health infrastructure.	Weak health systems; lack of mental health services post-disasters.	Awiti (2022) and Corvalan et al. (2022)

(Darjee et al., 2021). Additionally, empirical evidence from case studies highlights that locally produced knowledge, when recognized and leveraged, can lead to more effective and culturally relevant adaptation strategies (Boateng & Larbi, 2021). Subsequently, the meaningful incorporation of local knowledge into broader adaptation frameworks is essential to establishing equitable climate resilience.

Institutional strength and governance capacity are paramount for effective climate adaptation outcomes. Governance quality directly correlates with adaptive capacity, as strong institutional frameworks can facilitate collaborative efforts across scales, enabling communities to respond adeptly to climate challenges. However, fragmented governance structures often weaken coordination among different entities, hindering cumulative adaptation efforts. For instance, the complexity of multi-level governance systems in West Africa presents challenges to cohesive climate action, where overlapping responsibilities lead to inefficiencies and inertia (Brockhaus et al., 2012). The literature emphasizes that enhancing governance quality is imperative for fostering adaptive capacities, as governance truly underpins societal readiness to tackle climate change (Andrijevic et al., 2019). Equally notable is the role of political will; without committed leadership that prioritizes responsive adaptation strategies, efforts to combat climate inequalities become drastically undermined (Mwadzingeni et al., 2022).

Moreover, the intertwining factors of socio-demographic contexts and institutional frameworks significantly influence adaptive capacities in various geographical settings. Research shows that marginalized communities often face additional barriers, such as low literacy and limited access to financial resources, which impede their ability to engage in meaningful adaptation activities (Mwadzingeni et al., 2022). For instance, the circumstances of smallholder irrigators in Zimbabwe illustrate how socio-demographic variables shape adaptive responses, emphasizing the necessity for tailored governance that addresses the specific challenges faced by vulnerable populations. Robust institutions capable of mobilizing resources, fostering inclusivity, and coordinating policies across sectors can enhance the potential for adaptive success, particularly in regions most affected by climate change (Brockhaus et al., 2012). It is essential to recognize that global climate change adaptation strategies require transcending disciplinary boundaries and integrating diverse knowledge systems. As noted in the literature, vulnerabilities to climate change manifest differently across contexts, demanding localized knowledge alongside scientific inquiry to bolster adaptive frameworks (Wang et al., 2016). Greater alignment among diverse actors, including governmental bodies, NGOs, and local communities, offers a compelling avenue for developing innovative solutions to address the multifaceted nature of climate adaptation. Clearly, collaborative efforts combined with locally sourced insights fundamentally underpin robust adaptive capacities that hold the potential to bridge the growing chasm in global climate resilience (Gupta et al., 2010).

5.5 Climate Finance and Unequal Resource Allocation

Climate finance plays a vital role in addressing global inequality exacerbated by climate change, with crucial implications for resource allocation, particularly between developed and developing countries (Iyiola et al., 2024a, 2024b; Izah et al., 2023). Historical inequalities are stressed by a significant imbalance in greenhouse gas emissions contributions since the Industrial Revolution, with wealthier nations historically contributing disproportionately to the problem. In comparison, poorer nations bear a larger burden of the consequences. For instance, developed countries have emitted far more cumulative greenhouse gases, necessitating a discussion of climate debt and historical responsibility for climate finance allocations for mitigation and adaptation efforts (Bae et al., 2022; Ogwu et al., 2025a, 2025b, 2025c). Despite bearing the brunt of climate impacts, developing countries often encounter barriers in accessing mitigation and adaptation funds due to inequitable allocation mechanisms, governance challenges, and market-driven financing priorities. Figure 5.2 illustrates how this resource gap perpetuates global climate injustices, highlighting the need for a more equitable, inclusive, and reparative approach to climate finance that prioritizes the most vulnerable populations and regions.

The Green Climate Fund (GCF) and the Adaptation Fund are central to global funding mechanisms established to channel climate finance to vulnerable nations. However, critiques of these mechanisms highlight a troubling trend: significant discrepancies in disbursement patterns that favor wealthier nations or those with more robust governance infrastructure. For example, Doku et al. (2021) articulate that factors such as governance quality, social inequality, and economic stability significantly influence the level of climate finance that Sub-Saharan countries receive, often leaving the most vulnerable populations without necessary support. The GCF's allocation policies have faced criticism for potentially perpetuating

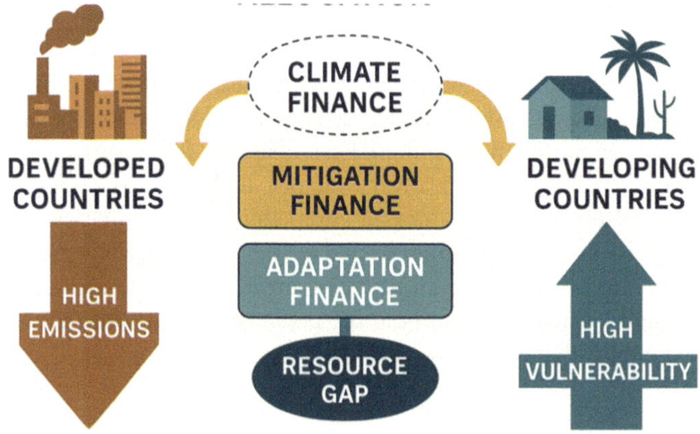

Fig. 5.2 Climate finance and unequal resource allocation: Disparities in global funding flows

these systemic disparities, particularly in how funding criteria are set and utilized, which can disadvantage countries that need it most, such as those in the Pacific region (Samuwai & Hills, 2019). When examining the dynamics of climate finance, it is critical to differentiate between mitigation and adaptation finance. The latter is particularly vital for developing nations facing immediate challenges posed by loss and damage from climate change. Unfortunately, as Chirisa et al. (2021) elucidate, these countries frequently struggle to access adequate funding, with the bureaucratic and technical requirements of international funding sources posing significant barriers. Furthermore, the historical neglect of loss-and-damage funding mechanisms reflects a broader trend of inequitable resource allocation in climate finance, leaving many nations feeling disenfranchised in global environmental negotiations (Perdana et al., 2023).

Moreover, the roles of multinational corporations and global financial institutions in shaping the landscape of climate financing cannot be overlooked. Studies stress that these entities often prioritize profit maximization over environmental justice, exacerbating inequality in financial distribution. Meyer (2019) highlights how corporate interests can dictate project outcomes in developing countries, with multinational corporations leveraging regulatory frameworks to their advantage, often at the expense of local communities. The relationship between private financiers and public policy entities can thus lead to a misalignment of climate projects that favor affluent stakeholders instead of the communities most adversely affected by climate impacts (Dikau & Volz, 2021).

Additionally, financing for climate adaptation has been hampered by prevailing dynamics that favor public investment over private funding. The emerging literature advocates an equitable approach in which investment channels are redirected to support local initiatives that build resilience against climate risks. This shift is necessary to close the accessibility gap and counterbalance the historical injustices of climate finance allocation (Sorg, 2025). By adopting a reparative finance model, there is potential for innovative funding solutions that prioritize historically marginalized groups while addressing both mitigation and adaptation necessities (Monasterolo et al., 2018).

The relationship between public and private investment flows complicates the landscape of climate finance. While public investments are essential for broad-scale projects, private capital, often driven by multinational corporations, is crucial for technology-driven solutions to climate issues. However, the tendency for private-sector investments to favor projects with quicker returns often sidelines critical adaptation initiatives essential to long-term sustainability in developing regions (Chirisa et al., 2021). This duality creates a challenging environment where the most immediate financial needs of vulnerable populations may be overlooked in favor of investments that cater to wealthier investors.

5.6 Colonial Legacies and Trade Dependencies

Colonial legacies profoundly influence modern economies, social structures, and environmental conditions. The historical intricacies of exploitation and resource extraction during the colonial era have contributed to significant environmental degradation and heightened social inequalities. Economic models prioritized the extraction of natural resources while neglecting indigenous ecological knowledge, thereby fostering an ongoing environmental crisis. The extraction processes often devastated large areas of their natural resources, such as timber and minerals, under the guise of economic development, leading to lasting ecological scars resulting from practices that were neither sustainable nor considerate of the communities affected (Mahony & Endfield, 2018; Masakure & Ndumeya, 2021; Zeng, 2023). Figure 5.3 illustrates how colonial-era resource extraction practices laid the groundwork for modern trade dependencies and environmental degradation. During the colonial period, vast territories were exploited for timber, minerals, and agricultural commodities, often at the expense of ecological sustainability and local livelihoods. The prioritization of export-oriented economies over subsistence and environmental stewardship entrenched patterns of ecological destruction and socio-economic inequality. Today, these historical practices persist through neo-colonial trade arrangements, ecological debt, and structural adjustment programs that limit the adaptive capacities of postcolonial nations. Due to the cyclical nature of resource extraction, environmental degradation, and primary commodity exports, colonial legacies continue to shape contemporary climate vulnerabilities and global inequalities (Fig. 5.3).

Extraction by colonial powers often disregarded environmental sustainability in favor of immediate economic gains, leading to accelerated deforestation and

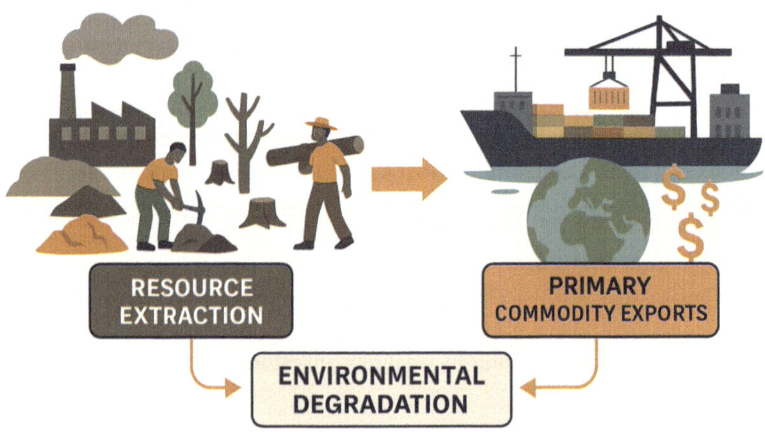

Fig. 5.3 Pathways from historical exploitation to modern climate vulnerabilities through colonial legacies and trade dependencies

soil depletion. For instance, in colonial Zimbabwe, authorities established a series of environmental regulations to mitigate the damaging effects of deforestation; however, these regulations were often ineffectively enforced, particularly against white settlers (Masakure & Ndumeya, 2021). Such practices exemplified a trend in which colonial economic interests took precedence over ecological concerns, exacerbating environmental degradation. Additionally, the ideologies promoting these economic pursuits justified environmentally damaging and socially disruptive acts (Mahony & Endfield, 2018; Musemwa, 2016).

As neo-colonial relations persist in contemporary trade frameworks, modern economies remain vulnerable to dependency on primary commodity exports. Neo-colonialism often manifests through unequal trade agreements and corporate land grabs, where indigenous peoples are systematically dispossessed of their lands by external entities pursuing agricultural or mineral exploitation. These relationships are characterized by an imbalance of power, with vast resources extracted from poorer nations while minimal benefits return (Bhambra & Newell, 2023; Hossen et al., 2025). This dynamic can create a concept termed "climate colonialism," reflecting how wealthier nations exploit poor countries' environmental resources while disproportionately contributing to global climate change. Such skewed systems result in marginalized populations facing significant socio-economic challenges, intensifying their vulnerabilities to climate impacts (Bhambra & Newell, 2023; van Ryneveld & Islar, 2023).

The concept of ecological debt is crucial for understanding the modern implications of historical exploitation. Ecological debt refers to the environmental degradation and biodiversity loss incurred by colonizing powers that have yet to be addressed by compensatory measures (Hossen et al., 2025). These debts significantly contribute to present-day climate vulnerabilities, especially within developing nations that disproportionately bear the brunt of climate change effects despite contributing minimally to historical emissions. This disparity raises ethical questions about climate justice, as nations striving for development often find themselves repaying debts accrued by their predecessors through resource depletion and economic sacrifices (Hossen et al., 2025; Resurrección, 2024).

Moreover, climate vulnerabilities are compounded by various structural challenges faced by postcolonial states, many of which struggle with historical debts and the legacy of structural adjustment programs. These programs, often imposed by international financial institutions, have constrained nations' capacities to develop effective, independent climate responses by mandating austerity measures and prioritizing debt repayment over environmental sustainability (van Ryneveld & Islar, 2023). As a result, countries may find themselves perpetually in cycles of debt that hinder sustainable development and the adaptation strategies essential to confronting modern climate challenges, driven by both internal and external factors. Postcolonial development struggles exhibit a complex interplay between historical context and contemporary realities. Many nations inherited frameworks ill-suited for fostering equitable and sustainable growth due to the exploitative dynamics of colonial governance, which prioritized foreign interests over local empowerment. The challenges intensify in contexts where historical injustices remain unaddressed, leading to social unrest and movements aimed at reclaiming agency in development planning and climate action initiatives (Bhambra & Newell, 2023; Resurrección, 2024).

Responses to these intertwined challenges have called for a re-evaluation of international approaches toward climate change and economic development frameworks. Proponents of decolonization argue that understanding historical context is vital in reshaping modern policy frameworks to prioritize local knowledge systems and equitable resource management (Hossen et al., 2025; van Ryneveld & Islar, 2023). Collaboration centered on justice, recognizing indigenous practices, acknowledging ecological debts, and developing reparative justice mechanisms can pave the way for more sustainable and inclusive development paths (Bhambra & Newell, 2023; Resurrección, 2024). Recognizing the interplay between colonial legacies and trade dependencies challenges the status quo of global inequality and environmental injustice. Comprehensive analyses and actionable frameworks should account for historical contexts and contemporary realities in policymaking. To transition toward a more equitable global society, nations should confront the environmental repercussions of their colonial pasts while actively seeking to dismantle the neo-colonial structures that perpetuate exploitation, inequality, and environmental degradation today (Junejo & Shaikh, 2022; van Ryneveld & Islar, 2023; Zeng, 2023). Policymakers should leverage historical insight to inform and shape future actions aimed at sustainability and social justice, while emphasizing ecological stewardship as both a local and global imperative (Bhambra & Newell, 2023; Resurrección, 2024).

5.7 Global Governance and Structural Inequities

The complex interplay of global governance, structural inequities, and climate change has increasingly become an area of significant examination within environmental and political sciences. The issues of unequal representation in climate institutions, critiques of multilateral agreements, tensions in North–South negotiation dynamics, and the geopolitics surrounding climate adaptation and mitigation all reflect an urgent need to address systemic disparities. Understanding these domains offers insight into how the global community can work toward more equitable climate governance. A significant aspect of climate governance is the unequal representation in institutions such as the United Nations Framework Convention on Climate Change (UNFCCC), where developed nations often wield disproportionate influence over their developing counterparts. This disparity is exacerbated by Northern leadership dominance, which often shapes the agenda and priorities of climate discussions. Studies indicate that agreements like the Kyoto Protocol and the Paris Agreement contain gaps in equity considerations that can marginalize voices from the Global South (Najam et al., 2003). Such inequity undermines the efforts of developing countries to reinforce their interests in these negotiations.

Attempts to engage various stakeholders through frameworks that encourage participation—such as the diagnostic strategic tools proposed by Hurlbert and Gupta (2015)—have not been uniformly successful, particularly in contexts marked by power imbalances. Issues regarding the efficacy of these participatory approaches are compounded by accusations that Northern-led initiatives often lack genuine consultation with affected communities, thereby perpetuating a cycle of exclusion. For instance, participatory gaps noted in contexts such as Chile reflect a systemic failure to equitably engage diverse actors, leading to skepticism about the intentions behind many climate initiatives (Hurlbert & Gupta, 2015).

Critiques of major multilateral climate agreements point to their inadequacies in addressing equity concerns. The Kyoto Protocol has faced criticism for failing to adequately involve the voices and interests of the Global South (Aldy et al., 2003). Countries like Canada, which initially participated in the Protocol, have since withdrawn, citing concerns about policy coherence, exemplifying the fragility of such agreements when critical equity dimensions are not addressed (Hurlbert & Gupta, 2015). Similarly, the Paris Agreement has been criticized for being more flexible and less binding, potentially exacerbating disparities as countries negotiate commitments (Mohanty, 2025). The dynamics observed during Conference of Parties (COP) summits reveal underlying tensions between developed and developing nations. Negotiations often become battlegrounds of conflicting priorities, with developed nations pushing for market-based solutions while developing countries advocate for mechanisms that emphasize historical responsibility and equitable resource allocation (Joshi, 2013; Mohanty, 2025). This tension is further complicated by geopolitical considerations, wherein economic and political interests shape the positions taken by various countries (Friman & Linnér, 2008). The structure of climate negotiations often systematically disadvantages Global South nations, which may lack the political clout to influence discussions in their favor (Mohanty, 2025).

Furthermore, the geopolitics surrounding climate adaptation and mitigation introduces additional complexities, especially considering the roles of militarization and energy security. The interests of developed countries frequently center around securing energy supplies and ensuring national security, often at the expense of Global South perspectives that emphasize sustainability and equity (Batel & Devine-Wright, 2017). The concept of "energy colonialism" elucidates how large-scale renewable energy projects may marginalize local voices and prioritize corporate interests over community needs (Batel & Devine-Wright, 2017). This manipulation of governance structures can lead to situations in which the adaptation and mitigation strategies promoted by Global North powerhouses fail to address the structural inequities faced by vulnerable communities worldwide.

Engaging with the lives and experiences of marginalized communities is crucial for any coherent climate strategy. Participatory approaches should incorporate local knowledge and expertise if they are to be effective in fostering resilience in climate-vulnerable communities (Gidley et al., 2009). Recognizing diverse worldviews and the knowledge systems associated with these communities is needed to create governance that is authentically representative and inclusive. Moreover, the participation of traditionally excluded voices can enhance the legitimacy and effectiveness of climate policies by aligning them more closely with the realities faced by those directly impacted by climate change. At the intersection of these discussions lies the urgent need for equitable climate finance that addresses past and present injustices. There is increasing recognition that financing should account for developing countries' historical emissions and socio-economic contexts while providing pathways for sustainable development (Carton et al., 2020). When financial assistance mechanisms are designed and implemented to prioritize the needs of the Global North, the stark discrepancies in global climate governance become even more pronounced, risking the marginalization of developing nations in these critical negotiations.

5.8 Emerging Patterns of Climate Inequality

Climate change represents a multifaceted challenge that amplifies existing inequalities, particularly evident through the lens of climate-induced migration, food and water security challenges, public health implications, and the intergenerational and intersectional dimensions of vulnerability (Iyiola et al., 2024a, 2024b). Each of these dimensions reveals a complex interplay of socio-economic factors exacerbated by climate destabilization, necessitating a nuanced understanding of how different communities are impacted.

5.8.1 Climate-Induced Migration

One of the most pressing consequences of climate change is climate-induced migration, which has gained attention in global policy discussions due to its implications for human security and rights. The conceptual shift from viewing individuals as "climate refugees" to recognizing them as "climate migrants" emphasizes the need for adaptive responses rather than merely security-based reactions (Jakobsson, 2021; Vydelingum, 2019). This shift highlights the socio-economic and environmental factors that push individuals to migrate; as reported in recent literature, many migrations are responses to failed livelihoods due to climatic pressures (Teye & Nikoi, 2022). For instance, in West Africa, governments' reluctance to encourage migration from rural areas experiencing agricultural failure often exacerbates local stress, leaving vulnerable populations with limited coping strategies (Sugg et al., 2023; Teye & Nikoi, 2022).

Empirical studies indicate an increasing trend of internal migration driven by climate impacts, such as those observed in Bangladesh, where millions are expected to be displaced by rising sea levels and extreme weather events (Bhardwaj et al., 2024; Faysal et al., 2023; Nayna Schwerdtle et al., 2021). The compounded effects of poverty and environmental degradation lead to a situation where already vulnerable populations face disproportionate pressures, often leading to urbanization as individuals migrate toward cities in search of better livelihoods, despite the risks of urban poverty and inadequate infrastructure (Ogwu & Kosoe, 2025; Oliveira & Pereda, 2020). Urban areas, while seen as havens for opportunities, often manifest their challenges, creating enclaves of precarious living conditions for migrants who have fled from rural calamities.

5.8.2 Food and Water Security Challenges

Simultaneously, climate-induced challenges to food and water security are paramount. Agricultural productivity losses driven by changing weather patterns and extreme conditions threaten the livelihoods of millions, particularly in impoverished regions. Reports suggest that projected increases in heat stress and disrupted rainfall patterns could significantly reduce crop yields, heightening food insecurity (Bhardwaj et al., 2024; Manni et al., 2020). For example, in many countries, especially those classified as small island developing states (SIDS), the intersection of rising sea levels and limited freshwater resources heightens vulnerabilities—agricultural systems falter, and communities are left facing famine risks (Thomas & Benjamin, 2018).

In terms of water security, changing climate impacts exacerbate competition for already limited resources; studies indicate that as climatic conditions worsen, water scarcity intensifies, posing significant threats to health and livelihoods across populations (Manni et al., 2020; Thomas & Benjamin, 2018). As climate conditions drive agricultural failures and reduce water availability, community resilience diminishes, further destabilizing food systems and heightening the potential for conflicts over resources that are increasingly contested due to climatic stresses (Boas et al., 2019).

5.8.3 Public Health Implications

The public health implications of climate change extend far beyond just food and water security. Rising temperatures and changing environmental conditions contribute to the proliferation of vector-borne diseases and alarming rates of malnutrition among vulnerable populations (Nayna Schwerdtle et al., 2021). For instance, tropical diseases such as malaria and dengue fever are expected to expand their range as changing climates create conditions suitable for vectors (Nayna Schwerdtle et al., 2021; Osei-Amponsah et al., 2023). Additionally, malnutrition rates can be directly linked to food insecurity exacerbated by climate-induced challenges, with marginalized communities facing the highest burdens due to inadequate diets rich in essential nutrients (Bhardwaj et al., 2024).

Moreover, heat stress presents a significant health hazard, particularly for those living in urban areas without adequate shelter or access to cooling systems. Vulnerable populations—such as children, the elderly, and individuals with pre-existing health conditions—are at increased risk of heat-related illnesses (Nayna Schwerdtle et al., 2021; Ogwu, 2019; Osei-Amponsah et al., 2023). The lack of infrastructure to support adapting to these changes exacerbates health disparities, perpetuating a cycle of disadvantage for already vulnerable groups.

5.8.4 Intergenerational and Intersectional Dimensions

The intergenerational and intersectional dimensions of climate change unveil the compounded vulnerabilities faced by specific demographics, such as youth, women, and indigenous communities. Women and girls, for instance, disproportionately bear the brunt of climate impacts due to entrenched social inequalities and roles that prioritize their responsibilities in food and water management (Bhardwaj et al., 2024). They face increased risks of migration and displacement, often resulting from a combination of socio-economic pressures and gender-based discrimination that limits their ability to adapt (Bettini et al., 2017; Ogwu et al., 2025c, 2025d). The voices of indigenous communities, often overlooked in climate discourses, highlight the need for inclusive policy frameworks that account for traditional knowledge and adaptive strategies that have sustained these communities for centuries. Their experiences illustrate the intersectionality of climate impacts, where cultural, economic, and environmental dimensions converge, necessitating targeted interventions that address both immediate needs and long-term resilience (Farbotko et al., 2018).

5.9 Toward Climate Justice and Equitable Futures

In recent years, discussions of climate justice have gained traction, emphasizing the need for reparations and compensation that acknowledge the historical injustices associated with climate change. These discussions are crucial because they aim to address the disproportionate impacts on marginalized communities resulting from both historical and ongoing environmental degradation. The necessity for loss-and-damage funds arises from the acknowledgment that developing countries, particularly those in the Global South, have contributed the least to climate change yet bear its brunt. This points to a pressing demand that polluting nations, primarily from the Global North, shoulder the burden of financial reparations for the damage inflicted on less polluting countries (Klein, 2024; Mohan, 2025).

Climate reparations are not merely financial transactions; they involve a broader spectrum of redress that ties back to historical accountability, reminding formerly colonial states of their lingering obligations toward nations that have been systematically exploited and disenfranchised (Malherbe & Oladejo, 2025; Mohan, 2025). For instance, the works of Klein (2024) highlight the necessity to transcend compensation into transformative practices that rectify colonial legacies, foster resilience, and contribute to sustainable living pathways. Achieving climate justice thus involves recognizing and addressing these historical grievances while navigating complex socio-political dynamics, calling for a robust reparatory framework grounded in ethics and justice (Malherbe & Oladejo, 2025; Page & Heyward, 2017).

Moreover, Indigenous and local knowledge systems play a vital role in the broader framework of climate equity. These systems offer unique perspectives on adaptation and mitigation that are often overlooked in mainstream climate strategies. By integrating community-led adaptation approaches, stakeholders can harness localized practices and values that reflect the nuances of specific socio-cultural contexts (Coger et al., 2022; Evans, 2012). Such collaborative frameworks have shown promise in empowering communities to take charge of their climate resilience strategies while ensuring the principles of equity and justice are embedded within decision-making processes (Bulkeley et al., 2013; Coger et al., 2022). The emphasis on participatory governance models, as suggested by various scholars, reinforces the need for inclusivity that allows those most affected to influence and lead climate-related planning and action (Bulkeley et al., 2013; Izah & Ogwu, 2025; Lori et al., 2025; Sultana, 2022).

Collaborative governance is further supported by a growing discourse on South-South cooperation, in which nations from the Global South mobilize resources and knowledge among themselves to enhance adaptive capacity and resilience without necessarily depending on external aid from the Global North (Malherbe & Oladejo, 2025; Webber, 2024). This dynamic not only fosters solidarity but also creates alternative development pathways that challenge the hierarchies imposed by traditional patterns of international relations (Gonzalez, 2021; Malherbe & Oladejo, 2025). The intricacies of these partnerships reflect a burgeoning understanding that global approaches to climate change can benefit significantly from grassroots initiatives that resonate with local realities, thus promoting a more equitable distribution of resources and responsibilities.

Research indicates that climate adaptation policies should account for socioeconomic disparities within and between communities, particularly in the Global South. The recognition of marginalized groups, especially those impacted by colonial histories, highlights the importance of equity in adaptation strategies (Evans, 2012; See & Wilmsen, 2022). The intersection of climate justice with issues of race, class, and historical exploitation facilitates a deeper inquiry into how climate policies can be restructured to avoid reinforcing existing vulnerabilities (Gonzalez, 2021; Grasso, 2024). Furthermore, highlighting the significance of procedural justice, which pertains to the inclusivity of policy processes, bolsters the argument for equitable access to decision-making (Domingue, 2023; Sultana, 2022). By prioritizing these marginalized perspectives, policymakers can aspire toward climate actions that undergo an ethical and just transformation, ensuring that no one is left behind (Coger et al., 2022; Sultana, 2022).

In practical terms, the development of localized, culturally relevant adaptation strategies requires verification from those with lived experiences and insights essential to addressing the unique challenges posed by climate change. Such an emphasis on agency shifts the paradigm from a top-down approach to one that prioritizes community aspirations and sovereignty, essential elements of climate justice (Coger et al., 2022; Sultana, 2022). It also reinforces the call for reparative investments to rebuild communities, particularly in regions such as Louisiana and the broader African context, where historical and contemporary inequities manifest in stark realities as a consequence of climate change (Domingue, 2023; Malherbe & Oladejo, 2025). Therefore, achieving climate justice necessitates a multifaceted approach that integrates loss-and-damage financing, acknowledgment of historical responsibilities, and a commitment to empowering marginalized knowledge systems through inclusive governance. The discussions around climate reparations extend beyond the financial realm into the moral and ethical dimensions of historical injustices, which require accountability frameworks that prioritize long-term change rather than temporary relief (Junker et al., 2022; Klein, 2024). For instance, issues of climate colonization and legacies of exploitation affect both local and global perspectives on sustainable practices. Addressing these intersections calls for transformative agreements and reparative justice mechanisms that engage communities and reflect their realities and needs (Junker et al., 2022; Malherbe & Oladejo, 2025; Mohan, 2025).

Furthermore, inclusion of marginalized groups in climate governance can catalyze transformative shifts in environmental policies and practices. By combining knowledge from Indigenous systems with scientific approaches, there is an opportunity to create a holistic climate adaptation response that embodies equity (Coger et al., 2022; Evans, 2012). This marks a significant pivot from conventional practices that often marginalize these crucial voices, thereby reaffirming the urgency for inclusive mechanisms and policies that embrace diverse societal inputs.

5.10 Conclusion

This chapter has demonstrated that climate change is not only an environmental crisis but also a profound amplifier of historical and structural inequalities across the global core, semi-periphery, and periphery. The mapping of differential exposure to climate risks revealed that communities already marginalized by centuries of colonization, resource extraction, and trade dependency face the most severe climate threats despite contributing least to greenhouse gas emissions. Moreover, the analysis of adaptive capacities and climate finance allocation highlighted how technological gaps, institutional weaknesses, and inequitable funding mechanisms perpetuate cycles of vulnerability in the Global South, while wealthier nations consolidate adaptive advantages. This chapter situated these disparities within the enduring legacies of colonialism and the persistent asymmetries of global governance, where decision-making power remains heavily skewed toward developed countries. Multilateral climate agreements, while symbolically inclusive, often reproduce geopolitical hierarchies, leaving periphery nations with limited voice in shaping climate agendas and accessing loss-and-damage resources. Yet, the findings also point toward possibilities for transformation. Calls for climate reparations, the strengthening of South–South cooperation, and the integration of indigenous and local knowledge systems offer pathways to reimagine climate governance beyond technocratic and market-driven paradigms. Addressing climate injustice thus requires dismantling structural inequities in global trade, finance, and political representation while fostering inclusive, participatory, and historically informed approaches to adaptation and mitigation. Achieving climate justice demands a radical shift in how the global community conceptualizes responsibility, equity, and sustainability—one that moves beyond incremental reforms toward systemic change capable of confronting both climate risks and the power imbalances that sustain them.

References

Akbulut, B. (2012). Community-based rmanagement in Turkey:Je participe, Tu participes, Il participe… Ils profitent1. *Development and Change, 43*(5), 1133–1158.

Aldy, J. E., Barrett, S., & Stavins, R. N. (2003). Thirteen plus one: A comparison of global climate policy architectures. *Climate Policy, 3*(4), 373–397.

Andrijevic, M., Crespo Cuaresma, J., Muttarak, R., & Schleussner, C. F. (2020). Governance in socioeconomic pathways and its role for future adaptive capacity. *Nature Sustainability, 3*(1), 35–41.

Andrijevic, M., Cuaresma, J., Muttarak, R., & Schleussner, C. (2019). Governance in socioeconomic pathways and its role for future adaptive capacity. *Nature Sustainability, 3*(1), 35–41. https://doi.org/10.1038/s41893-019-0405-0

Badolo, M. (2024). The Badolo health resilience scientific framework for health system resilience roadmaps.

Bae, S. M., Masud, M. A. K., Rashid, M. H. U., & Kim, J. D. (2022). Determinants of climate financing and the moderating effect of politics: Evidence from Bangladesh. *Sustainability Accounting, Management and Policy Journal, 13*(1), 247–272.

Baker, E. (2024). Ethical implications of environmental policies and practices. *International Journal of Philosophy, 3*(1), 37–40.

Batel, S., & Devine-Wright, P. (2017). Energy colonialism and the role of the global in local responses to new energy infrastructures in the UK: A critical and exploratory empirical analysis. *Antipode, 49*(1), 3–22.

Bettini, G., Nash, S. L., & Gioli, G. (2017). One step forward, two steps back? The fading contours of (in) justice in competing discourses on climate migration. *The Geographical Journal, 183*(4), 348–358.

Bhambra, G. K., & Newell, P. (2023). More than a metaphor: 'Climate colonialism' in perspective. *Global Social Challenges Journal, 2*(2), 179–187.

Bhardwaj, L. K., Rath, P., & Yadav, P. (2024). A comprehensive review on the impact of climate-induced livelihood, health and migration on women and girls.

Bigman, Y. E., Yam, K. C., Marciano, D., Reynolds, S. J., & Gray, K. (2021). Threat of racial and economic inequality increases preference for algorithm decision-making. *Computers in Human Behavior, 122*, 106859.

Boas, I., Farbotko, C., Adams, H., Sterly, H., Bush, S., Van der Geest, K., Wiegel, H., Ashraf, H., Baldwin, A., Bettini, G., & Blondin, S. (2019). Climate migration myths. Nature. *Climate Change, 9*(12), 901–903.

Boateng, K., & Larbi, R. T. (2021). Collaboration between chiefs and local government actors in combating climate change: Evidence from New Juaben. *Ghana. Contemporary Journal of African Studies, 8*(2), 25–41.

Boyce, J. K., Zwickl, K., & Ash, M. (2016). Measuring environmental inequality. *Ecological Economics, 124*, 114–123.

Brockhaus, M., Djoudi, H., & Kambire, H. (2012). Multi-level governance and adaptive capacity in West Africa. *International Journal of the Commons, 6*(2).

Brown, H. C. P., Smit, B., Somorin, O. A., Sonwa, D. J., & Ngana, F. (2013). Institutional perceptions, adaptive capacity and climate change response in a post-conflict country: A case study from Central African Republic. *Climate and Development, 5*(3), 206–216.

Buhaug, H. (2015). Climate–conflict research: Some reflections on the way forward. *Wiley Interdisciplinary Reviews: Climate Change, 6*(3), 269–275.

Bulkeley, H., Carmin, J., Broto, V. C., Edwards, G. A., & Fuller, S. (2013). Climate justice and global cities: Mapping the emerging discourses. *Global Environmental Change, 23*(5), 914–925.
Carton, W., Asiyanbi, A., Beck, S., Buck, H. J., & Lund, J. F. (2020). Negative emissions and the long history of carbon removal. *Wiley Interdisciplinary Reviews: Climate Change, 11*(6), e671.
Casson, N., Cameron, L., Mauro, I., Friesen-Hughes, K., & Rocque, R. (2023). Perceptions of the health impacts of climate change among Canadians. *BMC Public Health, 23*(1), 212.
Castells-Quintana, D., & McDermott, T. (2024). The far-reaching distributional effects of global warming: Evidence from half a century of climate and inequality data.
Chancel, L. (2022). Global carbon inequality over 1990–2019. *Nature Sustainability, 5*(11), 931–938.
Chirisa, I., Gumbo, T., Gundu-Jakarasi, V. N., Zhakata, W., Karakadzai, T., Dipura, R., & Moyo, T. (2021). Interrogating climate adaptation financing in Zimbabwe: Proposed direction. *Sustainability, 13*(12), 6517.
Climate Analytics. (2022). *COP participation and representation: Assessment report*. https://climateanalytics.org
Coger, T., Dinshaw, A., Tye, S., Kratzer, B., Aung, M. T., Cunningham, E., Ramkissoon, C., Gupta, S., Bodrud-Doza, M., Karamallis, A., Castells-Quintana, D., & McDermott, T. (2022). *Locally led adaptation: From principles to practice*. World Resources Institute.
Corvalan, C., Gray, B., Prats, E. V., Sena, A., Hanna, F., & Campbell-Lendrum, D. (2022). Mental health and the global climate crisis. *Epidemiology and Psychiatric Sciences, 31*, e86.
da Silva, H. G., & Requia, W. J. (2025). Racial and economic disparities in high-temperature exposure in Brazil. *International Journal of Environmental Research and Public Health, 22*(2), 200.
Darjee, K. B., Sunam, R. K., Köhl, M., & Neupane, P. R. (2021). Do national policies translate into local actions? Analyzing coherence between climate change adaptation policies and implications for local adaptation in Nepal. *Sustainability, 13*(23), 13115.
Di Napoli, C., McGushin, A., Romanello, M., Ayeb-Karlsson, S., Cai, W., Chambers, J., Dasgupta, S., Escobar, L. E., Kelman, I., Kjellstrom, T., & Kniveton, D. (2022). Tracking the impacts of climate change on human health via indicators: Lessons from the Lancet countdown. *BMC Public Health, 22*(1), 663.
Diffenbaugh, N. S., & Burke, M. (2019). Global warming has increased global economic inequality. *Proceedings of the National Academy of Sciences, 116*(20), 9808–9813.
Dikau, S., & Volz, U. (2021). Central bank mandates, sustainability objectives and the promotion of green finance. *Ecological Economics, 184*, 107022.
Doku, I., Ncwadi, R., & Phiri, A. (2021). Determinants of climate finance: Analysis of recipient characteristics in Sub-Sahara Africa. *Cogent Economics & Finance, 9*(1), 1964212.
Domingue, S. J. (2023). "Ground Zero" for climate crisis: Narratives about climate adaptation and implications for justice in coastal Louisiana. *Sociological Inquiry, 93*(2), 296–319.
Eskander, S. M., & Fankhauser, S. (2022). Income diversification and income inequality: Household responses to the 2013 floods in Pakistan. *Sustainability, 14*(1), 453.
Evans, E. C. (2012). *The right to reparation in international law for victims of armed conflict* (Vol. No. 91). Cambridge University Press.
FAO. (2023). *The state of food security and nutrition in the world 2023*. Food and Agriculture Organization of the United Nations..
Farbotko, C., McMichael, C., Dun, O., Ransan-Cooper, H., McNamara, K. E., & Thornton, F. (2018). Transformative mobilities in the Pacific: Promoting adaptation and development in a changing climate. *Asia & the Pacific Policy Studies, 5*(3), 393–407.
Friman, M., & Linnér, B. O. (2008). Technology obscuring equity: Historical responsibility in UNFCCC negotiations. *Climate Policy, 8*(4), 339–354.
Gidley, J. M., Fien, J., Smith, J. A., Thomsen, D. C., & Smith, T. F. (2009). Participatory futures methods: Towards adaptability and resilience in climate-vulnerable communities. *Environmental Policy and Governance, 19*(6), 427–440.

Gonzalez, C. G. (2021). Racial capitalism, climate justice, and climate displacement. *Oñati Socio-Legal Series, 11*(1), 108–147.

Grasso, M. (2024). The case for climate reparations by fossil fuel companies: Ethical foundations, monetary estimates and feasibility. *Development and Change, 55*(4), 727–751.

Guarino, R., Corsi, G., & Muñoz-Ulecia, E. (2024). Debunking competition-global ecologically unequal exchange explained by exploitation and control relations..

Gupta, J., Termeer, C., Klostermann, J., Meijerink, S., Van den Brink, M., Jong, P., Nooteboom, S., & Bergsma, E. (2010). The adaptive capacity wheel: A method to assess the inherent characteristics of institutions to enable the adaptive capacity of society. *Environmental Science & Policy, 13*(6), 459–471.

Håkon Inderberg, T. (2011). Institutional constraints to adaptive capacity: Adaptability to climate change in the Norwegian electricity sector. *Local Environment, 16*(4), 303–317.

Hannan, M. A. (2024). Climate migrants in Bangladesh: An overview of the legal protection mechanisms. *BAIUST Academic Journal.*

Hickel, J. (2020). Quantifying national responsibility for climate breakdown: An equality-based attribution approach. *The Lancet Planetary Health, 4*(9), e399–e404.

Sihotang, M.M, Hoddinott, H., & Rogers, R. (2022). Assessing vulnerability to climate-induced humanitarian crises: A global perspective. Sosateris, 10(2), 120–139.

Hossen, M. A., Benson, D., & Islam, M. (2025). Donor climate coloniality, ecological debt and vulnerabilities in coastal Bangladesh: Redesigning institutions for recognition justice. *Environmental Management*, 1–15.

Hurlbert, M., & Gupta, J. (2015). The split ladder of participation: A diagnostic, strategic, and evaluation tool to assess when participation is necessary. *Environmental Science & Policy, 50*, 100–113.

IPCC. (2021). *Climate change 2021: The physical science basis.* Intergovernmental Panel on Climate Change.

IPCC. (2022). *Climate change 2022: Impacts, adaptation and vulnerability.* Intergovernmental Panel on Climate Change.

IRENA. (2023). *Renewable energy investment trends 2023.* International Renewable Energy Agency.

Iyiola, A. O., Afolabi, O. A., Alimi, S. K., Akingba, O. O., Izah, S. C., & Ogwu, M. C. (2024a). Climate change and water crisis in the global south. In S. C. Izah, M. C. Ogwu, A. Loukas, & H. Hamidifar (Eds.), *Water crises and sustainable management in the global south* (pp. 111–140). Springer.

Iyiola, A. O., Afolabi, O. A., Alimi, S. K., Akingba, O. O., Izah, S. C., & Ogwu, M. C. (2024b). Climate change and water crisis in the global south. In S. C. Izah, M. C. Ogwu, A. Loukas, & H. Hamidifar (Eds.), *Water crises and sustainable management in the global south.* Springer. https://doi.org/10.1007/978-981-97-4966-9_4

Izah, S. C., Iyiola, A. O., Yarkwan, B., & Richard, G. (2023). Impact of air quality as a component of climate change on biodiversity-based ecosystem services. In A. L. Srivastav, A. K. Dubey, A. Kumar, S. K. Narang, & M. A. Khan (Eds.), (pp. 123–148). Visualization techniques for climate change with machine learning and artificial intelligence. isbn:978-0-323-99714-0.

Izah, S. C., & Ogwu, M. C. (2025). Modeling solutions for microbial water contamination in the global south for public health protection. *Frontiers in Microbiology, 16*, 1504829. https://doi.org/10.3389/fmicb.2025.1504829

Jakobsson, E. (2021). How climate-induced migration entered the UN policy agenda in 2007–2010: A multiple streams assessment. *Politics and Governance, 9*(4), 16–26.

Joshi, S. (2013). Understanding India's representation of North–South climate politics. Global Environmental Politics, 13(2), 128–147.

Junejo, M., & Shaikh, T. S. (2022). Land deterioration and environmental damage: A postcolonial eco-critical study of how beautiful we were. *University of Chitral Journal of Linguistics and Literature, 6*(I), 67–88.

Junker, K. W., Münster, S., & Shinde, M. (2022). Climate-induced migration: The need for a reparative legal regime. *Environmental Policy and Law, 52*(5–6), 473–485.

Khan, F., Shah, S. A. A., Shah, S. U. A., & Khan, A. (2024). Climate change and human health: impacts and vulnerability. *Journal of Health and Rehabilitation Research, 4*(1), 561–566.
Khine, M. M., & Langkulsen, U. (2023). The implications of climate change on health among vulnerable populations in South Africa: A systematic review. *International Journal of Environmental Research and Public Health, 20*(4), 3425.
Klein, E. (2024). Beyond compensation: Reparations and the ongoing colonization of Australia. *Development and Change, 55*(4), 830–854.
Kurani, S. S., McCoy, R. G., Lampman, M. A., Doubeni, C. A., Rutten, L. J. F., Inselman, J. W., Giblon, R. E., Bunkers, K. S., Stroebel, R. J., Rushlow, D., & Chawla, S. S. (2020). Association of neighborhood measures of social determinants of health with breast, cervical, and colorectal cancer screening rates in the US Midwest. *JAMA Network Open, 3*(3), e200618.
Landrigan, P. J., Rauh, V. A., & Galvez, M. P. (2010). Environmental justice and the health of children. *Mount Sinai Journal of Medicine: A Journal of Translational and Personalized Medicine: A Journal of Translational and Personalized Medicine, 77*(2), 178–187.
Levy, J. I., Chemerynski, S. M., & Tuchmann, J. L. (2006). Incorporating concepts of inequality and inequity into health benefits analysis. *International Journal for Equity in Health, 5*(1), 2.
Mahony, M., & Endfield, G. (2018). Climate and colonialism. *Wiley Interdisciplinary Reviews: Climate Change, 9*(2), e510.
Maina, P., & Parádi-Dolgos, A. (2024). The effectiveness of climate adaptation finance and readiness on vulnerability in African Economies. *Climate, 12*(5), 59.
Malherbe, N., & Oladejo, A. O. (2025). Against climate aid colonialism: The case for climate reparations and south-south solidarity. *Africa Spectrum, 60*(1), 84–105.
Malik, I. H., & Ford, J. D. (2024). Addressing the climate change adaptation gap: Key themes and future directions. *Climate, 12*(2), 24.
Manni, M., Di Giuseppe, A., Petrozzi, A., Nicolini, A., Rossi, F., & Cotana, F. (2020). High-reflective mulching membrane for a sustainable development: Monitoring campaign. In *E3S Web of Conferences* (Vol. 197, p. 08012). EDP Sciences.
Martenies, S. E., Milando, C. W., & Batterman, S. A. (2018). Air pollutant strategies to reduce adverse health impacts and health inequalities: A quantitative assessment for Detroit, Michigan. *Air Quality, Atmosphere & Health, 11*(4), 409–422.
Martenies, S. E., Milando, C. W., Williams, G. O., & Batterman, S. A. (2017). Disease and health inequalities attributable to air pollutant exposure in Detroit, Michigan. *International Journal of Environmental Research and Public Health, 14*(10), 1243.
Masakure, C., & Ndumeya, N. (2021). 'The trees do not belong to Chief Maranke but to the Native Reserves Trust': The politics of timber resource exploitation in African reserves, colonial Zimbabwe, 1924-1948. *Historia, 66*(1), 61–87.
Meyer, Y. (2019). The effectiveness of market-based initiatives for regulating development projects by multinational corporations in Africa with regard to human rights and environmental abuses. *African Human Rights Law Journal, 19*(1), 126–150.
Minas, S. (2022). Financing climate justice in the European Union and China: Common mechanisms, different perspectives. *Asia Europe Journal, 20*(4), 377–401.
Mohan, P. S. (2025). Climate reparations for a just response to climate change: A review of historical responsibility and future implications. *Wiley Interdisciplinary Reviews: Climate Change, 16*(2), e70007.
Mohanty, R. (2025). Need for fairness in climate change negotiations: A third world perspective. *Asian Journal of International Law, 15*(1), 157–189.
Monasterolo, I., Zheng, J. I., & Battiston, S. (2018). Climate transition risk and development finance: A carbon risk assessment of China's overseas energy portfolios. *China & World Economy, 26*(6), 116–142.
Musemwa, M. (2016). Sic utere tuo ut alienam non laedas. From wanton destruction of timber forests to environmentalism: The rise of colonial environmental and sustainability practices in colonial Zimbabwe, 1938-1961. *Environment and History, 22*(4), 521–559.

Mwadzingeni, L., Mugandani, R., & Mafongoya, P. L. (2022). Socio-demographic, institutional and governance factors influencing adaptive capacity of smallholder irrigators in Zimbabwe. *PLoS One, 17*(8), e0273648.

Nagy, G. (2021). Environmental justice and its geographical aspects in Hungary. *Tér és Társadalom, 35*(4), 76–103.

Najam, A., Huq, S., & Sokona, Y. (2003). Climate negotiations beyond Kyoto: Developing countries concerns and interests. *Climate Policy, 3*(3), 221–231.

Nayna Schwerdtle, P., Baernighausen, K., Karim, S., Raihan, T. S., Selim, S., Baernighausen, T., & Danquah, I. (2021). A risk exchange: Health and mobility in the context of climate and environmental change in Bangladesh—A qualitative study. *International Journal of Environmental Research and Public Health, 18*(5), 2629.

OECD. (2022). *Climate finance provided and mobilised by developed countries: Aggregate trends updated with 2020 data.* Organisation for Economic Co-operation and Development.

OECD. (2023). *Adaptation finance analysis report.* Organisation for Economic Co-operation and Development.

Ogwu, M. C. (2019). Towards sustainable development in Africa: The challenge of urbanization and climate change adaptation. In P. B. Cobbinah & M. Addaney (Eds.), *The geography of climate change daptation in urban Africa* (pp. 29–55). Springer Nature. https://doi.org/10.1007/978-3-030-04873-0_2

Ogwu, M. C., Izah, S. C., Merritt, H., Wise, R. D., & Jintrawet, A. (2025a). *Sustainable bioeconomy development in the global south: Volume I status and perspectives.* Springer. https://doi.org/10.1007/978-981-96-0640-5

Ogwu, M. C., Izah, S. C., Vazquez-Arenas, J. G., Feleke, S. T., & Wei, X. (2025b). Sustainable bioeconomy development in the global south: Volume II bioeconomy techniques. *Springer.* https://doi.org/10.1007/978-981-96-0305-3

Ogwu, M. C., & Kosoe, E. A. (2025). Integrating green infrastructure into sustainable agriculture to enhance soil health, biodiversity, and microclimate resilience. *Sustainability, 17*(9), 3838. https://doi.org/10.3390/su17093838

Ogwu, M. C., Kosoe, E. A., Alaka, A. C., & Xu, J. (2025c). Green infrastructure and environmental health security: The role of nature-based solutions in improving urban health. In S. C. Izah & M. C. Ogwu (Eds.), *Innovative approaches in environmental health management: Processes, technologies, and strategies for a sustainable future.* Environmental Science and Engineering Series. Springer Nature. https://doi.org/10.1007/978-3-031-81966-7_9

Oliveira, J., & Pereda, P. (2020). The impact of climate change on internal migration in Brazil. *Journal of Environmental Economics and Management, 103*, 102340.

Oramah, C. P., & Olsen, O. E. (2021). Equity and justice in climate change adaptation: Policy and practical implication in Nigeria. In W. Leal Filho, N. Oguge, D. Ayal, L. Adeleke, & I. da Silva (Eds.), *African handbook of climate change adaptation* (pp. 1767–1787). Springer International Publishing.

Osei-Amponsah, C., Quarmine, W., & Okem, A. (2023). Understanding climate-induced migration in West Africa through the social transformation lens. *Frontiers in Sociology, 8*, 1173395.

Oxfam. (2023). *Climate finance shadow report 2023.* Oxfam International.

Paavola, J. (2017). Health impacts of climate change and health and social inequalities in the UK. *Environmental Health, 16*(Suppl. 1), 113.

Page, E. A., & Heyward, C. (2017). Compensating for climate change loss and damage. *Political Studies, 65*(2), 356–372.

Pak, J., Le, N., Metwally, E.M., Wang, J., Planey, A.M., Lowman, A.M., Jackson, B.E., Butler, E.N. and Lund, J.L., 2025. Evaluating the intersection of climate vulnerability and cancer burden in North Carolina. JNCI Cancer Spectrum, 9(1), p.pkae124.

Palagi, E., Coronese, M., Lamperti, F., & Roventini, A. (2022). Climate change and the nonlinear impact of precipitation anomalies on income inequality. *Proceedings of the National Academy of Sciences, 119*(43), e2203595119.

Perdana, M. R., Sudiro, A., Ratnawati, K., & Rofiaty, R. (2023). Does sustainable finance work on banking sector in ASEAN?: The effect of sustainable finance and capital on firm value with

institutional ownership as a moderating variable. *Journal of Risk and Financial Management, 16*(10), 449.
Perry, B. L., Aronson, B., & Pescosolido, B. A. (2021). Pandemic precarity: COVID-19 is exposing and exacerbating inequalities in the American heartland. *Proceedings of the National Academy of Sciences, 118*(8), e2020685118.
Phuong, L. T. H., Biesbroek, G. R., & Wals, A. E. (2018). Barriers and enablers to climate change adaptation in hierarchical governance systems: The case of Vietnam. *Journal of Environmental Policy & Planning, 20*(4), 518–532.
Piyasena, N. M. P. M., & Bandara, S. L. (2025). Utilizing geospatial tools for assessing climate change vulnerability: A case study of the Ratnapura District. *Sri Lanka. Environmental Monitoring and Assessment, 197*(7), 774.
Reich, P. B., Grace, K., Agrawal, A., & Nagendra, H. (2022). *Mitigating climate change and climate injustice simultaneously*. Authorea Preprints.
Resurrección, B. P. (2024). Colonial erasures in gender and climate change solutions. *Wiley Interdisciplinary Reviews: Climate Change, 15*(5), e890.
Samuwai, J., & Hills, J. M. (2019). Gazing over the horizon: Will an equitable Green Climate Fund allocation policy be significant for the Pacific post-2020? *Pacific Journalism Review, 25*(1/2), 158–172.
Schlosberg, D., & Collins, L. B. (2014). From environmental to climate justice: Climate change and the discourse of environmental justice. *Wiley Interdisciplinary Reviews: Climate Change, 5*(3), 359–374.
See, J., & Wilmsen, B. (2022). A multidimensional framework for assessing adaptative justice: A case study of a small island community in the Philippines. *Climatic Change, 170*(1), 16.
SenGupta, S., & Atal, A. (2024). Income inequality in the face of climate change: An empirical investigation on unequal nations, vulnerable regions and India. *SN Business & Economics, 4*(8), 87.
Soergel, B., Kriegler, E., Bodirsky, B. L., Bauer, N., Leimbach, M., & Popp, A. (2021). Combining ambitious climate policies with efforts to eradicate poverty. *Nature Communications, 12*(1), 2342.
Song, K., Hsu, A., Peng, W., Yu, Y., & Kittner, N. (2023). Projecting future heat stress disparities to 2100 in the contiguous United States.
Sorg, C. (2025). Finance as a form of economic planning. *Competition & Change, 29*(1), 17–37.
Sugg, M. M., Ryan, S., Spurlock, T., Duncan, S., Hege, A., Ogwu, M. C., Tyson, J., Shay, E., Renwick, K., Hendren, C., Roy, M., & Runkle, J. D. (2023). Climate change and health in rural mountain environments: Summary of a workshop on knowledge gaps, barriers, and opportunities for action. *GeoJournal*. https://doi.org/10.1007/s10708-023-10916-4
Sultana, F. (2022). Critical climate justice. *The Geographical Journal, 188*(1), 118–124.
Swiss Re Institute. (2021). *The economics of climate change: No action not an option*. Swiss Re Institute.
Tesselaar, M., Botzen, W. W., Haer, T., Hudson, P., Tiggeloven, T., & Aerts, J. C. (2020). Regional inequalities in flood insurance affordability and uptake under climate change. *Sustainability, 12*(20), 8734.
Teye, J. K., & Nikoi, E. G. (2022). Climate-induced migration in West Africa. In *Migration in West Africa: IMISCOE regional reader* (pp. 79–105). Springer International Publishing.
Thomas, A., & Benjamin, L. (2018). Policies and mechanisms to address climate-induced migration and displacement in Pacific and Caribbean small island developing states. *International Journal of Climate Change Strategies and Management, 10*(1), 86–104.
Tony, M. (2023). Economic inequality and its effects on the decision-making of entrepreneurs in emerging economies. *Journal of Business Leadership and Management, 1*(1), 67–73.
UNDRR. (2022). *Global assessment report on disaster risk reduction 2022*. United Nations Office for Disaster Risk Reduction.
UN Women. (2023). *Gendered impacts of climate change: Global assessment*. UN Women.
UNEP. (2023). *Adaptation gap report 2023*. United Nations Environment Programme.

UNICEF & WHO. (2022). *Progress on household drinking water, sanitation and hygiene*. United Nations Children's Fund and the World Health Organization.
van Ryneveld, T. N., & Islar, M. (2023). Coloniality as a barrier to climate action: Hierarchies of power in a coal-based economy. *Antipode, 55*(3), 958–981.
Vydelingum, E. (2019). *National adaptation plans: A response to climate-induced migration?* (Doctoral dissertation). Toronto Metropolitan University.
Wang, C., Chen, X., & Jiang, T. (2025). Economic inequality brings about more inaction over climate change: The role of perception, discussion, and responsibility. *International Journal of Psychology, 60*(2), e70022.
Wang, J., Wang, Y., Li, S., & Qin, D. (2016). Climate adaptation, institutional change, and sustainable livelihoods of herder communities in northern Tibet. *Ecology and Society, 21*(1).
Wang, S., Zhang, B., & Slater, L. (2024). Larger contribution of anthropogenic climate change to the increasing frequency of compound droughts and heatwaves in poor countries compared to wealthy ones.
Webber, S. (2024). For and against climate capitalism. *Geographical Research, 62*(1), 14–27.
WHO. (2019). *Climate change and health: Key facts*. World Health Organization.
WHO. (2023). *Vector-borne diseases and climate change*. World Health Organization.
Whyte, K. (2020). Too late for indigenous climate justice: Ecological and relational tipping points. *Wiley Interdisciplinary Reviews: Climate Change, 11*(1), e603.
WMO. (2023). *State of the global climate 2023*. World Meteorological Organization.
World Bank. (2020). *Poverty and shared prosperity 2020*. The World Bank.
World Bank. (2021). *Groundswell part II: Acting on internal climate migration*. The World Bank.
Xing, Z., Guo, J., Zhang, Z., Xue, T., Yang, M., & Wu, W. (2024). Research on the impact of environmental inequality on labor mobility—A study based on the China General Social Survey (CGSS). *Sustainability (2071–1050), 16*(22), 9813.
Zeng, R. (2023). Challenging the growth paradigm: A multifaceted examination of development, disparity, and environmental concerns in modern economies. *Highlights in Business Economics and Management, 21*, 929–933.

Peripheral Ecologies: Environmental Degradation at the Margins of the World System

6

6.1 Introduction: Environmental Inequality in the World System

The concept of environmental inequality is a crucial focus in world-systems theory, necessitating a nuanced understanding of how global capitalism shapes ecological relations, particularly between core and peripheral regions. The core-periphery divide encapsulates an ongoing and complex interplay of economic dynamics where wealthier nations (the core) disproportionately exploit the natural resources and environments of poorer nations (the periphery) to sustain their levels of consumption and production. This exploitation is embedded in systemic processes that place environmental burdens not in affluent centers but at the fringes, where political and economic power is diminished (Faber, 2018; Rice, 2007). Research highlights that inequities in exposure to environmental stressors are deeply embedded in the structures underpinning capitalist economies, with peripheral nations bearing the brunt of ecological degradation. In contrast, core nations benefit from resource extraction (Briggs, 2003; Guarino et al., 2024).

Central to these challenges is the notion of ecological debt, which refers to the imbalance in the contributions different countries make to global environmental sustainability. Specifically, peripheral regions face significant environmental costs stemming from their roles as providers of raw materials and as waste-disposal sites for wealthier nations. This situation exemplifies the idea of "ecologically unequal exchange," where the environmental repercussions of production practices are externalized onto less developed regions (Newell, 2005; Rice, 2007). The burden on these peripheral areas intensifies, as they often lack the regulatory frameworks and technological advancements needed to effectively combat pollution and environmental degradation, further contributing to the cycle of environmental inequality (Newell, 2005).

M. C. Ogwu and S. Chibueze Izah, *Climate Frontiers and Social Transformation*, World-Systems Evolution and Global Futures,
https://doi.org/10.1007/978-3-032-17639-4_6

Moreover, socio-economic structures within peripheral countries typically do not allow for equitable benefits from the resources extracted. Instead, these benefits are disproportionately captured by global corporations often based in core nations, which maximize profits while minimizing environmental liabilities. Consequently, this creates a vicious cycle where underdevelopment is perpetuated by both economic exploitation and environmental degradation (Jorgenson, 2009). Studies indicate that individuals in peripheral nations are subjected to greater illness and shorter life expectancy due to these inequitable environmental practices, illustrating the dire human costs of such arrangements (Forouzanfar et al., 2016). As economies in core nations advance, often at the expense of peripheral economies, the disparities in environmental and health outcomes continue to widen (Millward-Hopkins, 2022; Shi et al., 2022).

Furthermore, the idea of ecological debt extends beyond economic considerations; it raises ethical questions about the responsibilities of core nations toward the environments and communities they adversely affect in the periphery. This ethical abstraction underlines contemporary debates around climate justice, suggesting that industries in the global south are compelled to pursue short-term gains at the expense of long-term ecological sustainability and human welfare (Faber, 2018; Newell, 2005). Economic policies that have historically favored core nations continue to dominate the global landscape, often constraining peripheral economies' ability to innovate and implement sustainable production processes, thereby deepening ecological inequalities linked to core-periphery dynamics (Briggs, 2003; Prüss-Ustün et al., 2017). The institutional frameworks governing environmental governance amplify these inequities. Existing international environmental agreements often fail to sufficiently address the unique circumstances of peripheral countries, leaving them vulnerable to the adverse effects of climate change and ecological exploitation. These nations frequently exhibit weakened institutional capacities to negotiate better terms for the extraction and use of their resources, leading to a situation in which foreign corporations dominate extraction processes while local communities gain minimal benefits (Guarino et al., 2024; Newell, 2005). Consequently, environmental inequalities become entangled with broader issues of social and economic power discrepancies, making them a significant area for policy intervention and academic inquiry. The challenge of ecological debt underscores a profound ethical and political imperative to reform international governance structures and economic practices to ensure a more equitable distribution of environmental responsibilities and benefits. Only through systemic change can the entrenched cycles of degradation be effectively addressed. Only when the global community recognizes and acts on the intricate injustices underlying environmental disparities can sustainable development and ethical accountability be realistically pursued.

This chapter aims to critically examine how environmental degradation in peripheral regions—particularly in Africa, Southeast Asia, and Latin America—reflects the structural inequalities embedded in the global world system. This scholarly assessment provides a comprehensive and multifaceted approach to understanding environmental inequalities as viewed through the lens of world-systems theory, amalgamating empirical evidence with theoretical insights to delineate the complexity of these issues. By focusing on extractive zones where deforestation, pollution, and biodiversity loss are most acute, the chapter interrogates how demands from core economies drive ecological harm and how these costs are systematically externalized to maintain global commodity and capital flows. Its contribution lies in connecting political economy, environmental justice, and ecological sociology to highlight the uneven geography of environmental burdens. Through comparative case studies and analysis of resistance movements, the chapter advances a nuanced understanding of how peripheral ecologies experience both exploitation and resilience, offering insights for rethinking global environmental governance toward equity and sustainability.

6.2 Peripheral Regions as Extractive Frontiers

The historical roots of resource extraction in peripheral regions can be traced back to colonial-era exploitation, characterized by the establishment of foreign rule and the extraction of human and natural resources to benefit imperial interests (Fig. 6.1). Beginning with colonial exploitation, Fig. 6.1 traces how imperial powers imposed systems of resource extraction that have evolved into modern industrial activities (including mining, fossil fuel production, agriculture, and logging) under neoliberal economic orders. This exploitation laid the groundwork for neo-colonial economic arrangements that continue to shape the dynamics of resource extraction in these regions, reinforcing existing inequalities and engendering ongoing struggles for Indigenous rights and environmental justice. Colonialism, defined as the domination of a territory and the subjugation of its peoples by distant powers, involved the systematic exploitation of resources, which has persisted into modern times through global economic practices (Adalikwu-Obisike & Obisike, 2019; Kroll-Zeldin, 2016).

As these peripheral regions are increasingly integrated into the global economy, a process termed "peripheralization," they are subjected to extractive roles supported by international trade agreements and financial institutions. Globalization has exacerbated resource dependency, as peripheral economies are often relegated to producing raw materials for consumption in more developed markets. According to recent studies, many conflicts surrounding extractive projects arise from such arrangements, particularly in sectors like mining, fossil fuels, and agriculture, significantly affecting local communities (Scheidel et al., 2023). This dynamic is further compounded by subsidies and externalities that favor extractive industries, allowing them to operate without fully accounting for their environmental impacts, thereby perpetuating patterns of environmental degradation and social inequity (Reyes-García et al., 2025).

The dominance of specific sectors (namely mining, logging, agriculture, and fossil fuel production) acts as a primary driver of environmental degradation in these peripheral regions. Mining represents a substantial source of conflict and

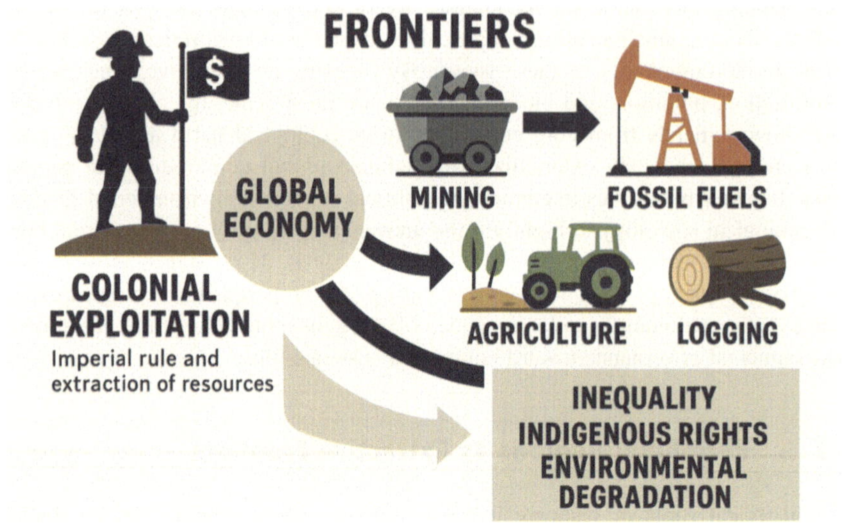

Fig. 6.1 Peripheral regions as extractive frontiers left behind colonial legacies and environmental inequality

environmental harm, as illustrated by case studies documenting the social and ecological impacts of mining activities (Cygańczuk & Roguski, 2024). The findings from diverse research indicate that the consequences of mining extend beyond economic benefits, often resulting in significant landscape loss, livelihood disruption, and land dispossession, which disproportionately affect Indigenous communities (Scheidel et al., 2023). The environmental legacies of such extractive activities include contamination of water sources, soil degradation, and loss of biodiversity, all of which contribute to broader ecological crises (Inatimi et al., 2022; Izah & Aigberua, 2023; Mahmood et al., 2024; Ogwu et al., 2022, 2023).

Agriculture, linked to deforestation and the intensification of fossil fuel use, represents another critical sector driving environmental change. The expansion of agricultural practices not only reduces carbon sinks but is often coupled with unsustainable practices leading to soil degradation and increased greenhouse gas emissions (Mahmood et al., 2024). Moreover, the heavy reliance on fossil fuels for agricultural operations exacerbates these issues, further entrenching the sector within a cycle of environmental harm and economic dependency (Mahmood et al., 2024). Beyond the immediate environmental impacts, the socio-economic fabric of peripheral regions is intertwined with these extractive activities. The neoliberal economic orders fostered by global capital often overlook local needs and priorities, emphasizing profit maximization over sustainable development. Thus, while resource extraction may generate short-term economic gains, it inevitably leads to long-term ecological and social costs that are disproportionately borne by marginalized communities (Kroll-Zeldin, 2016; Reyes-García et al., 2025). The need for policy reform and sustainable practices within these extractive industries is more urgent than ever as the consequences of environmental degradation manifest in increasingly severe forms (Reyes-García et al., 2025).

To mitigate the detrimental effects of industrial and extractive activities, reclamation and restoration practices are essential. Effective land reclamation and restoration strategies can help restore post-mining environments, paving the way for ecological recovery and sustainable use. These practices serve as crucial interventions that not only address environmental degradation but also provide social benefits by restoring livelihoods and lands to local communities (Cygańczuk & Roguski, 2024). Thus, comprehensive frameworks that incorporate both environmental imperatives and socio-economic needs are integral to creating sustainable futures for these peripheral regions.

The complexities of this extractive dynamic necessitate a holistic understanding of global relationships and local realities. The intersection of colonial legacies with contemporary environmental policies and practices highlights the urgent need for equitable frameworks that challenge existing power structures. The imbalance evident in the treatment of peripheral regions warrants a critical re-examination of extractive practices, not just for the sake of local communities but also for the broader health of the global ecosystem (Adalikwu-Obisike & Obisike, 2019; Kroll-Zeldin, 2016). It calls for a global community to adopt reforms such as reducing subsidies for environmentally damaging practices and investing in sustainable alternative industries that can thrive without compromising the ecological integrity of peripheral regions (Reyes-García et al., 2025).

Moreover, the drive toward climate justice highlights the importance of reparatory measures that acknowledge the historical injustices of colonial exploitation. This acknowledgment is vital to fostering a more equitable distribution of resources and responsibilities in mitigating climate change and its far-reaching impacts on vulnerable populations (Gürçam, 2025). By understanding colonial connections and establishing frameworks that prioritize Indigenous rights and environmental sustainability, it is possible to forge paths toward just, sustainable development in these regions (Adalikwu-Obisike & Obisike, 2019). Acknowledging this complexity is vital as the global community strives to address the multifaceted challenges posed by environmental degradation and social inequality. Table 6.1 summarizes the main drivers of resource extraction in peripheral regions, their associated environmental and social impacts, and potential policy responses. It highlights how colonial legacies, global economic structures, and industrial activities intersect to create complex challenges requiring sustainable and equitable solutions. The future requires a commitment to unified action that promotes sustainable practices across industries while simultaneously rectifying historical injustices through inclusive and equitable policies that empower local communities to reclaim agency over their resources and futures (Reyes-García et al., 2025).

Table 6.1 Key drivers, impacts, and responses in peripheral resource extraction

Drivers of extraction	Primary environmental & social impacts	Potential policy & practice responses
Colonial legacies and neo-colonial economic orders	Land dispossession, cultural erosion, and marginalization of Indigenous communities	Reparatory measures recognizing historical injustices; strengthening Indigenous land rights
Globalization and trade dependency	Resource dependency, unequal economic exchange, and social conflicts	Fair trade agreements, reduction of exploitative trade practices, and environmental justice frameworks
Mining and fossil fuel industries	Soil degradation, water contamination, greenhouse gas emissions, and biodiversity loss	Strict environmental regulations; investment in renewable energy; ecological remediation programs
Industrial agriculture and deforestation	Carbon sink loss, habitat destruction, increased emissions, and livelihood vulnerability	Promotion of sustainable agriculture; agroecological practices; forest restoration initiatives
Subsidies and externalities favoring industries	Corporate impunity, inadequate environmental accountability	Removal of harmful subsidies; carbon taxation; corporate responsibility mechanisms
Weak environmental governance	Persistent environmental degradation, failure to integrate local priorities	Inclusive policy reforms, participatory decision-making, and integration of local ecological knowledge into development planning

6.3 Drivers of Ecological Degradation

Ecological degradation is a multifaceted issue driven by several interlocked factors, including deforestation and land-use change, industrial pollution, biodiversity loss, and the exacerbating effects of climate change (Iyiola et al., 2024a, 2024b, Izah et al.,2023). Understanding these drivers is crucial for devising effective strategies to combat ecological degradation and foster sustainable practices across industrial and agricultural systems. Figure 6.2 illustrates the major drivers of ecological degradation across peripheral regions. It highlights how population growth, industrial activities, agricultural expansion, and fossil fuel use interact to intensify resource consumption, deforestation, pollution, and greenhouse gas emissions.

6.3.1 Deforestation and Land-Use Change

Deforestation is primarily driven by industrial agriculture, infrastructure expansion, and logging concessions. The expansion of agricultural land to support industrial farming practices, such as monoculture cultivation, poses significant risks to

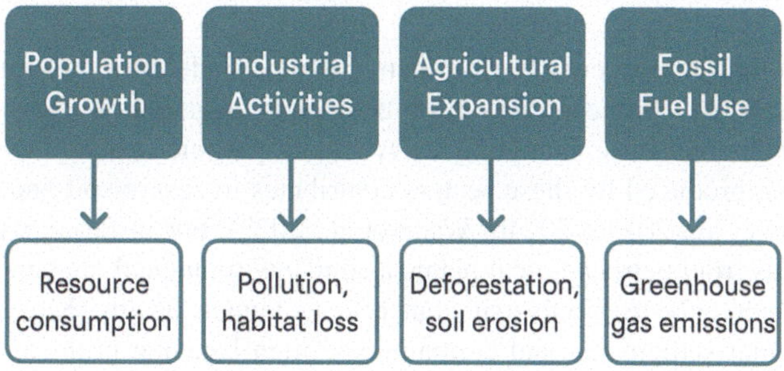

Fig. 6.2 Drivers of ecological degradation in ecological peripheries link human activities to environmental impacts

forests, particularly in developing regions where land is often converted without sustainable management practices (Geist & Lambin, 2002; Ogwu & Kosoe, 2025). This transformation of landscapes has been correlated with a loss of biodiversity, as natural habitats are destroyed to make way for plantations or urban development (Lee et al., 2016; Ogwu, 2019). For example, significant proportions of land previously covered by forests have been cut down for cash crops, significantly contributing to global deforestation rates (Lee et al., 2016). In many socio-economic contexts, local communities engaged in agriculture may be labeled as the principal culprits of deforestation. However, this view overlooks the systemic pressures from market demands and government policies that promote land conversion for economic gain (Asmin et al., 2016).

Infrastructure expansion exacerbates the situation by fragmenting habitats and directly leading to deforestation. Roads and urban areas expand at the expense of ecological integrity, creating pathways that further facilitate the invasion of non-native species and complicate ecosystem recovery (Evans et al., 2001). Additionally, logging concessions often operate in unsustainable ways, driven more by profit motives than by the health of forest ecosystems, leading to overexploitation and the degradation of valuable tree species that play critical roles in local and global carbon cycles (Geist & Lambin, 2002). Continuous logging contributes to the vulnerability of forests, making them more susceptible to fires and diseases that further degrade remaining ecosystems (Finlay, 2022).

Moreover, the industrial agriculture sector's contribution to deforestation is often intertwined with cultural and economic narratives that prioritize immediate gains over long-term ecological sustainability. It is crucial to incorporate traditional ecological knowledge (TEK) into land-use planning, which can shed light on sustainable practices that balance economic needs with conservation efforts (Asmin et al., 2016). This interplay between governmental policies and local practices is key in addressing the root causes of deforestation and formulating comprehensive conservation strategies that integrate local knowledge with scientific research (Chalmers & Fabricius, 2007).

6.3.2 Industrial Pollution and Toxic Waste

Industrial pollution, particularly from petrochemical industries and mining operations, is another significant driver of ecological degradation. The significant amount of industrial waste, including heavy metals, pesticides, and other hazardous contaminants, produced by these sectors contributes to severe soil and water pollution (Iyiola et al., 2024a, 2024b; Wang et al., 2022). For instance, studies show that emissions from petrochemical plants lead to environmental contamination that poses risks not only to biodiversity but also to human health (Kayembe et al., 2012). Informal settlements and communities often bear the brunt of this pollution, as hazardous waste is disposed of without proper treatment or regulation, exacerbating local health crises such as respiratory conditions and skin diseases (Lin et al., 2008).

Mining operations also generate toxic waste, especially through tailings, which often find their way into nearby waterways, leading to a variety of ecological impacts, including the destruction of aquatic habitats and species extinction (Bowman et al., 2020; Izah & Ogwu, 2025). The practice of shipping hazardous waste from developed countries to regions with less stringent environmental regulations further complicates the issue, creating a cycle of exploitation where vulnerable communities bear the environmental and health burdens of economic activities that largely benefit wealthier nations (Taufan & Purwanto, 2018). This environmental injustice highlights the need for international regulatory frameworks to manage waste disposal and ensure that pollution does not disproportionately impact marginalized communities (Belay, 2010).

Moreover, the concept of circular economics, which promotes the regeneration of materials within the economy, can serve as an innovative strategy to mitigate waste generation in industrial sectors. Strategies such as recycling and repurposing industrial by-products could lead to more sustainable production methods and lessen the demand for raw material extraction, further alleviating ecological strain (Gonul Kochan et al., 2016).

6.3.3 Biodiversity Loss and Habitat Fragmentation

The loss of biodiversity and habitat fragmentation are direct outcomes of many of the drivers discussed previously, particularly industrial agriculture, deforestation, and extractive projects. Monocropping practices often reduce plant diversity, which is critical for maintaining ecosystem services and resilience against diseases and pests (Geist & Lambin, 2002). The decline of native species not only impacts local ecosystems but can also lead to widespread ecological consequences, including the collapse of intricate food webs and significant biodiversity loss (Finlay, 2022).

Fragmentation of habitats, often instigated by roads and infrastructure development, isolates wildlife populations, making it challenging for them to migrate, breed, or adapt to changing environmental conditions (Ogwu et al., 2025c). This isolation can lead to inbreeding and a decrease in genetic diversity, making species more susceptible to extinction (Evans et al., 2001). Urban expansion and agricultural intensification have intensified these challenges, as natural corridors that wildlife previously relied upon are increasingly severed, severely limiting their habitats and contributing to a loss of ecosystem resilience (Chalmers & Fabricius, 2007; Ogwu, 2019).

Furthermore, the introduction of invasive species often follows habitat disturbance, further complicating conservation efforts by outcompeting native flora and fauna, altering nutrient cycles, and changing the landscape dynamics (Evans et al., 2001). As ecosystems lose their unique identities, the services they provide—such as water purification, pollination, and carbon sequestration—are diminished, ultimately affecting human livelihoods and well-being (Wang et al., 2022). It is therefore critical to consider a holistic approach to restoration that incorporates ecological and socio-economic dimensions to mitigate the impacts of fragmentation and promote biodiversity recovery (Hilderbrand et al., 2005; Ogwu et al., 2025a, 2025b).

6.3.4 Climate Change Linkages

The relationship between ecological degradation and climate change is intricate, with emissions from industrial practices in core economies exacerbating vulnerabilities in peripheral regions. The combustion of fossil fuels and other greenhouse gas (GHG)-emitting activities drives climate change, manifesting in extreme weather events, altered precipitation patterns, and changing temperature regimes, which, in turn, strain already vulnerable ecosystems (Finlay, 2022). These changes put additional pressure on flora and fauna, which may not have the adaptive capacity to cope with rapidly changing conditions, leading to increased species extinction rates and habitat loss (Peristiowati, 2023).

Moreover, climate change disproportionately affects nations that contribute the least to global emissions, illustrating environmental injustice and extending the cycle of ecological degradation. For instance, agricultural systems in developing countries, which are already under stress from industrial pollution and deforestation, face increased threats from extreme droughts and flooding linked to climate change (Peristiowati, 2023). This heightened vulnerability can exacerbate food insecurity and drive further land-use changes as communities struggle to adapt to new climate realities (Finlay, 2022). To address these linkages, a multifaceted approach is necessary, incorporating both mitigation strategies to reduce GHG emissions and adaptation efforts to enhance ecosystem resilience. Initiatives such as reforestation, sustainable agricultural practices, and wetland restoration can be effective tools for combating both climate change and ecological degradation (Lee et al., 2016). Additionally, incorporating indigenous knowledge into climate adaptation strategies could enhance the effectiveness of initiatives, ensuring they are culturally relevant and scientifically sound (Mwakaje, 2009).

6.4 Case Studies from the Global Peripheries

The examination of land-use policies and their ramifications in global peripheries reveals a complex interplay between economic development, environmental sustainability, and social justice. This synthesis focuses on three key regions: the Congo Basin in Africa, Southeast Asia, and the Amazon Basin in Latin America. Each case illustrates distinct patterns of resource exploitation, displacement of indigenous communities, and degradation of tropical ecosystems, highlighting the urgent need for effective governance and conservation strategies.

In the Congo Basin, dual pressures from mining and logging concessions exacerbate the vulnerabilities of indigenous communities and the rich biodiversity unique to tropical rainforests. Logging operations, focused on sustainable practices, often lead to significant land transformations. Such concessions can be reclassified for industrial use, including cash crops such as palm oil, thereby degrading the ecological integrity of these forests. In this context, it was noted that logging concessions meant for sustainable management have frequently been repurposed, contributing to forest loss (Wich et al., 2012). Meanwhile, mining activities displace local communities and cause long-term environmental degradation through soil erosion and water pollution, negatively impacting both human and wildlife health in the region (Wich et al., 2012).

Southeast Asia presents a similarly alarming scenario, driven primarily by the expansion of palm oil plantations and industrial shrimp aquaculture. The surge in palm oil production, particularly in Indonesia and Malaysia, is directly linked to widespread deforestation that threatens biodiversity and increases greenhouse gas emissions (Pacifici & Cristiano, 2025; Richards & Friess, 2016). Agricultural intensification, mainly through oil palm cultivation, remains a leading cause of deforestation and biodiversity loss in the region (Gibbs et al., 2010). The conversion of rich tropical forests into monoculture plantations has detrimental effects on various species, including those endemic to these habitats (Pacifici & Cristiano, 2025).

Moreover, the industrial shrimp aquaculture sector in Southeast Asia has contributed to significant mangrove loss. Research found that between 2000 and 2012, mangrove ecosystems declined primarily due to aquaculture expansion, threatening coastal biodiversity and the livelihoods of indigenous communities reliant on them (Richards & Friess, 2016). The combined impacts of oil palm and shrimp farming not only undermine ecosystem services but also create socio-economic pressures, as local communities' land tenure rights are increasingly challenged by corporate interests (Azhar et al., 2011). Similar dynamics occur in the Amazon Basin, where agricultural expansion, particularly cattle ranching and soybean cultivation, has driven alarming deforestation rates. Brazilian cattle ranching is a major driver of land-use change, with recent studies indicating that a substantial portion of new agricultural land in the Amazon originates from previously forested areas (Vijay et al., 2016). This degradation is not only ecological; it has severe human implications, as indigenous groups face encroachment and cultural dislocation (Vijay et al., 2016). The land transformation for commodity production poses significant risks to remaining primate habitats, impacting biodiversity and global ecological health(Estrada et al., 2019a, 2019b; Machado, 2019).

Mega-infrastructure projects, such as roads and dams in the Amazon, further exacerbate deforestation and habitat fragmentation, making already vulnerable ecosystems more susceptible to exploitation (Estrada et al., 2019a, 2019b). Recommendations suggest that enhancing protective measures for unique habitats, along with closely monitoring land-use changes associated with infrastructure projects, are essential to mitigate their negative impacts on biodiversity (Vijay et al., 2018). For effective policy responses, it is crucial to emphasize participatory governance that involves local communities in land-use and conservation decision-making (Lori et al., 2025). Strengthening protections for indigenous rights and ecosystem services can facilitate a transition toward more sustainable development practices that prioritize both the environment and the livelihoods of affected communities.

6.5 Externalizing Environmental Costs to the Periphery

The phenomenon of externalizing environmental costs to the periphery of the global economy has profound implications for capital accumulation, corporate strategies, and the environment itself. This issue is intricately linked to the dynamics of global commodity chains and the profit-maximization strategies of multinational corporations (MNCs). MNCs often exploit resource-rich regions under the guise of development, leaving significant ecological footprints while reaping the financial benefits of low-cost raw materials. Specifically, the models of capital accumulation used by companies frequently sidestep the hidden environmental costs associated with their production processes and financial flows. This allows central economies to benefit from cheap resources while externalizing the environmental burdens to the periphery, which bears the brunt of ecological degradation and resource depletion (Nababan & Hasyir, 2019). Table 6.2 summarizes the key mechanisms through which environmental costs are externalized from core to peripheral regions, linking industrial activities and global trade dynamics to their ecological and socio-economic consequences. It highlights how resource extraction, industrial relocation, market control, and subsidies disproportionately burden marginalized communities while sustaining economic growth in core economies.

Evidence indicates that the financial success of corporations is often contingent upon their exploitation strategies, including the use of tax havens and minimal adherence to environmental regulations in extraction areas. Such practices contribute to a vicious cycle in which MNCs maximize profits while local ecosystems degrade, exacerbating socio-economic disparities and environmental harm. These corporations' strategies epitomize a broader trend where ecological awareness is overshadowed by financial motives, often at the expense of local communities and biodiversity (Miladiasari et al., 2021). The environmental impact of such commodification extends beyond immediate ecological damage; it influences long-term sustainability strategies that are crucial in a climate-challenged world. Incorporating ecological costs into financial models should thus be a priority for

Table 6.2 Mechanisms and consequences of externalizing environmental costs to peripheral regions

Mechanism	Environmental impacts	Socio-economic consequences	Examples
Resource extraction for export	Deforestation, water pollution, soil degradation	Loss of livelihoods, displacement of communities	Mining in Congo Basin, palm oil in SE Asia
Industrial relocation from core economies	Air and water pollution, toxic waste dumping	Health crises, increased inequality	E-waste exports to West Africa
Commodity price control by core markets	Overexploitation of land, monocultures	Economic dependency, food insecurity	Coffee and cocoa sectors in Latin America
Subsidies for extractive industries	Habitat loss, biodiversity decline	Corporate profits over local welfare	Fossil fuel subsidies in Africa

policymakers to mitigate these externalities effectively (Hasibuan & Tinambunan, 2024).

The concept of ecological subsidies to core economies illustrates how central markets benefit from the peripheral regions' environmental resources without bearing the associated costs. By sourcing raw materials from less-regulated areas, companies can dramatically reduce production costs. This leads to an economic model in which financial benefits are concentrated in developed economies, while resource-extracting regions face increased environmental degradation, biodiversity loss, and socio-economic challenges (Djuitaningsih & Ristiawati, 2011). Ultimately, the real costs of extraction—such as the depletion of natural resources and the loss of ecosystem services—are often socialized onto society, creating a scenario in which profits are privatized while losses are externalized (Setiawan & Honesty, 2021).

Furthermore, the role of multinational corporations in shaping these dynamics cannot be overstated. MNCs deliberately structure their operations to benefit from regulatory arbitrage—taking advantage of lax environmental laws in certain countries. This not only maximizes profit margins but also undermines local governance and regulatory frameworks, often leading to an erosion of trust in institutions (Brady, 2025). Corporate practices such as strategic environmental management or corporate social responsibility are frequently employed as superficial measures, failing to alter the underlying profit-maximizing motives (Valentinov et al., 2021). The integration of environmental considerations into corporate strategies, though politically expedient, is often insufficient for genuine ecological restoration or sustainability unless paired with stringent regulatory frameworks and genuine corporate accountability.

Moreover, the implications of externalizing environmental costs extend into financial markets, where environmental performance can significantly influence corporate valuations. Investors are increasingly scrutinizing companies' sustainability practices, recognizing that firms with poor environmental records may face reputational damage and regulatory penalties that could impact their bottom line (Miladiasari et al., 2021). The connection between environmental and financial performance is not merely theoretical; empirical studies show that firms that proactively manage their environmental costs tend to outperform their less attentive counterparts (Prinz, 2019). This reality presents an opportunity for MNCs to reconsider their operating models and for consumers and investors to push for more accountability regarding ecological impacts.

The complexities surrounding environmental costs and their implications for financial performance suggest a need for systematic changes in how corporations operate within global commodity chains. Policies must focus on enforcing environmental accountability across borders, ensuring that corporations cannot simply shift their ecological burdens to the periphery without consequence. Sustainable practices should be incentivized through tax structures or tariffs that reflect environmental costs, thereby promoting a more equitable distribution of ecological responsibilities (Setiawan & Honesty, 2021). Additionally, stakeholders, including governments, NGOs, and local communities, are crucial for holding MNCs accountable and encouraging more sustainable production and consumption patterns. This complex arrangement has significant consequences for ecological integrity, social equity, and profitability. Effective interventions will require a multifaceted approach that balances profit maximization with ecological stewardship, ensuring that environmental costs are duly factored into financial models and corporate decision-making processes. This paradigm shift is not merely beneficial but essential for fostering resilience in both local and global ecosystems as we navigate the uncertainties of climate change and sustainability challenges in the twenty-first century (Kwak & Kim, 2017; Zheng, 2017).

6.6 Socio-Ecological Impacts on Peripheral Communities

The socio-ecological impacts on peripheral communities involve a range of issues, including health and well-being related to exposure to toxic substances, livelihood displacement resulting from forced evictions, and the cultural and spiritual erosion of indigenous identities tied to the environment. These multifaceted challenges demand a comprehensive examination through the lens of various environmental and socio-political frameworks. Figure 6.3 illustrates how exposure to pollution and toxic substances affects health and well-being, how forced evictions and economic displacements undermine livelihoods, and how the loss of land leads to cultural and spiritual erosion. Together, these dimensions capture the multilayered vulnerabilities facing marginalized populations in extractive and environmentally degraded regions.

Health and well-being are critical aspects of socio-ecological impacts, particularly for peripheral communities, which are often situated near industrial activities and agricultural lands treated with heavy metals and pesticides. Prolonged exposure to these toxic substances has been linked to various health issues,

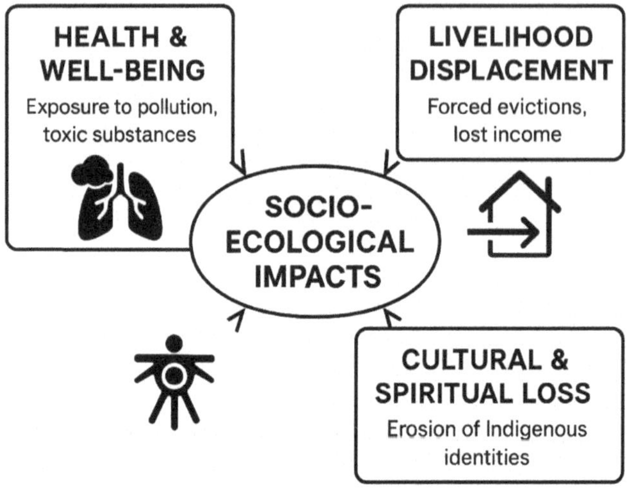

Fig. 6.3 Socio-ecological impacts on peripheral communities

including respiratory diseases, cancers, and developmental disorders among children. Research indicates that communities surrounding mining operations often face significant health risks due to contamination of air, soil, and water resources from industrial effluents and pesticide runoff (Agariga et al., 2021; Garvin et al., 2009). Moreover, significant environmental degradation exacerbates health disparities among vulnerable populations, as those with limited socio-economic resources further lack access to adequate healthcare and information, impeding their ability to respond to health crises effectively (Jelle et al., 2021; Osman & Abebe, 2023).

The issues of livelihood displacement through forced evictions highlight how socio-economic dynamics significantly alter the landscapes of peripheral communities. Studies identify that such displacements manifest through state-led interventions, often justified by development projects or conservation policies that prioritize economic growth over social equity (Lunstrum, 2016; Nitin et al., 2018;Waldorff, 2016). The uprooting of individuals from their agricultural lands leads to immediate economic disarray and a collapse of traditional subsistence economies reliant on agriculture and grazing, profoundly impacting food security. For instance, urbanization-induced evictions in regions like Addis Ababa have resulted in the loss of productive land for peri-urban households, thereby disrupting their food production systems and threatening their livelihoods (Gnamura et al., 2024). This phenomenon is not isolated, as various global experiences within agrarian economies highlight the adverse repercussions of forcible land acquisition and the erosion of agricultural viability due to industrial encroachments, evident in Ghana and parts of Uganda (Ogwang & Vanclay, 2019; Taer, 2024).

Culturally, the erosion of indigenous land and its associated ecosystems leads to the degradation of culturally significant landscapes tied deeply to community identities and cosmologies (Brigitte Chung, 2017; Murray & Agyare, 2018). The loss of access to these lands disrupts not only traditional practices and livelihoods but also obliterates the very spiritual and cultural frameworks that define community cohesion. The transformation of rural landscapes through enforced land seizures disrupts existing social fabrics, making the interplay among health, gender, and cultural practices increasingly precarious (Khan, 2025; Nitin et al., 2018; Sugg et al., 2023). The disconnect from land as a vital resource for spiritual and cultural expressions results in an identity crisis for displaced populations, as their historical narratives and survival strategies become fragmented and often forgotten (Busscher et al., 2020; Lunstrum, 2016).

Furthermore, it is essential to acknowledge the intersectionality of these socio-ecological issues, where health outcomes, economic displacement, and cultural degradation converge. Displaced individuals frequently experience heightened vulnerability to mental health issues, often stemming from the traumas of eviction and loss of home (Jelle et al., 2021; Osman & Abebe, 2023). The psychological toll, augmented by financial instability and health complications, fosters a climate of despair and hopelessness. The interplay between socio-economic fractures and deteriorating health conditions creates a feedback loop that further marginalizes peripheral communities as they navigate the aftermath of socio-ecological transformations (Busscher et al., 2020; Müller, 2023). By addressing these complex interrelations holistically, it is possible to formulate strategies that bolster community resilience and equity within the socio-ecological landscape (Bufebo & Elias, 2021; Islam et al., 2023; Murray & Agyare, 2018).

6.7 Resistance and Alternative Ecologies

The issues surrounding resistance and alternative ecologies have become increasingly relevant as advocates for environmental justice seek to document grassroots movements, indigenous activism, and the networks that effectively challenge mainstream development narratives. Fundamental to these undertakings are frameworks that support ecological citizenship and agency, which highlight community-led resource management and initiatives toward local sustainability. Furthermore, within these concepts lie alternative development pathways which involve agroecology, eco-tourism, and indigenous land rights, serving as counter-narratives to harmful extractivist practices. Figure 6.4 illustrates the interconnected dimensions of resistance and alternative ecologies in peripheral regions. It highlights grassroots movements, indigenous activism, ecological citizenship, and alternative development pathways such as agroecology and eco-tourism. Together, these elements showcase how local communities mobilize to defend their rights, preserve environmental integrity, and promote sustainable, community-centered development models in the face of extractivist pressures.

Fig. 6.4 Resistance and alternative ecologies pathways: Grassroots movements, indigenous activism, and sustainable pathways

One salient aspect of environmental justice movements is their grassroots nature, characterized by local communities coming together to oppose unwarranted exploitation of their land and resources. Research shows that grassroots resistance often manifests through diverse tactics, ranging from formal political engagement to disruptive protests, highlighting the multiplicity of strategies used to challenge institutional power structures. The collective mobilization of these grassroots efforts demonstrates their effectiveness in opposing elite interests and fostering community empowerment (Adams & Shriver, 2017). Such initiatives exemplify how local communities articulate their needs and values against the backdrop of global development models that prioritize profit over ecological sustainability.

Additionally, the concept of ecological citizenship encapsulates the role of individuals and communities in managing natural resources with a focus on sustainability and equity. This perspective emphasizes that citizenship extends beyond legal statuses to include active participation and agency in ecological stewardship (Maldonado-Mariscal, 2023). Engagement at this level fosters a sense of belonging and responsibility toward the environment, which is crucial for cultivating long-term sustainable practices. As evident in the literature, communities often develop alternative resource management strategies that pivot away from extractivist paradigms toward more sustainable approaches such as agroecology and eco-tourism, which not only protect ecological integrity but also enhance community welfare (Maldonado-Mariscal, 2023).

Indigenous activism plays a vital role in these environmental justice movements, as many indigenous communities are at the forefront of defending their rights against land dispossession and ecological degradation. Indigenous knowledge systems often provide valuable insights into sustainable practices that have been refined over generations, highlighting their essential contributions to biodiversity conservation. The rights to land and resources are central to these struggles, embodying the cultural, spiritual, and economic identities of indigenous peoples. The adoption of a framework that integrates indigenous knowledge with contemporary environmental management practices can lead to innovative solutions to pressing ecological issues, illustrating the efficacy of these alternative development pathways.

In addition, transnational advocacy networks have become instrumental in amplifying grassroots resistance. These networks facilitate the sharing of resources, information, and strategies among communities facing similar challenges across various geopolitical landscapes. The pooling of experiences from diverse movements strengthens efforts to influence policy changes from local initiatives to global dialogues on environmental governance. Research illustrates that such transnational grassroots coalitions can significantly influence public opinion and policy outcomes by framing their discourses around shared values of sustainability and justice (Apostolopoulou et al., 2022).

The exploration of agroecology as an alternative development pathway reveals how sustainable agricultural practices can contribute to food security while preserving environmental integrity. Studies illustrate that agroecological practices not only reduce dependency on chemical inputs but also foster resilience in the face of climate variability by promoting biodiversity. Community-based agroecological initiatives empower local farmers, align with traditional ecological knowledge, and facilitate the transition to sustainable food systems that prioritize collective health over individual profit. Furthermore, agroecological approaches often reinvigorate local economies, offering pathways to viable economic alternatives that challenge industrial agricultural models (Apostolopoulou et al., 2022). Eco-tourism emerges as another significant alternative to standard exploitative tourism models, which often exacerbate environmental and social injustices. When managed correctly, eco-tourism can promote conservation efforts, support local cultures, and provide economic benefits to communities by making them stakeholders in preserving their natural environments. The integration of indigenous perspectives in eco-tourism design and implementation ensures that local communities benefit from their heritage and environment rather than being marginalized within their landscapes. This approach promotes a more equitable distribution of resources and amplifies local agency in articulating their ecological rights.

The role of policy frameworks in shaping these movements cannot be underestimated. For grassroots activists and communities pursuing environmental justice, navigating local, national, and international policy landscapes is crucial. They often face significant institutional challenges, but successful movements have demonstrated that sustained grassroots activism can lead to policy reforms that recognize and institutionalize their rights. Moreover, grassroots movements must continuously adapt their strategies; when communities transform local struggles into broader societal campaigns, their chances of success significantly increase (Ju & Tang, 2011). The interplay between grassroots resistance, indigenous activism, and transnational advocacy networks has established a robust foundation for environmental justice movements. These entities challenge extractivist narratives by advocating for ecological citizenship and community agency in resource management, ultimately leading to sustainable, equitable, and deeply rooted alternative development pathways. Through their collective efforts, these movements not only resist encroachment on their rights and ecosystems but also actively reimagine development, prioritizing ecological health, social equity, and cultural integrity.

6.8 Conclusion: Rethinking Global Environmental Governance

The analysis of peripheral ecologies across Africa, Southeast Asia, and Latin America reveals that environmental degradation in these regions is not accidental, nor merely the byproduct of modernization. Instead, it is structurally embedded within the very architecture of the global world system. Core economies, sustained by relentless demands for commodities, energy, and capital accumulation, rely on peripheral regions to absorb the environmental costs of global production and consumption. Deforestation, biodiversity loss, and industrial pollution in these extractive frontiers reflect a pattern of ecological externalization that allows the global North to maintain economic growth while displacing its ecological footprint to the global South. It uncovered the paradox of peripheral ecologies: while they are sites of acute vulnerability, they are simultaneously spaces of resistance, resilience, and innovation. Grassroots environmental justice movements, indigenous knowledge systems, and alternative ecological practices challenge the dominant logic of extraction and open pathways toward more equitable and sustainable futures. Their struggles highlight the urgency of reimagining global environmental governance beyond technocratic solutions and toward frameworks that integrate ecological justice, indigenous sovereignty, and transnational accountability. Addressing ecological degradation in the peripheries demands systemic transformation. It requires dismantling the economic and political hierarchies that perpetuate unequal ecological exchange and advancing a paradigm where environmental sustainability is inseparable from social justice. By situating peripheral ecologies at the center of global environmental debates, this chapter calls for a radical rethinking of development, trade, and climate policies, thus one that recognizes the moral imperative of shared ecological responsibility in an interconnected world.

References

Adalikwu-Obisike, J., & Obisike, E. E. (2019). The legacy of imperialism in Nigeria. *European Scientific Journal, 15*(16), 229–246. https://doi.org/10.19044/esj.2019.v15n16p229

Adams, A. E., & Shriver, T. E. (2017). Tactics and targets: Explaining shifts in grassroots environmental resistance. *Social Currents, 4*(3), 265–281.

Agariga, F., Abugre, S., Siabi, E. K., & Appiah, M. (2021). Mining impact on livelihoods of farmers of Asutifi North District, Ghana. *Environmental Management and Sustainable Development, 10*(4), 29.

Apostolopoulou, E., Bormpoudakis, D., Chatzipavlidis, A., Cortés Vázquez, J. J., Florea, I., Gearey, M., et al. (2022). Radical social innovations and the spatialities of grassroots activism: Navigating pathways for tackling inequality and reinventing the commons. *Journal of Political Ecology, 29*(1), 144–188.

Asmin, F., Darusman, D., Ichwandi, I., & Suharjito, D. (2016). Local ecological knowledge on forest clearing: A case study of parak and rimbo practices in Simancuang Community, Indonesia. *Komunitas, 8*(2), 208–220.

Azhar, B., Lindenmayer, D. B., Wood, J., Fischer, J., Manning, A., McElhinny, C., & Zakaria, M. (2011). The conservation value of oil palm plantation estates, smallholdings and logged peat swamp forest for birds. *Forest Ecology and Management, 262*(12), 2306–2315.

Belay, A. A. (2010). Impacts of chromium from tannery effluent and evaluation of alternative treatment options. *Journal of Environmental Protection, 1*(1), 53–58.

Bowman, D. M., Kolden, C. A., Abatzoglou, J. T., Johnston, F. H., van der Werf, G. R., & Flannigan, M. (2020). Vegetation fires in the Anthropocene. *Nature Reviews Earth & Environment, 1*(10), 500–515.

Brady, M. (2025). Sales or profits? Managerial decision-making in a duopoly with symmetric and asymmetric costs. *Managerial and Decision Economics, 46*(6), 3318–3326.

Briggs, D. (2003). Environmental pollution and the global burden of disease. *British Medical Bulletin, 68*(1), 1–24.

Brigitte Chung, Y. (2017). Engendering the new enclosures: Development, involuntary resettlement and the struggles for social reproduction in coastal Tanzania. *Development and Change, 48*(1), 98–120.

Bufebo, B., & Elias, E. (2021). Land use/land cover change and its driving forces in Shenkolla watershed, south Central Ethiopia. *The Scientific World Journal, 2021*(1), 9470918.

Busscher, N., Parra, C., & Vanclay, F. (2020). Environmental justice implications of land grabbing for industrial agriculture and forestry in Argentina. *Journal of Environmental Planning and Management, 63*(3), 500–522.

Chalmers, N., & Fabricius, C. (2007). Expert and generalist local knowledge about land-cover change on South Africa's Wild Coast: Can local ecological knowledge add value to science? *Ecology and Society, 12*(1).

Cygańczuk, K., & Roguski, J. (2024). The consequences of mining damage and their impact on the natural environment—A Case Study of Trzebinia. *Safety & Fire Technology, 64*(2), 66–82.

Djuitaningsih, T., & Ristiawati, E. E. (2011). Pengaruh kinerja lingkungan dan kepemilikan asing terhadap kinerja finansial perusahaan. *Jurnal Akuntansi Universitas Jember, 9*(2), 31–54.

Estrada, A., Garber, P. A., & Chaudhary, A. (2019a). Expanding global commodities trade and consumption place the worlds primates at risk of extinction. *PeerJ, 7*, e7068.

Estrada, A., Garber, P. A., & Chaudhary, A. (2019b). Expanding global commodities trade and consumption place the world's primates at risk of extinction. *PeerJ, 7*, e7068.

Evans, R. D., Rimer, R., Sperry, L., & Belnap, J. (2001). Exotic plant invasion alters nitrogen dynamics in an arid grassland. *Ecological Applications, 11*(5), 1301–1310.

Faber, D. (2018). Global capitalism, reactionary neoliberalism, and the deepening of environmental injustices. *Capitalism Nature Socialism, 29*(2), 8–28.

Finlay, D. J. (2022). Climate change and ocean acidification: Are we just treating the symptoms?

Forouzanfar, M. H., Afshin, A., Alexander, L. T., Anderson, H. R., Bhutta, Z. A., Biryukov, S., Brauer, M., Burnett, R., Cercy, K., Charlson, F. J., & Cohen, A. J. (2016). Global, regional, and national comparative risk assessment of 79 behavioural, environmental and occupational, and metabolic risks or clusters of risks, 1990–2015: A systematic analysis for the global burden of disease study 2015. *The Lancet, 388*(10053), 1659–1724.

Garvin, T., McGee, T. K., Smoyer-Tomic, K. E., & Aubynn, E. A. (2009). Community–company relations in gold mining in Ghana. *Journal of Environmental Management, 90*(1), 571–586.

Geist, H. J., & Lambin, E. F. (2002). Proximate causes and underlying driving forces of tropical deforestation: Tropical forests are disappearing as the result of many pressures, both local and regional, acting in various combinations in different geographical locations. *BioScience, 52*(2), 143–150.

Gibbs, H. K., Ruesch, A. S., Achard, F., Clayton, M. K., Holmgren, P., Ramankutty, N., & Foley, J. A. (2010). Tropical forests were the primary sources of new agricultural land in the 1980s and 1990s. *Proceedings of the National Academy of Sciences, 107*(38), 16732–16737.

Gnamura, K., Antwi, M., & Abenet, B. (2024). The impact of urbanization-induced eviction on food security of peri-urban households in Addis Ababa, Ethiopia.

Gonul Kochan, C., Pourreza, S., Tran, H., & Prybutok, V. R. (2016). Determinants and logistics of e-waste recycling. *The International Journal of Logistics Management, 27*(1), 52–70.

Guarino, R., Corsi, G., & Muñoz-Ulecia, E. (2024). Debunking competition-global ecologically unequal exchange explained by exploitation and control relations. *Ecological Economics, 242*, 108909.

Gürçam, S. (2025). Colonial connections in the global climate crisis: Historical injustices and contemporary inequalities. *Ankara Üniversitesi Çevrebilimleri Dergisi, 12*(1), 19–36.

Hasibuan, D. H., & Tinambunan, M. S. M. (2024). Influence leverage, cost environment, and performance environment to financial performance. *Journal of Economics, Finance and Management Studies, 7*(02), 851–858.

Hilderbrand, R. H., Watts, A. C., & Randle, A. M. (2005). The myths of restoration ecology. *Ecology and Society, 10*(1).

Inatimi, S. A., Popoola, O. M., Yarkwan, B., Iyiola, A. O., & Izah, S. C. (2022). Therapeutic potentials of wildlife resources and options for conservation. In S. C. Izah (Ed.), *Biodiversity in Africa: Potentials, threats and conservation, sustainable development and biodiversity* (Vol. 29, pp. 143–174). Springer Nature Singapore Pte Ltd. https://doi.org/10.1007/978-981-19-3326-4_6

Islam, M. S., Crawford, T. W., & Shao, Y. (2023). Evaluation of predicted loss of different land use and land cover (LULC) due to coastal erosion in Bangladesh. *Frontiers in Environmental Science, 11*, 1144686.

Iyiola, A. O., Afolabi, O. A., Alimi, S. K., Akingba, O. O., Izah, S. C., & Ogwu, M. C. (2024a). Climate change and water crisis in the global south. In S. C. Izah, M. C. Ogwu, A. Loukas, & H. Hamidifar (Eds.), *Water crises and sustainable management in the global south* (pp. 111–140). Springer.

Iyiola, A. O., Afolabi, O. A., Alimi, S. K., Akingba, O. O., Izah, S. C., & Ogwu, M. C. (2024b). Climate change and water crisis in the global south. In S. C. Izah, M. C. Ogwu, A. Loukas, & H. Hamidifar (Eds.), *Water crises and sustainable management in the mlobal south*. Springer. https://doi.org/10.1007/978-981-97-4966-9_4

Izah, S. C., & Aigberua, A. O. (2023). Potential threats and possible conservation strategies of biodiversity in Niger Delta region of Nigeria. In S. C. Izah & M. C. Ogwu (Eds.), *Sustainable utilization and conservation of Africa's biological resources and environment. sustainable development and biodiversity* (Vol. 888). Springer. https://doi.org/10.1007/978-981-19-6974-4_6

Izah, S. C., Iyiola, A. O., Yarkwan, B., & Richard, G. (2023). Impact of air quality as a component of climate change on biodiversity-based ecosystem services. In A. L. Srivastav, A. K. Dubey, A. Kumar, S. K. Narang, & M. A. Khan (Eds.), *Visualization techniques for climate change with machine learning and artificial intelligence* (pp. 123–148). Elsevier. isbn:978-0-323-99714-0.

Izah, S. C., & Ogwu, M. C. (2025). Modeling solutions for microbial water contamination in the global south for public health protection. *Frontiers in Microbiology, 16*, 1504829. https://doi.org/10.3389/fmicb.2025.1504829

Jelle, M., Morrison, J., Mohamed, H., Ali,e R., Solomon, A., & Seal, A. J. (2021). Forced evictions and their social and health impacts in Southern Somalia: A qualitative study in Mogadishu Internally Displaced Persons (IDP) camps. Global Health Action, 14(1), 1969117.

Jorgenson, A. K. (2009). The sociology of unequal exchange in ecological context: A panel study of lower-income countries, 1975–2000. *Sociological Forum, 24*(1), 22–46.

Ju, C. B., & Tang, S. Y. (2011). External legitimacy, goal congruence and collective resistance: Environmental NGOs and land use politics in South Korea. *Urban Studies, 48*(4), 811–825.

Kayembe, K., Basosila, L., Mpiana, P. T., Makambo, L., Sikulisimwa, P. C., Tshibangu, D. S., Tshilanda, D. D., & Tati, R. K. (2012). The effect of the monosubstituted benzenes functional groups on the inhibition of methane gas biosynthesis. *Journal of Sustainable Bioenergy Systems, 2*(4), 92–96.

Khan, D. (2025). Reformulating theories of 'accumulation by dispossession': 'Contested accumulations through displacement' in postcolonial Punjab, Pakistan. *Environment and Planning A: Economy and Space, 57*(2–3), 203–222.

Kroll-Zeldin, O. (2016). Colonialism. Oxford bibliographies. doi:https://doi.org/10.1093/obo/9780199766567-0139.

Kwak, M., & Kim, H. (2017). Green profit maximization through integrated pricing and production planning for a line of new and remanufactured products. *Journal of Cleaner Production, 142*, 3454–3470.

Lee, K., Carrillo Botero, N., & Novotny, T. (2016). 'Manage and mitigate punitive regulatory measures, enhance the corporate image, influence public policy': Industry efforts to shape understanding of tobacco-attributable deforestation. *Globalization and Health, 12*(1), 55.

Lin, A. Y. C., Yu, T. H., & Lin, C. F. (2008). Pharmaceutical contamination in residential, industrial, and agricultural waste streams: Risk to aqueous environments in Taiwan. *Chemosphere, 74*(1), 131–141.

Lori, T., Osei-Kwarteng, M., Ojija, F., Ojo, A. O., & Ogwu, M. C. (2025). Participatory approaches to environmental health management. In S. C. Izah & M. C. Ogwu (Eds.), *Innovative approaches in environmental health management: Processes, technologies, and strategies for a sustainable future. Environmental science and engineering series*. Springer Nature. https://doi.org/10.1007/978-3-031-81966-7_11

Lunstrum, E. (2016). Green grabs, land grabs and the spatiality of displacement: Eviction from Mozambique's Limpopo National Park. *Area, 48*(2), 142–152.

Mahmood, H., Hassan, M. S., Meraj, G., Mahmood, H., Hassan, M. S., Meraj, G., & Furqan, M. (2024). Agriculture's role in environmental sustainability: A comprehensive review of challenges and solutions. *Challenges in Sustainability, 12*(3), 178–189.

Maldonado-Mariscal, K. (2023). Grassroots innovation and social innovation in perspective. *Frontiers in Sociology, 8*, 1247293.

Miladiasari, M., Agriyanto, R., Farida, D., Prasetyoningrum, A., & Muhlis, M. (2021, October). The effect of environmental performance and environmental cost on financial performance with good corporate governance as the moderating variable. In *Proceedings of the First International Conference on Islamic History and Civilization, ICON-ISHIC 2020, 14 October, Semarang, Indonesia* (pp. 29–42).

Millward-Hopkins, J. (2022). Inequality can double the energy required to secure universal decent living. *Nature Communications, 13*(1), 5028.

Müller, L. (2023). A socio-ecological approach to multi-dimensional inequality. *The Evolving Scholar*.

Murray, G., & Agyare, A. (2018). Religion and perceptions of community-based conservation in Ghana, West Africa. *PLoS One, 13*(4), e0195498.

Mwakaje, A. G. (2009). Wetlands, livelihoods and sustainability in Tanzania. *African Journal of Ecology, 47*, 179–184.

Nababan, L. M., & Hasyir, D. A. (2019). Pengaruh environmental cost dan environmental performance terhadap financial performance. *E-Jurnal Ekonomi Dan Bisnis Universitas Udayana, 3*(8), 259–286.

Ogwang, T., & Vanclay, F. (2019). Social impacts of land acquisition for oil and gas development in Uganda. *Land, 8*(7), 109.

Ogwu, M. C. (2019). Towards sustainable development in Africa: The challenge of urbanization and climate change adaptation. In P. B. Cobbinah & M. Addaney (Eds.), *The geography of climate change adaptation in urban Africa* (pp. 29–55). Springer Nature. https://doi.org/10.1007/978-3-030-04873-0_2

Ogwu, M. C., Iyiola, A. O., & Izah, S. C. (2023). Overview of African biological resources and environment. In S. C. Izah & M. C. Ogwu (Eds.), *Sustainable utilization and conservation of Africa's biological resources and environment. Sustainable development and biodiversity* (Vol. 888). Springer. https://doi.org/10.1007/978-981-19-6974-4_1

Ogwu, M. C., Izah, S. C., & Iyiola, A. O. (2022). An Overview of the potentials, threats and conservation of biodiversity in Africa. In S. C. Izah (Ed.), *Biodiversity in Africa: Potentials, threats and conservation, sustainable development and biodiversity* (Vol. 29, pp. 1–20). Springer Nature Singapore Pte Ltd. https://doi.org/10.1007/978-981-19-3326-4_1

Ogwu, M. C., Izah, S. C., Merritt, H., Wise, R. D., & Jintrawet, A. (2025a). *Sustainable bioeconomy development in the global south: Volume I status and perspectives.* Springer Singapore. https://doi.org/10.1007/978-981-96-0640-5

Ogwu, M. C., Izah, S. C., Vazquez-Arenas, J. G., Feleke, S. T., & Wei, X. (2025b). *Sustainable bioeconomy development in the Global South: Volume II bioeconomy techniques.* Springer Singapore. https://doi.org/10.1007/978-981-96-0305-3

Ogwu, M. C., & Kosoe, E. A. (2025). Integrating green infrastructure into sustainable agriculture to enhance soil health, biodiversity, and microclimate resilience. *Sustainability, 17*(9), 3838. https://doi.org/10.3390/su17093838

Ogwu, M. C., Kosoe, E. A., Alaka, A. C., & Xu, J. (2025c). Green infrastructure and environmental health security: The role of nature-based solutions in improving urban health. In S. C. Izah & M. C. Ogwu (Eds.), *Innovative approaches in environmental health management: Processes, technologies, and strategies for a sustainable future.* Environmental Science and Engineering Series. Springer Nature. https://doi.org/10.1007/978-3-031-81966-7_9

Osman, A. A., & Abebe, G. K. (2023). Rural displacement and its implications on livelihoods and food insecurity: The case of inter-riverine communities in Somalia. *Agriculture, 13*(7), 1444.

Pacifici, M., & Cristiano, A. (2025). Impacts of oil palm plantations expansion on the distribution of terrestrial mammals in South-East Asia. *bioRxiv*, 2025-05.

Peristiowati, Y. (2023). Analysis driving force, pressure, state, impact and regional environmental issues Kediri City. *Journal for Quality in Public Health, 7*(1), 104–109.

Prinz, A. (2019). Learning (not) to evade taxes. *Games, 10*(4), 38.

Prüss-Ustün, A., Wolf, J., Corvalán, C., Neville, T., Bos, R., & Neira, M. (2017). Diseases due to unhealthy environments: An updated estimate of the global burden of disease attributable to environmental determinants of health. *Journal of Public Health, 39*(3), 464–475.

Rai, N. D., Benjaminsen, T. A., Krishnan, S., & Madegowda, C. (2019). Political ecology of tiger conservation in India: Adverse effects of banning customary practices in a protected area. *Singapore Journal of Tropical Geography, 40*(1), 124–139.

Reyes-García, V., Villasante, S., Benessaiah, K., Pandit, R., Agrawal, A., Claudet, J., Garibaldi, L. A., Kabisa, M., Pereira, L., & Zinngrebe, Y. (2025). The costs of subsidies and externalities of economic activities driving nature decline. *Ambio, 55*, 1–14.

Rice, J. (2007). Ecological unequal exchange: Consumption, equity, and unsustainable structural relationships within the global economy. *International Journal of Comparative Sociology, 48*(1), 43–72.

Richards, D. R., & Friess, D. A. (2016). Rates and drivers of mangrove deforestation in Southeast Asia, 2000–2012. *Proceedings of the National Academy of Sciences, 113*(2), 344–349.

Scheidel, A., Fernández-Llamazares, Á., Bara, A. H., Del Bene, D., David-Chavez, D. M., Fanari, E., Garba, I., Hanaček, K., Liu, J., Martínez-Alier, J., & Navas, G. (2023). Global impacts of extractive and industrial development projects on indigenous peoples' lifeways, lands, and rights. *Science Advances, 9*(23), eade9557.

Setiawan, M. A., & Honesty, F. F. (2021, June). Environmental performance, environmental costs and financial performance. In *Sixth Padang International Conference on Economics Education, Economics, Business and Management, Accounting and Entrepreneurship (PICEEBA 2020)* (pp. 85–88). Atlantis Press.

Shi, J., Lu, C., & Wei, Z. (2022). Effects of social capital on pro-environmental behaviors in Chinese residents. *Sustainability, 14*(21), 13855.

Sugg, M. M., Ryan, S., Spurlock, T., Duncan, S., Hege, A., Ogwu, M. C., Tyson, J., Shay, E., Renwick, K., Hendren, C., Roy, M., & Runkle, J. D. (2023). Climate change and health in rural mountain environments: Summary of a workshop on knowledge gaps, barriers, and opportunities for action. *GeoJournal*. https://doi.org/10.1007/s10708-023-10916-4

Taer, A. (2024). Shrinking rice bowls: Tracing the decline of Philippine rice lands.

Taufan, H. S., & Purwanto, P. (2018). The management of toxic and hazardous waste materials in the food industry. In *E3S Web of Conferences* (Vol. 73, p. 07020). EDP Sciences.

Valentinov, V., Roth, S., & Pies, I. (2021). Social goals in the theory of the firm: A systems theory view. *Administration & Society, 53*(2), 273–304.

Vijay, V., Pimm, S. L., Jenkins, C. N., & Smith, S. J. (2016). The impacts of oil palm on recent deforestation and biodiversity loss. *PLoS One, 11*(7), e0159668.

Vijay, V., Reid, C. D., Finer, M., Jenkins, C. N., & Pimm, S. L. (2018). Deforestation risks posed by oil palm expansion in the Peruvian Amazon. *Environmental Research Letters, 13*(11), 114010.

Waldorff, P. (2016). 'The law is not for the poor': Land, law and eviction in Luanda. *Singapore Journal of Tropical Geography, 37*(3), 363–377.

Wang, Q., Wang, Q., Yang, Z., Wu, X., & Peng, Y. (2022). Quantitative analysis of industrial solid waste based on terahertz spectroscopy. *Photonics, 9*(3), 184.

Wich, S. A., Gaveau, D., Abram, N., Ancrenaz, M., Baccini, A., Brend, S., Curran, L., Delgado, R. A., Erman, A., Fredriksson, G. M., & Goossens, B. (2012). Understanding the impacts of land-use policies on a threatened species: Is there a future for the Bornean orang-utan? *PLoS One, 7*(11), e49142.

Zheng, F. (2017). The retailers' choices of profit strategies in a cournot duopoly: Relative profit and pure profit. *Modern Economy, 8*(2), 199–210.

Ecological Imperialism Revisited: Land Grabs, Resource Frontiers, and Global Demand

7

7.1 Introduction: Revisiting Alfred Crosby's Ecological Imperialism

Alfred Worchester Crosby (from the United States, 1931–2018) introduced the concept of ecological imperialism to describe how European colonial powers exerted control not only over land and people but also over the ecosystems of the regions they colonized. His work highlights how the transference of species, crops, and agricultural practices significantly altered ecological landscapes. This disruption was a strategic element of imperial expansion that facilitated resource extraction and ultimately shaped colonial economies. The ecological transformations brought about by colonialism had lasting effects, creating resource-dependent states and radically altering local ecosystems from the fifteenth to the twentieth centuries (Frame, 2022). The legacies of these practices can be observed today in modern agricultural practices and resource management approaches that continue to reflect historical patterns of exploitation and ecological imbalance (Abajue & Ogwu, 2024).

7.1.1 Historical Context of Ecological Imperialism: Colonial Legacies, Resource Extraction, and Ecological Transformations (Fifteenth–Twentieth Century)

Colonial legacies manifest in contemporary global industries, particularly agriculture and resource extraction, illustrating the continuity of ecological imperialism well into the twenty-first century. As noted by Tilman et al. (2001), demand for food and other natural resources is projected to surge due to population growth and rising consumption patterns. Land grabs, which typify modern ecological imperialism, echo historical practices whereby colonial powers appropriated vast areas

M. C. Ogwu and S. Chibueze Izah, *Climate Frontiers and Social Transformation*, World-Systems Evolution and Global Futures,
https://doi.org/10.1007/978-3-032-17639-4_7

of land for monoculture crop production under the guise of agricultural development. This not only devastated indigenous practices and ecosystems but also created dependencies that continue in today's global economic systems (Frame, 2022). The resulting pressures on land and ecological systems have led to increased greenhouse gas emissions and biodiversity loss, directly linked to persistent agricultural intensification strategies that trace their roots back to colonial resource extraction practices (Charles et al., 2010; Ogwu, et al., 2025).

As we critique these historical practices, it is essential to recognize that contemporary ecological transformations reflect the historical transformations instigated during imperial expansions. These transformations are evidenced by the increasing reliance on industrial agriculture, characterized by intensive monocropping and extensive use of fertilizers and pesticides, mirroring the ecological impositions made during the colonial era (Tilman et al., 2001). The technological innovations that emerged under colonialism were primarily designed for resource extraction, highlighting a pattern of ecological manipulation that still plagues modern societies (Charles et al., 2010; Frame, 2022).

7.1.2 Relevance in the Twenty-First Century: Globalized Capital and Climate Change

The relevance of ecological imperialism is starkly apparent in the contemporary setting, where globalization of capital and climate change reinforces new forms of ecological domination. Climate change is identified as a leading factor exacerbating food insecurity and resource scarcity, with trends suggesting that by 2050, food demand may increase significantly (potentially by as much as 56%) due to climate change impacts (Quan et al., 2023). This escalating demand draws parallels to historical crises of resource management and exploitation observed during the colonial era, necessitating a critical examination of current agricultural practices and their sustainability (Falcon et al., 2022). As highlighted by Costello et al. (2020), adaptive management of fisheries and aquaculture is fundamental to addressing the challenges posed by climate change and a growing global population. However, the continuation of imperialistic resource exploitation mechanisms (often dominated by corporations in the Global North seeking profit at the expense of ecological and social well-being in the Global South) replicates problems of the past (Frame, 2022). Corporations engage in land grabs that prioritize economic gains over sustainable practices, propelling a cycle of ecological degradation reminiscent of historical imperial practices (Obeng-Odoom, 2022).

This exploitative modality is intricately linked to technologies of agriculture and resource extraction that have, since colonial times, enabled the dominance of powerful entities over marginalized communities who are often left vulnerable and disenfranchised (Frame, 2022; Tilman et al., 2001). The socio-ecological impacts (ranging from biodiversity loss to increased greenhouse gas emissions) underscore the urgency of reconsidering our approach to global resource management in the context of ecological imperialism.

7.1.3 Linking Past and Present: From Territorial Conquests to Corporate Land Grabs

The historical linkage between territorial conquests and contemporary corporate land grabs becomes particularly evident when examining the resource frontiers exploited in the twenty-first century. The parallels between the past and present highlight how ecological imperialism has evolved but continues to manifest in similar exploitative frameworks. Today's land grabs are often justified under the guise of development and modernization, echoing colonial rhetoric that masked the exploitation of colonized lands (Frame, 2022). These practices result in the displacement of indigenous populations and the erosion of traditional ecological knowledge systems that sustained local environments.

In the twenty-first-century context, the globalization of agricultural practices reveals a persistent trajectory of ecological domination, deeply embedded within the mechanisms of capitalism. The statistics presented by Tilman et al. (2001) suggest that sustainable agricultural strategies can mitigate ecological damage while addressing impending food demands, underscoring the potential to diversify agricultural practices beyond monoculture. However, these alternative strategies often confront systemic barriers created by existing power dynamics within global food systems. The corporatization of agriculture and the capture of land by multinational entities result in increased competition for limited resources, drawing on historical exploitative tendencies that prioritize profit over environmental integrity (Charles et al., 2010; Frame, 2022).

Moreover, contemporary challenges, such as climate change, which compound food security issues (Iyiola et al., 2024; Izah et al., 2023), emphasize the necessity of effective and just resource management approaches that learn from past ecological imperialism. It is also essential to recognize the interdependencies of global food systems, as articulated by Falcon et al. (2022), who argue for rethinking food production in the face of growing demand. Solutions should embrace a holistic understanding of ecological systems and the historical context that shapes current agricultural policies, emphasizing equity and sustainability.

Revisiting A.W. Crosby's concept of ecological imperialism invites critical reflection on the historical practices of colonial powers and compels us to interrogate the enduring legacies within contemporary global dynamics. Understanding these connections helps frame the challenges we face today and encourages the exploration of pathways toward more equitable and sustainable ecological relations in the wake of perpetual global demands (Falcon et al., 2022; Frame, 2022).

This chapter aims to critically reassess Alfred Crosby's concept of ecological imperialism by situating it within the contemporary realities of global land grabs, resource frontiers, and intensifying competition over natural resources. Specifically, it seeks to analyze how foreign investments in agriculture, mining, and energy reshape rural ecologies, disrupt social relations, and generate new dependencies under conditions of neoliberal globalization and ecological crisis. This chapter also investigates how affected communities negotiate these transformations through resistance movements, alternative development practices, and emerging discourses on environmental justice. By integrating insights from political ecology, environmental history, and development studies, this chapter contributes to the growing body of scholarship on the socio-ecological implications of twenty-first-century resource extraction. It offers a conceptual bridge between historical forms of imperialism and present-day resource enclosures, highlighting the continuities and ruptures in global environmental governance. This chapter seeks to provide a nuanced understanding of how ecological imperialism has evolved in the Anthropocene and to identify pathways toward more equitable and sustainable resource management systems.

7.2 Global Land Grabs in the Twenty-First Century

Land grabbing, particularly in the twenty-first century, is a complex phenomenon with substantial implications for emerging economies, local communities, and ecological integrity. The concept of land grabbing revolves around large-scale land acquisitions, characterized by the political, economic, and symbolic dimensions that define contemporary property relations. The term "land grab" has increasingly become colloquial for any extensive acquisition or leasing of land by foreign investors, often resulting in detrimental consequences for local populations and their environments. Borras Jr and Franco (2012) argue that land grabbing signifies an increase in transnational land transactions focused predominantly on the production and export of essential commodities, including food and biofuels. Understanding land grabbing entails evaluating dimensions such as changes in land tenure and the consequential shifts in property relations within global capitalism (Margulis et al., 2013). Indicators of land grabs often refer to the area acquired relative to national cultivated land; thus, countries with notable land grabs frequently show discrepancies in their cultivated areas, leading to a drastic decline in per capita availability of arable land (Rulli et al., 2013).

This phenomenon is notably evident across regions, including Sub-Saharan Africa, Southeast Asia, and Latin America, each with unique characteristics pertinent to its socio-political context. Governments often participate directly or through intermediaries in facilitating these grabs, amplifying tensions between development needs and local community rights (Gemeda et al., 2019). Consequently, encapsulating land grabbing requires examining the underlying motivations—ranging from economic trajectories to land governance frameworks that empower or marginalize various stakeholders.

7.2.1 Drivers of Land and Resource Acquisitions

Several driving forces compel governments and corporations to pursue land acquisitions aggressively. Land and resource acquisitions are driven by multiple interrelated factors spanning economic, environmental, and social dimensions (Fig. 7.1). As populations grow and economies expand, demand for food, energy,

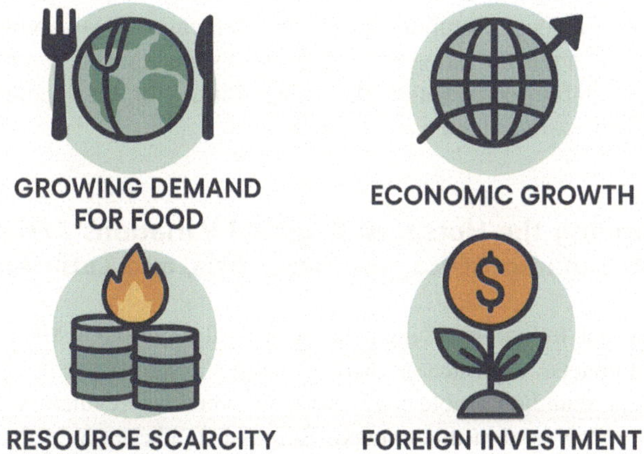

Fig. 7.1 Major drivers of land and resource acquisitions

water, and raw materials intensifies, while climate variability and resource scarcity heighten competition for these limited resources. Foreign investments and globalization further accelerate large-scale land deals, often reshaping local livelihoods and ecosystems. Food insecurity is a primary motivator, characterized by the pressing need for increased agricultural production to meet the dietary requirements of burgeoning populations. The global food crisis, visible from 2007 to 2008, exacerbated this urgency, leading various nations to secure land abroad to ensure their food sovereignty, as evidenced by extensive investments in agricultural land (Adam & Agegnehu, 2023). In Ethiopia, for instance, substantial investments have been made, primarily driven by Indian buyers seeking fertile land for export-oriented agriculture, reflecting the growing interdependencies in food supply chains (Teklemariam et al., 2017). Global market fluctuations further exacerbate these dynamics as nations grapple with rising food prices and discern supply-chain vulnerabilities that emphasize national interests in resource security and access.

As nations transition toward sustainable energy paradigms, the demand for biofuels has spurred a wave of land acquisitions aimed at establishing plantations for renewable energy production (Ogwu, et al., 2025). These shifts create a complex overlay with the ongoing global transition to green technologies, where agriculture is tightly intertwined with energy production and industrial growth (Margulis et al., 2013). The quest for alternative energy sources thus reinforces the trend of land grabs, particularly when countries seek to transition away from fossil fuel dependence, reflecting the interconnectedness between energy security and agricultural land use (Schoneveld et al., 2011).

Furthermore, speculative investments in land have emerged as a lucrative avenue for financial gains amidst fluctuating global economic scenarios. The anticipated returns on agricultural investments have attracted various actors, including multinational corporations and hedge funds, to secure land rights, often with minimal regard for local communities or environmental sustainability (Pistorius et al., 2017). This financialization of land leads to significant socio-environmental repercussions, notably exacerbating the vulnerabilities of local populations and intensifying conflicts over land as a resource.

7.2.2 Mapping the Hotspots: Regional Variations Across Sub-Saharan Africa, Southeast Asia, and Latin America

Regional variations in land-grabbing activities reveal profound distinctions in local contexts and impacts. Sub-Saharan Africa remains a significant hotspot for land grabs due to its abundant arable land and sometimes lax regulatory frameworks (Ogwu, et al. 2025). Within this context, Ethiopia stands out as a focal point for large-scale land acquisition, often described as the "epicenter of land grabs for food exports on the continent" (Wondimu & Gebresenbet, 2018). Diverse investments in biofuel and food production have led to significant land transfers, often displacing farming communities and causing widespread socio-economic disruptions, such as reduced access to essential resources and declining livelihoods among local populations.

In Southeast Asia, countries like Cambodia exemplify the challenges posed by land grabbing. Rural populations often face threats of displacement as agricultural land transitions into commercial estates or infrastructure projects aligned with national development goals (Muawanah, 2015). The Cambodian government's approach to land reform, often driven by capitalist interests, frequently results in the alienation of local communities from their ancestral lands, engendering conflict, undermining traditional livelihoods, and catalyzing further environmental degradation (Loehr, 2012). Consequently, understanding these regional variations requires detailed analyses of local political economies, which reveal how economic interests and regulatory frameworks shape land access and use.

Conversely, Latin America experiences a confluence of land acquisitions driven by the demands for commodities such as soy, palm oil, and sugarcane, which radically alter land-use patterns. The prioritization of export-oriented agriculture often aligns with government-backed initiatives that promote land investment through favorable legal conditions and incentives for foreign investors (Desalegn et al., 2014). The implications for local communities include loss of land rights, exposure to the dynamics of the global market, and diminished food security, underscoring the complex interplay between national development imperatives and local community rights.

7.2.3 Examples of Large-Scale Land Deals in Ethiopia, Brazil, and Cambodia

In Ethiopia, for instance, the surge in land leases to foreign investors has resulted in a reconfiguration of agricultural practices. The large-scale acquisition of land has frequently marginalized local farmers, disrupting traditional land-use practices and altering socio-cultural dynamics within communities. The shift toward monoculture and intensive farming methods driven by foreign investors raises alarms about long-term sustainability and local food security (Beyene, 2022). Furthermore, developments such as the Agricultural Commercialization Clusters represent state efforts to reconsolidate and commercialize land use, highlighting the tension between developing smallholder agriculture and promoting foreign investment (Wedekind & Chinigò, 2020).

In Brazil, the expansion of agribusiness has led to significant land seizures linked to soy production and cattle ranching. Large tracts of land are frequently converted for monoculture production, displacing indigenous communities and small farmers who rely on diverse agricultural systems for their livelihoods. The socio-environmental costs of such transformations include deforestation, biodiversity loss, and disruptions to the ecosystem services on which local populations depend (Perger et al., 2013). The intricate dynamics in Brazil showcase the perpetual conflict between economic growth driven by agribusiness and the preservation of local cultures and ecosystems, highlighting the broader implications of global land deals.

Similar narratives emerge in Cambodia, where contested land grabs have triggered widespread local resistance against government-sanctioned displacements. The formalization of land titles has favored elites and corporations, often at the expense of marginalized communities who are evicted from lands they have occupied for generations (Loehr, 2012). The resultant land conflicts in Cambodia underscore a systemic pattern of disenfranchisement that persists under contemporary governance structures aimed at integrating with global markets. A study by Muawanah (2015) reveals that resolving land tenure issues effectively contributes to improved poverty alleviation efforts, highlighting the need for policy interventions that prioritize local communities' rights and sustainable development.

Land grabbing in the twenty-first century reflects a myriad of interconnected factors, intertwining global and local interests that shape contemporary agricultural and investment landscapes. The intricate web of political economics, environmental sustainability, and social justice that frames the discourse on land grabs underscores the need for robust governance frameworks that uphold accountability and prioritize the rights of local communities over corporate profits. Addressing the implications of land grabbing requires comprehensive land reforms that resonate with both national aspirations and local realities within an increasingly interconnected global framework.

7.3 Rural Ecologies Under Global Capital

The transformations in agrarian landscapes under global capital have led to significant shifts from smallholder agriculture to industrial monocultures, with far-reaching implications for ecological and social systems. In various contexts, agricultural practices have been moving toward intensification and industrialization, often driven by global markets and the need for efficiency and productivity. Research indicates that countries like Uruguay have experienced a decrease in aboveground net primary production as a result of such agricultural transformations, in which increased dependency on inputs and the expansion of cultivated areas have been prioritized within agribusiness strategies (Gazzano et al., 2019). Shifting from smallholder farms, which traditionally sustained local communities and ecosystems, to large-scale industrial farming not only disrupts local economies but also intensifies resource appropriation, raising broader questions of equity and sustainability within rural ecologies (Gazzano et al., 2019).

Moreover, the concept of contract farming represents a noteworthy transformation in how smallholders engage with the agricultural economy, as it reflects a process where corporate entities control production while smallholders navigate severe economic dependencies. In India, for instance, companies involved in contract farming have been found to centralize capital without outright dispossession of farmers, indicating a complex interplay of power dynamics in agrarian contexts (Shrimali, 2016). While it may not displace farmers in the traditional sense, this centralization often comes at a cost, with labor conditions becoming strained and varied, exhibiting geographical differences based on local circumstances (Shrimali, 2016). As such, the agribusiness-led expansion into regions traditionally dominated by small-scale subsistence farming has transformed landscape dynamics and exacerbated socio-economic disparities, contributing to systemic vulnerabilities.

The environmental consequences of such industrial expansions cannot be overlooked. Large-scale agricultural practices have caused substantial ecological disruption, including habitat loss, soil degradation, and pollution. For instance, evidence from tropical regions indicates that the expansion of industries like oil palm has led to significant declines in biodiversity, with large-scale plantations threatening to displace native ecosystems and species (Azhar et al., 2011). Additionally, intensive resource exploitation has been linked with severe soil degradation, compounded by the demand for chemical use that ultimately depletes soil health and contributes to pollution in surrounding water systems (Cohn et al., 2017; Izah & Ogwu, 2025). In this light, extractive industries emerge as critical contributors to environmental costs in rural ecologies, prompting urgent calls for regulatory frameworks to mitigate these impacts (Azhar et al., 2011).

Water resources in particular face significant threats from agricultural intensification and extractive industrial practices. Pollution and direct alteration of water bodies from agricultural runoff and industrial waste discharge have reduced water quality and availability, posing risks to human health and ecological sustainability. In regions where intensive agriculture prevails, the overuse of fertilizers and pesticides degrades local water sources and heightens the risk of eutrophication, creating dead zones inhospitable to aquatic life (Cohn et al., 2017). This complex interplay of agricultural practices and resource exploitation encapsulates a broader pattern of ecological disruption wherein water scarcity and quality degradation serve as harsh reminders of the human-environment dynamic under global capitalism.

Climate change further complicates the predicament of rural ecologies, especially amid land grabs under the guise of "development" and agricultural expansion. The propensity for large-scale land acquisitions, often termed land grabs, has been shown to exacerbate vulnerability among local populations by stripping them of access to traditional lands and resources upon which they depend for their livelihoods (Cohn et al., 2017). Smallholder farmers, already facing the dual pressures of climate change and an evolving agricultural market, find themselves increasingly precarious as the climate crisis deepens and land becomes increasingly contested and commercialized (Cohn et al., 2017). The intersection of climate impacts and globalization underscores the vital need for a thorough understanding of how rural landscapes are reshaped, often at the expense of vulnerable communities and ecological integrity.

Moreover, the industrialization of agriculture is often accompanied by socio-political challenges, as land grabs may lead to violence and conflicts over land ownership and rights (Besley & Persson, 2011). Economic shocks induced by global market fluctuations can trigger disputes and multiply vulnerabilities among local populations. This situation reveals how ecological transformations are intertwined with social justice issues and highlights the necessity for policies that address both environmental sustainability and social inequalities (Besley & Persson, 2011). A holistic examination of agrarian landscapes under global capital underscores the urgent need for frameworks that balance ecological sustainability and community resilience. As ecological disruption unfolds, the prospect of long-term climatic shifts poses an additional layer of complexity to agricultural viability. Climate change has been correlated with increased variability in weather patterns, creating uncertainty in agricultural productivity, particularly for smallholder farmers (Cohn et al., 2017). These farmers, often among the most vulnerable, rely on stable climatic conditions that are increasingly unpredictable. Such dynamics can prompt further shifts toward monoculture practices to maintain yields amid climatic variability, perpetuating a cycle of ecological degradation and social vulnerability (Cohn et al., 2017).

The transformation of rural ecologies under global capital represents a multifaceted challenge, wherein the shift toward industrial agriculture and extractive industries reshapes landscapes and community dynamics while magnifying socio-economic disparities. The increasing prevalence of contract farming and large-scale agricultural practices redefines rural livelihoods, often with adverse environmental impacts. Furthermore, the intersection of climate change with industrialization underscores the urgency for innovative approaches to governance and resource management that prioritize ecological integrity alongside social equity. Table 7.1 summarizes the major transformations occurring in rural ecologies under global capital, highlighting the shift from smallholder agriculture to industrial

Table 7.1 Transformations in rural ecologies under global capital

Theme	Key Processes/ Trends	Implications for Rural Communities	Ecological Consequences	References
Shift to Industrial Agriculture	Expansion of monocultures, intensified input use, and agribusiness-driven land acquisitions.	Displacement of smallholders, loss of livelihoods, and growing socio-economic disparities	Habitat loss, reduced biodiversity, soil degradation, and water pollution	Gazzano et al., 2019; Azhar et al., 2011
Contract Farming	Centralization of production by corporate actors, and the economic dependence of smallholders	Limited land dispossession but heightened labor precarity and economic vulnerabilities	Resource overexploitation, and uneven rural development	Shrimali, 2016
Extractive Industries	Expansion of oil palm, mining, and industrial farming	Marginalization of indigenous communities, weakened land rights	Deforestation, soil erosion, and pollution of air and water systems	Azhar et al., 2011; Cohn et al., 2017; Ogwu & Izah, 2024
Water Resource Pressures	Agricultural runoff, industrial discharge, and chemical overuse	Reduced water quality and availability, health risks, and increased conflict over water	Eutrophication, decline in aquatic ecosystems, and long-term water scarcity	Cohn et al., 2017
Climate Change Intersections	Increased climatic variability, intensified land grabs under "development" projects.	Greater livelihood insecurity, land conflicts, and food insecurity	Accelerated ecosystem degradation, reduced resilience to climate extremes	Cohn et al., 2017; Besley & Persson, 2011

monocultures, the rise of contract farming, pressures on water resources, and the ecological and social implications of extractive industries.

7.4 Social Relations and New Dependencies

The contemporary discourse on social relations and new dependencies offers a multifaceted examination of rural livelihoods, gender dynamics, food sovereignty, and economic dependencies linked to neo-colonialism. The intricate interplay between these themes underscores the evolving paradigms of labor relations, socio-economic stratification, and cultural identities in rural contexts.

7.4.1 Rural Livelihoods and Labor Relations

In the wake of globalization, rural labor dynamics have undergone significant transformations. Wage labor and seasonal migration have become prevalent avenues for economic survival for many rural households, especially in developing economies. The rise of informal financing mechanisms, such as microcredit, has been posited as both a potential lifeline and a source of entrapment. For instance, Carswell et al. (2021) highlight that while microcredit can convert risky debts into more manageable forms, often allowing households to invest in essential aspects of their lives, it simultaneously introduces new forms of productive but hidden labor associated with debt repayment processes. Furthermore, in many contexts, fishermen and other laborers often find themselves in debt traps, leaving them vulnerable to ongoing cycles of repayment, as noted by Manzilati and Prestianawati (2022), who elucidate the paradoxical nature of informal financial contracts that can hinder genuine economic independence.

The interaction of socio-economic factors further complicates labor relations in rural settings. Households, particularly those belonging to wage worker classes, experience varying degrees of indebtedness shaped by their asset ownership and class positions. For example, research by Kandikuppa (2022) indicates that rural elites tend to engage in higher levels of borrowing than their less affluent counterparts, who become increasingly reliant on informal credit sources, often at exorbitant interest rates. This stratification not only affects access to capital but also influences the overall socio-economic mobility of such households, perpetuating cycles of poverty and dependence.

7.4.2 Gender and Generational Dimensions

The contemporary labor landscape is distinctly influenced by gender and generational viewpoints. Figure 7.2 highlights the dual challenges shaping rural economies: the undervaluation of women's labor and the growing disenfranchisement of youth. It illustrates how intergenerational knowledge loss and limited economic opportunities undermine traditional ecological practices and threaten the cultural and economic resilience of rural communities. Women in rural areas play crucial roles in household economies, often managing multiple responsibilities ranging from agricultural activities to informal labor. However, these roles are fraught with challenges; women's labor remains undervalued and usually unrecognized in formal economic analyses. Moreover, youth disenfranchisement poses a growing concern as younger generations face limited pathways to economic engagement, exacerbated by the loss of intergenerational knowledge transfer. As observed in various studies, the erosion of traditional ecological practices and localized knowledge systems threatens the cultural fabric and economic viability of rural communities.

Intergenerational knowledge loss is particularly salient in the context of agricultural practices. The displacement of traditional methods by modern technologies

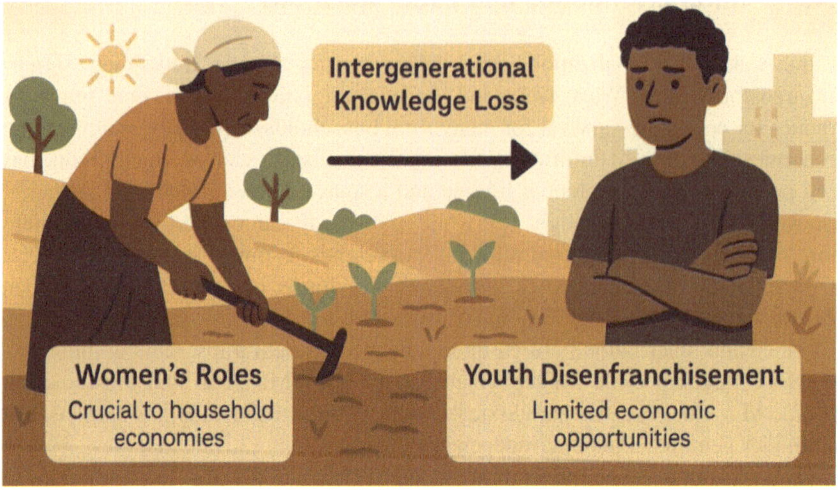

Fig. 7.2 Intergenerational knowledge loss and gendered labor dynamics in rural economies

reflects a broader trend in which young people may opt for wage labor in urban environments, seeking immediate financial returns that alienate them from traditional knowledge systems. As a consequence, the cultural narrative surrounding food sovereignty is at risk; modern agricultural models often prioritize yield over biodiversity and sustainable practices, sidelining indigenous techniques that have sustained communities for generations. Such dynamics not only affect food systems but also challenge cultural identities, as highlighted in anthropological discussions of cultural change related to agricultural transitions.

7.4.3 Food Sovereignty and Cultural Erosion

The decline of local food systems is intertwined with broader economic processes and cultural shifts. The commodification of agriculture has led to increased reliance on external markets, which adversely affects food sovereignty. As local food production decreases in favor of commercial agriculture, communities lose their food sources and the associated traditions surrounding food preparation and consumption. This phenomenon is exacerbated by the growth of multinational corporations that exert substantial influence over local markets, leading to the homogenization of food cultures and practices (Table 7.2).

Research on agricultural transitions indicates that prioritizing cash crops over subsistence farming often results in nutritional and cultural deficiencies within communities, as further demonstrated by Barbier et al. (2016), who outline how poor resource management within indebted smallholder economies can have long-term negative implications. The reliance on external support mechanisms, rather

Table 7.2 Economic transitions and cultural impacts on food sovereignty in rural communities

Issue	Major Drivers	Impacts on Food Systems	Cultural Implications
Commodification of Agriculture	Expansion of cash crops, external market dependency	Decline in local food production, loss of food diversity	Erosion of traditional food practices, homogenization of food cultures
Multinational Corporate Influence	Control over local markets, global supply chains	Prioritization of commercial agriculture over subsistence	Marginalization of indigenous food systems, weakening of cultural identity
Neo-colonial Economic Structures	Historical legacies, structural economic dependencies	Reliance on external inputs and markets	Continued cultural disempowerment, loss of sovereignty over food systems
Youth Disconnection from Tradition	Shift to wage labor, modernization pressures	Limited intergenerational knowledge transfer	Decline in cultural narratives tied to local food production

than enhancing local capacities for self-sufficiency, signals a worrying trend in how food systems are configured within rural landscapes.

Moreover, cultural erosion is not merely a byproduct of economic shifts; it actively contributes to the loss of identity for rural communities. As traditional practices lose importance, younger generations may struggle to connect with their cultural roots. This reflects broader narratives of neo-colonialism, in which former colonial relationships continue to shape economic and cultural realities, embedding systems of dependency. Table 7.2 summarizes how economic commodification, corporate influence, and modernization pressures reshape rural food systems and highlights the decline of subsistence farming, the erosion of traditional food practices, and the cultural disconnection emerging from global economic dependencies.

7.4.4 Economic Dependency and Neo-colonialism

Economic dependency in many developing countries mirrors aspects of neo-colonialism, where external influences and capital dominate local economies. This dependency perpetuates cycles of poverty and underdevelopment, restricting nations' ability to pursue independent development paths. The ongoing discussions around debt traps exemplify this phenomenon, particularly in countries engaging in significant borrowing from international creditors. As outlined by Al-Fadhat and Prasetio (2024), the repercussions of debt-trap diplomacy manifest in various contexts, wherein countries become excessively indebted and reliant on foreign powers, particularly evident in instances involving Chinese investments in Africa.

The ramifications of such dependencies extend beyond simple monetary exchanges; they encompass broader socio-political implications, perpetuating systems of control that hinder genuine sovereignty. Furthermore, economic interactions shaped by external debt often delay infrastructural development and social progress, keeping nations in economically subservient positions. It raises critical questions about the global economic order and its implications for self-determination and agency in resource-rich yet economically marginalized regions.

Notably, scholars have interrogated how economic policy decisions intersect with social inequities. Salti (2015) argues that the structure of debt and its management influence income distribution within economies, especially in low-income settings, where public debt primarily affects domestic stakeholders. Such dynamics reflect the broader inequities embedded in global capitalism and underscore the urgent need to re-evaluate how economic dependencies are structured.

7.5 Resistance and Resilience

The intertwined issues of resistance and resilience, particularly within the realms of local protests and land rights movements, indigenous knowledge and ecological stewardship, civil society and transnational advocacy networks, and alternative models of development, are crucial for understanding contemporary social and environmental justice movements. Grassroots mobilizations play a pivotal role in asserting justice and access, particularly in challenging cases exacerbated by systemic injustices and environmental degradation. Figure 7.3 illustrates the interconnected dimensions of resistance and resilience, highlighting how local protests, indigenous knowledge, civil society advocacy, and alternative development models collectively drive movements for social and environmental justice.

7.5.1 Local Protests and Land Rights Movements

Local protests, such as those advocating for land rights, arise from grassroots efforts driven by demands for justice and equitable access to resources. Communities frequently mobilize to defend their rights against encroachment by state-led development and corporate initiatives that threaten their land and livelihoods. Research highlights that these grassroots efforts can take various forms, from protests to legal battles, as communities seek to protect their ancestral lands against exploitation (Mayaputri et al., 2024; Sitthikriengkrai & Porath, 2017). Additionally, NGO involvement can strengthen these protests by providing legal support, raising awareness, and facilitating community organization.

Furthermore, local mobilizations around land rights underscore the importance of environmental justice and the intersection of socio-economic rights with indigenous knowledge systems (Carrington, 2020; Kuyper et al., 2018). Studies have shown that local protests challenge not only immediate socio-political contexts

Fig. 7.3 Resistance and resilience as pathways to social and environmental justice

but also broader structures of power, indicating a strategic evolution in global justice movements (Choudry, 2013; Bhandari et al., 2024). Moreover, NGOs act as allies in strategically framing these movements, enhancing their ability to connect local struggles to global frameworks of justice (Mosse & Nagappan, 2021).

7.5.2 Indigenous Knowledge and Ecological Stewardship

Indigenous knowledge systems offer vital insights into ecological stewardship and conservation ethics, both essential to sustainability. These systems, often transmitted across generations, emphasize a deep connection to the land and advocate sustainable practices that maintain ecological equilibrium (Brito & Rondon, 2024; Panday & Rabbani, 2017). The increasing recognition of the integration of indigenous knowledge into broader ecological governance frameworks highlights its importance for effective environmental management (Choir, 2022; Schlosberg & Collins, 2014).

Grassroots movements have been instrumental in advocating for the acknowledgment and incorporation of these knowledge systems within national policies, thereby enhancing cultural resilience amidst the pressures of globalization and industrialization (Anderson et al., 2019; Gottlieb & Joshi, 2010). By framing their heritage in the context of ecological stewardship, indigenous communities articulate their struggles in ways that resonate with broader audiences concerned with climate justice and human rights (Humphrey, 2012). This approach challenges the predominant reliance on Western scientific methods and advocates a more inclusive perspective in environmental governance (Brito & Rondon, 2024; Carrington, 2020).

7.5.3 Civil Society and Transnational Advocacy Networks

Civil society organizations, particularly NGOs, have become central actors in transnational advocacy networks, effectively amplifying local movements and generating global pressure for accountability (Carrington, 2020). These networks facilitate information exchange, strategic planning, and collective actions among various stakeholders committed to justice and human rights (Kuyper et al., 2018). The role of NGOs is crucial, as they often provide necessary resources, expertise, and platforms for advocacy that are essential for nurturing local activism (Allan, 2018).

In the realm of global governance, these advocacy networks challenge conventional structures that can hinder grassroots participation in policymaking (Atia & Herrold, 2018; Keddie, 2021). The dynamic interplay of local voices in transnational contexts can lead to policy changes that address diverse perspectives, particularly concerning environmental justice, where local communities have historically been marginalized (McGaughey et al., 2023). Consequently, NGOs not only operate at the national level but also harness international partnerships to enhance their effectiveness in countering corporate and state interests, solidifying their importance in the quest for justice (Anderson et al., 2019; Kuyper et al., 2018).

7.5.4 Alternative Models of Development

Alternative models of development, such as agroecology and cooperative governance, provide powerful counter-narratives to exploitative traditional development paradigms (Petrilla et al., 2020). These models emphasize sustainable practices grounded in local traditions and ecological principles, enhancing resilience in the face of climate change and social injustice (Van Huijstee et al., 2011). Community-based resource governance has shown promise in guaranteeing local populations control over their resources, circumventing the drawbacks of top-down development strategies that frequently disenfranchise local stakeholders (Choir, 2022).

Agroecology embodies a holistic approach that integrates scientific inquiry with traditional agricultural practices, thereby promoting not only food security but also social equity (Brito & Rondon, 2024; Panday & Rabbani, 2017). Research indicates that these approaches enhance sustainable agricultural practices and strengthen community cohesion and social capital, critical components when addressing external challenges such as climate change (Hansen & Umbreit, 2018). The transition toward these alternative paradigms reflects an increasing acknowledgment of the interconnectedness of environmental health, food justice, and social equity (Choir, 2022; Hansen & Umbreit, 2018).

7.6 Resource Competition and Geopolitical Tensions

The geopolitical landscape is intricately linked with the competition for critical resources, particularly in the context of energy and minerals essential for technological advancement and a sustainable green economy (Ogwu et al., 2025, d). The demand for vital resources such as oil, lithium, and rare earth elements is escalating globally, driven by the transition toward renewable energy and technological development (Izah, et al., 2025; Izah, Ogwu, & Hait, 2025a, 2025b). The shift from fossil fuels to renewable energy sources is fundamentally reshaping energy markets and the geopolitical dynamics governing access to these vital resources (Altiparmak, 2023; Yu & Wang, 2023).

Emerging economies and industrialized nations are vying for strategic control over energy supply routes and critical mineral reserves. This competition is intensified by geopolitical risks associated with ongoing tensions, as evidenced by the impact of the Russia–Ukraine conflict on foreign investment and corporate strategy (Wang, 2023; Meissner, 2024). These tensions significantly influence the strategy of transnational corporations (TNCs), as they seek stable environments for long-term investments, particularly in resource-intensive sectors (Husnain et al., 2022; Wang, 2023). Fluctuations in geopolitical stability not only affect investment decisions but also lead to hazardous economic consequences, creating ripple effects throughout global markets (Olasehinde et al., 2024).

Furthermore, the increasing awareness of resource security intersects with questions of sustainability and environmental stewardship. Strategic control over energy resources can be seen as a direct driver of geopolitical conflict, with major energy-producing regions such as the Middle East and Central Asia becoming focal points of international confrontation (Ibekwe et al., 2024). Concurrently, as countries grapple with the ramifications of climate change and strive for energy independence, there is a pressing need to reassess their energy policies. Countries like China, which depend heavily on the import of critical minerals, are re-evaluating their strategies to secure energy supply chains and mitigate risks stemming from geopolitical instability (Altiparmak, 2023; Yu & Wang, 2023). The convergence of agricultural policy with resource competition further complicates the geopolitics surrounding energy and minerals. Specifically, land grabs for biofuel production raise concerns about food security and the ethical implications of international trade policies (Meissner, 2024). Countries with significant agricultural investment are often those experiencing geopolitical instability, suggesting a prioritization of resource extraction over sustainable practices (Yang et al., 2024). These dynamic interrelations necessitate a foundational understanding of strategies for food and energy security amid rising global tensions.

Moreover, as climate-induced migration pressures escalate, the diplomacy surrounding environmental conflicts becomes increasingly critical. Water scarcity, exacerbated by geopolitical rivalries, is an urgent international concern that could lead to resource conflicts (Rabitz, 2023). Diplomatic efforts aimed at resolving conflicts over transboundary water resources reflect the intertwining of geopolitics, resource management, and environmental sustainability. Further analysis into these conflicts reveals how cooperative frameworks, or their absence, directly influence humanitarian outcomes and resource management strategies (Liu et al., 2021). Geopolitical risks extend into energy inflation concerns. Rising tensions linked to significant geopolitical events can precipitate fluctuations in energy prices, affecting economies worldwide (Olasehinde et al., 2024). The Geopolitical Risk Index reveals periodic spikes aligned with historical geopolitical disruptions, subsequently impacting energy markets globally (Caldara & Iacoviello, 2022). Understanding the nexus between geopolitical risk and energy pricing is crucial for policymakers and economists as they design strategies to address the challenges posed by geopolitical competition over diminishing natural resources.

The interrelation between resource competition and geopolitical tensions necessitates a multifaceted approach. The global demand for energy and critical minerals underscores the urgency of strategic investment and the need for sustainable, politically stable environments. Additionally, the implications of geopolitical dynamics extend beyond energy supply chains, influencing food security, environmental policy, and international cooperation in resource management. As we confront these challenges, it is paramount that scholars, policymakers, and global leaders adopt holistic strategies that address the economic intricacies and ethical realities of resource security in a rapidly changing geopolitical landscape.

7.7 Rethinking Ecological Imperialism in the Anthropocene

7.7.1 The Anthropocene as a Conceptual Lens

The Anthropocene is fundamentally characterized by human dominance over planetary processes, as reflected in significant alterations to Earth's systems. Richardson et al. (2023) note that humanity has already transgressed several of the nine planetary boundaries that, if crossed, may lead to catastrophic ecological thresholds. They describe how critical boundaries concerning biosphere integrity were surpassed in the late nineteenth century, coinciding with the onset of intensified land-use practices that drastically reshaped ecosystems (Richardson et al., 2023). This historical context aligns with the broader understanding that human activities have escalated disruption across multiple environmental parameters, including climate change and biodiversity loss (Lade et al., 2019). Consequently, the Anthropocene era presents a critical juncture at which cumulative anthropogenic pressures risk driving the Earth system into non-reversible states, challenging the very conditions that facilitated the rise of human civilization (Donges et al., 2017).

Within this framework, the notion of "ecological tipping points" emerges, underscoring the need for systematic recognition of the interconnectedness of various biophysical phenomena beyond isolated assessments. For instance, increased water extraction, a key component of the water planetary boundary, actively contributes to changes in the hydrological cycle, affecting broader ecological interactions (Gleeson et al., 2020). These modifications, representative of the Anthropocene's complex socio-ecological fabric, necessitate thoughtful consideration of human-environment relationships and policies that judiciously navigate these boundaries. Moreover, El-Sayed and Cloutier (2022) advocate a regenerative approach to food systems to recognize these interdependencies, stressing the importance of integrating traditional ecological knowledge with modern scientific understanding. Thus, the conceptualization of the Anthropocene requires an interdisciplinary lens, wherein ecological imperialism is reexamined under the pressing context of global sustainability challenges (Ogwu & Izah, 2025a, 2025b). The recognition of planetary boundaries serves as a critical reference point for evaluating human influence and developing frameworks to mitigate detrimental impacts, underscoring the need for a proactive stance in environmental governance and resource management.

7.7.2 From Imperialism to Extractivism

The transition from traditional imperialism to contemporary extractivism embodies a paradigm shift that reflects ongoing forms of exploitation in the Anthropocene. This shift recognizes how the extraction of natural resources has been foundational to economic practices but has increasingly prioritized short-term gains over long-term ecological considerations. Cianchi et al. (2024) highlight the challenges inherent in employing mechanistic methodologies for environmental assessments, with traditional frameworks often ill-equipped to encompass the systemic complexity of socio-ecological interactions. This limitation complicates efforts to reconcile objectives of economic growth with the imperatives of ecological sustainability and justice.

Moreover, contemporary extractivism often perpetuates colonial legacies, wherein marginalized communities bear disproportionate burdens of resource extraction, facing environmental degradation and socio-economic inequities. Political ecology perspectives emphasize how power dynamics shape the governance of natural resources, arguing that failure to address these imbalances exacerbates vulnerability among disenfranchised populations (Cianchi et al., 2024). This perspective resonates with Thomson and Newman (2020) analysis, which underscores the role of urban governance in mitigating ecological risks within rapidly urbanizing contexts. They contend that cities, empowered by evidence-based policies, have the potential to redefine relationships with the environment, advocating for regenerative urban strategies that counteract extractivist practices.

In facilitating a transition from extractivism to more sustainable paradigms of resource use, it is imperative to employ inclusive and equitable frameworks that integrate local knowledge systems and foster participatory governance models. This approach not only challenges hegemonic narratives of development but also positions local communities as active agents in ecological restoration efforts, promoting resilience and adaptability within the socio-ecological fabric of their environments (Ikhajiagbe et al., 2022).

7.7.3 Political Ecology Perspectives

Political ecology offers a robust analytical lens for unpacking the intersections of power, justice, and environmental governance in an era marked by globalization. Central to this approach is the understanding that environmental issues are inextricably linked to socio-economic structures and dominant political narratives. Haselsteiner et al. (2021) stress that a paradigm shift toward regenerative neighborhoods requires examining societal frameworks holistically rather than isolating variables such as energy efficiency or material use. This reflects an essential critique of conventional sustainability models, which often fail to account for the intricate interdependencies present within socio-ecological systems.

Furthermore, the political ecology framework necessitates an engagement with concepts of justice, particularly in the context of environmental governance (Thomson & Newman, 2020). As Gleeson et al. (2020) articulate, demographic shifts and the acceleration of urbanization challenge traditional governance structures, compelling urban centers to explore innovative solutions that align resource management with principles of equity and sustainability (Gleeson et al., 2020). This perspective aligns with broader discussions on how environmental degradation disproportionately affects marginalized communities, necessitating governance systems that prioritize justice and inclusivity. As practitioners and policymakers navigate the complexities of globalization, fostering environments that promote resilience and equitable resource distribution remains paramount. The need to synthesize insights from political ecology into practical governance frameworks can pave the way for systems-level changes that prioritize both ecological integrity and social justice, ultimately transcending historical patterns of exploitation and fostering a more equitable and sustainable future.

7.7.4 Toward a Post-Imperial Ecology

The pursuit of a post-imperial ecology necessitates a fundamental rethinking of sustainability ethics and the adoption of decolonial frameworks that challenge existing paradigms of knowledge and practice. This shift calls for the integration of diverse worldviews and ontologies, as highlighted by Gibbons (2020), who argues for advancing regenerative sustainability through integrative approaches that harness community knowledge and experiences. By acknowledging the complexities of local ecological contexts and their inherent knowledge systems, regenerative practices can emerge that not only address contemporary environmental challenges but also actively engage with historical injustices rooted in colonial practices.

Moreover, regenerative approaches emphasize the need for a systemic understanding of ecological interactions, moving beyond reductive methodologies that have historically dominated environmental assessments. Pistore et al. (2023) develop a framework for promoting regenerative environments that prioritize holistic relationships between human health and ecosystems. This necessitates fostering designs that not only minimize harm but also actively contribute to the rejuvenation of social-ecological systems, aligning with the broader call for sustainability that transcends mere maintenance of existing conditions.

Additionally, embracing principles of regenerative design promotes an ethical obligation to restore ecological integrity while fostering community empowerment. By adopting frameworks that prioritize co-evolution with natural systems, stakeholders can catalyze shifts toward practices that celebrate biodiversity and environmental resilience. Such efforts are further supported by the idea of a circular economy, as outlined by Khan (2024), which advocates for high human development within ecological limits and frames the transition to sustainability as intrinsically connected to the cultural and social dimensions of local practices. Rethinking ecological imperialism in the Anthropocene entails a thorough reassessment across canonical narratives, underpinned by an emphasis on justice, equity, and regenerative processes (Ogwu & Izah, 2025b). This approach not only acknowledges the historical roots of environmental exploitation but actively engages with diverse voices to shape more sustainable futures, fostering a deeply interwoven relationship between human societies and the broader biosphere.

7.8 Conclusion: Lessons for Sustainability and Equity

The re-examination of ecological imperialism in the context of twenty-first-century land and resource grabs reveals the persistence of unequal power relations in shaping human–environment interactions. What began as a colonial project of territorial conquest and ecological domination has evolved into complex regimes of global capital, transnational governance, and corporate extraction that transcend national boundaries. The expansion of industrial agriculture, mining, and energy infrastructures into rural landscapes underscores how ecological imperialism has adapted to contemporary demands for food, fuel, and raw materials, often deepening vulnerabilities among local communities while eroding environmental resilience. The analysis presented here also illuminates the agency and resilience of affected populations. Across the Global South, grassroots resistance movements, indigenous land rights campaigns, and transnational advocacy networks are challenging the legitimacy of exploitative resource regimes and demanding alternative pathways to development. These responses underscore that ecological imperialism is neither totalizing nor uncontested; instead, it is met with visions of environmental justice, food sovereignty, and climate equity that envision more democratic and sustainable resource futures. Conceptually, this chapter advances the dialogue between environmental history and political ecology by framing land and resource grabs as a continuation (and transformation) of ecological imperialism in the Anthropocene. For policymakers and scholars, the findings signal the urgency of embedding equity, local knowledge systems, and environmental sustainability into global resource governance. Future trajectories should therefore move beyond extractivist paradigms toward regenerative, decolonial, and climate-resilient approaches that honor both human and ecological well-being on a rapidly changing planet.

References

Abajue, M. C., & Ogwu, M. C. (2024). Sustainable alternative land management strategies and their impacts on soil arthropod diversity. In M. C. Ogwu, S. C. Izah, J. Dessureault-Rompré, & D. Gasparatos (Eds.), *Sustainable soil systems in global south*. Springer. https://doi.org/10.1007/978-981-97-5276-8_18

Adam, A. G., & Agegnehu, A. W. (2023). Contract farming as an alternative to large-scale land acquisition and promoting inclusive and responsible agricultural investment: Evidence from Ethiopia. *Corporate Social Responsibility and Environmental Management, 30*(6), 2840–2851.

Al-Fadhat, F., & Prasetio, H. (2024). How China's debt-trap diplomacy works in African countries: Evidence from Zimbabwe, Cameroon, and Djibouti. *Journal of Asian and African Studies, 59*(5), 1463–1483.

Allan, J. I. (2018). Seeking entry: Discursive hooks and NGOs in global climate politics. *Global Policy, 9*(4), 560–569.

Altiparmak, S. O. (2023). China and lithium geopolitics in a changing global market. *Chinese Political Science Review, 8*(3), 487–506.

Anderson, C. R., Bruil, J., Chappell, M. J., Kiss, C., & Pimbert, M. P. (2019). From transition to domains of transformation: Getting to sustainable and just food systems through agroecology. *Sustainability, 11*(19), 5272.

Atia, M., & Herrold, C. E. (2018). Governing through patronage: The rise of NGOs and the fall of civil society in Palestine and Morocco. *VOLUNTAS: International Journal of Voluntary and Nonprofit Organizations, 29*(5), 1044–1054.

Azhar, B., Lindenmayer, D. B., Wood, J., Fischer, J., Manning, A., McElhinny, C., & Zakaria, M. (2011). The conservation value of oil palm plantation estates, smallholdings and logged peat swamp forest for birds. *Forest Ecology and Management, 262*(12), 2306–2315.

Barbier, E. B., López, R. E., & Hochard, J. P. (2016). Debt, poverty and resource management in a rural smallholder economy. *Environmental and Resource Economics, 63*(2), 411–427.

Besley, T., & Persson, T. (2011). The logic of political violence. *The Quarterly Journal of Economics, 126*(3), 1411–1445.

Beyene, T. (2022). Current land tenure and households' preferences to ownership of farmland in South East Ethiopia. *Journal of African Development Studies, 9*(1), 41–55.

Bhandari, S., Yadav, B., Yadav, P. K., Lahutiya, V., & Koirala, S. (2024). Exploring agroecological approaches for sustainable agriculture and rural development: A comprehensive. *Asian Journal of Research in Agriculture and Forestry, 10*(1), 79–90.

Borras, S. M., Jr., & Franco, J. C. (2012). Global land grabbing and trajectories of agrarian change: A preliminary analysis. *Journal of Agrarian Change, 12*(1), 34–59.

Brito, L., & Rondon, G. (2024). Reproductive justice lessons for catalyzing advocacy strategies during Zika Crisis in Brazil. *Physis: Revista de Saúde Coletiva, 34*, e34SP107.
Caldara, D., & Iacoviello, M. (2022). Measuring geopolitical risk. *American Economic Review, 112*(4), 1194–1225.
Carrington, J. M. (2020). Rethinking nongovernmental organizations: Neoliberalism, "nonstate" actors, and the politics of recognition in the United States. *Law & Policy, 42*(4), 344–364.
Carswell, G., De Neve, G., & Ponnarasu, S. (2021). Good debts, bad debts: Microcredit and managing debt in rural south India. *Journal of Agrarian Change, 21*(1), 122–142.
Charles, S. T. (2010). Strength and vulnerability integration: A model of emotional well-being across adulthood. *Psychological Bulletin, 136*(6), 1068–1091. https://doi.org/10.1037/a0021232
Choir, M. (2022). Rights to free from corruption: A vernacular mobilization of anti-corruption movement in Indonesia. *Simulacra, 5*(2), 113–128.
Choudry, A. (2013). Struggles against bilateral FTAs: Challenges for transnational global justice activism. *Studies in Social Justice, 7*(1), 7–25.
Cianchi, B., Everard, M., Gething, B., Cooke, R., & Ginepro, M. (2024). The efficacy of biodiversity and ecosystem assessment approaches for informing a regenerative approach to built development. *Integrated Environmental Assessment and Management, 20*(1), 248–262.
Cohn, A. S., Newton, P., Gil, J. D., Kuhl, L., Samberg, L., Ricciardi, V., Manly, J. R., & Northrop, S. (2017). Smallholder agriculture and climate change. *Annual Review of Environment and Resources, 42*(1), 347–375.
Costello, C., Cao, L., Gelcich, S., Cisneros-Mata, M. Á., Free, C. M., Froehlich, H. E., Golden, C. D., Ishimura, G., Maier, J., Macadam-Somer, I., & Mangin, T. (2020). The future of food from the sea. *Nature, 588*(7836), 95–100.
Desalegn, T., Cruz, F., Kindu, M., Turrión, M. B., & Gonzalo, J. (2014). Land-use/land-cover (LULC) change and socioeconomic conditions of local community in the central highlands of Ethiopia. *International Journal of Sustainable Development & World Ecology, 21*(5), 406–413.
Donges, J. F., Winkelmann, R., Lucht, W., Cornell, S. E., Dyke, J. G., Rockström, J., Heitzig, J., & Schellnhuber, H. J. (2017). Closing the loop: Reconnecting human dynamics to earth system science. *The Anthropocene Review, 4*(2), 151–157.
El-Sayed, S., & Cloutier, S. (2022). Weaving disciplines to conceptualize a regenerative food system. *Journal of Agriculture, Food Systems, and Community Development, 11*(2), 23–51.
Falcon, W. P., Naylor, R. L., & Shankar, N. D. (2022). Rethinking global food demand for 2050. *Population and Development Review, 48*(4), 921–957.
Frame, M. L. (2022). Ecological imperialism: A world-systems approach. *American Journal of Economics and Sociology, 81*(3), 503–534.
Gazzano, I., Achkar, M., & Díaz, I. (2019). Agricultural transformations in the southern cone of Latin America: agricultural intensification and decrease of the aboveground net primary production, Uruguay's case. *Sustainability, 11*(24), 7011.
Gemeda, B. S., Abebe, B. G., & Cirella, G. T. (2019). The role of land speculators around the urban edge of Shashemene City, Ethiopia. *European Journal of Sustainable Development Research, 4*(1), em0108.
Gibbons, L. V. (2020). Regenerative—The new sustainable? *Sustainability, 12*(13), 5483.
Gleeson, T., Wang-Erlandsson, L., Porkka, M., Zipper, S. C., Jaramillo, F., Gerten, D., Fetzer, I., Cornell, S. E., Piemontese, L., Gordon, L. J., & Rockström, J. (2020). Illuminating water cycle modifications and Earth system resilience in the Anthropocene. *Water Resources Research, 56*(4), p.e2019WR024957.
Gottlieb, R., & Joshi, A. (2010). *Food justice*. Mit Press.
Hansen, T., & Umbreit, M. (2018). State of knowledge: Four decades of victim-offender mediation research and practice: The evidence. *Conflict Resolution Quarterly, 36*(2), 99–113.
Haselsteiner, E., Rizvanolli, B. V., Villoria Sáez, P., & Kontovourkis, O. (2021). Drivers and barriers leading to a successful paradigm shift toward regenerative neighborhoods. *Sustainability, 13*(9), 5179.

Humphrey, M. (2012). Victims, civil society and transitional justice in Bosnia and Herzegovina. *Temida, 15*(1), 59–75.

Husnain, M. I. U., Syed, Q. R., Bashir, A., & Khan, M. A. (2022). Do geopolitical risk and energy consumption contribute to environmental degradation? Evidence from E7 countries. *Environmental Science and Pollution Research, 29*(27), 41640–41652.

Ibekwe, K. I., Etukudoh, E. A., Nwokediegwu, Z. Q. S., Umoh, A. A., Adefemi, A., & Ilojianya, V. I. (2024). Energy security in the global context: A comprehensive review of geopolitical dynamics and policies. *Engineering Science & Technology Journal, 5*(1), 152–168.

Ikhajiagbe, B., Chukwujama, C. A., & Ogwu, M. C. (2022). Intrinsic restoration of a ferruginous utisol using goosegrass (*Eleusine indica* L. [Gaertn.]) obtained from different land use areas. *Journal of Natural Resources and Development, 12*, 1–18. https://doi.org/10.18716/ojs/jnrd2022.12.01

Iyiola, A. O., Afolabi, O. A., Alimi, S. K., Akingba, O. O., Izah, S. C., & Ogwu, M. C. (2024). Climate change and water crisis in the Global South. In S. C. Izah, M. C. Ogwu, A. Loukas, & H. Hamidifar (Eds.), *Water crises and sustainable management in the Global South* (pp. 111–140). Springer.

Izah, S. C., Iyiola, A. O., Yarkwan, B., & Richard, G. (2023). Impact of air quality as a component of climate change on biodiversity based ecosystem services. In A. L. Srivastav, A. K. Dubey, A. Kumar, S. K. Narang, & M. A. Khan (Eds.), *Visualization techniques for climate change with machine learning and artificial intelligence* (pp. 123–148). Published by Elsevier.

Izah, S. C., & Ogwu, M. C. (2025). *Innovative approaches in environmental health management: Processes, technologies, and strategies for a sustainable future*. Springer Nature. https://doi.org/10.1007/978-3-031-81966-7

Izah, S. C., Ogwu, M. C., & Akram, M. (2025). *Bioactive compounds in edible mushrooms: Sustainability and health applications*. Springer. https://doi.org/10.1007/978-3-031-52642-8

Izah, S. C., Ogwu, M. C., & Hait, M. (2025a). Clean energy solutions and public health: Innovations in renewable energy for a healthier environment. In S. C. Izah & M. C. Ogwu (Eds.), *Innovative approaches in environmental health management: Processes, technologies, and strategies for a sustainable future*. Environmental Science and Engineering Series. Springer Nature. https://doi.org/10.1007/978-3-031-81966-7_7

Izah, S. C., Ogwu, M. C., & Hait, M. (2025b). Waste management and health: Addressing the processes behind hazardous waste and pollution. In S. C. Izah & M. C. Ogwu (Eds.), *Innovative approaches in environmental health management: Processes, technologies, and strategies for a sustainable future*. Environmental Science and Engineering Series. Springer Nature. https://doi.org/10.1007/978-3-031-81966-7_6

Kandikuppa, S. (2022). Class and vulnerability to debt in rural India: A \ statistical overview. *Rural Sociology, 87*(2), 454–488.

Keddie, A. (2021). NGOs working for gender justice with boys and men: Exploring challenges of accountability. *Gender, Work and Organization, 28*(4), 1461–1474.

Khan, T. (2024). Circular-ESG Model for Regenerative Transition. *Sustainability, 16*, 7549. https://doi.org/10.3390/su16177549

Kuyper, J. W., Linnér, B. O., & Schroeder, H. (2018). Non-state actors in hybrid global climate governance: Justice, legitimacy, and effectiveness in a post-Paris era. *Wiley Interdisciplinary Reviews: Climate Change, 9*(1), e497.

Lade, S., Steffen, W., Vries, W., Carpenter, S., Donges, J., Gerten, D., et al. (2019). Human impacts on planetary boundaries are amplified by Earth system interactions. *Nature Sustainability, 3*(2), 119–128. https://doi.org/10.1038/s41893-019-0454-4

Liu, Y., Yu, L., Yang, C., & Li, Z. (2021). Heterogeneity of the impact of geopolitical events on energy trade: An empirical study based on regression discontinuity design. *Frontiers in Environmental Science, 9*, 722910.

Loehr, D. (2012). Land reforms and the tragedy of the anticommons—A case study from Cambodia. *Sustainability, 4*(4), 773–793.

Manzilati, A., & Prestianawati, S. A. (2022). Informal financing or debt traps: are the UN sustainable development goals being met in emerging economies? *Review of International Business and Strategy, 32*(1), 132–145.

Margulis, M. E., McKeon, N., & Borras, S. M., Jr. (2013). Land grabbing and global governance: critical perspectives. *Globalizations, 10*(1), 1–23.

Mayaputri, A. P., Paramitha, G., & Isnaeni, N. (2024). The Island's call for justice: NGO advocacy in the climate litigation case of Pari Island in Indonesia. *Technium Social Sciences Journal, 56*, 282.

McGaughey, F., Rafferty, R., & Maguire, A. (2023). Transitional justice from above and below: exploring the potential glocalising role of non-governmental organisations through a Northern Ireland case study. *Northern Ireland Legal Quarterly, 74*, 472.

Meissner, K. (2024). Caught up by geopolitics: sanctions and the EU's response to Russia's war against Ukraine. *The Economists' Voice, 20*(2), 275–283.

Mosse, D., & Nagappan, S. B. (2021). NGOs as social movements: Policy narratives, networks and the performance of Dalit rights in South India. *Development and Change, 52*(1), 134–167.

Muawanah, U. (2015). Does land conflict matter to farm productivity? A case study of Cambodia. *Asian Journal of Agriculture and Development, 15*(2), 77–92.

Obeng-Odoom, F. (2022). Mainstream economics and conventional environmental policies. *American Journal of Economics and Sociology, 81*(3), 443–472.

Ogwu, M. C., Iyiola, A. O., Fawole, W. O., & Izah, S. C. (2025). Biobased resource recovery techniques in the Global South. In M. C. Ogwu, S. C. Izah, J. G. Vazquez-Arenas, S. T. Feleke, & X. Wei (Eds.), *Sustainable bioeconomy development in the global south.* Springer. https://doi.org/10.1007/978-981-96-0305-3_17

Ogwu, M. C., & Izah, S. C. (2024). *Sustainable strategies for air pollution mitigation: Development, economics, and technologies.* Springer. https://doi.org/10.1007/978-3-031-74157-9

Ogwu, M. C., & Izah, S. C. (2025a). *Edible insects: Nutritional benefits, culinary innovations and sustainability.* Springer Nature. https://doi.org/10.1007/978-3-031-90087-7

Ogwu, M. C., & Izah, S. C. (2025b). *Evaluating environmental processes and technologies.* Springer Nature. https://doi.org/10.1007/978-3-031-85327-2

Ogwu, M. C., Izah, S. C., Merritt, H., Wise, R. D., & Jintrawet, A. (2025). *Sustainable bioeconomy development in the Global South: Volume I status and perspectives.* Springer. https://doi.org/10.1007/978-981-96-0640-5

Ogwu, M. C., Izah, S. C., Vazquez-Arenas, J. G., Feleke, S. T., & Wei, X. (2025). *Sustainable bioeconomy development in the Global South: Volume II bioeconomy techniques.* Springer. https://doi.org/10.1007/978-981-96-0305-3

Ogwu, M. C., Ojo, A. O., & Alaka, A. C. (2025). Biodiversity and human health: The interconnections of species loss and ecosystem services. In S. C. Izah & M. C. Ogwu (Eds.), *Innovative approaches in environmental health management: Processes, technologies, and strategies for a sustainable future.* Environmental Science and Engineering Series. Springer Nature. https://doi.org/10.1007/978-3-031-81966-7_5

Olasehinde-Williams, G., Olanipekun, I., & Usman, O. (2024). Does geopolitics trigger energy inflation in the European economic area? Evidence from a panel time-varying regression. *International Journal of Energy Sector Management, 18*(5), 941–955.

Panday, P. K., & Rabbani, G. (2017). Do legal empowerment activities of NGOs reduce gender-based violence in Bangladesh? *International Journal of Gender Studies in Developing Societies, 2*(1), 1–18.

Perger, C., McCallum, I., Albrecht, F., See, L., Fritz, S., & Thau, D. (2013). *Improved cropland mapping in Ethiopia.* GI Forum.

Petrilla, C. M., Silva, D. J., Huggins, S. L., McNamara, R., & Sylk, G. (2020). Healing lives and communities: How can NGOs support restorative justice involving youth. *Journal of Applied Juvenile Justice Services*, 1–27.

Pistore, L., Konstantinou, T., Pasut, W., & Naboni, E. (2023). A framework to support the design of a regenerative indoor environment. *Frontiers in Built Environment, 9*, 1225024.

Pistorius, T., Carodenuto, S., & Wathum, G. (2017). Implementing forest landscape restoration in Ethiopia. *Forests, 8*(3), 61.

Quan, N., Lee, S. C., Chopra, C., Nesic, Z., Porto, P., Pow, P., Jassal, R. S., Smukler, S., Krzic, M., Knox, S. H., & Black, T. A. (2023). Estimating net carbon and greenhouse gas balances of potato and pea crops on a conventional farm in Western Canada. *Journal of Geophysical Research: Biogeosciences, 128*(3), e2022JG007113.

Rabitz, F. (2023). Space resources and the politics of international regime formation. *International Journal of the Commons, 17*(1), 243–255.

Richardson, K., Steffen, W., Lucht, W., Bendtsen, J., Cornell, S. E., Donges, J. F., Drüke, M., Fetzer, I., Bala, G., Von Bloh, W., & Feulner, G. (2023). Earth beyond six of nine planetary boundaries. *Science Advances, 9*(37), eadh2458.

Rulli, M. C., Saviori, A., & D'Odorico, P. (2013). Global land and water grabbing. *Proceedings of the National Academy of Sciences, 110*(3), 892–897.

Salti, N. (2015). Income inequality and the composition of public debt. *Journal of Economic Studies, 42*(5), 821–837.

Schlosberg, D., & Collins, L. B. (2014). From environmental to climate justice: climate change and the discourse of environmental justice. *Wiley Interdisciplinary Reviews: Climate Change, 5*(3), 359–374.

Schoneveld, G. C., German, L. A., & Nutakor, E. (2011). Land-based investments for rural development? A grounded analysis of the local impacts of biofuel feedstock plantations in Ghana. *Ecology and Society, 16*(4), 160410.

Shrimali, R. (2016). Accumulation by dispossession or accumulation without dispossession: The case of contract farming in India. *Human Geography, 9*(3), 77–88.

Sitthikriengkrai, M., & Porath, N. (2017). 'Lead-polluted water changed our lives' a Thai-Karen village's quest for environmental justice. *South East Asia Research, 25*(2), 139–156.

Teklemariam, D., Azadi, H., Nyssen, J., Lanckriet, S., Asfaha, T. G., Haile, M., Taheri, F., & Witlox, F. (2017). Dynamics of the North–South capital flows or rise of South–South land deals? features of land acquisition in Ethiopia. *Land Degradation & Development, 28*(8), 2389–2407.

Thomson, G., & Newman, P. (2020). Cities and the Anthropocene: Urban governance for the new era of regenerative cities. *Urban Studies, 57*(7), 1502–1519.

Tilman, D., Fargione, J., Wolff, B., D'antonio, C., Dobson, A., Howarth, R., Schindler, D., Schlesinger, W. H., Simberloff, D., & Swackhamer, D. (2001). Forecasting agriculturally driven global environmental change. *Science, 292*(5515), 281–284.

Van Huijstee, M., Pollock, L., Glasbergen, P., & Leroy, P. (2011). Challenges for NGOs partnering with corporations: WWF Netherlands and the environmental defense fund. *Environmental Values, 20*(1), 43–74.

Wang, R. (2023). The impact of geopolitical stability on foreign investment and transnational corporations: a case study of the Russia-Ukraine conflict. *Advances in Economics Management and Political Sciences, 63*(1), 141–147.

Wedekind, J., & Chinigò, D. (2020). Contract and control: Agrarian labour mobilisation and resistance under large-scale land investments for biofuel crop production in Ethiopia. *Annales d'Éthiopie, 33*, 49–78.

Wondimu, T. & Gebresenbet, F. (2018). Resourcing land, dynamics of exclusion and conflict in the Maji area, Ethiopia. *Conflict Security and Development, 18*(6), 547–570. https://doi.org/10.1080/14678802.2018.1532644

Yang, Z., Yuan, X., Hu, J., Liu, D., & Tang, W. (2024). The impact of the proportion of renewable energy consumption on geopolitical risks in the United States and the United Kingdom. *Energy Exploration & Exploitation, 42*(6), 1958–1986.

Yu, M., & Wang, N. (2023). The influence of geopolitical risk on international direct investment and its countermeasures. *Sustainability, 15*(3), 2522.

Urban Metabolism and the Global South: Climate Change and Infrastructural Fragility

8

8.1 Introduction

The intersection of urban metabolism and climate change resilience represents a critical area of inquiry, particularly within the Global South, where the challenges of rapid urbanization compound vulnerabilities to climate-related hazards. Urban areas in these regions are increasingly recognized as potential hotspots for the impacts of climate change, warranting an integrated understanding of how infrastructural fragility affects urban longevity and quality of life. Factors such as socio-economic disparities, governance challenges, and limited financial resources frequently intersect, exacerbating the consequences of climate change in these urban environments.

8.1.1 Background and Rationale

Climate change poses a multitude of challenges for cities in the Global South, characterized by significant socio-economic inequities and underdeveloped infrastructure (Iyiola et al., 2024; Izah et al., 2023). As noted by the Intergovernmental Panel on Climate Change (IPCC), the impacts of climate change in these countries are particularly acute, often leading to increased urban vulnerability as cities expand and populations surge (Wilcocks & Laubscher, 2021). Urban growth in the Global South is predicted to facilitate the migration of 2.5 billion people into cities by 2050, adding to already stretched urban infrastructure that is ill-equipped to handle such demographic surges (Miklian & Hoelscher, 2020). Infrastructural fragility results from inadequate planning and investment, leading to a vicious cycle where climate impacts further expose these deficiencies, perpetuating a vulnerable urban existence.

M. C. Ogwu and S. Chibueze Izah, *Climate Frontiers and Social Transformation*, World-Systems Evolution and Global Futures,
https://doi.org/10.1007/978-3-032-17639-4_8

The increasing frequency and severity of climate-related disasters underscore the urgent need for resilience strategies, highlighting the connections between climate change and urban energy systems, waste management, and social services (Bigger & Webber, 2021). Thus, this chapter aims to articulate the urgent need to reconcile urban metabolism with climate resilience through a robust understanding of the interdependencies between climate and the urban world system (Bulkeley & Castán Broto, 2013). By emphasizing the unique socio-political and economic challenges of the Global South, this chapter offers critical commentary on how localized responses to climate change can shape the future of urban sustainability.

8.1.2 The Climate–World-System Nexus in the Global South

The climate–world-system nexus highlights the complexities of governance and adaptation strategies, particularly as they apply to cities in the Global South. The governance landscape in these urban spaces varies considerably, influenced by actors such as governmental bodies, non-governmental organizations, and community groups (O'Donoghue et al., 2022). These diverse stakeholders are increasingly under pressure to mobilize resources and implement strategies to address the onset of climate crises. However, the disparity in capacity between the Global North and the South complicates these efforts. For instance, while the Global North has historically accounted for the majority of emissions, the resulting climatic impacts are predominantly felt in the South, raising questions of equity and justice in global climate governance (Malherbe & Oladejo, 2025).

Moreover, as urban environments grapple with the immediacy of climate threats, the role of planning becomes central. The concept of "urban resilience" emphasizes the need to endure, adapt, and thrive amid climate change-induced challenges. However, such resilience cannot solely depend on technocratic solutions; it requires a nuanced understanding of localized vulnerabilities and adaptive capacities (Sareen & Waagsaether, 2023; Ogwu et al., 2025b, 2025f). Efforts should extend beyond infrastructure upgrades to involve community engagement and resource allocation, addressing the fundamental social injustices that compound climatic impacts.

This chapter seeks to elucidate the pressing agenda of climate change adaptation within the Global South, detailing the climatic vulnerabilities that afflict urban populations. This chapter critically examines the interconnectedness of urban metabolism, climate change, and infrastructural fragility in the Global South, emphasizing how cities function as vital nodes in the climate–world-system nexus. By applying the urban metabolism framework, this chapter analyzes energy and material flows within urban environments to reveal the underlying vulnerabilities driving urban heat islands, water scarcity, informal settlements, and infrastructure breakdowns under climate stress. Beyond diagnosing challenges, this chapter contributes to knowledge by integrating systems thinking with climate adaptation discourse, offering new insights into how urban metabolism can inform resilience planning, circular economy transitions, and sustainable infrastructure design. Through case studies, theoretical synthesis, and policy-oriented analysis, it positions cities in the Global South not only as sites of vulnerability but also as laboratories for transformative climate solutions, providing a conceptual and practical bridge between urban sustainability scholarship and climate governance imperatives.

8.2 Theoretical Foundations of Urban Metabolism

8.2.1 Origins and Evolution of Urban Metabolism as a Concept

The concept of urban metabolism has its roots in ecological and biological modeling, deriving its framework from analogies drawn between cities and living organisms. This comparison allows cities to be understood through their resource and energy consumption patterns—much like biological systems that rely on inputs and outputs for sustenance. Early models of urban metabolism, as illustrated by Abel Wolman in his 1965 article, "*The Metabolism of Cities*," published in *Scientific American*, aimed to quantify the energy and material flows into hypothetical urban environments, particularly focusing on large cities' dependencies on their surrounding ecosystems for materials and energy (Fig. 8.1). Figure 8.1 illustrates the cyclical flows of energy, water, resources, and emissions within urban systems that represent atypical urban metabolism, which highlight how cities function as dynamic hubs of consumption and transformation. It emphasizes the interconnected processes driving environmental impacts and the need for sustainable urban planning. Over the past few decades, urban metabolism research has evolved significantly. Initial studies emphasized linear models of resource flows, which have now transitioned to more dynamic, cyclical, and network models that reflect the complex interactions within urban systems (Bai, 2016; Zhang et al., 2015).

The field has expanded from simple energy input–output analyses to incorporate a multidisciplinary approach that includes insights from economics, sociology, and environmental science. This evolution traces back over half a century, highlighting the growing complexity and interdependence of urban areas (Galychyn et al., 2020; Wiesmann et al., 2011). Notably, researchers have identified the need for improved methodologies to assess urban metabolism, emphasizing the development of spatially resolved models and material flow analyses to understand urban ecosystems and their sustainability better (Liu et al., 2022a, 2022b; Wiesmann et al., 2011).

8.2.2 Linking Urban Metabolism to Climate Change Research

The relationship between urban metabolism and climate change research is increasingly vital for addressing the sustainability challenges cities face. Urban areas are significant contributors to greenhouse gas emissions due to their high energy consumption and resource use. Understanding urban metabolism thus

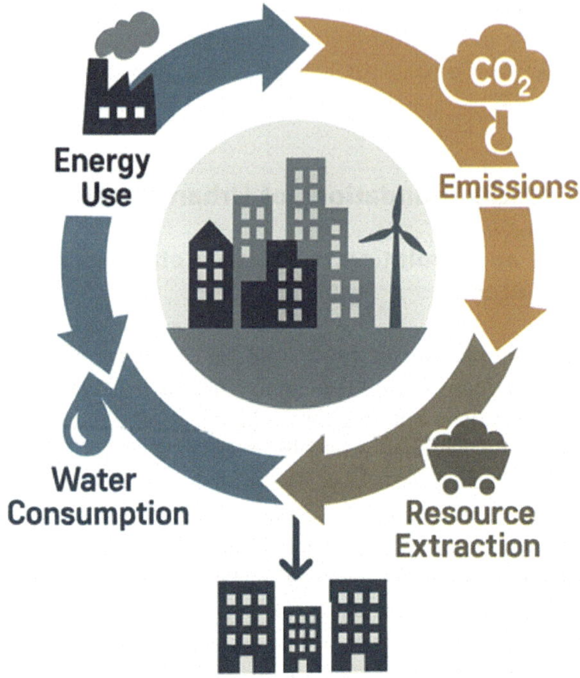

Fig. 8.1 Illustration of urban metabolism: resource flows and environmental interactions

becomes essential for developing strategies that mitigate climate change impacts and enhance urban resilience (Ziebell & Singh, 2018). Recent research emphasizes the role of urban metabolism in assessing the sustainability of cities and guiding policy decisions toward climate-responsive urban planning (Pardo Martínez et al., 2021).

The exploration of energy and material flows within urban systems illuminates patterns that directly correlate with climate resilience. For instance, urban heat islands (UHIs) significantly affect energy use and demand, which, in turn, influence greenhouse gas emissions. Understanding these interactions underscores the need to incorporate climate adaptation strategies into urban metabolism studies, such as integrating urban green infrastructure to mitigate heat impacts while enhancing ecological quality (Borah & Sarma, 2025). Moreover, cities that effectively manage their metabolic processes can unlock potential for a circular economy, where waste is minimized and materials are reused, thus supporting broader climate goals (Perrotti & Stremke, 2020).

8.2.3 Energy and Material Flows in Urban Systems

Energy and material flows represent the lifeblood of urban systems, facilitating the functioning and growth of cities. Different urban ecosystems exhibit distinct characteristics shaped by their energy and material practices, often reflecting their socio-economic context. Recent investigations into the energy profiles of urban areas reveal stark contrasts in efficiency and sustainability outcomes across cities (Song et al., 2016). Urban metabolism is now viewed as a critical framework for understanding how cities can optimize resource use while minimizing environmental impacts, reinforcing the interplay between material consumption, waste production, and economic growth (Ziebell & Singh, 2018).

Furthermore, cities are increasingly recognized as complex networks where energy and material flow dynamics influence urban form and infrastructure (Bai, 2016). Advanced studies employ Geographic Information Systems (GIS) to provide a spatial dimension to material flow analysis, thereby capturing the intricate patterns of interactions among various urban systems (Liu et al., 2022a, 2022b). These methodologies reveal how urban metabolism varies spatially and how interconnected flows can support targeted interventions that enhance sustainability and ecological health in urban areas (Zhen et al., 2014).

8.2.4 System Dynamics and Feedback Loops in Cities

System dynamics and feedback loops are integral to understanding urban metabolism, as they elucidate how different components of the urban system interact and influence one another. Cities function as complex adaptive systems, in which changes in one aspect of urban metabolism can cascade throughout the entire system (Lu & Chen, 2014). Feedback mechanisms—both positive and negative—can amplify or mitigate impacts of urban processes, particularly in response to external pressures such as rapid population growth or environmental changes. For instance, an increase in urban density might enhance energy efficiency through shared infrastructure. Yet, it may also exacerbate pressure on local resources, leading to increased waste production and environmental degradation (Patrakeyev et al., 2017). By analyzing these dynamics, urban planners and policymakers can better anticipate unintended consequences of urban development strategies and design interventions that promote resilience and sustainability.

The cyclical nature of urban metabolism underscores the need for cities to adapt to changes in their infrastructure and resource flows continuously. This adaptability is crucial for fostering long-term sustainability and managing the environmental footprints associated with urban expansion (Moreau & Massard, 2017). Urban metabolism thus presents a valuable lens for examining how cities can evolve sustainably while addressing the pressing challenges posed by climate change and resource scarcity.

8.2.5 Comparative Perspectives: Global North vs. Global South

The comparative study of urban metabolism between global northern and southern contexts reveals significant disparities that merit further exploration. Urbanization patterns in the Global South are often driven by rapid population growth and economic transitions, leading to informal settlements and inadequate infrastructure that create distinct metabolic challenges (Duman & Dinçer, 2024). Research suggests that while cities in the Global North may focus on the efficiency of energy and material flows, those in the Global South are frequently grappling with developmental issues and infrastructural deficits that exacerbate environmental vulnerabilities (Li & Yue, 2012).

Scholarly discourse increasingly recognizes the need to contextualize urban metabolism within specific socio-economic frameworks, highlighting how conditions in the Global South necessitate tailored approaches to metabolism studies. This includes recognizing the role of informal economies and undervalued resources in shaping urban metabolic patterns (Duman & Dinçer, 2024). Furthermore, the interplay between urbanization and economic growth in these regions underscores the complexities of achieving sustainability amidst ongoing inequalities and varying governance structures (Lemanski, 2019). Urban metabolism provides a robust framework for assessing these differences and guiding research initiatives to foster inclusive, sustainable growth paths across diverse urban landscapes. By integrating lessons from both the Global North and South, future research can contribute to a more nuanced understanding of urban metabolism as a global phenomenon while emphasizing the need for local solutions that respect cultural and environmental contexts (Bi & Wang, 2020).

8.3 Urban Heat Islands and Thermal Vulnerabilities

8.3.1 Defining Urban Heat Islands and Their Characteristics

Urban Heat Islands (UHIs) are areas that experience significantly higher temperatures than their surrounding rural areas, especially at night. This phenomenon is predominantly attributed to anthropogenic influences, as urbanization alters land use, replacing natural landscapes with impervious surfaces like buildings and roads that absorb and retain heat. The urbanization process exacerbates local climate by altering surface materials, vegetation cover, and increased human activities that contribute to heat generation and retention (Keeratikasikorn & Bonafoni, 2018; Ye et al., 2012). For instance, Liu and Zhang (2011) noted a direct relationship between increased urbanization and greater temperature disparities, with urban areas exhibiting notably warmer conditions than adjacent rural areas. Impervious surfaces, waste heat, and reduced vegetation in urban areas lead to the formation of urban heat islands (Fig. 8.2).

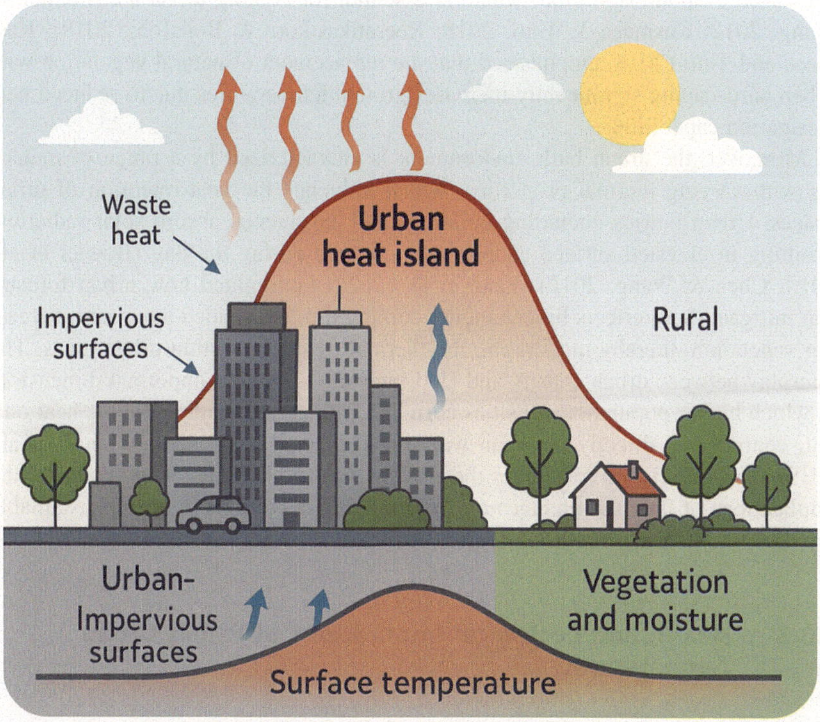

Fig. 8.2 Temperature dynamics in urban vs. rural landscapes

The characteristics of UHIs can vary based on multiple factors, including geographic location, time of year, and the surface materials present in the urban environment. Urban areas typically feature more heat-absorbing infrastructure, such as asphalt and concrete, which have lower albedo than vegetated regions. Moreover, the intensity and spatial distribution of UHIs can vary widely across cities due to differences in urban design, population density, and local climate (Huang et al., 2024; Ye et al., 2012). According to a systematic review by Stewart (2011), these spatial characteristics necessitate precise methodological approaches in UHI studies to identify and quantify their presence and intensity accurately.

8.3.2 Land-Use Change, Built Environments, and Thermal Dynamics

Land-use change plays a critical role in the development and intensity of UHIs. As cities expand, the transformation of natural landscapes into urban environments creates a variety of thermal dynamics that result in higher ambient temperatures. The conversion of green spaces and pervious surfaces into built structures leads to decreased evapotranspiration, which is essential for cooling urban areas (Chen & Wang, 2012; Kasmaee & Tinti, 2018; Keeratikasikorn & Bonafoni, 2018). Kasmaee and Tinti (2018) emphasized that the replacement of natural vegetation with urban hardscaping significantly increases ground temperatures due to reduced heat dissipation capabilities.

Moreover, the urban built environment is characterized by a range of materials with varying thermal properties, which influence the heat retention of urban spaces. Urban fabrics, including buildings and pavements, absorb solar radiation, resulting in elevated surface and air temperatures during the day (Bowler et al., 2010; Chen & Wang, 2012). Yang et al. (2022) highlighted how urban forestry can mitigate these effects by enhancing cooling through shaded areas and increasing vegetation, thereby moderating the thermal dynamics within urban areas. The interplay between urban density and UHI intensity is another important dimension, in which higher population densities correlate with greater anthropogenic heat output, contributing directly to urban warming (Maheng et al., 2019; Manoli et al., 2019). This correlation reinforces the idea that urban planning should consider the implications of land-use changes on local thermal conditions to create sustainable urban environments (Abajue & Ogwu, 2024).

8.3.3 Health and Ecological Implications of Rising Urban Temperatures

The health and ecological implications of rising urban temperatures due to the UHI effect are profound. Elevated temperatures can exacerbate heat-related health issues, with urban populations facing increased risks of heat stress, cardiovascular diseases, and exacerbation of chronic illnesses (Maheng et al., 2019; Stone Jr et al., 2014; Ye et al., 2012). The phenomenon is particularly severe in vulnerable populations, including the elderly, children, and those with pre-existing health conditions, highlighting the urgent need for public health interventions and urban design strategies that prioritize heat resilience (Manoli et al., 2019; Stone Jr et al., 2014). Ecologically, higher temperatures can alter urban biodiversity, affecting species composition and distribution. As urban areas become increasingly inhospitable due to extreme heat, many native plant and animal species face increased stress and potential decline (Huang et al., 2024; Jiang & Yang, 2024). The thermal stress experienced by ecosystems can disrupt biological processes such as reproduction and growth, reducing resilience to other environmental stressors, including pollution and habitat destruction.

Furthermore, as found by Mittermüller et al. (2021), the combined impacts of climate change and UHI may exacerbate urban heat stress, underscoring the need for effective mitigation strategies to improve the overall urban climate. The ecological consequences underscore the interconnected nature of urban heat dynamics and public health, necessitating comprehensive approaches that integrate ecological preservation with human health considerations.

8.3.4 UHI Impacts in Major Global South Cities

Several case studies illustrate the severe impacts of UHI effects in major cities of the Global South, where rapid urbanization is often unchecked. For instance, in Colombo, Sri Lanka, increasing urban green space has been shown to influence UHI patterns significantly, demonstrating a strong relationship between vegetation cover and ambient temperature reduction (Maheng et al., 2019). Additionally, research conducted in contemporary and historical metropolitan areas across China has revealed that cities classified as "stove cities" exhibit more severe UHI effects, leading to significant health risks for urban dwellers (Huang et al., 2024).

Similarly, in Delhi NCR, studies indicate that rapid urban sprawl has contributed to escalating UHI effects, with cities experiencing a rise in temperatures that far exceeds surrounding rural areas (Kushwaha & Nithiyanandam, 2019). These findings are corroborated by Narvaez et al. (2024), emphasizing that urbanization significantly enhances localized warming and alters atmospheric conditions, making urban centers more susceptible to extreme weather events. Analyzing the UHI effects in Bangkok has uncovered critical insights into how different land-use practices contribute to temperature variations, underscoring the necessity for integrated urban planning that considers land-use allocation as a means to mitigate these effects (Keeratikasikorn & Bonafoni, 2018). Collectively, these case studies serve to reinforce the pressing need for targeted interventions that address the unique challenges posed by urbanization in the Global South.

8.3.5 Strategies for Mitigation

Effectively mitigating the effects of urban heat islands requires a multifaceted approach, with green infrastructure and cool roofs emerging as vital strategies (Fig. 8.3). Green infrastructure involves a range of practices, including the creation of urban green spaces, parks, and green roofs, all of which foster cooling through increased vegetation and improved air quality (Bruns & Simko, 2017; Koc et al., 2018; Mittermüller et al., 2021; Ogwu & Izah, 2024). Such strategies not only help reduce ambient temperatures but also offer ecological benefits, such as enhancing biodiversity and improving stormwater management.

Research conducted by Koc et al. (2018) supports the notion that urban greening not only cools cities through shading and evapotranspiration but also promotes mental well-being and enhances the aesthetic value of urban environments, creating more livable cities. Moreover, a systematic review of methods used to evaluate green infrastructure highlights the need for robust assessment frameworks to gauge its effectiveness in mitigating UHI (Mittermüller et al., 2021).

Cool roofs represent another effective strategy, utilizing reflective materials to reduce surface temperatures and improve energy efficiency in buildings (Izah et al., 2025a, 2025b; Mittermüller et al., 2021). These roofs reflect more sunlight and absorb less heat, thereby mitigating the overall temperature in urban areas. In conjunction with urban greening, they can play a crucial role in managing heat

Fig. 8.3 Nature-based and technological approaches for mitigating urban heat islands

extremes, particularly in densely populated areas where heat retention is more pronounced (Jiang & Yang, 2024; Wang & Yang, 2024). Mitigation efforts should also be complemented by policy frameworks that encourage sustainable urban development and promote practices that prioritize ecological health and community resilience in the face of climate challenges. By integrating green infrastructure with innovative building solutions, cities can effectively combat the UHI effect, thus safeguarding public health and preserving ecological integrity.

8.4 Informal Settlements and Infrastructural Fragility

The phenomenon of informal settlements is a pressing issue that speaks to the broader contexts of urbanization, socio-economic inequality, and environmental vulnerability. As cities grow, informal settlements emerge in response to rapid migration and the inadequacies of formal housing markets, underscoring significant challenges in urban planning and public health (Kosoe & Ogwu, 2025; Ogwu & Kosoe, 2025). Table 8.1 summarizes the key challenges facing informal settlements, including rapid urbanization, climate risks, inadequate services, and governance gaps. It also highlights the impacts on communities and outlines critical policy needs for building resilience and equity in urban planning.

Table 8.1 Challenges and policy needs for informal settlements and infrastructural fragility

Key aspect	Challenges	Impacts	Policy needs
Growth & distribution	Rapid urbanization, weak housing markets	Hazard-prone settlements, planning exclusion	Integrated urban planning
Climate vulnerabilities	Exposure to floods, heatwaves, and poor infrastructure	Disaster risks, low adaptive capacity	Climate-resilient housing & risk planning
Housing & public health	Inadequate water, sanitation, and overcrowding	Disease outbreaks, economic burdens	Universal services & health safeguards
Social inequality	Exclusion from decision-making, resource biases	Poverty, environmental injustice	Community participation & equity policies
Governance gaps	Reactive policies, top-down governance	Limited resilience measures	Inclusive, long-term urban reforms

8.4.1 Growth and Spatial Distribution of Informal Settlements

Informal settlements are often viewed as manifestations of urban inequality and as the failure of formal housing systems to keep pace with rapid urbanization. They are characterized by inadequate housing and lack of access to basic services such as water, sanitation, and electricity. A study by Anierobi et al. (2023) detailing the demographics of informal settlements in Enugu Metropolis, Nigeria, revealed a significant population residing in these areas, often unaccounted for in official census records. This underrepresentation underscores the neglect these communities face within formal urban planning processes, illustrating how spatially concentrated inequalities can develop (Anierobi et al., 2023). Informal settlements represent a substantial portion of the urban populace grappling with resource scarcity and social exclusion.

The spatial distribution of these settlements often correlates with environmental vulnerabilities. For instance, the characteristics of urban land use heavily influence settlement patterns, leading residents to occupy marginal areas that are prone to flooding and other hazards, exacerbated by climate change (Kihato et al., 2021). By clustering in high-risk zones, these communities face compounded vulnerabilities as they are often unable to evacuate or secure resources during climate-induced disasters. The intricate relationship between environmental risk and social vulnerability reveals how systemic inequities shape the existence of informal settlements, necessitating a critical reassessment of urban policy to include these realities.

8.4.2 Climate Vulnerabilities in Urban Informality

The intersecting challenges of climate change and urbanization highlight the precariousness of life in informal settlements. Residents are frequently positioned in harm's way due to inadequate infrastructure and services that fail to shield them from extreme weather events. This underscores the concept of "environmental justice," in which marginalized communities suffer disproportionately from environmental hazards due to socio-economic and political factors (Kekana et al., 2023). The COVID-19 pandemic further exposed these vulnerabilities, demonstrating how informal settlements are ill-equipped to manage public health crises due to poor sanitation and overcrowding, which are characteristic of such living environments (Dalasile et al., 2024).

Understanding how climate change impacts informal settlements requires a consideration of socio-economic dynamics that perpetuate poverty and limit adaptive capacity. For example, without secure land tenure or adequate financial resources, residents are less likely to invest in resilient housing or infrastructure (Kihato et al., 2021). Climate adaptation strategies should therefore be inclusive and tailored to the unique needs and circumstances of informal settlement residents to enhance their resilience against climate risks.

8.4.3 Housing, Sanitation, and Public Health Risks

The health consequences of living in informal settlements are profound. Access to clean water and sanitation is a critical determinant of health, with poor sanitation leading to high rates of communicable diseases such as diarrhea, which disproportionately affect women and children (Corburn & Hildebrand, 2015). Studies in Kenya highlight that women living in informal contexts report an array of health problems, including violence and respiratory illness, rooted in inadequate living conditions (Winter et al., 2019).

Furthermore, inadequate sanitation not only affects health but also compounds economic burdens by increasing healthcare costs and lost productivity. Research from informal settlements in South Africa confirms that residents face a cycle of health-related issues intertwined with the socio-environmental contexts of their living environments (Anierobi et al., 2023). These findings advocate for ensuring comprehensive sanitation infrastructure as a fundamental right, essential for improving health outcomes in informal settlements and breaking the cycle of poverty.

8.4.4 Social Inequality and Environmental Justice Dimensions

Social inequalities manifest starkly within urban informal settlements, where marginalized populations are often excluded from political processes that determine resource allocation and urban planning. The environmental justice literature highlights the need for equitable access to resources and decision-making processes as vital to addressing systemic inequities in informal settlements (Kekana et al., 2023). The lack of representation often means that policies fail to address the specific needs of these communities, perpetuating cycles of disadvantage and environmental degradation.

Consequently, the discourse around environmental justice should involve not only the physical and logistical aspects of urban planning but also the socio-political structures that govern access to resources. Studies have shown that when community voices are neglected in urban governance, the resultant policies tend to overlook adaptive capacity and resilience, leading to infrastructural fragility and increased vulnerability (Howard et al., 2025). Thus, the integration of social equity principles (by recognizing the environmental and health risks associated with informal settlements) should guide policy frameworks aimed at achieving urban sustainability and justice.

8.4.5 Policy Gaps and Adaptive Governance Mechanisms

Despite acknowledging the critical issues surrounding informal settlements, policy frameworks often remain inadequate. Many urban areas have reactive rather than proactive policies that fail to incorporate long-term solutions to address both climate change and housing needs. Governance mechanisms should shift from top-down approaches to inclusive practices that empower residents of informal settlements (Howard et al., 2025). This adaptive governance involves co-creating solutions with community stakeholders, ensuring policies are reflective of the needs and experiences of marginalized populations living in informal contexts.

Furthermore, achieving just sustainability requires an intersectional approach that incorporates diverse perspectives, particularly those of women and other marginalized groups who often experience the brunt of environmental vulnerability (Kekana et al., 2023). Policies should also focus on resilience-building interventions that prioritize access to essential services and secure housing, emphasizing community-driven strategies to foster socio-economic improvement and environmental sustainability (Howard et al., 2025). This transformation in governance structures holds the potential to mitigate the fragility of informal settlements while enhancing the quality of life for millions living in these environments.

8.5 Water Scarcity and Climate-Induced Urban Crises

Water scarcity, exacerbated by climate change and urbanization, is a critical issue facing cities globally, particularly in the Global South (Jacob et al., 2024a, 2024b). In this comprehensive exploration of urban water metabolism, the interdependencies between water demand, supply, and wastewater flows are critically examined. Urban water systems are under increasing stress from rising populations and

Fig. 8.4 Urban water metabolism: interactions between demand, supply, and wastewater systems

changing climates, making it imperative to understand their hydrological dynamics to devise effective management strategies (Izah et al., 2024; Numbere et al., 2024). Figure 8.4 illustrates the dynamic flows of water within urban environments that characterize urban water metabolism, linking sources, consumption, and wastewater treatment. It highlights how increasing urban demand, limited supply, and inadequate treatment infrastructure interact under climate and population pressures, shaping water sustainability and resilience challenges in cities.

8.5.1 Urban Water Metabolism: Demand, Supply, and Wastewater Flows

The concept of urban water metabolism encapsulates the intricate interactions among various components of the urban water cycle, notably water demand, supply, and wastewater flows. Urbanization has significantly altered these interactions, leading to a decline in the natural filtering capability of watersheds due to increased impervious surfaces (e.g., roads and buildings) (O'Driscoll et al., 2010). Haddeland et al. (2014) indicate that human interventions, including urban development and reservoir operations, have a profound impact on global water resources, further complicating water management in urban contexts. The growth of metropolitan regions exacerbates water consumption patterns, leading to unsustainable demand relative to supply (Aragay & Ostos, 2011). Studies have shown that average water consumption per capita in urban areas has risen markedly with increasing urbanization (Wei et al., 2019). Compounding these challenges is the prevalence of combined sewer overflows (CSOs), which often result from insufficient stormwater infrastructure, further degrading urban water quality (Sun et al., 2016). Thus, urban water metabolism should be viewed as a situated challenge, responding not only to population pressures but also to infrastructural dynamics.

The supply side also suffers significantly due to climate-induced changes, which have been documented to affect precipitation patterns globally (Wang & Liu, 2023). The transformation of natural water systems into more regulated, human-centered utility frameworks has led to decreased groundwater recharge and increased surface runoff, consequently diminishing the resilience of urban areas to extreme weather events (Ogwu et al., 2025a; Valeo et al., 2021). Therefore, integrating sustainable water practices and technologies into urban water systems is essential to enhance their capacity to handle variability in water supply and quality.

8.5.2 Climate Change Effects on Urban Hydrological Cycles

Climate change is fundamentally reshaping urban hydrological cycles, leading to increased variability in both the frequency and intensity of precipitation events. This increased variability can manifest as more severe flooding and prolonged droughts, which pose considerable challenges to urban water management (Valeo et al., 2021; Wang & Liu, 2023). The impacts of climate change are particularly pronounced where urban expansion has reduced the natural landscape's ability to absorb rainfall due to increased impervious surfaces. Studies by Valeo et al. (2012) show that metropolitan areas experience intensified peak flows during extreme precipitation events, a direct consequence of climate change, reinforcing the critical need for adaptive management strategies.

Moreover, the interaction between climate change and urban land use further complicates hydrological responses. Liu et al. (2023) highlight that urbanization alters land cover, exacerbating runoff and diminishing the natural water-retention capacity of ecosystems. The resulting changes in hydrological cycles often require cities to invest in infrastructure improvements to handle increased runoff and flooding risks effectively. Research indicates that areas especially vulnerable to climate impacts experience both increased water demand due to higher evaporation rates and a decline in adequate water supply due to increased contamination in urban environments (Wang & Liu, 2023). Thus, the interplay between urbanization and climate change necessitates comprehensive planning that prioritizes resilient water management techniques, including low-impact development and green infrastructure, to mitigate adverse outcomes (Izah et al., 2025a; Ogwu et al., 2025a, 2025b, 2025c, 2025d, 2025e, 2025f).

8.5.3 Water Scarcity and Governance Challenges in the Global South

The Global South faces unique governance challenges in managing water scarcity as climate impacts worsen existing vulnerabilities. Cities in these regions often lack the robust institutions and infrastructure necessary to effectively manage water resources under shifting climatic conditions. The urban poor are particularly vulnerable to water scarcity, experiencing disproportionate effects of diminishing water availability (Kaykhosravi et al., 2020; Putro et al., 2016).

The interplay between ineffective governance and systemic inequalities often results in marginalized communities facing water insecurity, perpetuating cycles of poverty and socio-economic disenfranchisement (Nunno et al., 2024; Ogwu, 2019). Governance frameworks in many Global South cities inadequately address contemporary water management challenges, such as the need for integrated urban planning that considers climate change implications. Research suggests that developing inclusive governance frameworks that incorporate localized water management strategies is essential for improving urban water resilience (Jacob et al., 2024a, 2024b).

Furthermore, existing governance models often prioritize economic growth over sustainable water management practices, leading to increased pressures on already scarce water resources (Putro et al., 2016). The lack of participatory approaches in water governance results in policies that are misaligned with community needs and environmental sustainability, further compounding the challenges faced in water-scarce urban areas (Nunno et al., 2024). A paradigm shift toward more equitable governance frameworks that emphasize stakeholder involvement and adaptive strategies is crucial for effectively addressing urban water scarcity.

8.5.4 Socio-economic Implications for Marginalized Populations

The socio-economic implications of water scarcity run deep through the fabric of urban societies, particularly affecting marginalized populations. Limited access to clean water exacerbates health disparities and discourages educational and economic opportunities, leading to further marginalization (Nunno et al., 2024). Urban areas experiencing climate-induced water stress often see social tensions escalate due to competition for dwindling resources (Kaykhosravi et al., 2020; Putro et al., 2016). Marginalized groups, often situated in vulnerable locations, suffer disproportionately from inadequate water supply and increased flooding risks, highlighting the nexus between social equity and water management (Askarizadeh et al., 2015). For example, urban areas with prevalent informal settlements typically lack adequate infrastructure for stormwater management, leading to widespread flooding during heavy rain events and subsequent health crises (Nunno et al., 2024).

Table 8.2 Socio-economic implications of urban water scarcity for marginalized populations

Key dimension	Drivers of water scarcity	Socio-economic implications	Policy/planning needs
Access to clean water	Limited supply, unequal distribution, aging infrastructure	Health risks, higher water costs, increased time spent collecting water	Equitable water allocation, infrastructure upgrades
Livelihoods & employment	Industrial demand, agricultural overuse	Loss of income in water-dependent sectors, reduced productivity	Sustainable water use in economic planning
Gender inequalities	Household water management burdens on women	Increased labor time, reduced education, and economic opportunities	Gender-sensitive water policies
Urban poverty & inequality	Informal settlements, poor governance	Exacerbation of poverty cycles, heightened social vulnerabilities	Inclusive urban planning, social safety nets
Public health & sanitation	Inadequate wastewater treatment, climate shocks	Spread of waterborne diseases, economic costs of healthcare	Investments in sanitation and climate resilience

Research has shown that addressing water scarcity through targeted policies and resource allocation can significantly improve the quality of life for disadvantaged populations (Nunno et al., 2024). Implementing community-centric water management strategies that prioritize equity can not only help alleviate water scarcity but also enhance social cohesion and community resilience to climate impacts (Putro et al., 2016). Thus, tackling water scarcity should be coupled with a focus on social equity and on empowering marginalized communities through improved water governance and resource management. Table 8.2 shows how urban water scarcity disproportionately impacts marginalized populations by linking limited access, gendered burdens, and livelihood disruptions to broader social and economic vulnerabilities, and outlines the policy actions needed to promote fair water management and resilience in cities.

8.5.5 Technological and Nature-Based Solutions for Urban Water Resilience

In light of the multifaceted challenges posed by climate change and urbanization, cities are increasingly turning to technological and nature-based solutions to enhance their water resilience. Nature-based solutions, such as green roofs, permeable pavements, and urban wetlands, have gained traction as effective means to restore natural hydrological processes (Fig. 8.5; Palermo et al., 2023). These solutions are not only cost-effective but also provide ecosystem services such as biodiversity enhancement and urban cooling (Palermo et al., 2023). Technological advancements in data analytics, sensor networks, and smart water management

Fig. 8.5 Integrating technological and nature-based solutions for urban water resilience

systems have the potential to optimize water use, reduce waste, and improve service delivery in urban areas (Liu et al., 2017). Integrating these technologies into existing urban infrastructure can help cities adapt to climate variability by enabling more efficient prediction and management of water demand.

Moreover, the adoption of low-impact development (LID) strategies has emerged as a promising approach to mitigate urban runoff and improve water quality by emulating natural hydrological patterns (Askarizadeh et al., 2015). These sustainable practices aim to lessen the burden on conventional drainage systems, thereby enhancing water quality and fostering community engagement and environmental stewardship (Askarizadeh et al., 2015). Nonetheless, the application of these solutions is often limited by institutional barriers and a lack of supportive policy frameworks. To maximize their benefits, cities should invest in capacity-building initiatives and promote interdisciplinary approaches that harness both technological innovation and ecological restoration (Ikhajiagbe et al., 2022;

Palermo et al., 2023). Urban water resilience can be significantly improved through a holistic approach that combines modern technology with nature-based strategies, ensuring sustainable and equitable water management practices.

8.6 Infrastructure Under Climate Stress

Climate change has emerged as a critical global challenge, significantly impacting infrastructure across multiple sectors. This discourse will explore vulnerabilities and resilience pathways in energy infrastructure, transportation systems, communication and digital infrastructures, cross-sectoral interdependencies, and lessons from climate-related infrastructure failures. Understanding these elements is vital for enhancing adaptive capacities and ensuring sustainable growth in the face of intensifying climate stressors.

8.6.1 Energy Infrastructure: Vulnerabilities and Resilience Pathways

The vulnerabilities of energy infrastructure to climate stress are profound, necessitating robust assessments and adaptive measures. Critical infrastructure, such as power plants, transmission networks, and distribution systems, is increasingly susceptible to disruptions caused by climate events like floods, hurricanes, and extreme heat. Studies indicate that climate-induced hazards can directly impair energy generation and supply, thereby exacerbating societal vulnerabilities and economic losses (Becher et al., 2023; Verschuur et al., 2024). For instance, inundation of coastal energy facilities from rising sea levels and storm surges poses a significant risk to energy production capabilities across many regions (Becker et al., 2015).

To foster resilience in energy infrastructure, strategies should integrate adaptive planning frameworks that are responsive to predicted climate scenarios. Implementing nature-based solutions (NbS) presents a promising approach for enhancing infrastructure resilience. These solutions can help reduce vulnerability while offering ancillary benefits such as improved biodiversity and social equity. Furthermore, community-led initiatives observed in Southeast Asia emphasize decentralized energy models and local stakeholder engagement, underscoring the importance of collective efforts in building climate resilience (Kurniawan et al., 2024). Investments in sustainable energy technologies and infrastructure, such as microgrids and renewable energy sources, offer not only mitigation against climate impacts but also economic and social benefits (Cortesão & Copeland, 2023).

8.6.2 Transportation Systems: Disruptions and Climate-Induced Risks

Transportation systems represent another critical sector facing unprecedented climate risks. Flooding, rising sea levels, and extreme weather events can cause significant disruptions to both land and maritime transport (Rokooei et al., 2023). Research highlights those systemic vulnerabilities in transportation infrastructure—such as roads, railways, and ports—expose communities to severe risks during climate events, leading to service interruptions and potentially cascading failures across interconnected systems (Gallagher & Cruickshank, 2016). For example, recent incidents in major urban centers demonstrated how transportation system failures can precipitate public safety issues and hinder emergency response activities, thus amplifying the impact of climate threats (Mühlhofer et al., 2023).

Addressing these vulnerabilities requires a comprehensive understanding of interdependencies among transportation nodes and other infrastructure systems. For instance, transportation disruptions can significantly impact the operational effectiveness of energy systems, creating a feedback loop of stress across infrastructure (Guerini & Mertens, 2023). Solutions like adaptive roadway designs, improved drainage systems, and enhanced emergency preparedness protocols can bolster resilience (Dunn et al., 2013). Moreover, integrating advanced modeling techniques to understand better risk scenarios and infrastructure interactions will be essential for developing effective resilience pathways (Johnson, 2017).

8.6.3 Communication and Digital Infrastructures in a Changing Climate

Communication and digital infrastructures are crucial for societal functionality, particularly during climate-induced emergencies. The interdependence of these infrastructures with energy and transportation systems underscores their vulnerability. Climate events can disrupt communication networks, thereby inhibiting effective emergency response and coordination efforts. Studies emphasize the importance of maintaining redundant systems within telecommunications networks to ensure continuity during critical climate events (Bayili & Badolo, 2024).

To bolster resilience, enhancing cybersecurity measures alongside traditional physical infrastructure improvements is essential. The increasing integration of smart technologies and digital tools enables better data collection and dissemination, fostering informed decision-making in crisis scenarios. Furthermore, stakeholder engagement, particularly involving local communities, can play a vital role in building adaptive capacity within communication infrastructures (Ray et al., 2022; Verschuur et al., 2024). Such engagement is integral not only for bolstering awareness but also for incorporating inclusive design practices that account for the diverse needs of vulnerable populations (Silvast et al., 2021).

8.6.4 Cross-Sectoral Interdependencies and Systemic Risks

Understanding cross-sectoral interdependencies is vital for addressing systemic risks posed by climate change. Infrastructure systems, including energy, water, transportation, and digital platforms, function within a complex web of interactions that can compound vulnerabilities during climate events. The operational capacity of one infrastructure sector often depends on the status of others. For example, power outages can compromise water supply systems, triggering cascading failures across interdependent infrastructures (Verschuur et al., 2024). In recognizing these interdependencies, resilience strategies should transcend sectoral lines and promote a holistic view of infrastructure development. Emphasizing integrated planning approaches that consider how multiple infrastructures interact and influence one another can better prepare systems for climate-induced disruptions (Sumardjo et al., 2023). Urban planners and policymakers should devise frameworks that enhance collaboration across sectors, ensuring collective resilience improves community sustainability (Okvat & Zautra, 2011).

8.6.5 Lessons from Climate-Related Infrastructure Failures

Historical instances of infrastructure failures during climate events offer invaluable lessons for future resilience planning. For instance, the impacts of Hurricane Katrina emphasized the critical need for robust levee systems and the role of social dynamics in shaping vulnerabilities (Becker et al., 2015). Such incidents reveal that climate-resilient infrastructure should incorporate not only engineering solutions but also social and economic considerations to ensure equitable access to adaptive benefits (Silvast et al., 2021). Understanding the social dimensions of climate impacts, particularly their impacts on marginalized populations, can inform intervention strategies to enhance resilience (Nijhawan et al., 2022).

Additionally, fostering a culture of preparedness and community engagement around infrastructure resilience is paramount. Lessons from urban centers that experienced extreme weather underscore the importance of public awareness initiatives that educate communities about climate risks and adaptive measures (Kurniawan et al., 2024; Ray et al., 2022). Implementing adaptive management frameworks provides opportunities for continual learning and iterative improvement of infrastructure resilience strategies, thereby enhancing systems' ability to respond to evolving climate challenges (Habibian & Minaie, 2018).

Addressing the vulnerabilities of infrastructure under climate stress requires an integrated approach that accounts for energy, transportation, communication, and cross-sectoral interdependencies. Enhancing resilience through adaptive planning, stakeholder engagement, and innovative solutions, such as nature-based approaches, will mitigate climate risks and improve infrastructure performance during extreme events. Drawing on historical failures, it is imperative to adopt a holistic approach that integrates social dynamics into infrastructure resilience planning while fostering community engagement and awareness. Future infrastructure strategies should prioritize not only robustness and efficiency but also inclusivity to ensure equitable outcomes in the face of an increasingly volatile climate.

8.7 Urban Metabolism as a Framework for Climate Adaptation

Urban metabolism is increasingly recognized as a crucial framework for understanding urban systems, particularly concerning climate adaptation. By adopting a systems perspective and applying methodologies such as Material Flow Analysis (MFA), urban planners and policymakers can assess resource flows, energy consumption, and waste management within cities. These methodologies not only enhance resilience to climate change but also facilitate the integration of circular economy principles that aim to minimize waste and optimize resource use. Cities that implement robust MFA frameworks are better equipped to respond to climate-related challenges through effective adaptation strategies focused on resource efficiency (Rivas et al., 2021; Urrutia-Azcona et al., 2020).

Material Flow Analysis serves as a key tool for quantifying and managing material and energy flows in urban environments, providing vital insights for developing sustainable urban systems. Urban metabolism analysis enables cities to track resources from extraction through production to consumption and waste generation, highlighting inefficiencies and areas for intervention. By identifying patterns of resource consumption and waste production, planners can develop strategies to improve resource use while enhancing resilience to climate impacts (Tatar et al., 2020; Urrutia-Azcona et al., 2021a). This analysis underscores the need for an integrated approach that blends energy and urban planning to address climate risks effectively, particularly as urban areas contribute significantly to global carbon emissions (Urrutia-Azcona et al., 2021b).

Furthermore, urban metabolism indicators can play a critical role in climate planning. By measuring variables such as energy consumption, waste generation, and carbon emissions, urban decision-makers can create informed adaptation plans. These metrics can guide policies promoting low-carbon urban development, aligning with global climate policy targets, such as the Paris Agreement (DeLosRíos-White et al., 2020). Indicators derived from MFA can enhance understanding of urban resilience, demonstrating how cities can adapt to climate change and thrive amid its challenges by utilizing circular economy practices that emphasize resource recovery and sustainability (Urrutia-Azcona et al., 2021a).

The successful implementation of climate adaptation strategies also hinges on engaging diverse stakeholders throughout the planning process (Ogwu & Izah, 2025). Inclusive stakeholder participation leads to more equitable and effective adaptation goals. Communities that actively participate in climate adaptation planning are more likely to incorporate diverse perspectives and address social equity issues, fostering community buy-in and promoting shared ownership of adaptation measures (Guenat et al., 2021; Hügel & Davies, 2020; Lioubimtseva, 2022). Recognizing that diverse stakeholder input significantly enhances the effectiveness of adaptation plans underscores the need for robust participation frameworks that accommodate diverse voices in the planning process (Kujala et al., 2022). Participatory planning fosters a shared understanding and trust among community members, enhancing the legitimacy of urban policies. Engaging stakeholders from diverse sectors and demographics ensures that adaptation strategies reflect the community's priorities and needs, especially in vulnerable urban areas (Kujala et al., 2022). Research suggests that tailored engagement methods, such as workshops and online platforms, can substantially increase participation rates and improve the quality of community members' input, leading to more actionable urban plans (Ju & Kim, 2023; Yu & Ahlgren, 2023).

Scenario modeling emerges as a critical tool in crafting climate-resilient urban futures. By employing scenario analysis frameworks, cities can visualize potential futures, assess risks, and evaluate the adaptability of different strategies across different climate scenarios. These models elucidate the cascading effects of climate impacts on urban systems, including infrastructure, health, and economic stability (Izah & Ogwu, 2025; Izah et al., 2025a). Moreover, scenario modeling helps stakeholders understand complex interactions within urban systems, fostering informed decision-making grounded in rigorous analysis (Scarpellini et al., 2018; Urrutia-Azcona et al., 2021b). As communities cope with uncertainties posed by climate change, robust scenario modeling becomes essential for articulating pathways to resilience (Tatar et al., 2020).

Integrating circular economy principles into urban systems is vital for adapting to climate challenges, emphasizing sustainability through resource efficiency and waste reduction. The circular economy paradigm aims to maintain the utility of products, materials, and resources for as long as possible (He et al., 2022; Oguw et al., 2025c, 2025d). This integration can lead to improved resource management and reduced carbon footprints within urban contexts. Policymakers increasingly view the circular economy as a critical component of urban adaptation strategies, promoting sustainable consumption and production in line with climate adaptation goals (Tatar et al., 2020; Urrutia-Azcona et al., 2020).

Furthermore, the intersection of urban metabolism and circular economy principles positions cities to become proactive rather than reactive in addressing climate change. By focusing on how materials and energy flow through urban settings, cities can identify opportunities to reduce dependence on finite resources and mitigate adverse environmental impacts. This approach involves rethinking production and consumption patterns, encouraging recycling and promoting the reuse of materials, thereby enhancing resilience in urban systems (Rivas et al., 2021; Urrutia-Azcona et al., 2021b). As cities model future scenarios alongside innovative economic principles, the path to climate resilience becomes increasingly evident, emphasizing proactive, integrated approaches as crucial for navigating the complexities of future urban challenges (DeLosRíos-White et al., 2020; Scarpellini et al., 2018; Urrutia-Azcona et al., 2020).

8.8 Transformative Potentials and Policy Implications in Urban Design and Climate Governance

8.8.1 Toward Climate-Resilient Urban Design and Governance

Climate change poses severe risks to urban environments worldwide, necessitating the integration of resilience into urban design and governance frameworks. Research indicates that successful climate-resilient urban design should incorporate multifaceted approaches that leverage technology, community engagement, and regulatory frameworks. Dąbrowski (2018) explored how urban areas can span governance boundaries to facilitate effective climate adaptation, emphasizing that even in developed regions like the Dutch city of Rotterdam, cross-jurisdictional cooperation for climate action poses significant challenges. This underscores the need for continuous learning and adaptation within governance systems to address complexities arising from climate impacts.

Furthermore, the need to incorporate hybrid frameworks into urban climate governance cannot be overstated. A holistic view should encompass socio-political dimensions, as well as the infrastructural and economic aspects of cities. For instance, Long and Rice (2019) illustrate the concept of "climate urbanism," in which urban spaces become the frontline for climate action, prioritizing infrastructure protection and community resilience. Such an approach requires a re-evaluation of existing urban policies to ensure they foster sustainable practices while addressing social inequalities and their impacts (Xu et al., 2023).

8.8.2 Linking Urban Metabolism with Sustainable Development Goals

Linking urban metabolism—the way cities utilize resources—with the SDGs is fundamental for sustainable urban development. As global trends indicate an increase in urbanization, frameworks that assess urban metabolism through a sustainability lens can prove instrumental in advancing the SDGs. Efficient resource management in urban areas can facilitate economic, environmental, and social benefits, thus enhancing overall urban resilience and sustainability (Sun & Cui, 2018). The interplay between economic, social, and ecological systems is especially crucial, as demonstrated in Beijing, where studies highlight the coordination among these domains to optimize urban infrastructure for sustainability (Tao, 2019).

Moreover, the financing of sustainable urban infrastructure is vital to achieving the SDGs. White and Wahba's (2019) work emphasizes the critical role of private funding in the Global South, particularly given the constraints on government budgets to provide climate-resilient infrastructure. Investments should not only fund necessary projects but also support the institutional capacity to navigate the complexities of urban systems. Hence, engaging various stakeholders across levels of governance and sectors is essential to link urban metabolism to the broader SDG agenda.

8.8.3 Multi-level Governance and Regional Climate Cooperation

Implementing a multi-level governance framework is vital for effective urban climate action. It enables local, national, and international cooperation in addressing climate challenges that transcend political boundaries. The analysis by Opdyke et al. (2017) underscores the increasing recognition of disaster resilience as an essential aspect of climate governance. This framework fosters necessary synergies among different levels of government, enabling the creation of policies that enhance urban infrastructure resilience.

Studies indicate that robust multi-level governance structures facilitate the sharing of resources and knowledge, which is essential for coordinating effective climate responses across regions. For example, Tozer et al. (2023) found that city climate action in Africa is significantly strengthened by constructive relationships between local and national stakeholders, enabling systemic changes in legislative and operational frameworks. Collaborative efforts not only bolster the efficacy of climate strategies but also promote the equitable distribution of resources across urban populations, which is crucial to avoid the exacerbation of socio-economic inequities during climate-related vulnerabilities.

8.8.4 Financing Resilient Urban Infrastructure in the Global South

The financing of resilient urban infrastructure, particularly in the Global South, faces numerous challenges that should be strategically addressed. The complexity of urban infrastructure demands substantial investments to mitigate the adverse effects of climate change, with developing countries anticipated to require over a trillion dollars to meet their infrastructure needs (Fekete et al., 2023). Consequently, it is imperative to explore innovative financing mechanisms that leverage non-state actors and blended finance models.

Moreover, assessing barriers to private investment in climate-resilient projects can provide insights into tailoring financial strategies that suit local contexts. Experts suggest engaging stakeholders at all levels, fostering partnerships with the private sector, and aligning financial incentives with sustainability objectives (White & Wahba, 2019). This combined effort can lead to the development of urban infrastructure that is not only resilient but also economically viable for the communities it serves.

8.8.5 Research, Policy, and Practice Interfaces for Transformative Action

The intersection of research, policy, and practice is crucial for informing transformative climate action in urban contexts. Scholarly investigations can catalyze policy innovation by identifying climate adaptation needs and resilience strategies. For instance, findings by Shaker et al. (2019) stress the importance of infrastructure resilience as a research imperative that should influence urban planning and policy frameworks—integrating empirical evidence into policy design aids in crafting strategies that are both informed by current realities and responsive to future climate scenarios.

To transition effectively toward sustainable urban environments, it is essential to foster a dialogue among researchers, policymakers, and practitioners. This collaboration can lead to the co-production of knowledge and shared learning experiences vital for advancing resilience in urban settings. Niamir and Pachauri (2023) emphasize the need for comprehensive transformative actions that incorporate behavioral, cultural, and institutional mechanisms, highlighting the importance of bridging adaptation and mitigation for coastal cities' resilience.

8.9 Conclusion

This chapter has underscored the critical role of urban metabolism in understanding and addressing the climate-induced vulnerabilities of cities in the Global South. By conceptualizing cities as dynamic systems of energy, material, and information flows, we revealed how urban heat islands, water scarcity, informal settlements, and infrastructural breakdowns are not isolated challenges but interlinked manifestations of socio-ecological and infrastructural fragility. The urban metabolism framework provided a lens to move beyond fragmented analyses, allowing for an integrated understanding of how climate stress disrupts essential urban functions while opening pathways for transformative adaptation. One of the key insights emerging from this analysis is that infrastructural resilience in the Global South cannot be decoupled from questions of social equity, governance, and ecological sustainability. Informal settlements, frequently excluded from formal urban planning, epitomize the intersection of vulnerability and adaptive capacity, demonstrating both the risks of exclusionary policies and the potential for community-driven climate resilience. Similarly, the urban water and energy crises highlight the need for cross-sectoral integration, technological innovation, and participatory governance to build adaptive infrastructures capable of withstanding climate shocks.

This chapter contributes to ongoing climate governance debates by positioning urban metabolism as both an analytical tool and a policy framework for achieving climate-resilient, low-carbon, and socially inclusive urban futures. As the Global South faces escalating climate uncertainties, the capacity to integrate material flow analysis, circular economy principles, and multi-level governance will determine whether its cities remain sites of vulnerability or become incubators for transformative climate solutions. Future research should therefore focus on refining urban metabolism models, expanding empirical datasets, and deepening collaborations between scientists, policymakers, and local communities to ensure that urban adaptation strategies are both evidence-based and socially just.

References

Abajue, M. C., & Ogwu, M. C. (2024). Sustainable alternative land management strategies and their impacts on soil arthropod diversity. In M. C. Ogwu, S. C. Izah, J. Dessureault-Rompré, & D. Gasparatos (Eds.), *Sustainable soil systems in Global South*. Springer. https://doi.org/10.1007/978-981-97-5276-8_18

Anierobi, C. M., Nwalusi, D. M., Efobi, K. O., Nwosu, K. I., Nwokolo, N. C., & Ibem, E. O. (2023). Urban housing inequality and the nature of relationship between formal and informal settlements in Enugu Metropolis, Nigeria. *SAGE Open, 13*(3), 21582440231192390.

Askarizadeh, A., Rippy, M. A., Fletcher, T. D., Feldman, D. L., Peng, J., Bowler, P., Mehring, A. S., Winfrey, B. K., Vrugt, J. A., AghaKouchak, A., & Jiang, S. C. (2015). From rain tanks to catchments: Use of low-impact development to address hydrologic symptoms of the urban stream syndrome. *Environmental Science & Technology, 49*(19), 11264–11280.

Bai, X. (2016). Eight energy and material flow characteristics of urban ecosystems. *Ambio, 45*(7), 819–830.

Bay, B. D., & Dinçer, F. (2024). Turkey's urbanization within the context of the Global South/North divide: Continuities and differences. *Research Studies, 4*(11), 1229–1234.

Bayili, B. R., & Badolo, M. (2024). *Cities resilience to flood risks communication: The B2CresCom scientific framework*. Center for Open Science.

Becher, O., Pant, R., Verschuur, J., Mandal, A., Paltan, H., Lawless, M., Raven, E., & Hall, J. (2023). A multi-hazard risk framework to stress-test water supply systems to climate-related disruptions. *Earth's Future, 11*(1), e2022EF002946.

Becker, A. H., Matson, P., Fischer, M., & Mastrandrea, M. D. (2015). Towards seaport resilience for climate change adaptation: Stakeholder perceptions of hurricane impacts in Gulfport (MS) and Providence (RI). *Progress in Planning, 99*, 1–49.

Bi, X., & Wang, D. (2020). Research on the path of low-carbon city system construction mode based on energy consumption per unit output value method. In *IOP conference series: Earth and environmental science* (Vol. 450(1), p. 012045). IOP Publishing.

Bigger, P., & Webber, S. (2021). Green structural adjustment in the World Bank's resilient city. *Annals of the American Association of Geographers, 111*(1), 36–51.

Borah, S., & Sarma, K. (2025). Temporal assessment of land surface temperature and urban heat island effect in Guwahati City: A geospatial analysis. *Ecology, Environment & Conservation, 31*, S26–S33.

Bowler, D. E., Buyung-Ali, L., Knight, T. M., & Pullin, A. S. (2010). Urban greening to cool towns and cities: A systematic review of the empirical evidence. *Landscape and Urban Planning, 97*(3), 147–155.

Bruns, J., & Simko, V. (2017). Stable hotspot analysis for intra-urban heat islands. *GI_Forum: Journal for Geographic Information Science, 1*, 79–92.

Bulkeley, H., & Castán Broto, V. (2013). Government by experiment? Global cities and the governing of climate change. *Transactions of the Institute of British Geographers, 38*(3), 361–375.

Chen, H., & Wang, K. (2012). The study for the influencing factors of urban heat island development. *Advanced Materials Research, 524*, 3524–3529.

Chirambo, D. (2020). Improving climate change resilience in Global South cities through South-South climate finance. *International Journal of Entrepreneurship and Governance in Cognitive Cities (IJEGCC), 1*(2), 1–8.

Corburn, J., & Hildebrand, C. (2015). Slum sanitation and the social determinants of women's health in Nairobi, Kenya. *Journal of Environmental and Public Health, 2015*(1), 209505.

Cortesão, J., & Copeland, S. (2023). Urban climate resilience. In R. C. Brears (Ed.), *The Palgrave encyclopedia of urban and regional futures* (pp. 1962–1967). Springer International Publishing.

Dąbrowski, M. (2018). Boundary spanning for governance of climate change adaptation in cities: Insights from a Dutch urban region. *Environment and Planning C: Politics and Space, 36*(5), 837–855.

Dalasile, S., Itoba Tombo, E., Madonsela, B. S., Mpungose, P. P., Mshicileli, N., & Menziwa, M. (2024). Alcohol-based hand sanitizers used for COVID-19 prevention in the informal settlements of Cape Town, South Africa. *COVID, 4*(10), 1655–1675.

DeLosRíos-White, M. I., Roebeling, P., Valente, S., & Vaittinen, I. (2020). Mapping the life cycle co-creation process of nature-based solutions for urban climate change adaptation. *Resources, 9*(4), 39.

Dunn, S., Fu, G., Wilkinson, S., & Dawson, R. (2013). Network theory for infrastructure systems modelling. *Proceedings of the Institution of Civil Engineers-Engineering Sustainability, 166*(5), 281–292.

Firmansyah, A., & Dharmawan, L. (2023). Peri-urban community resilience in facing the dynamics of global climate-change impacts toward sustainable development. In *IOP conference series: Earth and environmental science* (Vol. 1266(1), p. 012045). IOP Publishing.

Gallagher, D., & Cruickshank, H. (2016). Planning under new extremes: Resilience and the most vulnerable. *Proceedings of the Institution of Civil Engineers-Municipal Engineer, 169*(3), 127–137.

Galychyn, O., Buonocore, E., & Franzese, P. P. (2020). Exploring the global scientific literature on urban metabolism. *Ecological Questions, 31*(4), 89–99.

Gaum, T., & Laubscher, J. (2022). Building energy codes: Reviewing the status of implementation strategies in the Global South. *International Journal of Built Environment and Sustainability, 9*(1), 39–53.

Guenat, S., Porras Lopez, G., Mkwambisi, D. D., & Dallimer, M. (2021). Unpacking stakeholder perceptions of the benefits and challenges associated with urban greenspaces in sub-Saharan Africa. *Frontiers in Environmental Science, 9*, 591512.

Guerini, C. J., & Mertens, J. B. (2023). RESIDENT: A neighborhood-level framework for measuring and visualizing social capital and community resilience in dense urban terrain. In *Disruptive technologies in information sciences VII* (Vol. 12542, pp. 17–24). SPIE.

Habibian, A., & Minaie, E. (2018). Achieving lifeline infrastructure resilience through an adaptive and risk-based approach. *Journal AWWA, 110*(8), 42–49.

Haddeland, I., Heinke, J., Biemans, H., Eisner, S., Flörke, M., Hanasaki, N., Konzmann, M., Ludwig, F., Masaki, Y., Schewe, J., & Stacke, T. (2014). Global water resources affected by human interventions and climate change. *Proceedings of the National Academy of Sciences, 111*(9), 3251–3256.

Howard, B. C., Moulds, S., Agyei-Mensah, S., Nahin, K. T. K., Quayyum, Z., Robinson, B. E., & Buytaert, W. (2025). Four principles of transformative adaptation to climate change-exacerbated hazards in informal settlements. *Wiley Interdisciplinary Reviews: Climate Change, 16*(3), e70008.

Huang, M., Zhong, S., Mei, X., & He, J. (2024). Spatiotemporal patterns in the urban heat island effect of several contemporary and historical Chinese "stove cities". *Sustainability, 16*(7), 3091.

Hügel, S., & Davies, A. R. (2020). Public participation, engagement, and climate change adaptation: A review of the research literature. *Wiley Interdisciplinary Reviews: Climate Change, 11*(4), e645.

Ikhajiagbe, B., Chukwujama, C. A., & Ogwu, M. C. (2022). Intrinsic restoration of a ferruginous utisol using goosegrass (*Eleusine indica* L. [Gaertn.]) obtained from different land use areas. *Journal of Natural Resources and Development, 12*, 1–18. https://doi.org/10.18716/ojs/jnrd2022.12.01

Iyiola, A. O., Afolabi, O. A., Alimi, S. K., Akingba, O. O., Izah, S. C., & Ogwu, M. C. (2024). Climate change and water crisis in the Global South. In S. C. Izah, M. C. Ogwu, A. Loukas, &

H. Hamidifar (Eds.), *Water crises and sustainable management in the Global South* (pp. 111–140). Springer.

Izah, S. C., Iyiola, A. O., Yarkwan, B., & Richard, G. (2023). Impact of air quality as a component of climate change on biodiversity-based ecosystem services. In A. L. Srivastav, A. K. Dubey, A. Kumar, S. K. Narang, & M. A. Khan (Eds.), *Visualization techniques for climate change with machine learning and artificial intelligence* (pp. 123–148). Elsevier. isbn:978-0-323-99714-0.

Izah, S. C., Jacob, D. E., Nelson, I. U., Daniel, K. S., & Ogwu, M. C. (2024). Exploring the influence of protected areas on water crises in the Global South: A balancing act. In S. C. Izah, M. C. Ogwu, A. Loukas, & H. Hamidifar (Eds.), *Water crises and sustainable management in the Global South* (pp. 625–654). Springer. https://link.springer.com/chapter/10.1007/978-981-97-4966-9_20

Izah, S. C., & Ogwu, M. C. (2025). *Innovative approaches in environmental health management: Processes, technologies, and strategies for a sustainable future.* Springer Nature. https://doi.org/10.1007/978-3-031-81966-7

Izah, S. C., Ogwu, M. C., & Akram, M. (2025a). *Bioactive compounds in edible mushrooms: Sustainability and health applications.* Springer Cham. https://doi.org/10.1007/978-3-031-52642-8

Izah, S. C., Ogwu, M. C., & Hait, M. (2025b). Clean energy solutions and public health: Innovations in renewable energy for a healthier environment. In S. C. Izah & M. C. Ogwu (Eds.), *Innovative approaches in environmental health management: Processes, technologies, and strategies for a sustainable future. Environmental science and engineering series.* Springer Nature. https://doi.org/10.1007/978-3-031-81966-7_7

Izah, S. C., Ogwu, M. C., & Hait, M. (2025c). Waste management and health: Addressing the processes behind hazardous waste and pollution. In S. C. Izah & M. C. Ogwu (Eds.), *Innovative approaches in environmental health management: Processes, technologies, and strategies for a sustainable future. Environmental science and engineering series.* Springer Nature. https://doi.org/10.1007/978-3-031-81966-7_6

Jacob, D. E., Nelson, I. U., & Izah, S. C. (2024a). Indigenous water management strategies in the Global South. In S. C. Izah, M. C. Ogwu, A. Loukas, & H. Hamidifar (Eds.), *Water crises and sustainable management in the Global South* (pp. 487–525). Springer. https://link.springer.com/chapter/10.1007/978-981-97-4966-9_16

Jacob, D. E., Nelson, I. U., Eniang, E. A., & Izah, S. C. (2024b). Advancing water security and resilience in the Global South through recreational development. In S. C. Izah, M. C. Ogwu, A. Loukas, & H. Hamidifar (Eds.), *Water crises and sustainable management in the Global South* (pp. 527–554). Springer. https://link.springer.com/chapter/10.1007/978-981-97-4966-9_17

Jiang, X., & Yang, H. (2024). Study on the mitigation effect of urban landscape in Hangzhou City on the urban heat island phenomenon. In *2023 5th International Conference on Hydraulic, Civil and Construction Engineering (HCCE 2023)* (pp. 283-291). Atlantis Press.

Johnson, B. R. M. (2017). *Urban resilience: What role can business improvement areas play in increasing climate change and extreme weather resilience in cities?* (Doctoral dissertation). Toronto Metropolitan University.

Ju, J., & Kim, J. (2023). Applying the Delphi approach to incorporate voiceless stakeholders in community planning. *Land, 12*(10), 1941.

Kasmaee, S., & Tinti, F. (2018). Metoda za procjenu utjecaja urbanizacije na regionalnu promjenu potpovršinske temperature. *Rudarsko-geološko-naftni zbornik, 33*(5), 12–12.

Kaykhosravi, S., Khan, U. T., & Jadidi, M. A. (2020). The effect of climate change and urbanization on the demand for low impact development for three Canadian cities. *Water, 12*(5), 1280.

Keeratikasikorn, C., & Bonafoni, S. (2018). Urban heat island analysis over the land use zoning plan of Bangkok by means of Landsat 8 imagery. *Remote Sensing, 10*(3), 440.

Kekana, H. N., Ruhiiga, T. M., Ndou, N. N., & Palamuleni, L. G. (2023). Environmental justice in South Africa: The dilemma of informal settlement residents. *GeoJournal, 88*(4), 3709–3725.

Kihato, C. W., De Villiers, S., Mohamed, S., & Mohulatsi, B. (2021). Spatial injustice in Johannesburg in the time of COVID-19. *Current History, 120*(826), 178–182.

Koc, C. B., Osmond, P., & Peters, A. (2018). Evaluating the cooling effects of green infrastructure: A systematic review of methods, indicators and data sources. *Solar Energy, 166*, 486–508.

Kosoe, E. A., & Ogwu, M. C. (2025). Sustainable urban planning and environmental management. In M. C. Ogwu & S. Chibueze Izah (Eds.), *Evaluating environmental processes and technologies. Environmental science and engineering series*. Springer. https://doi.org/10.1007/978-3-031-85327-2_14

Kujala, J., Heikkinen, A., Jokinen, A., & Tapaninaho, R. (2022). Stakeholder engagement in the generation of urban ecosystem services: The case of the Vuores stormwater system. *South Asian Journal of Business and Management Cases, 11*(1), 49–63.

Kurniawan, T. A., Goh, K. C., Goh, H. H., Zhang, D., Jiang, M., Dai, W., Ali, I., Othman, M. H. D., Aziz, F., Anouzla, A., & Haddout, S. (2024). Strengthening climate resilience: Urban water technologies for heat-resilient physical infrastructure in Southeast Asia cities amidst extreme temperature events and El Niño challenges. *ACS ES&T Water, 4*(9), 3713–3724.

Kushwaha, S., & Nithiyanandam, Y. (2019). The study of heat island and its relation with urbanization in Gurugram, Delhi NCR for the period of 1990 to 2018. *The International Archives of the Photogrammetry, Remote Sensing and Spatial Information Sciences, 42*, 49–56.

Leal, O. J. U., Fekete, A., Eudave, R. R., Matos, J. C., Sousa, H., & Teixeira, E. R. (2024). A systematic review of integrated frameworks for resilience and sustainability assessments for critical infrastructures. *Structural Engineering International, 34*(2), 266–280.

Li, H. Y., & Yue, X. P. (2012). Study on energy-saving of urban water supply system variable frequency pumping station. *Advanced Materials Research, 383*, 254–259.

Lioubimtseva, E. (2022). The role of inclusion in climate vulnerability assessment and equitable adaptation goals in small American municipalities. *Discover Sustainability, 3*(1), 3.

Liu, H., Li, X., Tian, S., & Guan, Y. (2022a). Research on the evaluation of resilience and influencing factors of the urban network structure in the three provinces of Northeast China based on multiple flows. *Buildings, 12*(7), 945.

Liu, W., Chang, A. C., Chen, W., Zhou, W., & Feng, Q. (2017). A framework for the urban eco-metabolism model-Linking metabolic processes to spatial patterns. *Journal of Cleaner Production, 165*, 168–176.

Liu, Y., Guan, Z., Huang, T., Wang, C., Guan, R., & Ma, X. (2023). Combined effects of land use/cover change and climate change on runoff in the Jinghe River Basin, China. *Atmosphere, 14*(8), 1237.

Liu, Y., Li, J., Chen, W. Q., Song, L., & Dai, S. (2022b). Quantifying urban mass gain and loss by a GIS-based material stocks and flows analysis. *Journal of Industrial Ecology, 26*(3), 1051–1060.

Long, J., & Rice, J. L. (2019). From sustainable urbanism to climate urbanism. *Urban Studies, 56*(5), 992–1008.

Lu, Y., & Chen, B. (2014). Urban studies based on emergy–a review in perspective of causality. *Energy Procedia, 61*, 2546–2549.

Maheng, D., Ducton, I., Lauwaet, D., Zevenbergen, C., & Pathirana, A. (2019). The sensitivity of urban heat island to urban green space—a model-based study of City of Colombo, Sri Lanka. *Atmosphere, 10*(3), 151.

Malherbe, N., & Oladejo, A. O. (2025). Against climate aid colonialism: The case for climate reparations and South-South solidarity. *Africa Spectrum, 60*(1), 84–105.

Maskun, M., Al Mukarramah, N. H., & Angi, J. C. W. (2023). Policy strategies for climate-resilient infrastructure in climate change adaptation: Jurisdictional comparison between Chile and Indonesia. In *E3S web of conferences* (Vol. 467, p. 05001). EDP Sciences.

Miklian, J., & Hoelscher, K. (2020). Entrepreneurial strategies to address rural-urban climate-induced vulnerabilities: Assessing adaptation and innovation measures in Dhaka, Bangladesh. *Sustainability, 12*(21), 9115.

Mittermüller, J., Erlwein, S., Bauer, A., Trokai, T., Duschinger, S., & Schönemann, M. (2021). Context-specific, user-centred: Designing urban green infrastructure to effectively mitigate urban density and heat stress. *Urban Planning, 6*(4), 40–53.

Moreau, V., & Massard, G. (2017). Material and energy flow analysis. In *Oxford research encyclopedia of environmental science*. Oxford University Press.

Mühlhofer, E., Koks, E. E., Kropf, C. M., Sansavini, G., & Bresch, D. N. (2023). A generalized natural hazard risk modelling framework for infrastructure failure cascades. *Reliability Engineering & System Safety, 234*, 109194.

Narvaez, J. P., Yanoviak, S. P., Bitzer, P. M., Burchfield, J. C., & Gora, E. M. (2024). Effects of urbanization on cloud-to-ground lightning strike frequency: A global perspective. *Journal of the Royal Society Interface, 21*(218), 20240257.

Niamir, L., & Pachauri, S. (2023). From social and natural vulnerability to human-centered climate resilient coastal cities. *Frontiers in Sustainable Cities, 5*, 1137641.

Nijhawan, A., Howard, G., Poudel, M., Pregnolato, M., Eunice Lo, Y. T., Ghimire, A., Baidya, M., Geremew, A., Flint, A., & Mulugeta, Y. (2022). Assessing the climate resilience of community-managed water supplies in Ethiopia and Nepal. *Water, 14*(8), 1293.

Numbere, A. O., Kika, E. P., & Izah, S. C. (2024). Challenges and mitigation of water resource management in rural areas in Nigeria. In S. C. Izah, M. C. Ogwu, A. Loukas, & H. Hamidifar (Eds.), *Water crises and sustainable management in the Global South* (pp. 373–390). Springer. https://doi.org/10.1007/978-981-97-4966-9_12

Nunno, F., Diodato, N., Bellocchi, G., Tricarico, C., Marinis, G., & Granata, F. (2024). Evapotranspiration analysis in central Italy: A combined trend and clustering approach. *Climate, 12*(5), 64. https://doi.org/10.3390/cli12050064

O'Donoghue, S., Morgan, D., Leck, H., & Haydvogl, K. (2022). The Durban Climate Change Strategy: Lessons learnt from the 2021 strategy review and implementation plan. *Town and Regional Planning, 81*, 84–96.

O'Driscoll, M., Clinton, S., Jefferson, A., Manda, A., & McMillan, S. (2010). Urbanization effects on watershed hydrology and in-stream processes in the southern United States. *Water, 2*(3), 605–648.

Ogwu, M. C. (2019). Towards sustainable development in Africa: The challenge of urbanization and climate change adaptation. In P. B. Cobbinah & M. Addaney (Eds.), *The geography of climate change adaptation in Urban Africa* (pp. 29–55). Springer Nature. https://doi.org/10.1007/978-3-030-04873-0_2

Ogwu, M. C., El Malahi, S., & Izah, S. C. (2025a). Policy and regulatory frameworks for sustainable environmental practices. In M. C. Ogwu & S. Chibueze Izah (Eds.), *Evaluating environmental processes and technologies. Environmental science and engineering series*. Springer. https://doi.org/10.1007/978-3-031-85327-2_16

Ogwu, M. C., Iyiola, A. O., Fawole, W. O., & Izah, S. C. (2025b). Biobased resource recovery techniques in the global South. In M. C. Ogwu, S. C. Izah, J. G. Vazquez-Arenas, S. T. Feleke, & X. Wei (Eds.), *Sustainable bioeconomy development in the Global South*. Springer. https://doi.org/10.1007/978-981-96-0305-3_17

Ogwu, M. C., & Izah, S. C. (2024). *Sustainable strategies for air pollution mitigation: Development, economics, and technologies*. Springer Cham. https://doi.org/10.1007/978-3-031-74157-9

Ogwu, M. C., & Izah, S. C. (2025). *Evaluating environmental processes and technologies*. Springer Nature. https://doi.org/10.1007/978-3-031-85327-2

Ogwu, M. C., Izah, S. C., Merritt, H., Wise, R. D., & Jintrawet, A. (2025c). *Sustainable bioeconomy development in the Global South: Volume I status and perspectives*. Springer Singapore. https://doi.org/10.1007/978-981-96-0640-5

Ogwu, M. C., Izah, S. C., Vazquez-Arenas, J. G., Feleke, S. T., & Wei, X. (2025d). *Sustainable bioeconomy development in the Global South: Volume II bioeconomy techniques*. Springer Singapore. https://doi.org/10.1007/978-981-96-0305-3

Ogwu, M. C., & Kosoe, E. A. (2025). Environmental economics: Valuing ecosystem services and green technologies. In M. C. Ogwu & S. Chibueze Izah (Eds.), *Evaluating environmental processes and technologies. Environmental science and engineering series*. Springer. https://doi.org/10.1007/978-3-031-85327-2_15

Ogwu, M. C., Kosoe, E. A., Alaka, A. C., & Xu, J. (2025e). Green infrastructure and environmental health security: The role of nature-based solutions in improving urban health. In S. C. Izah & M. C. Ogwu (Eds.), *Innovative approaches in environmental health management: Processes, technologies, and strategies for a sustainable future. Environmental science and engineering series*. Springer Nature. https://doi.org/10.1007/978-3-031-81966-7_9

Ogwu, M. C., Ojo, A. O., & Alaka, A. C. (2025f). Biodiversity and human health: The interconnections of species loss and ecosystem services. In S. C. Izah & M. C. Ogwu (Eds.), *Innovative approaches in environmental health management: Processes, technologies, and strategies for a sustainable future. Environmental science and engineering series*. Springer Nature. https://doi.org/10.1007/978-3-031-81966-7_5

Okvat, H. A., & Zautra, A. J. (2011). Community gardening: A parsimonious path to individual, community, and environmental resilience. *American Journal of Community Psychology, 47*(3), 374–387.

Opdyke, A., Javernick-Will, A., & Koschmann, M. (2017). Infrastructure hazard resilience trends: An analysis of 25 years of research. *Natural Hazards, 87*(2), 773–789.

Palermo, S. A., Turco, M., Pirouz, B., Presta, L., Falco, S., De Stefano, A., Frega, F., & Piro, P. (2023). Nature-based solutions for urban stormwater management: An overview. In *IOP Conference Series: Earth and Environmental Science* (Vol. 1196(1), p. 012027). IOP Publishing.

Pardo Martínez, C. I., Alfonso Piña, W., Facchini, A., & Cotte Poveda, A. (2021). Trends and dynamics of material and energy flows in an urban context: A case study of a city with an emerging economy. *Energy, Sustainability and Society, 11*(1), 24.

Patrakeyev, I., Ziborov, V., & Lazorenko-Hevel, N. (2017). Methods of evaluation of the state and efficiency of the urban environment. *ISPRS Annals of the Photogrammetry, Remote Sensing and Spatial Information Sciences, 4*, 43–49.

Perrotti, D., & Stremke, S. (2020). Can urban metabolism models advance green infrastructure planning? Insights from ecosystem services research. *Environment and Planning B: Urban Analytics and City Science, 47*(4), 678–694.

Putro, B., Kjeldsen, T. R., Hutchins, M. G., & Miller, J. (2016). An empirical investigation of climate and land-use effects on water quantity and quality in two urbanising catchments in the southern United Kingdom. *Science of the Total Environment, 548*, 164–172.

Ray, S., Goronga, T., Chigiya, P. T., & Madzimbamuto, F. D. (2022). Climate change, disaster management and primary health care in Zimbabwe. *African Journal of Primary Health Care & Family Medicine, 14*(1), 3684.

Rivas, S., Hernandez, Y., Urraca, R., & Barbosa, P. (2021). A comparative analysis to depict underlying attributes that might determine successful implementation of local adaptation plans. *Environmental Science & Policy, 117*, 25–33.

Rokooei, S., Vahedifard, F., & Belay, S. (2023). Gender-based perception of civil engineering and construction students towards infrastructure and community resilience. *International Journal of Disaster Resilience in the Built Environment, 14*(1), 40–52.

Sareen, S., & Waagsaether, K. L. (2023). New municipalism and the governance of urban transitions to sustainability. *Urban Studies, 60*(11), 2271–2289.

Shaker, R. R., Rybarczyk, G., Brown, C., Papp, V., & Alkins, S. (2019). (Re) emphasizing urban infrastructure resilience via scoping review and content analysis. *Urban Science, 3*(2), 44.

Silvast, A., Kongsager, R., Lehtonen, T. K., Lundgren, M., & Virtanen, M. (2021). Critical infrastructure vulnerability: A research note on adaptation to climate change in the Nordic countries. *Geografisk Tidsskrift-Danish Journal of Geography, 121*(1), 79–90.

Song, T., Yang, Z., & Chahine, T. (2016). Efficiency evaluation of material and energy flows, a case study of Chinese cities. *Journal of Cleaner Production, 112*, 3667–3675.

Stewart, I. D. (2011). A systematic review and scientific critique of methodology in modern urban heat island literature. *International Journal of Climatology, 31*(2), 200–217.

Stone, B., Jr., Vargo, J., Liu, P., Habeeb, D., DeLucia, A., Trail, M., Hu, Y., & Russell, A. (2014). Avoided heat-related mortality through climate adaptation strategies in three US cities. *PLoS One, 9*(6), e100852.

Sun, N., Yearsley, J., Baptiste, M., Cao, Q., Lettenmaier, D. P., & Nijssen, B. (2016). A spatially distributed model for assessment of the effects of changing land use and climate on urban stream quality. *Hydrological Processes, 30*(25), 4779–4798.

Sun, Y., & Cui, Y. (2018). Analyzing the coupling coordination among economic, social, and environmental benefits of urban infrastructure: Case study of four Chinese autonomous municipalities. *Mathematical Problems in Engineering, 2018*(1), 8280328.

Tao, Z. (2019). Research on the degree of coupling between the urban public infrastructure system and the urban economic, social, and environmental system: A case study in Beijing, China. *Mathematical Problems in Engineering, 2019*(1), 8206902.

Tatar, M., Kalvet, T., & Tiits, M. (2020). Cities4ZERO approach to foresight for fostering smart energy transition on municipal level. *Energies, 13*(14), 3533.

Tello, E., & Ostos, J. R. (2012). Water consumption in Barcelona and its regional environmental imprint: A long-term history (1717–2008). *Regional Environmental Change, 12*(2), 347–361.

Tozer, L., Mayr, M., Greenwalt, J., Nadi, G., & Runhaar, H. (2023). Mobilizing infrastructure investments for urban climate action in Africa: Enabling factors for multilevel action. *Local Environment, 28*(7), 867–881.

Urrutia-Azcona, K., Molina-Costa, P., Muñoz, I., Maya-Drysdale, D., Garcia-Madruga, C., & Flores-Abascal, I. (2021a). Towards an integrated approach to urban decarbonisation in practice: The case of vitoria-gasteiz. *Sustainability, 13*(16), 8836.

Urrutia-Azcona, K., Tatar, M., Molina-Costa, P., & Flores-Abascal, I. (2020). Cities4ZERO: Overcoming carbon lock-in in municipalities through smart urban transformation processes. *Sustainability, 12*(9), 3590.

Urrutia-Azcona, K., Usobiaga-Ferrer, E., De Agustín-Camacho, P., Molina-Costa, P., Benedito-Bordonau, M., & Flores-Abascal, I. (2021b). ENER-BI: Integrating energy and spatial data for cities' decarbonisation planning. *Sustainability, 13*(1), 383.

Valeo, C., He, J., & Kasiviswanathan, K. S. (2021). Urbanization under a changing climate–impacts on hydrology. *Water, 13*, 393.

Valero-Cabré, A., Sanches, C., Godard, J., Fracchia, O., Dubois, B., Levy, R., Truong, D. Q., Bikson, M., & Teichmann, M. (2019). Language boosting by transcranial stimulation in progressive supranuclear palsy. *Neurology, 93*(6), e537–e547.

Valero-Gil, J., Allué-Poc, A., Ortego, A., Tomasi, F., & Scarpellini, S. (2018). What are the preferences in the development process of a sustainable urban mobility plan? New methodology for experts involvement. *International Journal of Innovation and Sustainable Development, 12*(1-2), 135–155.

Verschuur, J., Fernández-Pérez, A., Mühlhofer, E., Nirandjan, S., Borgomeo, E., Becher, O., Voskaki, A., Oughton, E. J., Stankovski, A., Greco, S. F., & Koks, E. E. (2024). Quantifying climate risks to infrastructure systems: A comparative review of developments across infrastructure sectors. *PLoS Climate, 3*(4), e0000331.

Wang, G., & Yang, H. (2024). Optimization of green space pattern for alleviating the urban heat island effect in Qiantang District. In *Proceedings of the 2023 5th International Conference on Hydraulic, Civil and Construction Engineering (HCCE 2023)* (Vol. 26, p. 274). Springer Nature.

Wang, X., & Liu, L. (2023). The impacts of climate change on the hydrological cycle and water resource management. *Water, 15*(13), 2342.

Wei, H., Zhang, J. E., Singh, S. K., & Wang, M. (2019). Response of hydrological system to urbanization: A case study in Tianjin City, China. *Desalination and Water Treatment, 149*, 305–314.

White, R., & Wahba, S. (2019). Addressing constraints to private financing of urban (climate) infrastructure in developing countries. *International Journal of Urban Sustainable Development, 11*(3), 245–256.

Wiesmann, D., Ferrão, P., Niza, S., Rosado, L., & Brattebø, H. (2011). A spatially-resolved material flow model: Of the Lisbon metropolitan area. In *International Conference Virtual City and Territory (7è: 2011: Lisboa)* (pp. 103–106). Department of Civil Engineering of the University of Coimbra and e-GEO, Research Center in Geography and Regional Planning of the Faculty of Social Sciences and Humanities of the Nova University of Lisbon.

Winter, S. C., Obara, L. M., & Barchi, F. (2019). Environmental correlates of health-related quality of life among women living in informal settlements in Kenya. *International Journal of Environmental Research and Public Health, 16*(20), 3948.

Xu, W., Cong, J., & Proverbs, D. G. (2023). Evaluation of infrastructure resilience. *International Journal of Building Pathology and Adaptation, 41*(2), 378–400.

Yang, Y., Song, F., Ma, J., Wei, Z., Song, L., & Cao, W. (2022). Spatial and temporal variation of heat islands in the main urban area of Zhengzhou under the two-way influence of urbanization and urban forestry. *PLoS One, 17*(8), e0272626.

Ye, X., Wolff, R., Yu, W., Vaneckova, P., Pan, X., & Tong, S. (2012). Ambient temperature and morbidity: A review of epidemiological evidence. *Environmental Health Perspectives, 120*(1), 19–28.

Yu, H., & Ahlgren, E. O. (2023). Enhancing urban heating systems planning through spatially explicit participatory modeling. *Energies, 16*(11), 4264.

Zhang, Y., Yang, Z., & Yu, X. (2015). Urban metabolism: A review of current knowledge and directions for future study. *Environmental Science & Technology, 49*(19), 11247–11263.

Ziebell, A. C., & Singh, V. K. (2018). Energy indicator in sustainable urban energy metabolism and challenges. *Proceedings of the Institution of Civil Engineers-Energy, 171*(1), 26–31.

Ziervogel, G., Cowen, A., & Ziniades, J. (2016). Moving from adaptive to transformative capacity: Building foundations for inclusive, thriving, and regenerative urban settlements. *Sustainability, 8*(9), 955.

Part III

Actors, Institutions, and Movements

Hegemony and Climate Governance: Core States, Agreements, and the Limits of Multilateralism

9

9.1 Introduction

Global climate change emerges not only as an environmental crisis (Iyiola et al., 2024; Izah et al., 2023) but also as a profound challenge for governance. With the ongoing rise in global temperatures and the resulting extreme weather events, countries and communities are navigating a complex landscape where climate change intersects with political, economic, and social dynamics. Governance frameworks at all levels should evolve to recognize the multifaceted nature of climate challenges. The integration of climate policies into national, local, and municipal plans has become essential, as responses to climate change are increasingly seen as vital not only for environmental sustainability but also for safeguarding public health and societal stability (Fox et al., 2019; Khanal et al., 2024). Hegemony plays a critical role in international relations and global governance, notably within the climate change discourse. The hegemony of certain countries, particularly those with significant economic and military power, influences global climate agreements and defines the terms of climate governance. Emerging powers such as China also contribute to this dynamic, reflecting a shift in climate leadership that includes both traditional powerhouses and new contenders (Ju et al., 2022). The Paris Agreement, as a landmark framework, demonstrates how hegemonic processes shape international accords by allowing nations substantial leeway in their climate pledges, while hoping that this flexibility will spur broader collective action against climate change (Falkner, 2016).

Despite these advancements in global governance frameworks, significant power asymmetries persist in climate governance. These disparities often become evident in both nations' capacities to adapt to climate impacts and their political influence in international negotiations (Gong et al., 2020; Khanal et al., 2024). For instance, developing countries frequently face greater challenges in implementing climate policies due to resource constraints and limited institutional capacities,

M. C. Ogwu and S. Chibueze Izah, *Climate Frontiers and Social Transformation*,
World-Systems Evolution and Global Futures,
https://doi.org/10.1007/978-3-032-17639-4_9

which stand in stark contrast to the capabilities of developed nations. This reality underscores a fundamental inequity that threatens the effectiveness of global climate action, as wealthier nations have the resources to adapt and innovate. In comparison, poorer nations struggle to meet their developmental needs amid climate pressures (Eissa and Khalil 2022). The governance of climate change is further complicated by the diverse roles that non-state actors increasingly play in this domain. Urban areas, particularly cities, serve as critical actors in climate governance due to their concentration of emissions and vulnerability to climate impacts (Gordon, 2018; Ogwu, 2019; Sareen & Waagsaether, 2023). Cities often pursue innovative climate strategies, but their experiences also reflect broader governance challenges, in which local actions may be hampered by inadequate support from state entities or global frameworks. The role of cities underscores the importance of multi-level governance, in which power dynamics and resource allocation are crucial for effective climate action (Aylett, 2015; Xu, 2024).

Moreover, the politics of climate governance invariably reflect broader global power dynamics and the historical context of environmental policy formation. The legacy of colonialism and the current economic structures influence how climate policies are constructed and enacted across different regions (Gong et al., 2020; Ndambwa & Moonga, 2024). For example, developed nations tend to push frameworks that prioritize market-based solutions such as carbon trading. In contrast, nations in the Global South advocate for approaches that address development needs and acknowledge historical emissions. This divergence becomes a critical point of contention in international negotiations and highlights a persistent challenge to achieving consensus on climate action.

Given the interconnectedness of health and climate, governance frameworks should incorporate public health considerations in climate change policies, recognizing that climate change disproportionately impacts marginalized communities, particularly those in low-income scenarios (Fox et al., 2019; Khanal et al., 2024). Incorporating health perspectives into climate discussions could utilize existing frameworks in health governance, leading to more integrated and effective responses to climate-induced challenges. Institutions should evolve to unify efforts in climate change mitigation and public health, rather than treating them as isolated areas of concern (Izah & Ogwu, 2025).

The evolution of climate governance should also consider technological advancements and their potential for bridging governance gaps, particularly through digital tools that enhance transparency and accountability in climate action (Engvall & Flak, 2022). The use of digital platforms for reporting and monitoring climate actions can foster greater compliance and promote collaborative initiatives among nations. Nevertheless, these technologies should be implemented with caution, given the varying capacities of countries to leverage them. The interplay of technology and governance necessitates careful consideration of how power is distributed and exercised within the climate governance system. Addressing the structural imbalances in global climate governance requires a reframing of global priorities that recognize the shared nature of climate challenges while respecting national interests. Initiatives that promote mutual benefits (such as technology transfer, capacity building, and financial support for adaptation efforts) are essential for overcoming power disparities. Establishing a framework that prioritizes collaborative problem-solving over competitive posturing will enhance the legitimacy and effectiveness of global climate governance (Bäckstrand et al., 2017; Zhang & Bai, 2023). Such an approach could pave the way for a more equitable and effective international response to climate change, fostering greater shared responsibility among nations.

This chapter focuses on the role of hegemonic powers, particularly core states such as the United States, the European Union, and China, in shaping the structures and outcomes of global climate governance. It examines how international climate agreements, including the Kyoto Protocol and the Paris Agreement, have been influenced by the interests, strategies, and dominance of these states. The chapter further explores the limits of multilateralism by highlighting the power asymmetries that shape negotiations, the difficulties of enforcing global commitments, and the persistent tensions between equity, justice, and effectiveness in climate action. In doing so, it emphasizes how hegemony simultaneously enables progress while constraining genuine collective action on climate change. The work is beneficial as it provides a deeper theoretical understanding of how hegemony intersects with climate governance, thereby bridging international relations theory with environmental politics. It offers policy-relevant insights by showing how power dynamics shape climate agreements, helping negotiators and decision-makers adopt fairer, more effective strategies. The work also highlights issues of equity and justice by examining how hegemonic dominance affects developing countries, small island states, and other vulnerable groups. Additionally, it evaluates the strengths and weaknesses of multilateral climate frameworks and explores alternative governance mechanisms such as climate clubs, transnational networks, and private initiatives. The study further contributes to discussions on reforming global climate governance by proposing ways to balance the influence of core states with broader participation, accountability, and equity. Finally, it serves as a practical resource for scholars, activists, and policymakers seeking to understand the dynamics of power, cooperation, and resistance in international climate politics.

9.2 Theoretical and Conceptual Framework

Hegemony and climate governance are critical in understanding the interplay of power dynamics among states and their implications for global climate action (Table 9.1). Theoretical frameworks elucidate the nature of hegemony, ranging from its definition in international relations to its manifestation in climate governance. Hegemony, in this context, transcends mere military or economic dominance; it involves powerful states' ability to shape global norms, ideas, and values, particularly regarding climate responsibilities and strategies. The Gramscian perspective illustrates this notion, positing that hegemony entails the creation of consensus among weaker states through ideological leadership, often observed in how developed nations construct narratives about climate accountability and solutions. This framing is critical because it shapes perceptions of responsibility for climate action and the commitments made in international agreements such as the Paris Accord (Shen & Zou, 2024).

In contrast, the realist approach to hegemony emphasizes the concentration of power in certain states, which often use coercive measures to maintain order in international relations. Under this lens, climate treaties may be seen as instruments through which hegemonic states assert their strategic interests, rather than as collaborative efforts aimed at achieving global justice (Besombes et al., 2021). This dichotomy raises substantial questions about the nature and efficacy of international climate agreements, often revealing a tension between the ideals of global responsibility and the realpolitik of state interactions.

Furthermore, the neoliberal perspective adds complexity, suggesting that hegemony can facilitate international cooperation by providing a framework for establishing rules, reducing transaction costs, and enabling states to engage in multilateral negotiations. In this view, dominant states may play a vital role in

Table 9.1 Theoretical and conceptual perspectives on hegemony and climate governance

Concept	Major ideas	Implications for climate governance
Hegemony in IR	*Gramscian*: dominance through ideas, norms, and consent. *Realist*: power through coercion and strategic interests. *Neoliberal*: leadership that enables cooperation and rule-setting.	Shapes how climate issues are framed, negotiated, and enforced; hegemonic states drive or block agreements.
Core vs. periphery (world-systems theory)	Core states control resources, technology, and agenda-setting. Periphery states face vulnerabilities but limited influence. Semi-periphery states push for equity but remain constrained.	Explains global climate divides: those most responsible for emissions dominate, while vulnerable states struggle for justice.
Multilateralism in climate governance	Based on collective action, inclusivity, and consensus under UNFCCC. Gives all states formal voice in negotiations. Weaknesses: slow progress, weak enforcement, and dominance of powerful states.	Provides legitimacy but limited effectiveness; sovereignty vs. collective responsibility remains unresolved.
Legitimacy, soft power, and coercion	*Legitimacy*: fairness and inclusivity boost compliance. *Soft power*: diplomacy, finance, and technology transfer build alliances. *Coercion*: sanctions, trade restrictions, or imposed standards.	Demonstrates that climate governance is shaped by unequal power relations, not purely cooperation.

fostering agreements like the Paris Accord, thereby positioning themselves as essential players in achieving collective climate goals (Shen & Zou, 2024). This perspective highlights the potential for hegemonic states to orchestrate cooperative frameworks that account for the complex realities of global climate governance, even though their motivations may not align exclusively with altruism or the global good. Central to understanding the implications of hegemony in climate governance is the application of World-Systems Theory, which categorizes states into core, semi-periphery, and periphery nations. Core states, such as the United States, the European Union, and China, typically have substantial economic and technological advantages, enabling them to set the parameters for climate governance operations. Their dominance allows them to influence the discourse and direction of climate policies through negotiations and agreements that may favor their interests over those of less powerful nations (Shen & Zou, 2024). Conversely, peripheral states, which include a range of developing countries and small island states, frequently bear the gravest impacts of climate change yet possess limited negotiating power in these discussions. This neglect underscores a critical imbalance in which those least responsible for emissions bear the harshest consequences of policies shaped by the interests of powerful nations (Calmon et al., 2021). Semi-periphery states, including India, Brazil, and South Africa, find themselves in a transitional position, advocating for equitable terms in climate negotiations while grappling with constraints imposed by structural inequalities. These states often push for reforms that highlight their vulnerability to climate effects while simultaneously seeking to enhance their political agency within international frameworks. This ongoing struggle illustrates the entrenched divide between those who contribute disproportionately to climate problems and those who suffer from their ramifications, thereby complicating the narratives around global climate justice and responsibility (Nguyen et al., 2024).

In exploring multilateralism in international climate governance, it is essential to examine both its foundational principles and inherent limitations. On the one hand, the multilateral approach is grounded in collective action, inclusivity, and consensus-driven decision-making, often facilitated by institutions such as the United Nations Framework Convention on Climate Change (UNFCCC). This system provides legitimacy by ensuring that all states, irrespective of size or power, have a voice in climate negotiations (Pujiningsih et al., 2023). The quest for global consensus amplifies representation of diverse interests and enriches decision-making, fostering an inclusive dialogue that acknowledges varying circumstances among nations. On the other hand, the limitations of this multilateral approach are notable, particularly its slow decision-making and inadequate enforcement mechanisms. The influence of powerful states often dilutes bold commitments, leaving efforts to achieve more ambitious climate targets vulnerable to the strategic interests of dominant players (Ogwu & Izah, 2025). This imbalance poses significant challenges to effective climate governance, as powerful nations may resist policies that threaten their economic or political stability, thereby prioritizing their sovereignty over collective responsibility (Bahmanteymouri and Mohammadzadeh 2024). The need to harmonize national interests with global imperatives poses an ongoing dilemma in climate negotiations, shedding light on the intricacies of governance in an interconnected world confronting pressing environmental crises.

Furthermore, the complexities of legitimacy in international climate governance cannot be overlooked, as they play a paramount role in shaping alliances and negotiating dynamics among nations. Agreement legitimacy derives from its perceived fairness, inclusivity, and alignment with global norms. The widely recognized Paris Agreement aims to establish a substantive and inclusive framework that encourages universal participation (Akhtar, 2024). However, legitimacy is closely tied to the power dynamics at play; countries with greater political and economic might leverage their influence to cultivate supportive narratives or coerce compliance through soft power, including diplomacy, technology transfer, and climate financing. Coercion emerges as another mechanism through which power is exerted in climate governance, revealing the multilayered approaches that hegemonic states may employ to uphold their climate policies. Instances of coercion can manifest in various forms, including threats of sanctions, trade restrictions, or the imposition of rigorous environmental standards. Such measures underscore the complex interplay between cooperation and coercion in climate governance, showing that the collaboration often touted in international agreements can coexist with power imbalances and hegemonic tactics.

Throughout this discourse on hegemony and climate governance, there exists a pronounced need to address the structural inequalities that fundamentally characterize global climate negotiations. The differing levels of influence between core and periphery states articulate a broader narrative about climate justice and responsibility. Discussions should pivot toward acknowledging the disproportionate burdens borne by developing nations, enhancing their capability to advocate for their interests, and fostering equitable and effective climate governance structures (Nguyen et al., 2024). This nuanced understanding of hegemony not only amplifies the voices of the marginalized but also explores pathways for restructuring global climate governance to promote justice and equity. To consider the role of cooperation in addressing climate change, it is vital to recognize the interconnectedness of states in our global ecosystem. Successful climate action requires collaboration across borders, underscoring the need for national interests to consider global environmental imperatives increasingly. For instance, nations such as China and India, despite their rising status as major economies, highlight the intricate balance between competition and cooperation in climate discussions, where collective action should prevail over unilateral pursuits for real progress (Calmon et al., 2021).

In synthesizing these perspectives, a robust framework emerges to critically evaluate hegemonic practices in climate governance alongside their implications for both national and global landscapes. By understanding how dominant states frame climate discourse, communities on the periphery can strategize to ensure their needs and rights are recognized in international frameworks. Through this, it is possible to envision a more equitable future in which climate governance transcends power disparities. Instead, it fosters collaborations that address the myriad challenges posed by climate change.

In chronologically mapping the evolution of climate policies and international cooperation, it is paramount to identify moments of progressive engagement, particularly in contexts marked by contention and conflict. Such analysis enables scholars and policymakers alike to discern patterns that elucidate potential alliances and avenues for future action. The analysis of BRICS nations' cooperative efforts underscores the need to address not only climate-related development but also the political realities shaping these alliances (Shen & Zou, 2024). The emergence of new global coalitions represents both opportunities for reform and sources of contention within existing power structures.

This theoretical and conceptual framework underscores the centrality of hegemony in understanding the dynamics of climate governance. As hegemonic states continue to shape climate policies, addressing the profound inequalities present in international negotiations becomes paramount. By examining the prismatic notions of hegemony through various theoretical lenses (notably Gramscian thought, realism, and neoliberal perspectives), scholars can better comprehend the intricate relationships that inform climate governance. Acknowledging these complexities can pave the way for more inclusive dialogues and reformative initiatives equipped to tackle climate challenges effectively.

Figure 9.1 illustrates how global climate governance power is distributed across the world-system hierarchy. Core states—represented by the United States, the European Union, and China—hold the greatest agenda-setting authority, shape negotiation outcomes, and dominate climate finance and technology flows. Semi-peripheral states, such as India, Brazil, and South Africa, occupy an intermediate position: they possess growing political influence and developmental aspirations but remain constrained by structural inequalities within global institutions. Peripheral states, including Small Island Developing States (SIDS) and Least Developed Countries (LDCs), bear the highest climate vulnerability while exercising the least negotiation power. Arrows indicate the dominant direction of rule-setting, financial conditionalities, and technology transfer, visually highlighting the asymmetries that underpin climate governance and climate justice debates.

9.3 Historical Evolution of Climate Governance

The historical evolution of climate governance represents a complex tapestry woven from significant international moments, agreements, and shifting political dynamics. Initially, the Stockholm Conference in 1972 served as a vital global meeting, framing environmental issues as urgent international concerns. This landmark event catalyzed the establishment of the United Nations Environment Programme (UNEP), which coordinated global environmental efforts and marked the beginning of structured international dialogue on climate issues (Held & Roger, 2018). The resulting focus laid the groundwork for subsequent negotiations and discussions surrounding climate action and sustainable development (Lövbrand et al., 2017). Fast forward to the influential 1987 Brundtland Report, which popularized the concept of sustainable development and intricately

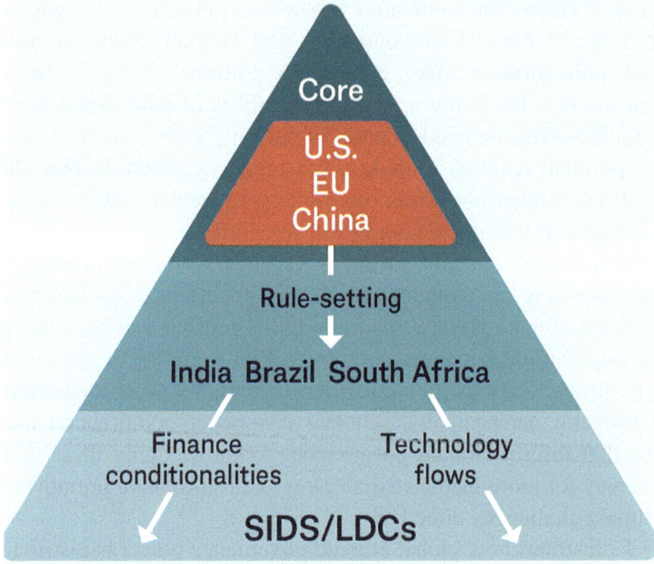

Fig. 9.1 Core–semi-periphery–periphery governance influence map

linked it to the need for comprehensive environmental stewardship. This report recognized the interconnectedness of economic development and environmental sustainability, emphasizing that effective environmental policies should also consider socio-economic factors (Held & Roger, 2018). By the time the 1992 Rio Earth Summit was convened, the momentum had built to a point where international agreements on climate governance could finally take shape. The Earth Summit spurred the adoption of the UN Framework Convention on Climate Change (UNFCCC), as well as Agenda 21 and the Rio Declaration, establishing a framework for multilateral climate negotiations and showcasing a global commitment toward environmental issues (Lövbrand et al., 2017).

The Kyoto Protocol, established in 1997, marked a significant differentiation in responsibilities among nations within the climate governance domain. It introduced the principle of Common but Differentiated Responsibilities (CBDR), stipulating that industrialized nations bear greater historical responsibility for greenhouse gas emissions than developing nations (Hale, 2016). This created binding emissions reduction targets for developed countries while exempting developing countries from such obligations. The introduction of market-based mechanisms, such as the Clean Development Mechanism and emissions trading, was a revolutionary step that facilitated more flexible approaches to reducing greenhouse gas emissions. However, various challenges soon emerged, including the United States' withdrawal from the Protocol and structural weaknesses in compliance and enforcement mechanisms, highlighting the complexities in coordinating global action on climate change (Hale, 2016). As discourse around climate governance matured, the Paris Agreement in 2015 introduced further nuanced approaches to managing national climate pledges. Transitioning from binding emissions targets to voluntary commitments known as Nationally Determined Contributions (NDCs), the Paris Agreement achieved nearly universal participation. This marked a shift toward a more inclusive, bottom-up approach that respects national sovereignty while fostering collective ambitions to mitigate climate change (Milkoreit, 2019). Despite its inclusive nature, the Agreement was not without weaknesses, as the lack of rigorous enforcement mechanisms and reliance on peer pressure for compliance raised questions about its effectiveness in achieving substantial climate action (Hale, 2016; Milkoreit, 2019).

The changing dynamics of global climate governance can also be understood through shifting hegemonic influences and the roles of prominent global actors. Initially, the United States wielded considerable influence as a leader in early climate negotiations. However, its retreat from commitments, such as withdrawal from the Kyoto Protocol and later the Paris Agreement under the Trump administration, represented a significant shift in leadership roles (Blok, 2020). In contrast, the European Union has emerged as a normative climate leader, advocating for ambitious climate policies through initiatives such as the EU Green Deal and mechanisms like carbon border adjustments to address its climate responsibilities (Krogmann, 2025). At the same time, China has become a critical player in global climate governance, representing both the world's largest emitter of greenhouse gases and a leader in renewable energy technologies (Izah et al., 2025). Its role complicates existing narratives on responsibility and power dynamics, as the Global South increasingly voices demands for justice and equity, particularly regarding climate finance and the recognition of "loss and damage" from climate impacts (Bäckstrand et al., 2017). The interplay among the United States, the EU, China, and the demands of developing nations exemplifies the intricate power struggles that characterize climate negotiations and the ongoing push for a fairer distribution of responsibilities (Abbott et al., 2016).

Significant developments, such as the 2009 Copenhagen COP15, underscored these tensions, as major powers sidelined broader negotiations in favor of limited bilateral accords, underscoring the need for an integrated multilateral approach (Abbott et al., 2016). The later Durban Platform in 2011 charted a path to the Paris Agreement, showcasing the critical role of multilateral talks in advancing global climate governance (Milkoreit, 2019). The Glasgow COP26 in 2021 further solidified commitments to coal phase-down and climate finance, yet it illustrated persistent limitations in ambition and the need for stronger enforcement measures (Milkoreit, 2019). At COP27 in 2022, the establishment of a Loss-and-Damage Fund marked a milestone for vulnerable states, highlighting the ongoing demand for justice in climate action (Hale, 2016). An increasingly vital role has emerged for non-state actors, including cities, corporations, and non-governmental organizations (NGOs), in pushing for climate action. Their participation supplements formal multilateral processes, underscoring the importance of local initiatives and grassroots movements alongside high-level negotiations (Bäckstrand et al., 2017).

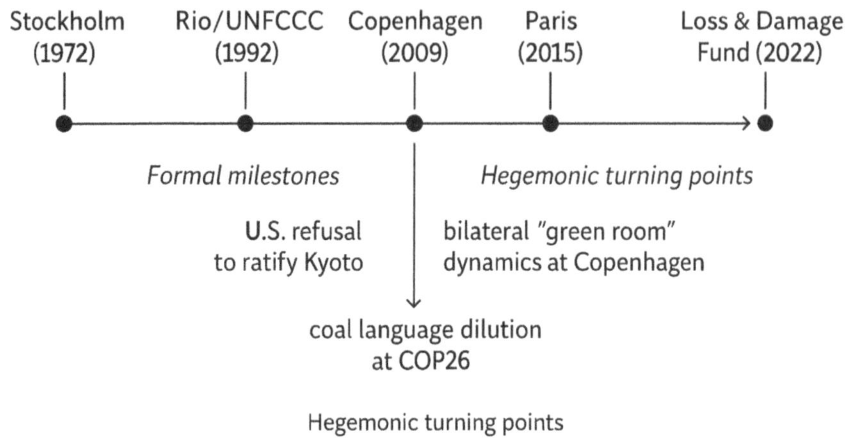

Fig. 9.2 Historical timeline of climate governance under hegemony

This citizen engagement enhances both traditional governmental frameworks and the broader landscape of climate governance, demonstrating the power of collective action to pursue ambitious goals (Depledge, Viñuales, Lees, and Reiner 2022).

Figure 9.2 provides a historical overview of the major moments that shaped global climate governance and highlights the critical hegemonic interventions that influenced the direction and ambition of international agreements. The timeline traces key diplomatic milestones—from Stockholm in 1972 to the establishment of the Loss-and-Damage Fund in 2022—and situates them alongside pivotal turning points where core states steered negotiations, blocked commitments, or reshaped institutional designs.

9.4 Core States and Hegemony in Climate Governance

The dynamics of climate governance involve various frameworks, institutions, and actors spanning both state and non-state entities that mediate and forge responses to climate change. Central to this governance structure are core states (primarily the United States, the European Union, and China) each playing distinct and influential roles that reflect their own geopolitical interests and histories.

9.4.1 The Role of Core States in Climate Governance

9.4.1.1 The United States

The United States has historically occupied a crucial role in shaping global climate governance, maneuvering between leadership and obstruction depending on the prevailing political climate. A salient example is the Kyoto Protocol of 1997, which the United States notably refused to ratify. The administration at that time cited economic costs and perceived inequities, particularly regarding developing countries' exemptions from binding commitments, as justification for its stance (Roelfsema et al., 2020). This refusal marked a significant moment in international climate negotiations, illustrating how domestic politics can jeopardize cooperative global efforts.

Conversely, the United States demonstrated leadership under the Obama administration in 2015 through the Paris Agreement, pushing for nationally determined contributions (NDCs). This participatory approach encouraged nations to declare their climate pledges while framing them within flexible agreements to promote compliance without coercion (Jordan et al., 2015). Nevertheless, the transition to the Trump administration marked a notable retreat from global climate leadership as the United States withdrew from the Paris Agreement. This withdrawal not only weakened immediate global climate momentum but also sent a disheartening signal regarding multilateral cooperation, a cornerstone of effective climate governance (Malini & Janardhana, 2024).

Under the Biden administration, the United States has re-engaged with the Paris Agreement and pledged to achieve net-zero emissions by 2050, indicating a renewed focus on climate action. Initiatives such as the Methane Pledge, launched in collaboration with the EU, seek to reduce methane emissions and are part of a broader strategy to align U.S. policy with global climate goals (Ju et al., 2022). However, the oscillating nature of U.S. climate policy underscores a troubling undercurrent in global climate negotiations: credibility suffers as international partners question the reliability of U.S. commitments amid partisan domestic politics (Lockwood, 2021).

9.4.1.2 The European Union

In stark contrast, the European Union is often seen as a normative leader in climate governance, advocating for sustainability and collective action through various legislative frameworks. One of the EU's landmark initiatives has been the European Green Deal, which integrates climate goals into economic policy and emphasizes the interconnected nature of sustainability and economic transformation (Earsom and Delreux 2023). This ambitious package not only aims to reduce carbon emissions but also aligns with broader social and economic objectives, exemplifying how climate policy can drive holistic societal change.

Furthermore, the EU employs impactful tools such as the Carbon Border Adjustment Mechanism (CBAM), designed to combat carbon leakage by imposing tariffs on imports from countries with less stringent climate regulations. This regulatory framework not only puts pressure on trade partners to decarbonize but also establishes the EU's position as a climate governance leader (Zhang et al., 2025). Notably, while the EU frequently compensates for U.S. policy vacuums, the internal divisions among its Eastern and Western members can hinder cohesive action, reflecting the complexities of multi-level governance (Jordan et al., 2015).

9.4.1.3 China

China, currently the most significant global greenhouse gas emitter, presents a nuanced profile in the context of climate governance. While it actively positions itself as a leader in renewable energy investment and advocates for developing countries at international forums, it is also cautious about committing to binding targets that may impede its development (Dubash, 2021). The pledge for carbon neutrality by 2060 and significant investments in solar and wind technologies highlight China's dual focus: balancing economic growth with climate responsibilities (Ju et al., 2022).

China's strategic behavior in international negotiations has made it a "swing state" regarding climate action—crucial to the success of global agreements, yet protective of its economic development narrative. Initiatives like the Belt and Road Initiative (BRI) afford China greater influence. However, concerns about continued fossil fuel financing underscore the complexities of its role in global climate governance (Ju et al., 2022). By negotiating through frameworks that prioritize mutual development and equity, China aims to defend the interests of the Global South while augmenting its rising global stature (Zhang et al., 2025).

9.4.2 Emerging Powers in Climate Governance

Emerging powers like India, Brazil, and South Africa are increasingly asserting their presence in global climate negotiations. India's climate strategy, grounded in the principles of "common but differentiated responsibilities," seeks to balance national interests with global solidarity, advocating poverty reduction while investing significantly in solar energy initiatives through platforms such as the International Solar Alliance (Lueong, 2024). Brazil's climate governance is profoundly affected by political leadership (including the environmental policies of past and current administrations), illustrating the volatility of climate commitments in political contexts (Eissa and Khalil 2022).

South Africa has emerged as a vocal proponent for climate justice, framing its climate policies around the need for stronger climate finance and equitable energy transitions for the Global South (Cihon et al., 2020). Collectively, countries within groupings such as BASIC (Brazil, South Africa, India, China) leverage their collective power to counterbalance Western dominance in climate negotiations, highlighting the evolving multipolar dynamics of climate governance (Chikosi et al., 2019).

9.4.3 The Role of Hegemonic Groupings

The Group of Seven (G7) and the Group of Twenty (G20) serve as critical platforms for climate governance among the world's most influential economies. The G7, primarily composed of wealthy industrialized democracies, sets important climate finance agendas and benchmarks for emission reduction, shaping global discourse on sustainability (Malini & Janardhana, 2024). However, the G20, which includes emerging economies among major emitters, has become increasingly central in climate cooperation albeit weakened by divergent national interests and priorities within the group (Malini & Janardhana, 2024).

These hegemonic groupings often use smaller forums like the Climate Club and the Major Economies Forum to expedite decision-making, circumventing the often slow and cumbersome United Nations Framework Convention on Climate Change (UNFCCC) process. This highlights an inherent tension between the inclusivity of the UNFCCC and the efficiency of elite gatherings, calling into question the legitimacy of negotiations that may exclude less powerful but highly vulnerable states and communities (Bansard, Pattberg, and Widerberg 2017).

9.4.4 Non-State Actors in Climate Governance

The landscape of climate governance is further enriched by the roles of non-state actors, including multinational corporations, financial institutions, philanthropic organizations, and subnational entities, which increasingly drive climate action. Corporations influence climate governance through innovation, lobbying for policy change, and establishing sustainability standards in global supply chains (Bäckstrand et al., 2017). Financial institutions such as the World Bank and the International Monetary Fund advocate green financing initiatives, facilitating investments in climate resilience that align with sustainable development goals (Chou 2021). Moreover, philanthropic networks such as the Gates Foundation and advocacy groups contribute to agenda-setting within climate negotiations, reinforcing norms around sustainability and equitable resource distribution. At the local level, cities and regions organize into transnational networks, exemplified by initiatives such as C40 Cities and the Under2 Coalition, which have become vital players in the climate governance arena, often acting more swiftly than national governments due to their capacity for localized decision-making (Bansard et al., 2017).

9.5 Multilateral Agreements and Their Limits

Multilateral agreements significantly shape the landscape of climate governance, yet these frameworks face substantial limitations stemming from hegemonic power and structural weaknesses. The United Nations Framework Convention on Climate Change (UNFCCC) is a foundational element of international climate governance; however, its consensus-based decision-making process often hampers swift, effective agreement. This method frequently leads to negotiations that yield outcomes at the lowest common denominator, inhibiting the ambitious climate action needed to address the escalating climate crisis (Aklin & Mildenberger, 2020; Falkner, 2016). The UNFCCC's reliance on consensus can result in the sidelining of the needs and priorities of less powerful states, as more dominant nations exert considerable influence over agenda-setting, often prioritizing their interests (Zhang & Li, 2024). This structural weakness translates into a deficit of capability in binding commitments, in which countries may interpret or dilute their obligations due to the absence of strong enforcement mechanisms (Burke, 2022). This lack of stringent obligations is concerning, as states tend to prioritize voluntary compliance, leading to inconsistencies in accountability and eroding collective action (Selby, 2018).

Multilateral climate negotiations under the UNFCCC present an inherent tension between formal inclusivity and constrained decision-making (Figure 9.3). While all states participate equally at the outset, the consensus rule narrows the range of viable outcomes and introduces layers of informality that privilege powerful actors as negotiations move from broad participation to "green room" processes and final text drafting, hegemonic influence intensifies, compressing ambition even as legitimacy is maintained. The emergence of voluntary Nationally Determined Contributions (NDCs) functions as a release valve, prioritizing participation over enforceability. Figure 9.3 visually summarizes how multilateralism is simultaneously broad in structure yet constrained in practice.

The principle of "Common but Differentiated Responsibilities" (CBDR) recognizes the historical emissions and responsibilities of developed countries. However, tensions arise when developed nations advocate for broader universal commitments, potentially diluting the equity mechanisms intended to safeguard the interests of developing countries (Hochstetler, 2021). The redefinition or reinterpretation of CBDR by more powerful states can minimize their obligations

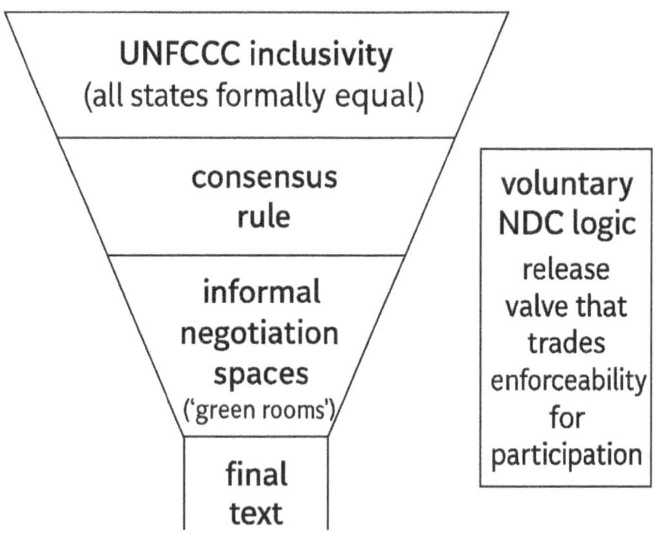

Fig. 9.3 The decision funnel of global climate negotiations

while disproportionately increasing expectations from developing countries, creating an environment that impedes progress in global climate negotiations (Bechtel et al., 2022). This situation exemplifies broader patterns of environmental injustice, contributing to the North-South divide in climate governance (Rietig 2021).

The Paris Agreement exemplifies both a significant step forward and the embodiment of these limitations, as it introduced a model of Nationally Determined Contributions (NDCs). While this system improves inclusivity by allowing nations to set self-defined targets for emissions reduction, it also risks diluting commitments. The flexibility inherent in voluntary pledges can result in lower ambition levels, undermining the collective goals intended to combat climate change (Falkner, 2016). The absence of binding commitments has already led powerful states to treat pledges as symbolic rather than substantive, largely due to insufficient legal enforcement mechanisms that allow substantial non-compliance (Burke, 2022; Liang & Liu, 2020).

A critical limitation becomes evident in the monitoring and compliance frameworks set forth under current multilateral agreements. Many reporting mechanisms are perceived as weak, leading to situations where states can underreport or misrepresent their climate progress without facing significant repercussions (Jaeger, 2023). This accountability gap creates a disparity in which less powerful states feel pressure to comply with international norms. In contrast, more powerful states can evade their commitments, perpetuating inequities in climate governance (Tache 2022). These power imbalances not only compromise the credibility of agreements but also reveal how the interests of dominant states overshadow the voices of vulnerable nations (Hochstetler, 2021; Zhang & Li, 2024).

Case studies, such as the Paris Agreement and the COP26 negotiations in Glasgow, highlight the persistent challenges posed by these hegemonic dynamics. While the Paris Agreement has been praised for its inclusiveness, it has also faced criticism regarding its weak enforcement mechanisms; the 2021 Glasgow meeting starkly illustrated these issues when major emitters like China and India successfully pressured negotiators to dilute commitments concerning coal phase-out, further demonstrating how power dynamics skew outcomes in favor of the powerful (Elsler et al., 2022; Zhang & Li, 2024). The U.S. withdrawal from the Paris Agreement under the Trump administration exemplifies how hegemonic states can withdraw from commitments without facing consequences, thereby undermining global climate governance and stability (Selby, 2018; Tache 2022).

The North-South divide is particularly pronounced in discussions surrounding climate finance, where developing nations demand essential changes, including financial assistance, technology transfer, and recognition of historical responsibilities for emissions (Hochstetler, 2021; Voigt, 2023). Despite these demands, developed nations frequently fail to fulfill their pledges, such as the long-promised \$100 billion annually for climate finance, exacerbating inequities and highlighting the unjust distribution of responsibilities (Oberthür and Groen 2017; Selby, 2018). Ongoing negotiations regarding loss and damage remain hindered by core states' hesitation to accept structural liabilities, underscoring the persistence of historical inequalities within contemporary governance frameworks (Burke, 2022; Hochstetler, 2021).

In addition to state-centric frameworks, the influence of non-state actors, including corporations and lobby groups, complicates the pursuit of effective climate action. These entities often align with powerful states, leading to outcomes that undermine ambitious climate objectives (Zhang & Li, 2024). Consequently, political discourse and financial flows related to climate change are vulnerable to manipulation by interests prioritizing short-term gains over long-term, global climate goals (Bechtel et al., 2022; Rietig 2021). The increasing reliance on "minilateral" arrangements and bilateral agreements reflects a diminishing faith in multilateralism, as states increasingly seek independent pathways to fulfill their climate obligations (Elsler et al., 2022).

9.6 Beyond Multilateralism: Alternative Governance Mechanisms

The contemporary dynamics of climate governance are increasingly characterized by mechanisms that extend beyond traditional multilateral frameworks (Table 9.2). The emergence of alternative governance structures, particularly bilateral and minilateral arrangements, signifies a shift in the paradigm of climate action. Such agreements, often formed among core states, provide avenues for more responsive decision-making than the cumbersome processes associated with global negotiations, such as those under the United Nations Framework Convention on Climate Change (UNFCCC). This trend can be illustrated through various high-profile collaborations, such as the EU–China cooperation on renewable energy technology and the U.S.–China climate dialogue, as well as commitments made by summits like the G7 and G20 that prioritize swift, actionable responses to climate change challenges (Bansard et al., 2017; Kaplan & Marantz, 2025).

This minilateralism offers a double-edged sword; while it empowers a small group of states to act more decisively, it can inadvertently reinforce hegemony. Core states, by establishing rules and norms within these informal networks, may exert pressure on less developed nations to conform to their frameworks, potentially marginalizing their voices in global discourse (Erbil et al., 2022; Kaplan & Marantz, 2025). This hierarchical structure raises concerns about inclusivity and the legitimacy of outcomes generated outside the universal frameworks of multilateralism. The challenge remains to ensure that such frameworks do not merely replicate existing power asymmetries but instead foster a more equitable approach to global climate governance. In the context of transnational networks, cities have emerged as vital actors in the fight against climate change. Initiatives such as C40 Cities and the Global Covenant of Mayors illustrate how municipal alliances pursue climate agendas that often surpass those of national governments. Through these networks, cities pursue ambitious climate actions and share innovative strategies that demonstrate their local commitment to sustainability (Bansard et al., 2017; Hsu et al., 2019). Studies indicate that urban alliances play a significant

Table 9.2 Alternative governance mechanisms beyond multilateralism in climate governance

Mechanism	Major features	Implications for hegemony and climate governance
Bilateral and minilateral climate clubs	Small-scale cooperation (e.g., EU–China, US–China, G7/G20); faster and more flexible than UN multilateralism.	Core states set agendas, enabling quicker progress but reinforcing hegemonic dominance.
Transnational networks	Cities (C40, Global Covenant of Mayors), NGOs, corporations, and civil society movements.	Decentralizes climate action, challenges hegemonic control, and pushes for greater ambition.
Private governance initiatives	Carbon markets (EU ETS), corporate frameworks (SBTi, RE100), ESG finance, and voluntary offsets.	Market-driven climate action that can complement or bypass states, often reflecting corporate or financial hegemony.
Finance and technology transfer	Climate funds, green bonds, renewable energy tech sharing, and adaptation programs.	Enables support for vulnerable states but is often shaped by conditions that reflect hegemonic interests.
Regional governance mechanisms	AU, ASEAN, EU, and MERCOSUR environmental policies and agreements.	Provides region-specific approaches that balance sovereignty with collective action.
Judicial and legal mechanisms	Domestic and international climate litigation (e.g., Uganda case, youth-led lawsuits).	Holds states and corporations accountable, reinforcing climate justice outside hegemonic negotiations.
Other emerging mechanisms	Climate security frameworks, digital monitoring (AI, blockchain), philanthropy, and knowledge coalitions (IPCC, universities).	Expands governance beyond traditional politics; can democratize action but risks new power asymmetries.

role in enhancing civic participation in climate policymaking, creating new opportunities for stakeholders that may feel excluded from formal negotiations (Arango et al., 2024; Salazar, 2023; Ogwu et al., 2025). The unique character of these initiatives lies in their ability to take swift, results-oriented actions tailored to local needs and conditions, which are essential for effective climate strategy implementation.

Moreover, non-governmental organizations (NGOs), corporations, and grassroots movements have begun to shape climate governance outside traditional multilateral frameworks. Environmental advocacy groups like Greenpeace and the Climate Action Network exert pressure on governments to enhance accountability and transparency in climate commitments. These organizations work collectively to hold states and corporations accountable, pushing for binding agreements rather than voluntary pledges (Tosun & Schoenefeld, 2017; Widerberg and Pattberg 2017). On the corporate front, many global firms have committed to sustainability through pledges to utilize renewable energy and reduce emissions, thus influencing climate norms through operational practices (Charalampaki, 2022). Civil society initiatives, such as Fridays for Future and Extinction Rebellion, further exemplify how grassroots movements are challenging the prevailing discourse on climate governance and advocating for more radical reforms based on principles of equity and justice (Wamsler, 2016).

Private governance initiatives also play a crucial role in shaping climate action by establishing market-driven frameworks that operate outside formal treaty regimes. Mechanisms like the EU Emissions Trading System (ETS) and voluntary carbon offset schemes illustrate how private governance creates economic incentives for emissions reduction (Budak, 2025; Tosun & Schoenefeld, 2017). This market-oriented approach not only facilitates greenhouse gas emission reductions but also drives investment flows toward a greener economy, as seen with the rise of green bonds and environmental, social, and governance (ESG) criteria in investment decision-making (Carbune 2023; Lee, 2018; Ogwu et al., 2025c, d). Each of these private initiatives can bypass bureaucratic inertia and accelerate climate action, though existing inequities should be addressed to ensure a just transition. Climate finance and technology transfer frameworks also illustrate alternative governance mechanisms relevant to climate governance in developing countries. Instruments such as the Green Climate Fund aim to provide the resources needed for mitigation and adaptation efforts in vulnerable regions. However, the effectiveness of these mechanisms is often tempered by the conditions imposed by donor states, which can create dependencies rather than empowering sustainable climate action (Budak, 2025; Erbil et al., 2022). The approach to climate finance should promote recipient nations' agency in decision-making, ensuring they are not merely passive beneficiaries but active agents shaping their climate trajectories (Huang et al., 2019; Liu et al., 2023).

At the regional governance level, various initiatives undertaken by entities such as the African Union, ASEAN, and regional blocs in Latin America highlight the importance of localized responses to global climate challenges. These governance frameworks often permit context-specific solutions that align with broader international commitments, fostering a collective response addressing localized climate vulnerabilities while promoting sustainable development (Fenton, 2017; Widerberg and Pattberg 2017). Judicial and legal mechanisms have also gained prominence as alternative methods of climate governance. Increasingly, courts are being used to compel states and corporations to meet their climate obligations. Landmark cases, such as the Urgenda ruling in the Netherlands and various youth-led lawsuits worldwide, have demonstrated the potential of legal action to drive significant policy change (Charalampaki, 2022; Lee, 2018). Such legal avenues not only bypass stagnation associated with multilateral negotiations but also enhance demands for accountability and climate justice, particularly for marginalized communities most affected by climate change (Karhinen, Peltomaa, Riekkinen, and Saikku 2021; Tosun and Schoenefeld 2017).

Emerging mechanisms in climate governance are also observed in how security frameworks, digital governance tools, and philanthropic initiatives reshape the landscape of climate action. For instance, linking climate action with global security agendas has gained traction among certain governments and organizations emphasizing the interconnectedness of climate stability and global peace (Fenton, 2017; Karhinen, Peltomaa, Riekkinen, and Saikku 2021). Additionally, the application of digital technologies, such as artificial intelligence and blockchain, is beginning to enhance compliance tracking and emissions monitoring, transforming governance modalities for climate commitments (Fenton, 2017; Gesing, 2018). Philanthropic efforts by major foundations to fund clean energy projects demonstrate how resources can support innovative solutions outside traditional public funding structures, prompting discussions on accountability and influence over climate priorities (Liu et al., 2023; Mai, 2019).

9.7 Hegemony, Inequality, and Climate Justice

The intricate relationship between hegemony, inequality, climate governance, and climate justice presents a multifaceted challenge (Table 9.3), particularly for vulnerable nations such as Small Island Developing States (SIDS) and Least Developed Countries (LDCs). Developing nations contribute the least to global emissions but disproportionately bear the consequences of climate change, raising ethical and moral concerns regarding climate justice. These vulnerabilities arise from systemic inequalities embedded in the international climate governance framework, where hegemonic actors often dominate decision-making, sidelining the voices of marginalized countries.

SIDS, including those in the Pacific and Caribbean, face existential threats from rising sea levels, intensifying storms, and coastal erosion, despite minimal contributions to global greenhouse gas emissions. These nations are often in precarious positions highlighted by their environmental fragility and economic constraints that limit their adaptive capacities. It has been documented that the institutional and financial capabilities required to combat and adapt to climate impacts are severely lacking in these regions (Newell et al., 2021). The stark reality is that while SIDS typically lack the resources necessary for effective adaptation or resilience-building efforts, they confront escalating climate threats, highlighting a profound imbalance within global climate governance systems (Sultana 2022).

Similarly, LDCs face the dual burdens of economic vulnerability and environmental impact as they engage in climate governance arenas dominated by wealthier, high-emitting states. These countries often lack the robust economic structures and institutional capacities necessary to participate effectively in international negotiations, leading to their marginalization in discussions that significantly affect their futures (Sasser 2024). The structural inequalities prevalent in these negotiations mean that critical adaptation measures for LDCs are frequently overlooked or inadequately funded. This reflects a broader trend in which negotiations are heavily influenced by powerful economies that frame the agenda to prioritize their interests over those of developing nations.

Current debates surrounding "loss-and-damage" mechanisms further complicate the landscape of international climate governance. The discourse often centers on whether wealthier, historically high-emission states should bear the financial

Table 9.3 Hegemony, inequality, and climate justice in global climate governance

Dimension	Key issues	Implications for climate governance
Vulnerabilities of SIDS and LDCs	Existential threats from sea-level rise, storms, and weak adaptive capacity. Moreover, contribute the least to emissions but face the most significant risks.	Limited ability to shape negotiations; survival needs are often sidelined by hegemonic priorities.
Loss-and-damage debates	Calls for finance, compensation, and responsibility from high-emission states and underfunded mechanisms with weak accountability.	Highlights the justice gap; resistance from core states stalls progress on binding commitments.
Power asymmetries in decision-making	Core states dominate agenda-setting and outcomes. Developing states lack resources, expertise, and representation in informal negotiation spaces.	Agreements often reflect hegemonic interests rather than equitable solutions.
Equity, representation, and inclusivity	Contestation of "common but differentiated responsibilities" (CBDR) and marginalized groups (youth, women, Indigenous communities) seek greater voice.	Inclusivity often remains symbolic; systemic inequalities persist in governance structures.
Other critical dimensions	Climate finance gaps (e.g., unmet $100bn pledge), barriers to technology transfer due to intellectual property regimes, "carbon colonialism" via offsets and resource dependence. Geopolitical rivalries (US–China, EU carbon borders), and justice movements mobilizing from below.	Deepens distrust between North and South; civil society and grassroots movements push for transformative justice.

responsibility for compensating vulnerable nations for irreversible climate damage. Despite establishing mechanisms under the United Nations Framework Convention on Climate Change (UNFCCC), significant issues persist related to funding shortfalls and accountability (Haque, 2024). Many developing nations contend that compensation should align with principles of historical responsibility and justice; however, core states frequently resist binding financial commitments, illustrating a significant divergence in interests between hegemonic actors and vulnerable nations (Leroux & Mintz-Woo, 2023). This ongoing debate reflects the tensions inherent in climate governance and broader global inequalities marked by power asymmetries and ethical considerations in climate justice (Okereke, 2024). Loss and damage have become one of the most politically charged dimensions of global climate governance, revealing sharp divides between highly vulnerable states and major emitters. The pathway shown in Figure 9.4 illustrates the sequence through

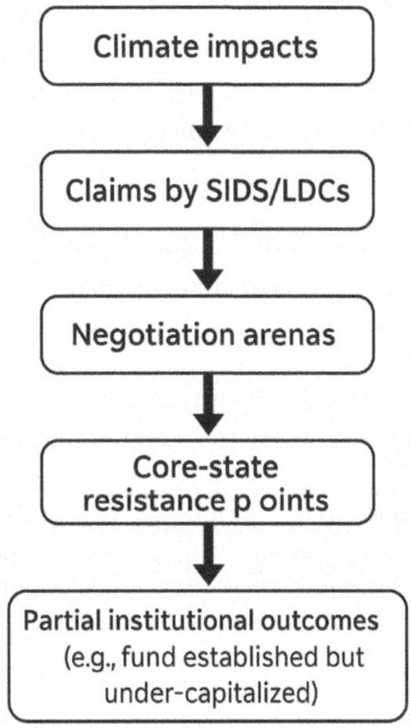

Fig. 9.4 Loss-and-Damage Politics Pathway

which climate impacts generate compensation claims, enter negotiation arenas, encounter resistance from core states, and ultimately produce partial institutional outcomes.

Power asymmetries in decision-making processes within the UNFCCC and other climate conferences illustrate a pattern of exclusion. Hegemonic states often dominate agenda-setting and negotiations, diluting more ambitious climate action proposals that could benefit the most vulnerable (Pruett & Hill, 2024). Developing countries typically lack the resources, technical expertise, and negotiating leverage to ensure a level playing field, undermining their potential contributions to climate governance discussions. Informal spaces, such as "green rooms," often exclude smaller delegations from critical discussions, perpetuating an environment where comprehensive and inclusive dialogue remains a challenge (Miner & Thurman, 2024). Climate governance structures tend to mirror global power dynamics, where the strategic and economic interests of dominant actors overshadow pressing needs for equity and inclusivity in policymaking.

Discussions surrounding equity and representation highlight the contested nature of "common but differentiated responsibilities" (CBDR), as developed nations often push for universal commitments that do not account for disparities in capabilities and responsibilities. Historically marginalized groups, including Indigenous peoples and vulnerable communities, stress the necessity of their participation in climate governance to ensure that their rights and perspectives are acknowledged (Watts et al., 2019). Furthermore, movements advocating for climate justice intrinsically link climate governance with broader social justice issues, necessitating a holistic examination of how climate policies intersect with economic inequalities and human rights (Jenkins, 2018; Wagle & Philip, 2022). While inclusivity is a common aspiration in climate governance forums, hegemonic structures often constrain meaningful participation from representatives of developing nations, reducing inclusivity to a symbolic gesture rather than fostering genuine engagement (Hölscher et al., 2019).

One of the critical dimensions of these discussions is climate finance, particularly the unmet pledge to mobilize $100 billion annually for climate action in developing countries. This financial shortfall exacerbates mistrust among countries of the Global South toward their Northern counterparts and reinforces existing inequalities within climate governance frameworks (Pham & Saner, 2021). Additionally, barriers to technology transfer, driven by restrictive patent regimes and intellectual property rights, inhibit developing nations from accessing the clean technologies necessary for their transition to sustainable practices and resilience-building (Gupta et al., 2024). Criticism has also been directed at "carbon colonialism," in which carbon offset markets can lead to the expropriation of land and resources in the Global South, perpetuating dependency and exploitation while undermining efforts to achieve true climate justice (Stephens, 2021).

Geopolitical rivalries, such as tensions between the United States and China, intricately shape the climate governance landscape, often overshadowing environmental priorities and steering international discourse toward strategic competition rather than collaborative climate action (Bellamy 2016). However, grassroots activism and the efforts of Indigenous leaders highlight the potential for transformative change driven by those most affected by climate change (Bailey, 2017). These movements illustrate marginalized societies' capacity to challenge hegemonic frameworks, advocate for their rights, and demand a more equitable distribution of power in climate governance, aligning with broader justice movements that recognize the intersectionality of climate issues (Yan and Luo 2024).

9.8 Future Prospects: Reforming Climate Governance

The prospects of climate governance, particularly regarding issues of hegemony and inequality, necessitate a thorough reevaluation of existing frameworks and practices. As nations grapple with the effects of climate change, it becomes increasingly evident that effective governance should transcend national boundaries and embark on cooperative strategies that engage both developed and developing nations. This reformation of climate governance should empower historically marginalized voices from the Global South, navigate the complexities of national sovereignty, and reinforce accountability and enforcement mechanisms across multilateral agreements. One proposed strategy in this transformative process is the establishment of climate clubs and coalitions of the willing. These informal agreements can foster a group of forward-thinking countries committed to ambitious climate policies, thereby circumventing the inertia often found in more traditional multilateral negotiations (Hale, 2020). Such clubs not only hold member states accountable but also set benchmarks for policy implementation and progress. Research has shown that these catalytic institutions can provide a framework for cooperation similar to that of the Paris Agreement and cultivate transnational networks that enhance the overall efficacy of climate action (Hale, 2020). This approach could serve as a mechanism to mitigate the tensions associated with climate unilateralism, which often emerges when powerful nations prioritize their interests at the expense of global cooperation (Zuk, Conversi, and Zuk 2024). By facilitating collaboration among like-minded states, climate clubs would also accelerate the diffusion of innovative technologies and sustainable practices, enhancing the collective ability to respond to global climate imperatives.

A critical element of reforming climate governance is strengthening accountability and enforcement mechanisms. Current frameworks often suffer from weak compliance and insufficient mechanisms to address breaches of commitments. The effectiveness of international climate accords hinges not only on the commitment of signatory states but also on the real capacity to monitor and enforce compliance. Enhanced accountability can be achieved by integrating civil society organizations (CSOs) into governance structures. CSOs have demonstrated the ability to advocate for marginalized communities and hold states accountable for their climate actions or inactions (Stepanova, 2024). By leveraging grassroots mobilization, civil society can amplify the voices of those in the Global South who are disproportionately affected by climate change but frequently excluded from vital decision-making processes. The experiences of communities in places like Mexico indicate that collaborative efforts between local stakeholders and governmental entities can lead to more effective environmental governance (Cotler et al., 2022).

Moreover, the democratization of climate governance is integral to embedding equity in the climate response. This entails recognizing the voices of the Global South and ensuring they play a substantial role in shaping policy frameworks that affect their nations. The literature emphasizes that inclusive governance structures enable better outcomes and enhance legitimacy (Stepanova, 2024). By actively involving diverse groups in policy discussions, states create an environment in which climate policies reflect the needs and priorities of various populations rather than the interests of elites alone. This participatory approach aligns with the concept of ecological sovereignty, which advocates collaborative efforts that transcend mere state interests (Azis & Sukaraja, 2024).

Balancing national sovereignty with global climate imperatives poses significant challenges for governance structures. The rise of energy nationalism, characterized by political decisions that prioritize national interests over global responsibilities, complicates multilateral cooperation (Zuk, Conversi, and Zuk 2024). For example, energy policies should be navigated carefully to avoid undermining national autonomy while adhering to global climate objectives. National sovereignty has historically been seen as fundamental to state behavior. Yet, the pressing need for coordinated action has highlighted tensions in unilateral approaches that neglect the collective nature of climate change (Shafi et al., 2021). A nuanced understanding of sovereignty that balances national context and global obligations is crucial to advancing effective climate policies.

The role of science and technology in reshaping climate governance cannot be overstated. Initiatives such as the Brazilian Earth System Model illustrate how Southern nations strive to enhance their representation in global climate discussions (Miguel et al., 2019). Such efforts can bridge the North-South knowledge divide, enabling the integration of localized climate data into international discourses. This incorporation not only democratizes knowledge production but also values the unique climatic and socio-economic conditions across diverse countries, thereby enhancing the effectiveness of policy recommendations and actions. Technology transfer processes that account for local contexts and are supported by international frameworks can lead to more sustainable and equitable outcomes in climate governance (Palley, 2021). The incorporation of civil society into climate discussions is paramount. Through their grassroots efforts, CSOs can empower community members to become advocates for policy reforms and contributors to climate solutions (Stepanova, 2024). The lessons learned from various community-based initiatives suggest that cooperation among diverse stakeholders significantly enhances governance efficacy and builds networks of trust (Cotler et al., 2022). This highlights an essential factor for governing climate change effectively: a broader coalition that embeds collective interests and responsibilities across differentiated contexts.

9.9 Conclusion

The evolution of climate governance highlights the deep entanglement of power, politics, and environmental urgency. Hegemonic states (particularly the United States, the European Union, and China) have consistently shaped global climate outcomes by setting agendas, financing initiatives, and driving technological innovation. However, these same actors have also served as obstacles, often prioritizing national interests over collective global responsibilities, thereby exposing the structural weaknesses of multilateral frameworks such as the UNFCCC. Consensus-based decision-making, voluntary pledges, and the erosion of the principle of Common but Differentiated Responsibilities illustrate the limits of multilateralism under unequal power relations. The historical trajectory from Stockholm through Kyoto to Paris demonstrates both adaptation and stagnation, revealing how core states and shifting political contexts continue to define progress. Finally, the intersection of hegemony, inequality, and climate justice underscores the need for a more inclusive and equitable global response. The challenge of climate governance is not merely technical but deeply political, requiring structures that recognize power asymmetries while amplifying the voices of vulnerable states.

To move toward a more effective and just climate governance framework, the international community should pursue reforms that enhance inclusivity, accountability, and resilience. This entails embedding equity and climate justice at the core of global agreements, ensuring that Small Island Developing States (SIDS), Least Developed Countries (LDCs), and marginalized communities participate meaningfully in decision-making processes. Stronger compliance and enforcement mechanisms are needed to hold core states accountable for emissions reductions and climate finance commitments, particularly in adaptation and loss-and-damage support. At the same time, alternative governance models (such as minilateral climate clubs, regional alliances, transnational networks, and grassroots initiatives) should be leveraged to complement multilateralism while guarding against new forms of exclusion. Recognizing the interconnectedness of climate change with public health, technological advancement, and sustainable development, governance structures should remain flexible, cooperative, and adaptive. Only by rebalancing power and embedding principles of fairness can global climate governance transcend symbolic agreements and respond effectively to the existential threat posed by climate change.

References

Abbott, K. W., Green, J. F., & Keohane, R. O. (2016). Organizational ecology and institutional change in global governance. *International Organization, 70*(2), 247–277.

Akhtar, I. (2024). The Significance of International Cooperation on Climate Change Mitigation in Afghanistan. *Nangarhar University International Journal of Biosciences*, 456–459.

Aklin, M., & Mildenberger, M. (2020). Prisoners of the wrong dilemma: why distributive conflict, not collective action, characterizes the politics of climate change. *Global Environmental Politics, 20*(4), 4–27.

Angel, P. A., & S. (2025). Empowering change: The crucial role of civil society in climate change mitigation. *American Journal of Economics and Sociology, 84*(1), 21–32.

Aylett, A. (2015). Institutionalizing the urban governance of climate change adaptation: Results of an international survey. *Urban Climate, 14*, 4–16.

Azis, F., & Sukaraja, D. (2024). Ecological sovereignty: beyond state sovereignty and legal responsibility. *Jurisprudensi Jurnal Ilmu Syariah Perundang-Undangan Ekonomi Islam, 16*(1), 195–209.

Bäckstrand, K., Kuyper, J. W., Linnér, B. O., & Lövbrand, E. (2017). Non-state actors in global climate governance: from Copenhagen to Paris and beyond. *Environmental Politics, 26*(4), 561–579.

Bahmanteymouri, E., & Mohammadzadeh, M. (2024). 'Neoliberalism is dead': Traversing neoliberal planning education is an exigency. *Policy Futures in Education, 22*(4), 625–641.

Bailey, I. (2017). Spatializing climate justice: Justice claim making and carbon pricing controversies in Australia. *Annals of the American Association of Geographers, 107*(5), 1128–1143.

Bansard, J. S., Pattberg, P. H., & Widerberg, O. (2017). Cities to the rescue? Assessing the performance of transnational municipal networks in global climate governance. *International Environmental Agreements: Politics, Law and Economics, 17*(2), 229–246.

Bechtel, M. M., Scheve, K. F., & van Lieshout, E. (2022). Improving public support for climate action through multilateralism. *Nature Communications, 13*(1), 6441.

Bellamy, R. (2016). A sociotechnical framework for governing climate engineering. *Science, Technology, & Human Values, 41*(2), 135–162.

Besombes, C., Pannekoucke, O., Lapeyre, C., Sanderson, B., & Thual, O. (2021). Producing realistic climate data with GANs. *Nonlinear Processes in Geophysics Discussions, 2021*, 1–39.
Blok, A. (2020). Urban green gentrification in an unequal world of climate change. *Urban Studies, 57*(14), 2803–2816.
Budak, T. G. (2025). Exploring Alternative Governance Structures for Climate Cooperation. In *Beyond Treaties: Rethinking Legal Mechanisms for International Climate Governance* (pp. 79-101).
Burke, A. (2022). An architecture for a net zero world: Global climate governance beyond the epoch of failure. *Global Policy, 13*, 24–37.
Calmon, D., Jacovetti, C., & Koné, M. (2021). Agrarian climate justice as a progressive alternative to climate security: Mali at the intersection of natural resource conflicts. *Third World Quarterly, 42*(12), 2785–2803.
Carbune, A. (2023). Shaping Interstate Decision-Making? Framing as a Persuasive Tactic in Transnational Climate Governance. *JCMS: Journal of Common Market Studies, 61*(5), 1143–1160.
Charalampaki, E. (2022). Minilateralism for Multilateralism: What Role for the EU in the Instrumentalization of the Japan-Australia Strategic Partnership. *HAPSc Policy Briefs Series, 3*(2), 103–116.
Chikosi, E. S., Mugambiwa, S. S., Tirivangasi, H. M., & Rankoana, S. A. (2019). Climate change and variability perceptions in Ga-Dikgale community in Limpopo Province, South Africa. *International Journal of Climate Change Strategies and Management, 11*(3), 392–405.
Chou, M. (2021). Australian local governments and climate emergency declarations: Reviewing local government practice. *Australian Journal of Public Administration, 80*(3), 613–623.
Cihon, P., Maas, M. M., & Kemp, L. (2020). Fragmentation and the future: Investigating architectures for international AI governance. *Global Policy, 11*(5), 545–556.
Cotler, H., Cuevas, M. L., Landa, R., & Frausto, J. M. (2022). Environmental governance in urban watersheds: the role of civil society organizations in Mexico. *Sustainability, 14*(2), 988.
Depledge, J., Viñuales, J. E., Lees, E., & Reiner, D. M. (Eds.). (2022). *Climate Policy after the 2015 Paris Climate Conference*. Routledge, Taylor & Francis Group.
Dubash, N. K. (2021). Varieties of climate governance: the emergence and functioning of climate institutions. *Environmental Politics, 30*(sup1), 1–25.
Earsom, J., & Delreux, T. (2023). One big conversation: the EU's climate diplomacy across the international regime complex on the Paris agreement negotiations. *JCMS: Journal of Common Market Studies, 61*(1), 198–214.
Eissa, Y., & Khalil, H. A. E. E. (2022, June). Urban climate change governance within centralised governments: A case study of Giza, Egypt. In *Urban forum* (Vol. 33, No. 2, pp. 197-221). Dordrecht: Springer Netherlands.
Elsler, L. G., Oostdijk, M., Levin, L. A., Satterthwaite, E. V., Pinsky, M. L., Crespo, G. O., & Wisz, M. S. (2022). Protecting ocean carbon through biodiversity and climate governance. *Frontiers in Marine Science, 9*, 880424.
Engvall, T. S., & Flak, L. S. (2022). The state of information infrastructure for global climate governance. *Transforming Government: People, Process and Policy, 16*(4), 436–448.
Erbil, A. Ö., Eroğlu, M., & Türk, E. G. (2022). Bridging climate change science and policy through TMNs in Turkey: CoM as a boundary-object. *Science and Public Policy, 49*(6), 830–842.
Falkner, R. (2016). The Paris Agreement and the new logic of international climate politics. *International Affairs, 92*(5), 1107–1125.
Fenton, P. (2017). The role of port cities and transnational municipal networks in efforts to reduce greenhouse gas emissions on land and at sea from shipping–An assessment of the World Ports Climate Initiative. *Marine Policy, 75*, 271–277.
Fox, M., Zuidema, C., Bauman, B., Burke, T., & Sheehan, M. (2019). Integrating public health into climate change policy and planning: state of practice update. *International journal of environmental research and public health, 16*(18), 3232.

Gesing, F. (2018). Transnational municipal climate networks and the politics of standardisation: The contested role of climate data in the new global covenant of mayors for climate and energy. *Politics and Governance, 6*(3), 126–135.

Gong, X., Liu, Y., & Sun, T. (2020). Evaluating climate change governance using the "polity–policy–politics" framework: A comparative study of China and the United States. *Sustainability, 12*(16), 6403.

Gordon, D. J. (2018). Global urban climate governance in three and a half parts: Experimentation, coordination, integration (and contestation). *Wiley Interdisciplinary Reviews: Climate Change, 9*(6), e546.

Gupta, V., Jadhav, D., & Ronald, B. (2024). Transfer of ESTs in international law: A climate justice approach. *Environmental Law Review, 26*(2), 117–140.

Hale, T. (2016). "All hands on deck": The Paris agreement and nonstate climate action. *Global environmental politics, 16*(3), 12–22.

Haque, M. A. (2024). Conceptualizing Climate Justice. *Journal of Climate Policy, 3*(1), 12–30.

Held, D., & Roger, C. (2018). Three models of global climate governance: from Kyoto to Paris and beyond. *Global Policy, 9*(4), 527–537.

Hochstetler, K. (2021). Climate institutions in Brazil: three decades of building and dismantling climate capacity. *Environmental Politics, 30*(sup1), 49–70.

Hölscher, K., Frantzeskaki, N., McPhearson, T., & Loorbach, D. (2019). Tales of transforming cities: Transformative climate governance capacities in New York City, US and Rotterdam, Netherlands. *Journal of environmental management, 231*, 843–857.

Hsu, A., Höhne, N., Kuramochi, T., Roelfsema, M., Weinfurter, A., Xie, Y., Lütkehermöller, K., Chan, S., Corfee-Morlot, J., Drost, P., & Faria, P. (2019). A research roadmap for quantifying non-state and subnational climate mitigation action. *Nature Climate Change, 9*(1), 11–17.

Huang, L., Kelly, S., Lv, K., & Giurco, D. (2019). A systematic review of empirical methods for modelling sectoral carbon emissions in China. *Journal of cleaner production, 215*, 1382–1401.

Iyiola, A. O., Afolabi, O. A., Alimi, S. K., Akingba, O. O., Izah, S. C., & Ogwu, M. C. (2024). Climate Change and Water Crisis in the Global South. In S. C. Izah, M. C. Ogwu, A. Loukas, & H. Hamidifar (Eds.), *Water Crises and Sustainable Management in the Global South* (pp. 111–140). Springer. https://doi.org/10.1007/978-981-97-4966-9_4

Izah, S. C., Iyiola, A. O., Yarkwan, B., & Richard, G. (2023). Impact of air quality as a component of climate change on biodiversity-based ecosystem services. In A. L. Srivastav, A. K. Dubey, A. Kumar, S. K. Narang, & M. A. Khan (Eds.), *Visualization Techniques for Climate Change with Machine Learning and Artificial Intelligence* (pp. 123–148). Published by Elsevier. isbn:978-0-323-99714-0.

Izah, S. C., & Ogwu, M. C. (2025). *Innovative Approaches in Environmental Health Management: Processes, Technologies, and Strategies for a Sustainable Future*. Springer Nature. https://doi.org/10.1007/978-3-031-81966-7

Izah, S. C., Ogwu, M. C., & Hait, M. (2025). Clean Energy Solutions and Public Health: Innovations in Renewable Energy for a Healthier Environment. In S. C. Izah & M. C. Ogwu (Eds.), *Innovative Approaches in Environmental Health Management: Processes, Technologies, and Strategies for a Sustainable Future. Environmental Science and Engineering Series*. Springer Nature. https://doi.org/10.1007/978-3-031-81966-7_7

Jaeger, W. K. (2023). Climate change and the problem of social cost. *PLOS Climate, 2*(9), e0000287.

Jenkins, K. (2018). Setting energy justice apart from the crowd: Lessons from environmental and climate justice. *Energy Research & Social Science, 39*, 117–121.

Jordan, A. J., Huitema, D., Hildén, M., Van Asselt, H., Rayner, T. J., Schoenefeld, J. J., Tosun, J., Forster, J., & Boasson, E. L. (2015). Emergence of polycentric climate governance and its future prospects. *Nature Climate Change, 5*(11), 977–982.

Ju, B., Shi, X., & Mei, Y. (2022). The current state and prospects of China's environmental, social, and governance policies. *Frontiers in Environmental Science, 10*, 999145.

Kaplan, R., & Marantz, E. A. (2025). Glocal Capitalist Class: How Same-Nation Interlocks Facilitate Transnational Corporate Political Unity in Global Environmental Politics. *Global Networks, 25*(1), e12522.

Karhinen, S., Peltomaa, J., Riekkinen, V., & Saikku, L. (2021). Impact of a climate network: The role of intermediaries in local level climate action. *Global Environmental Change, 67*, 102225.

Khanal, S., Bolte, G., & Boeckmann, M. (2024). Sector silos in climate action-a missed opportunity to prioritize health equity. *PLoS Climate, 3*(2), e0000349.

Krogmann, D. (2025). Liberal environmentalism and climate change in the polycrisis. *Global Sustainability, 8*, e8.

Lee, T. (2018). Local energy agencies and cities' participation in translocal climate governance. *Environmental Policy and Governance, 28*(3), 131–140.

Liang, C., & Liu, B. (2020). Challenge or opportunity of climate financial fragmentation: evidence from China-initiated cooperation with emerging multilateral institutions. *International Journal of Climate Change Strategies and Management, 12*(3), 289–303.

Liu, W., Tao, Y., & Bi, K. (2023). An Empirical Enquiry into Transnational Co-Patent Networks and their Characteristics, Evolution: The Case of Electric Vehicle Technology. *International Journal of Innovation and Technology Management, 20*(05), 2350030.

Lockwood, M. (2021). A hard act to follow? The evolution and performance of UK climate governance. *Environmental Politics, 30*(sup1), 26–48.

Lövbrand, E., Hjerpe, M., & Linnér, B. O. (2017). Making climate governance global: how UN climate summitry comes to matter in a complex climate regime. *Environmental Politics, 26*(4), 580–599.

Lueong, L. A. (2024). Role of Governance in Implementing Climate Change Adaptation Strategies in the Congo Basin Case of CCameroon and the DDemocratic Republic of Congo. *Journal of Geography, Environment and Earth Science International, 28*(2), 70–80.

Mai, L. (2019). Governing Climate Change: Global Cities and Transnational Lawmaking, by Jolene Lin Cambridge University Press, 2018, 222 pp,. £ 74.99 hb,. £ 24.99 pb,. £ 21.50 e-bk ISBN 9781108424851 hb, 9781108440981 pb, 9781108620963 e-bk. *Transnational Environmental Law, 8*(1), 193–197.

Malini N, and Janardhana, V. (2024). An analysis of the role of g20 in global climate governance: navigating the climate maze. In: *A Decade of Shaping the Future: Global Harmony, Co-operation and G20* E-ISBN: 978-93-6252-893-3 426-437.

Miguel, J. C. H., Mahony, M., & Monteiro, M. S. A. (2019). A "geopolítica infraestrutural" do conhecimento climático: o Modelo Brasileiro do Sistema Terrestre e a divisão Norte-Sul do conhecimento. *Sociologias, 21*, 44–75.

Milkoreit, M. (2019). The Paris agreement on climate change—made in USA? *Perspectives on Politics, 17*(4), 1019–1037.

Miner, S. A., & Thurman, T. J. (2024). Genomics and Biodiversity: Applications and Ethical Considerations for Climate-Just Conservation. *Hastings Center Report, 54*, S114–S119.

Ndambwa, B. J., & Moonga, G. (2024). Decentralising climate governance in the Global South: lessons from Itezhi-Tezhi and the Kafue Wetlands, Zambia. *Journal of Contemporary Governance and Public Policy, 5*(1), 33–54.

Newell, P., Srivastava, S., Naess, L. O., Torres Contreras, G. A., & Price, R. (2021). Toward transformative climate justice: An emerging research agenda. *Wiley Interdisciplinary Reviews: Climate Change, 12*(6), e733.

Nguyen, D., Wang, B., & Mutsvairo, B. (2024). (Pragmatic) collaboration for progress or threat to autonomy? African news discourses about Chinese technology in Nigeria and Ghana. *Global Media and China, 9*(4), 469–490.

Oberthür, S., & Groen, L. (2017). The European Union and the Paris Agreement: leader, mediator, or bystander? *Wiley Interdisciplinary Reviews: Climate Change, 8*(1), e445.

Ogwu, M. C. (2019). Towards Sustainable Development in Africa: The Challenge of Urbanization and Climate Change Adaptation. In P. B. Cobbinah & M. Addaney (Eds.), *The Geography of Climate Change Adaptation in Urban Africa* (pp. 29–55). Springer Nature. https://doi.org/10.1007/978-3-030-04873-0_2

Ogwu, M. C., El Malahi, S., & Izah, S. C. (2025c). Policy and Regulatory Frameworks for Sustainable Environmental Practices. In M. C. Ogwu & S. Chibueze Izah (Eds.), *Evaluating Environmental Processes and Technologies. Environmental Science and Engineering Series*. Springer. https://doi.org/10.1007/978-3-031-85327-2_16

Ogwu, M. C., & Izah, S. C. (2025). *Evaluating environmental processes and technologies*. Springer Nature. https://doi.org/10.1007/978-3-031-85327-2

Ogwu, M. C., Izah, S. C., Merritt, H., Wise, R. D., & Jintrawet, A. (2025d). *Sustainable Bioeconomy Development in the Global South: Volume I Status and Perspectives*. Springer Singapore. https://doi.org/10.1007/978-981-96-0640-5

Ogwu, M.C., Izah, S.C., Vazquez-Arenas, J.G., Feleke, S.T., and Wei, X. (2025e). *Sustainable Bioeconomy Development in the Global South: Volume II Bioeconomy Techniques*. Springer Singapore. doi:https://doi.org/10.1007/978-981-96-0305-3

Okereke, C. (2024). Degrowth, green growth, and climate justice for Africa. *Review of International Studies, 50*(5), 910–920.

Palley, T. (2021). National policy space: reframing the political economy of globalization and its implications for national sovereignty and democracy. *Brazilian Journal of Political Economy, 41*, 447–465.

Pham, H., & Saner, M. (2021). A systematic literature review of inclusive climate change adaption. *Sustainability, 13*(19), 10617.

Pruett, D., & Hill, C. (2024). *Climate Plans for the People: Civil society and community participation in national action plans on climate change.*

Pujiningsih, S., Suryani, A. W., Larasati, I. P., & Yusuf, S. N. S. (2023). Political hegemony and accounting discourse: valuing nationalization. *Asian Review of Accounting, 31*(4), 559–582.

Qi, S., & Xiaolong, C. (2024). Эволюция сотрудничества по глобальному климатическому управлению между странами БРИКС: от РКИК до Парижского соглашения. *MGIMO Review of International Relations, 17*(1), 65–85.

Rietig, K. (2021). Multilevel reinforcing dynamics: Global climate governance and European renewable energy policy. *Public Administration, 99*(1), 55–71.

Roelfsema, M., van Soest, H. L., Harmsen, M., van Vuuren, D. P., Bertram, C., den Elzen, M., Höhne, N., Iacobuta, G., Krey, V., Kriegler, E., & Luderer, G. (2020). Taking stock of national climate policies: Future impact of climate policies in the context of the Paris Agreement climate goals. *Nature Communications, 11*, 1–12.

Salazar, V. (2023). From practices to praxis: ASEAN's transnational climate governance networks as communities of practice. *Journal of Current Southeast Asian Affairs, 42*(2), 190–215.

Sareen, S., & Waagsaether, K. L. (2023). New municipalism and the governance of urban transitions to sustainability. *Urban Studies, 60*(11), 2271–2289.

Sasser, J. S. (2024). At the intersection of climate justice and reproductive justice. *Wiley Interdisciplinary Reviews: Climate Change, 15*(1), e860.

Selby, J. (2018). The trump presidency, climate change, and the prospect of a disorderly energy transition. *Review of International Studies, 45*(3), 471–490. https://doi.org/10.1017/s0260210518000165

Shafi, K. M., Khan, A. U., & Islam, R. (2021). Climate Change Action and State Sovereignty. *Margalla Papers, 25*(2), 98–108.

Stephens, J. C. (2021). Electrification: Opportunities for social justice and social innovation. *MRS Bulletin, 46*(12), 1205–1209.

Sultana, F. (2022). Critical climate justice. *The Geographical Journal, 188*(1), 118–124.

Tache, C. E. P. (2022). The EU-China road to the Comprehensive Agreement on Investment. *Tribuna Juridică, 12*(4), 476–494.

Tosun, J., & Schoenefeld, J. J. (2017). Collective climate action and networked climate governance. *Wiley Interdisciplinary Reviews: Climate Change, 8*(1), e440.

Voigt, C. (2023). The power of the Paris Agreement in international climate litigation. *Review of European, Comparative & International Environmental Law, 32*(2), 237–249.

Wagle, P., & Philip, K. (2022). Climate justice is social justice: articulating people's rights to the city in Mumbai. *Environment and Urbanization, 34*(2), 331–348.

Wamsler, C. (2016). From risk governance to city–citizen collaboration: Capitalizing on individual adaptation to climate change. *Environmental Policy and Governance, 26*(3), 184–204.

Watts, N., Amann, M., Arnell, N., Ayeb-Karlsson, S., Belesova, K., Boykoff, M., et al. (2019). The 2019 report of the lancet countdown on health and climate change: ensuring that the health of a child born today is not defined by a changing climate. *The Lancet, 394*(10211), 1836–1878. https://doi.org/10.1016/s0140-6736(19)32596-6

Widerberg, O., & Pattberg, P. (2017). Accountability challenges in the transnational regime complex for climate change. *Review of Policy Research, 34*(1), 68–87.

Yan, C., & Luo, Y. (2024). Local government-led climate governance and social inclusion: the case study of J County in China. *Social Inclusion, 12.*

Zapata Arango, M., Hoppe, T., Itten, A., & Blok, K. (2024). The role of City Climate Networks in Promoting Citizen Participation in Municipalities: A Critical Multi-Case Analysis. *Energy, Sustainability and Society, 14*(1), 5.

Zhang, D., Zhang, S., Ding, H., Liang, M., & Cao, Y. (2025). Comparative analysis of climate governance in China and Germany. *Energy and Climate Management, 1*(2).

Zhang, L., & Bai, E. (2023). The regime complexes for global climate governance. *Sustainability, 15*(11), 9077.

Zhang, S., & Li, N. (2024). Addressing Climate Change through International Investment Agreements: Obstacles and Reform Options. *Sustainability, 16*(4), 1471.

Zuk, P., Conversi, D., & Zuk, P. (2024). Conceptualising energy nationalism in the context of climate change: Framework and review. *Frontiers in Energy Research, 12.*

Resistance from the Periphery: Indigenous, Semi-Free or Unfree Farmers, and Climate Justice Movements

10

10.1 Introduction

The contemporary landscape of global climate politics displays a vivid panorama marked by stark contrasts between the periphery and the center, manifesting in a persistent cycle of marginalization and resistance. The narratives surrounding climate action are often influenced by powerful states and corporations that promote a hegemony largely disconnected from the realities faced by local communities disproportionately affected by climate change. For instance, it has been documented that after three decades of political efforts to combat climate change, global carbon dioxide emissions have risen significantly, being approximately 60% higher today than they were in 1990, suggesting that prevailing governance structures fail to address essential underlying issues of power dynamics and inequity in climate governance (Stoddard et al., 2021). This systemic marginalization often sidelines voices that can effectively articulate local needs, pushing grassroots movements to the periphery of climate policy discussions.

Grassroots, bottom-up strategies have emerged as counter-hegemonic forces advocating for a reconfiguration of power dynamics in climate governance to include perspectives often neglected by hegemonic institutions. The structural inequalities that drive environmental degradation are no longer merely abstract issues; they are intensely localized struggles facing communities, particularly in the Global South (Bouman & Steg, 2019). The interplay between grassroots activism and broader climate governance can be traced to the recognition that authentic pro-environmental actions are often stifled by a collective underestimation of public concern and commitment (Bouman & Steg, 2019). By fostering a more accurate societal narrative about environmental values, these bottom-up initiatives can potentially galvanize widespread participation in climate action that effectively challenges dominant power structures.

M. C. Ogwu and S. Chibueze Izah, *Climate Frontiers and Social Transformation*, World-Systems Evolution and Global Futures,
https://doi.org/10.1007/978-3-032-17639-4_10

Framing this discourse on resistance requires a nuanced understanding of justice and alternative futures, in which communities redefine resilience and sustainability through their localized experiences. Bottom-up assessments of climate risks and adaptation strategies are essential in this regard, allowing for local knowledge to inform broader policy frameworks (Conway et al., 2019). Significantly, local adaptation initiatives, through frameworks such as Local Adaptation Plans of Action (LAPAs), demonstrate that integrating multiple sectors can facilitate participatory governance and enhance community capacity (Ghimire & Chhetri, 2022). These frameworks characterize a forward-looking paradigm that moves away from an exclusive reliance on top-down directives that often exacerbate historical inequities, channeling instead the emergent local wisdom toward sustainable climate strategies.

Within the discourse of resistance in climate politics, there is a critical need to underscore the significance of social movements and collective actions arising from local disenfranchisement. Insights from studies on citizen participation show that positioning climate change within community contexts fosters greater engagement and mobilization around climate initiatives (Chitsa et al., 2022). In various case studies, local communities that are willing to adapt continue to face barriers rooted in structural inequalities and limited access to resources (Doust et al., 2021). Such observations starkly reflect the broader dynamics of climate justice, where marginalized populations not only bear the brunt of climate change impacts but also serve as crucial actors in shaping adaptive strategies (Sekine, 2021).

The entanglement of grassroots movements with local governance models reveals the paths toward alternative, equitable futures. A polycentric approach to governance suggests that multitiered systems incorporating both top-down policies and grassroots initiatives may yield more effective climate action outcomes (Cole, 2015). By acknowledging and integrating the insights from local actors, governance systems can foster greater inclusivity and responsiveness to diverse climate challenges, thereby dismantling long-standing power imbalances. Through participatory mechanisms that engage local voices, cities and communities can contribute to systemic changes aligned with the Sustainable Development Goals (SDGs), demonstrating the multifaceted potential of bottom-up approaches (Mondal et al., 2024).

Documenting the socio-political context in which these movements operate, such as the anger toward climate inaction and distrust in policymakers, emphasizes a critical driving force behind collective action (Wang et al., 2025). These emotions are not merely incidental; they form the bedrock on which communities rally to demand accountability and tangible, transformative action from those in power. Moreover, the role of communication and the articulation of ecological concerns underscore the importance of bridging narratives from local experiences to global discourses in climate politics (Caggiano et al., 2023). To galvanize effective climate responses, it is imperative to address perception gaps about climate change's urgency and to recognize that much of its impact is a byproduct of structural challenges inherent in current governance frameworks.

This chapter challenges dominant structures of climate governance. It examines the historical legacies of colonialism, dispossession, and global capitalism that have placed indigenous peoples, semi-free or unfree farming communities, and climate-vulnerable populations at the periphery of decision-making while exposing them to the harshest impacts of environmental degradation. By exploring indigenous resistance to extractivism, semi-free or unfree farmers' struggles for food sovereignty, and the rise of climate justice movements, the chapter highlights how these actors not only defend their lands and livelihoods but also propose alternative models of sustainability rooted in equity, cultural identity, and ecological balance. Attention is also given to the convergences and tensions within these movements, showing how solidarity and conflict coexist in the fight against structural injustice. The benefit of this chapter lies in its ability to broaden the reader's understanding of climate politics beyond state-centered and technocratic solutions. It demonstrates how peripheral movements, often dismissed as reactive or local, are in fact global drivers of justice-oriented change. Readers will see how traditional ecological knowledge, agroecology, and grassroots mobilizations provide viable alternatives to dominant "green growth" narratives. By analyzing their impact on international negotiations and local governance, as well as their visions of decolonial futures and just transitions, the chapter equips readers with both critical insight and hopeful possibilities. Finally, it shows that resistance from the periphery is not merely defensive but transformative, pointing toward more inclusive, equitable, and sustainable pathways in global climate governance.

10.2 Historical and Structural Context: Colonial Legacies, Capitalism, and Climate Inequality

Colonial legacies continue to shape modern socio-economic and environmental inequalities. The historical backdrop of colonialism is marked by dispossession, extractive economies, and racial hierarchies that have systematically marginalized indigenous and semi-free or unfree farmers communities (Table 10.1). The process of land dispossession was frequently accompanied by the violent displacement of these communities, effectively severing their ties to ancestral lands and resources. This violation of land rights did not merely result in physical displacement but also dismantled the cultural and spiritual connections these communities had with their environments, leading to a profound loss of indigenous knowledge systems that have sustained these communities for millennia (Bolt & Hillbom, 2016).

In the colonial experience, extractive economies, such as mining, plantations, and logging, were designed to benefit colonial powers at the expense of local sustainability. These systems prioritized resource extraction for export rather than fostering local economic resilience or integrating sustainable practices. This legacy persists today, as many postcolonial states grapple with economic models that still heavily rely on resource extraction, often under the control of local elites or foreign entities. The commodification associated with these practices frequently fails to recognize and integrate traditional ecological knowledge systems, which

Table 10.1 Historical and structural foundations of climate inequality

Theme	Major issues	Details/examples
Colonial legacy: dispossession, extractivism, racial hierarchies	Land dispossession and forced displacement	Indigenous and semi-free or unfree farmers communities were pushed off their ancestral lands during colonial expansion
	Extractive economies	Mining, plantations, and logging were structured to serve colonial powers
	Racial hierarchies	Indigenous knowledge is dismissed while Western science/governance is privileged
	Continuities in postcolonial states	Elite capture of land and natural resources after independence
Capitalism & globalization: land grabs, resource exploitation	Neoliberal globalization	Export-oriented economies are deepening dependency
	Land grabs	Foreign investors, agribusiness, and infrastructure projects are driving dispossession
	Fossil fuel extraction & industrial agriculture	Intensified exploitation is causing ecological degradation
	Commodification of nature	Carbon markets, biodiversity offsets reproducing inequalities
North–south divide: unequal responsibilities, climate debt	Historical emissions	Industrialized nations as major contributors to climate change
	Climate debt	Wealthy nations owe reparations for historical exploitation.
	Unequal financing	Limited adaptation/mitigation funds for Global South, often conditional
	Exclusion from governance	Marginalization of Global South voices in UNFCCC, COP negotiations
Other structural dimensions	Neocolonial relations	Dominance of multinational corporations in local economies
	Gendered impacts	Women in semi-free or unfree farmers /indigenous communities are disproportionately affected
	Urban–rural divide	Rural communities are neglected in adaptation/resilience planning
	Security & militarization	Environmental defenders criminalized; militarized responses to protests
	Debt dependency	Structural adjustment programs forcing extractivist development

could offer alternative pathways toward sustainability and resilience (Hamouchene, 2023). Racial hierarchies established during colonial rule not only justified dispossession but also embedded social stratification within the fabric of society. Indigenous and semi-free or unfree farmers knowledge was systematically devalued, aligning governance and scientific paradigms with Eurocentric standards that often disregarded local expertise (Bolt & Hillbom, 2016). This epistemic violence laid the groundwork for persistent inequalities, where indigenous voices are frequently sidelined in environmental governance discussions. The structured exclusion of these communities from meaningful participation in decision-making processes exacerbates their vulnerability in the face of contemporary issues such as climate change, demonstrating a continuity of colonial patterns in the postcolonial context (MacKay et al., 2020; Ogwu & Izah, 2025).

As neoliberal globalization progresses, it reinforces existing inequalities through practices such as land grabs and intensified resource exploitation. Agribusiness and large-scale infrastructure projects often lead to significant land acquisitions by foreign investors, resulting in the displacement of local populations and the erosion of subsistence livelihoods (Adow, 2024). This form of economic colonization reflects a continuity of historical practices that extract value from the Global South, consolidating profit while perpetuating ecological degradation and social fragmentation. The commodification of nature (as seen in the development of carbon markets and biodiversity offsets) poses environmental justice at risk by reproducing existing inequalities rather than rectifying them (Hamouchene, 2023). In analyzing the North–South divide, it becomes evident that industrialized nations have historically contributed the most to greenhouse gas emissions, while the Global South disproportionately bears the brunt of climate impacts. This dynamism raises critical issues surrounding climate debt, the idea that wealthier nations owe reparations to those they have historically exploited and who now face the consequences of environmental degradation (Alvarez et al., 2021). The wealth generated by fossil fuels and extractive industries in the Global North often shields its citizens from the immediate impacts of climate change, which are acutely felt in poorer Southern nations, underscoring the need for equitable climate action. Moreover, access to adaptation and mitigation financing remains unequal, with development aid often tied to conditionalities that reflect neoliberal priorities rather than addressing the genuine needs of vulnerable communities (Adow, 2024).

The ongoing exclusion of voices from the Global South in multilateral negotiations furthers this inequity. International discussions, such as those held at the United Nations Framework Convention on Climate Change (UNFCCC) and Conference of the Parties (COP) processes, frequently marginalize indigenous perspectives and experiences (MacKay et al., 2020). This exclusion is compounded by a neocolonial structure in which multinational corporations operate with little regard for local governance and environmental sustainability, thereby exacerbating climate insecurities in affected regions (Lee et al., 2023). Examining other structural dimensions shaping climate inequality reveals further complexities that intertwine with issues of militarization and environmental governance. The environmental impacts of militarized states and multinational corporations are substantial, as their operations often lead to increased ecological footprints while simultaneously neglecting the welfare of local communities. Military expenditures directed at securing resources can create tensions and conflict while contributing to an eco-political crisis in which environmental degradation and social injustice are inseparable (Rothe, 2017; Verma & Sharma, 2023). Indigenous people, particularly women in semi-free or unfree conditions in farmers' and rural communities, often bear the brunt of these impacts, disproportionately affected by climate change due to their roles in agriculture, water collection, and traditional ecological stewardship (Alvarez et al., 2021; O'Connell & Botchwey, 2023).

As the urban-rural divide widens, rural communities are often overlooked in infrastructure and resilience planning compared to their urban counterparts. This neglect further entrenches inequalities, leaving rural populations increasingly vulnerable to climate impacts while urban areas receive more attention and resources. The differential distribution of climate adaptation resources highlights how historical and structural inequalities persist in climate policies today, often echoing colonial patterns of exploitation and marginalization (Adow, 2024; Verma & Sharma, 2023).

Militarization also plays a dual role in climate injustice: it contributes to environmental degradation through military activities and criminalizes environmental defenders in affected regions (Na'puti & Frain, 2023; Rothe, 2017). Communities resisting exploitation often face violent crackdowns, further illustrating the intersection of militarization and environmental degradation in the context of climate change and its impacts on local ecosystems. The pressures of external debt and structural adjustment programs compel many postcolonial states to prioritize extractive industries over sustainable development, worsening both environmental and social inequalities (Adow, 2024; Na'puti & Frain, 2023).

10.3 Indigenous Resistance: Defending Land, Knowledge, and Sovereignty

Indigenous resistance primarily involves defending land and knowledge systems and asserting sovereignty. Each of these facets represents a critical dimension of the struggles faced by Indigenous communities across the globe, particularly in light of the historical and ongoing impacts of colonialism, capitalism, and climate change (Table 10.2).

Territorial Defense: Sovereignty, Land Rights, and Resource Protection

Territorial defense emerges as a crucial element in Indigenous peoples' struggle for recognition and autonomy. Legal battles to secure recognition of ancestral territories are a primary avenue for Indigenous communities to assert their rights. This quest for recognition often intersects with national and international legal frameworks, including the United Nations Declaration on the Rights of Indigenous Peoples (UNDRIP), which calls for the recognition and respect of Indigenous land rights (Bravo Espinosa & Sierra Camacho, 2024). Studies indicate that legal recognition of ancestral territories not only empowers communities but also plays

Table 10.2 Dimensions of indigenous resistance in climate and environmental struggles

Dimension	Narrative/Examples
Territorial defense: sovereignty, land rights, and resource protection	Struggles for recognition of ancestral territories; defense against mining, logging, agribusiness, and oil extraction; assertion of sovereignty and sacred ties to land
Knowledge systems: traditional ecological knowledge (TEK) and sustainable stewardship	Indigenous ecological practices as alternatives to extractivism; emphasis on reciprocity, balance, and interconnection; contributions to biodiversity, forest conservation, and climate adaptation
Case studies: indigenous movements across regions	Amazon defenders (anti-deforestation, oil resistance); Standing Rock Sioux (Dakota Access Pipeline protests); Mapuche (resistance to forestry industries); Ogoni (struggles against oil exploitation in Nigeria); Pacific Islanders (defending against rising seas)
Conflict and repression: confronting state and corporate power	Criminalization and violence against activists; militarization of indigenous territories; displacement and human rights violations; corporate–state collusion undermining autonomy
Other dimensions: cultural, legal, and transnational resistance	Cultural revitalization (languages, rituals, spirituality); legal tools (ILO Convention 169, UNDRIP); transnational alliances (Indigenous Environmental Network, Global Alliance of Territorial Communities); climate justice leadership in global summits; intergenerational and women's leadership

a critical role in environmental conservation. For instance, research has shown that when Indigenous communities are granted formal legal titles to their lands, deforestation and degradation are substantially reduced, thereby enhancing biodiversity preservation (Blackman et al., 2017; Correia et al., 2025).

Moreover, the resistance against land grabs for mining, logging, oil extraction, and agribusiness has intensified as Indigenous territories become targets for exploitation. The struggle against such incursions is rooted in an understanding of land that transcends mere property; for many Indigenous groups, land is woven into the fabric of cultural identity and spirituality (Barbosa & Rosset, 2024; Townsend et al., 2020). The ongoing conflicts over natural resources have led to self-organization among communities as they mobilize to protect their territories. The concept of hydrosocial territories exemplifies how Indigenous groups articulate their connection to water resources and manage them amid external pressures from urban expansion and extractive industries (Boelens et al., 2016; Manosalvas et al., 2021).

Indigenous sovereignty increasingly encapsulates the assertion of self-determination over the governance of resources. Movements across Latin America illustrate various strategies to regain control over essential natural resources while promoting governance models that align with traditional knowledge systems (Copeland, 2024; Herencia-Carrasco, 2022). The emphasis on autonomy allows Indigenous peoples not only to reclaim land but also to redefine governance and environmental stewardship in line with their cultural values and practices.

10.3.1 Knowledge Systems: Traditional Ecological Knowledge and Sustainable Stewardship

Traditional Ecological Knowledge (TEK) offers a unique lens through which to understand Indigenous stewardship of the environment. The preservation and revitalization of TEK are vital for sustaining ecological balance and addressing contemporary environmental challenges linked to climate change (Schroeder & González, 2019; Townsend et al., 2020). An emerging body of scholarship highlights TEK as a crucial counter-model to modern extractivist development paradigms, emphasizing principles such as reciprocity, balance, and interdependence with nature (Cariño & González, 2022). TEK practices, often grounded in centuries of observation and adaptation, can meaningfully contribute to contemporary biodiversity conservation efforts.

Moreover, the emphasis on reciprocity in TEK fosters a sustainable relationship not merely focused on resource extraction but on stewardship and conservation (Adawiah, 2023). Indigenous practices that prioritize biodiversity management have significant implications for climate adaptation, impacting regional resilience to climate variances. For instance, Indigenous fire management practices, as highlighted in various studies, have been shown to reduce wildfire risks and enhance ecosystem health effectively (Izah & Ogwu, 2025; Mistry et al., 2019; Oliveira et al., 2022). The role of TEK extends beyond mere environmental management; it embodies an epistemological framework that challenges dominant scientific paradigms. This alternative knowledge system stresses the need for collaborative approaches, integrating Indigenous insights with contemporary environmental science to develop more holistic solutions to climate issues (Boelens et al., 2016; Schroeder & González, 2019). The potential convergence of TEK with modern scientific approaches offers a pathway to address the climate crisis while respecting Indigenous rights and worldviews.

10.3.2 Case Studies: Indigenous Movements Across Regions

Indigenous movements worldwide provide poignant illustrations of resistance and struggle against colonial and capitalist forces. In the Amazon rainforest, defenders have mobilized against deforestation fueled by oil drilling and hydroelectric projects, emphasizing their role as guardians of biodiversity and environmental integrity (Copeland, 2024; Garnett et al., 2018). The case of the Standing Rock Sioux's opposition to the Dakota Access Pipeline poignantly illustrates the power of Indigenous activism, encapsulated in their mantra that "Water is Life," thereby highlighting the interconnection between resource protection and cultural survival (Gottardi, 2020; Herencia-Carrasco, 2022). Similarly, the Mapuche peoples in Chile and Argentina have fought to reclaim their territories amidst neocolonial pressures from forestry companies, asserting their rights to self-governance and resource management (Altamirano-Jiménez, 2021; Bravo Espinosa & Sierra Camacho, 2024). In Nigeria, the Ogoni people's resistance against the environmental devastation wrought by oil exploitation reflects the intertwined nature of social justice and environmental issues, showcasing how extractive industries disproportionately affect Indigenous communities (Hope, 2022).

Furthermore, Indigenous nations in the Pacific are increasingly vocal in defending against the impacts of climate change, such as rising seas. Their movements advocate assertive action against dispossession and environmental degradation, reinforcing the importance of Indigenous knowledge in mitigating the impacts of climate change (Barbosa & Rosset, 2024; Townsend et al., 2020). These case studies highlight the diversity and resilience of Indigenous struggles while illuminating the broader systemic challenges they confront.

10.3.3 Conflict and Repression: Confronting State and Corporate Power

The struggles faced by Indigenous peoples are often met with violent repression and systemic challenges. The criminalization of Indigenous leaders and activists seeking to protect their lands and rights reflects a broader strategy of state and corporate interests to undermine resistance (Gottardi, 2020; Townsend et al., 2020). The militarization of Indigenous territories under the guise of development or national security further exacerbates tensions, as communities find themselves navigating hostile conditions imposed by state policies (Bravo Espinosa & Sierra Camacho, 2024; Lavoie et al., 2021). Violence, forced displacement, and human rights violations are prevalent tactics utilized to suppress Indigenous movements. These tactics not only violate fundamental rights but also seek to erase Indigenous identities tied to land and culture (Altamirano-Jiménez, 2021; Cariño & González, 2022). The collusion between corporations and state agencies often complicates these dynamics, where Indigenous autonomy and governance are systematically undermined through legal and institutional mechanisms designed to prioritize economic advancement and extractivist agendas (Herencia-Carrasco, 2022; Huo et al., 2016; Ogwu et al., 2025c,d).

Furthermore, the impacts of colonial legacies remain evident as new forms of coloniality manifest in contemporary environmental governance, often marginalizing Indigenous perspectives while promoting top-down approaches that disregard local knowledge systems (Barbosa & Rosset, 2024; Hope, 2022). This resistance against systemic oppression and violence forms a critical part of Indigenous struggles, emphasizing the need for recognition, autonomy, and justice.

10.3.4 Others: Expanding Dimensions of Indigenous Resistance

Indigenous resistance extends beyond land and sovereignty to encompass broader cultural, legal, and social dimensions. Cultural revitalization is a powerful act of resistance, in which Indigenous communities actively preserve and revitalize languages, rituals, and spiritual traditions as a form of defiance against colonial narratives (Adawiah, 2023; Xu & Li, 2022). These acts serve not only to preserve culture but also to reinforce identity and community cohesion in the face of external pressures. Legal strategies have emerged as critical instruments in Indigenous resistance, with groups increasingly leveraging international courts and human rights instruments to assert their land and resource rights (Briseño-Roa, 2023; Herencia-Carrasco, 2022). Landmark legal victories, as seen in cases such as Júba Wajiín in Mexico, illustrate how strategic litigation can confront and overturn unjust decisions that impact Indigenous lands (Herencia-Carrasco, 2022).

Transnational alliances play an instrumental role in strengthening Indigenous resistance globally by fostering solidarity networks that transcend borders and share knowledge and strategies (Mistry et al., 2019; Xu & Li, 2022). The Indigenous Environmental Network and the Global Alliance of Territorial Communities exemplify collaborative efforts that amplify Indigenous voices in various international fora, including critical climate negotiations (Hope, 2022; Townsend et al., 2020). The leadership of Indigenous youth and women is crucial in driving contemporary movements, reflecting intergenerational and gendered dimensions of resistance. Empowering youth through education about their rights and cultural heritage, alongside the recognition of the vital role of women in activism, shapes a more inclusive and resilient future for Indigenous resistance (Cariño & González, 2022; Schroeder & González, 2019). The integration of youth and women's perspectives not only enriches the discourse but also fosters sustainable practices rooted in traditional knowledge.

10.4 Semi-Free or Unfree Farmers Movements: Struggles for Food Sovereignty and Against Agribusiness

The discourse surrounding semi-free or unfree farmers' movements, particularly in the context of food sovereignty, is framed by a complex interplay of historical colonial legacies, capitalist exploitation, and climate inequality (Table 10.3). La Vía Campesina, a robust global network advocating for farmers' rights, emerges as an essential force in this struggle, championing the concept of food sovereignty. Thus, the right of people to define their own food systems. This movement emphasizes the importance of agroecology and seed autonomy, recognizing traditional agricultural practices and cultural heritage as essential components of sustainable food systems. The principles of food sovereignty, as articulated by La Vía Campesina, challenge dominant agro-industrial narratives and advocate for local

Table 10.3 Semi-free or unfree farmers movements-struggles for food sovereignty and against agribusiness

Theme	Key elements	Focus/benefit
Food sovereignty	La Vía Campesina, agroecology, seed autonomy, local food systems	Empowers communities to control food systems and preserve traditional practices
Struggles against agribusiness	Opposition to GMOs, monocultures, pesticides, and land grabs	Protects biodiversity, farmer health, and ecological sustainability
Resistance to neoliberalism	Critique of WTO, trade liberalization, and structural adjustment policies	Defends small farmers from unfair global trade policies
Transnational solidarity	Rural-urban alliances, global semi-free or unfree farmers' networks, climate justice forums	Builds global solidarity and amplifies semi-free or unfree farmers' voices in climate debates
Other dimensions	Gender roles, youth engagement, climate adaptation, land rights, repression, alternative economies	Highlights inclusivity, resilience, and justice in the struggles of semi-free or unfree farmers

control over food production, pointing to the need for systems that are ecologically sound and culturally appropriate (Clendenning et al., 2016; Nyholm et al., 2024).

Agroecology represents a transformative approach that aligns closely with the philosophy of food sovereignty. It seeks to create sustainable, low-input agricultural practices that maximize the beneficial interactions among communities, ecosystems, and food systems (Anderson et al., 2022; Pimbert, 2015). The infusion of traditional knowledge within agroecological practices not only enhances food security but also restores ecological balance. Additionally, preserving seed autonomy is highlighted as a critical endeavor, allowing communities to maintain their seed varieties, which are often threatened by corporate patents and genetically modified organisms (GMOs) (Jefferson & Adhikari, 2019). Resistance against the encroachment of agribusiness encompasses not only the rejection of GMOs but also a broader critique of monoculture farming, which compromises biodiversity and soil health (Blue Bird Jernigan et al., 2021; Hoover, 2021; Ogwu et al., 2025e).

The opposition to agribusiness extends into broader socio-political struggles against neoliberal economic policies, particularly those advanced by institutions such as the World Trade Organization (WTO) and structural adjustment programs (SAPs). These entities promote trade liberalization measures that favor industrial agriculture and undermine smallholder farmers by flooding markets with cheap imports (Cerkleski, 2023; Henderson, 2017). Protests and advocacy from semi-free or unfree farmers' movements underscore the adverse effects of such policies on local farming systems and highlight the urgent need to prioritize food sovereignty in agricultural policy (Carlile et al., 2021; Hamouchene, 2023). The call for recognition of farmers' rights within national and international frameworks arises from these ongoing struggles against economic structures that perpetuate inequality and food insecurity.

Transnational solidarity plays an essential role in fortifying these movements. Global networks, particularly La Vía Campesina, facilitate collaboration across borders, resulting in robust rural-urban alliances between semi-free or unfree farmers, workers, and consumers (Clendenning et al., 2016; Ogwu, 2019). This synergy not only enhances the collective power of movements advocating for food sovereignty but also ensures that issues of ecological justice are interconnected with struggles against corporate greed and environmental degradation. The participation in international protest movements against neoliberal trade agreements illustrates how local struggles are articulated within a global framework, allowing for the sharing of experiences and strategies that transcend national boundaries.

Youth engagement within semi-free or unfree farmers movements has grown increasingly crucial as the demographic of farmers shifts and the need for sustainable agricultural practices becomes paramount. The incorporation of younger generations is vital to ensure the continuation of cultural practices related to food sovereignty and to challenge existing power dynamics in the agricultural sector (Eide & Kunelius, 2021; MacKay et al., 2020). Educational initiatives and advocacy are designed to foster a fresh understanding of agricultural practices that respect both traditional knowledge and contemporary ecological insights, thereby empowering youth to remain on the land and contribute to the resilience of local food systems (O'Connell & Botchwey, 2023; Ogwu et al., 2025a).

The dimension of gender within semi-free or unfree farmers movements cannot be overlooked, as women play a significant role in nurturing local food systems and leading advocacy efforts for food sovereignty (Motta, 2021). Feminist movements intersect with food sovereignty to highlight the specific challenges women face in agriculture—from limited access to resources to social and economic disempowerment (Ogwu et al., 2025b). By foregrounding gender in the discourse of food sovereignty, these movements seek to dismantle patriarchal structures that sustain inequality while promoting the rights and contributions of women farmers (Siebert, 2020).

Furthermore, addressing climate adaptation within the framework of food sovereignty emphasizes the need for resilient agricultural practices in the face of climate change (Iyiola et al., 2024; Izah et al., 2023). Semi-free or unfree farmers often possess intimate knowledge of their local ecosystems, allowing them to develop adaptive strategies that minimize vulnerability to climatic impacts, such as droughts and floods (Blue Bird Jernigan et al., 2021; Seminar et al., 2018). These practices not only ensure food security but also help mitigate the effects of climate change through sustainable agricultural methods. In tandem, land rights campaigns underscore the urgency of resisting privatization and advocating for collective ownership. The criminalization of semi-free or unfree farmers' activists poses a significant threat, as many face repression and violence in their struggle to defend their land against corporate interests and state encroachments (Henderson,

2017). The defense of land is not only about securing livelihoods but also about asserting dignity, autonomy, and cultural identity (Hudson et al., 2023; Kapała, 2023).

10.5 Climate Justice Movements: Equity, Reparations, and Grassroots Mobilizations

Climate justice movements represent a crucial intersection within the broader discourse on environmental change, emphasizing equity, reparations, and grassroots mobilizations that address the socio-economic disparities inherent to climate crises (Table 10.4). The principles underpinning these movements center on the acknowledgment of climate debt—an obligation for historical emissions predominantly borne by industrialized nations. This acknowledgment is essential, as the impacts of climate change are often most acutely experienced by vulnerable nations and communities in the Global South, who have contributed minimally to the origins of the crisis. The concept of equity encompasses not only the rectification of past injustices through reparations but also the provision of climate-related financial support, technology transfers, and the establishment of loss-and-damage funds to facilitate adaptation and mitigation efforts in these vulnerable regions (Malherbe & Oladejo, 2025; Ogunbode, 2022).

Table 10.4 Dimensions of climate justice movements—equity, reparations, and grassroots mobilizations

Category	Major focus areas
Equity & reparations	- Climate debt and historical responsibility of industrialized nations - Financial reparations and technology transfers - Climate finance debates: adaptation, mitigation, and loss & damage fund - "Polluter pays" principle and accountability mechanisms
Frontline communities	- Small island states facing sea-level rise - Global South youth reframing climate discourse - Women activists linking gender and climate justice - Indigenous and rural populations resisting extractivism
Grassroots mobilization	- Fridays for Future (South) adapting strikes to local contexts - Extinction Rebellion (periphery chapters) engaging in local struggles - Pacific Climate Warriors' resistance to fossil fuels - Local campaigns: anti-coal (India), oil resistance (Nigeria), anti-mining (Latin America)
Strategies	- Protests, strikes, and mass mobilizations Litigation against states and corporations - Civil disobedience (sit-ins, blockades)- Direct action to halt extractive projects - Policy advocacy at UNFCCC and national levels
Others	- Narrative shift: from "carbon problem" to "justice problem"- Intersectionality: racial, economic, gender justice - Digital activism and online mobilization - Cultural resistance through art, music, and storytelling - Transnational solidarity with labor, feminist, and indigenous groups - Community resilience projects and grassroots innovation

The ongoing debates within climate finance highlight a dichotomy between adaptation and mitigation strategies, which can perpetuate inequities if not managed carefully. The historical context demands a concerted emphasis on the "polluter pays" principle, underscoring the need for accountability mechanisms to ensure that countries and corporations that are disproportionately responsible for emissions contribute equitably to solutions. Establishing these frameworks can foster a sense of justice and moral responsibility while rebalancing power dynamics within the global economic landscape (Klinsky et al., 2017; O'Connell & Botchwey, 2023). Advocates for climate justice argue that extending reparations and support mechanisms should address not only the immediate impacts of climate change but also dismantle the systemic and structural inequalities that contribute to ongoing environmental degradation (O'Connell & Botchwey, 2023).

Frontline communities and tiny island states face existential threats from rising sea levels, underscoring the urgent need for climate justice. This urgency is echoed among youth in the Global South, who increasingly shift the climate discourse from abstract future risks to the tangible present injustices faced by their communities. The involvement of youth activists has been essential in framing climate justice as a justice issue linked to broader narratives of social, racial, and economic inequality (Han & Ahn, 2020; Sloan Morgan et al., 2024). Women's voices within these movements further amplify calls to link climate justice to gender justice, highlighting how climate change disproportionately affects women, especially in their unpaid care and survival labor (Garoupa et al., 2023).

Indigenous and rural populations face direct threats from extractivism, which endangers traditional territories and undermines cultural and social structures (Whyte, 2020). Addressing these challenges requires a framework that recognizes the rights of indigenous peoples and advocates for their participation in land-use and resource management decision-making (Abajue & Ogwu, 2024). In this context, climate justice movements should prioritize indigenous knowledge systems and practices, which have been effective in managing and conserving ecosystems (Whyte, 2020). The call for inclusive dialogue and mutual respect for diverse knowledge systems is essential to achieving lasting and equitable climate solutions (Garoupa et al., 2023).

Grassroots mobilization is a cornerstone of climate justice movements, with various organizations engaging in direct actions and protests tailored to local contexts. Movements like Fridays for Future have adapted the climate strike model to resonate with diverse socio-economic realities while merging local poverty issues with the global climate agenda, thereby creating compelling narratives for action (Neas et al., 2022). Extinction Rebellion's chapters exemplify these adaptive strategies, utilizing civil disobedience to spotlight local environmental issues and engage communities more deeply in climate justice dialogues. Such localized approaches not only address the climate crisis but also serve as platforms for community empowerment and social justice, reinforcing the interconnectedness of environmental and socio-economic issues (Córdova et al., 2023; O'Brien et al., 2018).

The Pacific Climate Warriors symbolize resilience and resistance against the expansion of fossil fuels, declaring, "We are not drowning, we are fighting." This sentiment encapsulates the collective struggle of vulnerable communities facing climate change challenges and underscores the need for both local and global efforts to combat fossil fuel dependency (Bowman & Buckley, 2022). Various local campaigns, such as anti-coal struggles in India, oil resistance in Nigeria, and anti-mining protests in Latin America, showcase the diverse strategies grassroots groups employ to challenge extractive industries while proposing alternative visions for sustainable futures (McGregor & Christie, 2021). Each instance underscores the urgency of solidarity among affected communities and the critical role of grassroots activism in reshaping narratives of climate justice (Verlie & Flynn, 2022).

Discussions of strategies within climate justice movements include a range of tactics, from protests to litigation and direct actions against systems that perpetuate environmental degradation. Mass mobilizations, such as strikes and marches, serve as collective calls for systemic change, asserting that climate action should prioritize justice over technological advancements. Litigation has become an increasingly utilized strategy, with activists pursuing legal avenues to hold governments and corporations accountable for climate inaction, emphasizing the need for legal frameworks that recognize the ethical dimensions of climate justice (Malherbe & Oladejo, 2025; Neas et al., 2022).

Civil disobedience has gained traction in the climate justice context, reflecting communities' determination to halt extractive projects perceived as detrimental to the environment and local communities. Disruptive tactics, including sit-ins and blockades, have proven effective in drawing media attention to urgent climate issues, thereby championing the voices and rights of frontline communities (McGregor & Christie, 2021). Direct action initiatives, from dismantling pipelines to preventing deforestation, signal a shift toward more radical approaches that emphasize the need for significant changes to existing destructive practices. Policy advocacy is a critical component, as climate justice activists continuously push for justice-oriented frameworks within international environmental negotiations like the United Nations Framework Convention on Climate Change (UNFCCC), advocating for the integration of social justice considerations into climate action (Córdova et al., 2023; Klinsky et al., 2017). An integral aspect of climate justice movements is the narrative framing that shifts the focus from viewing climate change merely as a "carbon problem" to redefining it as a "justice problem." This reframing underscores the moral obligations that societies have to address the disproportionate impacts of climate change on vulnerable populations (Klinsky et al., 2017). The intersectionality of climate justice with various social movements (whether racial, economic, or social) richly enhances the discourse and underscores the need to tackle these intersecting injustices simultaneously for more comprehensive and effective solutions (Ogunbode, 2022).

Moreover, digital activism plays a transformative role within climate justice movements, facilitating online campaigns and social media mobilization while forming global solidarity networks. These digital platforms enable marginalized voices to reach broader audiences, fostering transnational connections among various movements advocating for climate justice. This burgeoning solidarity, particularly in the Global South, transcends regional barriers, allowing diverse groups to unite around shared goals of environmental and social justice (O'Connell & Botchwey, 2023; Ogunbode, 2022). Cultural resistance through art, storytelling, and music has become a vital tool in mobilizing communities and raising awareness around climate justice issues. Such creative expressions resonate deeply within local contexts, inspiring collective action and fostering resilience in the face of the environmental crisis (McGregor & Christie, 2021). Transnational solidarity further strengthens these movements, calling for collaborative efforts across borders, particularly among Global South nations, amplifying voices that might otherwise remain unheard in global climate dialogue (Bowman & Buckley, 2022).

Resilience-building strategies are increasingly emphasized in climate justice agendas, as communities develop adaptation projects and cooperative economic initiatives to enhance sustainability. Grassroots innovations rooted in environmental justice principles address both the urgent needs imposed by environmental degradation and provide frameworks for sustainable development based on local knowledge and practices (Córdova et al., 2023; O'Connell & Botchwey, 2023). These initiatives represent a robust response to climate change, creating pathways that prioritize community needs while integrating broader climate justice objectives.

10.6 Convergences and Tensions: Shared Struggles, Divergent Strategies, and the Politics of Solidarity

Colonial legacies that have shaped contemporary landscapes of capitalism and climate inequality reveal complex relationships among various communities affected by the ongoing environmental crisis. At the forefront of this discourse are indigenous and semi-free or unfree farmers communities deeply entrenched in struggles over land autonomy and ecological justice (Table 10.5). These communities resist the encroachments of state and corporate powers, often leading to significant tensions in their narratives and practices. For instance, in Bolivia, attempts by the government to centralize control over resource extraction have been perceived as a direct threat to indigenous autonomy, suggesting that historical patterns of dispossession continue to operate within modern governance frameworks (Postero, 2017). Such dynamics highlight a continuous thread linking colonial histories to present-day struggles for land and governance, emphasizing the crucial need for autonomy in community-led initiatives.

These communities engage in shared struggles against extractivism. This term relates to the exploitation of natural resources by large corporations and states that prioritize profit over environmental and social well-being. The defense against deforestation and unsustainable agribusiness practices highlights a collective ambition toward ecological justice. This ambition links human rights to environmental rights, advocating their recognition within legal and policy frameworks. Thus,

Table 10.5 Convergences and tensions among indigenous, semi-free or unfree farmers, and climate justice movements

Theme	Major issues and illustrations
Shared struggles: land, autonomy, ecological justice	Resistance to land dispossession and extractivism; defense of forests, rivers, and territories; emphasis on ecological justice linking human rights and environmental rights
Convergences: alliances between semi-free or unfree farmers, indigenous groups, and climate justice activists	Joint campaigns against pipelines, mining, and agribusiness; solidarity through La Vía Campesina, Indigenous Environmental Network, Friends of the Earth; South–South and South–North collaborations
Tensions: representation, strategy, differing priorities	Disputes over who represents grassroots voices; differences in strategy (legal advocacy vs. direct action); resource imbalances favoring NGOs; clash between local territorial defense and global climate policy agendas
Bridging divides: transnational networks and solidarity platforms.	Platforms like the World Social Forum and Peoples' Climate Summit; digital activism linking movements; intercultural exchanges; framing struggles under a global justice agenda
Other dynamics: underlying challenges and opportunities	Risk of co-optation by states/donors; intersectionality of gender, youth, and marginalized voices; resilience in the face of repression; opportunities for pluriversal, justice-centered climate futures

the quest for ecological justice is framed not merely as an environmental issue but as a foundational aspect of human rights, reflecting the inseparable nature of humans and the environment (Sankey, 2022). The fight against environmental degradation foregrounds local knowledge and practices as vital to sustainable governance, advocating for systems that recognize and integrate these voices in decision-making processes (Boucquey & Fly, 2024).

The convergences between indigenous communities, semi-free or unfree farmers groups, and climate justice activists further illustrate a broad coalition in opposition to corporate greed and environmental exploitation. Joint campaigns aimed at halting mining operations, pipeline constructions, and industrial agricultural expansion reflect an increasing solidarity among these disparate groups. Organizations like La Vía Campesina and the Indigenous Environmental Network exemplify the importance of collaborative frameworks that center on equity and historical responsibilities while resisting the hegemony of capitalistic agendas (Boucquey & Fly, 2024). These alliances are not only a form of resistance but also a proactive engagement in the political discourse surrounding climate change, where strategies are crafted to amplify the local realities of marginalized communities on a global stage.

However, the relationships between these groups are not without tensions. A critical challenge is the representation and voice within transnational advocacy arenas. Conflicts often arise over who articulates the narratives of affected communities, which can lead to further marginalization of grassroots movements in favor of more established non-governmental organizations that may have greater access to resources and funding (Howard-Wagner et al., 2023; Milan & Treré, 2024). As these organizations navigate the landscape of climate advocacy, disparities in strategy become apparent; some prefer negotiation and legal frameworks, while others engage through more confrontational direct actions. Such differing methodologies can create friction within movements striving for common goals, where priorities based on local context can diverge significantly from global agendas focused on large-scale policy negotiations (Kaban & Ramadani, 2023). Figure 10.1 illustrates the intersecting agendas and points of friction among Indigenous movements, semi-free/unfree farmers, and climate justice movements. By mapping their shared commitments—such as land autonomy, ecological justice, and anti-extractivism—alongside areas of divergence.

Moreover, the unequal distribution of resources has profound implications for the effectiveness of various movements. Better-funded NGOs often overshadow grassroots voices, which exacerbates tensions between those with access to capital and those striving for recognition through local engagement. This discrepancy poses a significant challenge, especially when local priorities (such as territorial defense) clash with broader global agendas oriented toward climate policy negotiations (Orankiewicz & Turała, 2024). The political landscape should address these inequities if it seeks to create a truly inclusive platform for collective action against

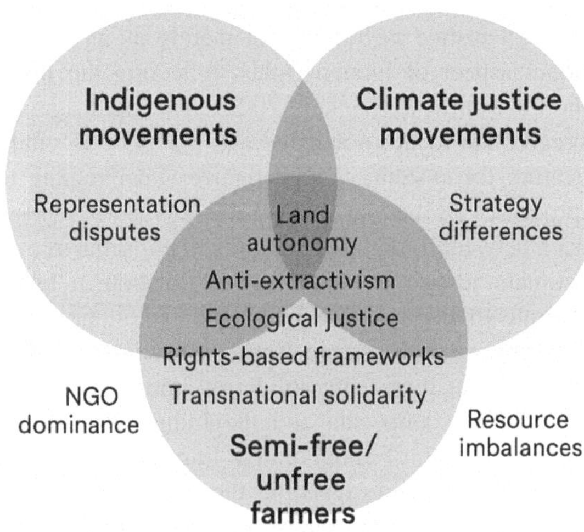

Fig. 10.1 Convergences and tensions among three movements

climate injustice. In the face of these disparities, transnational networks and solidarity platforms emerge as crucial conduits for bridging divides. Events such as the World Social Forum and the People's Climate Summit highlight the necessity of global solidarity in addressing climate challenges while allowing diverse struggles to converge on shared platforms. Digital activism further enhances connectivity among movements, enabling the sharing of strategies, success stories, and resources across vast distances. Intercultural exchanges encourage mutual understanding and knowledge sharing across movements, fostering a richer tapestry of resistance against diverse forms of oppression (Kupers & Hendriks, 2023). Shared framing of struggles within a global justice agenda fosters collaboration and solidarity rather than viewing these conflicts as isolated incidents.

Despite the apparent risks, including potential co-optation by state or donor agendas, the resilience of these movements is notable. The integration of intersectionality into broader resistance efforts enables the inclusion of voices often silenced in traditional environmental discourses, particularly those of women, youth, and other marginalized groups (Masquelier, 2022). Acknowledging the intersectional nature of struggles enriches the discourse on climate justice by expanding its focus beyond singular narratives of oppression. Furthermore, the potential for new political imaginaries to emerge becomes conceivable as these movements envision pluriversal futures that reject homogenizing climate solutions. The concept of pluriversalism (as articulated in contemporary decolonial thought) insists that multiple realities coexist, each shaped by distinct cultural, historical, and ecological contexts (Hann, 2024). This framework invites diverse communities to imagine and enact their pathways toward sustainability, resisting the oversimplified narratives often propagated by dominant paradigms. As such, the diverse strategies employed by various movements reflect a foundational belief in self-determination and local governance, enabling communities to tailor solutions to their unique circumstances.

10.7 Impact on Global Climate Governance: Disrupting Narratives and Creating Alternative Pathways

Climate governance is increasingly recognized as a critical arena where diverse narratives are challenged and alternative pathways for addressing climate change are forged. This discourse often centers on rejecting market-based and technocratic solutions, such as carbon trading and geoengineering, which have been criticized for perpetuating existing inequalities and failing to address the root causes of environmental degradation. As Kim (2023) articulates, market-based approaches are often shaped by capitalist interests, rendering them ineffective for meaningful climate action and unjust in their implementation. Furthermore, these approaches have been linked to green gentrification, where environmental improvements disproportionately benefit affluent communities while marginalizing lower-income and minority populations.

Advocacy spaces, such as the United Nations Framework Convention on Climate Change (UNFCCC) and events surrounding the Conference of the Parties (COP), have become essential platforms for mobilization and protest against these market-centric narratives. Young activists, in particular, have shown remarkable resilience and creativity in challenging dominant discourses through grassroots movements (Han & Ahn, 2020). For example, the People's Summit organized alongside COP gatherings demonstrates the potential for alternative frameworks that center on justice and equity, rejecting technocratic models in favor of community-led initiatives and recognition of rights for marginalized groups. The intersection of these advocacy spaces with local movements has led to recognition of rights and localized victories, as seen in various global contexts where communities have fought for their rights in climate governance.

Despite these achievements, significant limits exist within the climate governance landscape. There is often a backlash against progressive movement actors, leading to repression and co-optation as state and corporate interests seek to control and undermine grassroots initiatives (Daly, 2024). This opposition is compounded by structural inequalities within global governance frameworks, such as the UNFCCC, where negotiating power is unevenly distributed, often favoring wealthier nations (Borras & Franco, 2020). Recognizing climate injustice is crucial for understanding these dynamics, as marginalized communities often face heightened vulnerability to climate impacts yet have insufficient representation in decision-making processes (Chu & Michael, 2019). The ramifications of these limitations extend to the broader socio-political spheres in which claims for climate justice are made. For instance, in urban contexts like Mumbai, social and climate justice movements highlight the interconnections between environmental degradation and the rights of urban inhabitants to access clean living conditions and public services (Wagle & Philip, 2022). This unearths the multifaceted nature of climate governance, along with the necessity for comprehensive strategies that tackle not just environmental outcomes but also social inequities entangled with climate issues.

Therefore, alternative pathways for climate governance must be developed, focusing on grassroots mobilization and the inclusion of diverse voices in decision-making. This includes a shift toward more just adaptation planning practices that prioritize affected communities (Fünfgeld & Schmid, 2020). Moreover, intellectual discourses surrounding climate justice should evolve to consider the unique challenges faced by vulnerable populations, integrating concepts from social justice movements into climate action frameworks (Khadse & Srinivasan, 2022; Kim, 2023). Ultimately, the call for climate justice should resonate across multiple scales of governance, creating a synergistic relationship between local movements and global climate policy initiatives.

Figure 10.2 illustrates the multi-scalar pathways through which peripheral actors actively reshape global climate governance. Beginning with localized forms of resistance—such as territorial defense, TEK, and agroecological practices—the diagram shows how these struggles generate national policy shifts that challenge extractivist and inequitable development models. As these changes accumulate,

Fig. 10.2 How peripheral resistance shapes global climate governance

they catalyze the formation of regional South–South alliances that amplify shared demands for climate justice, technological equity, and financial reform.

10.8 Toward an Alternative Future: Decolonial Visions, Community-Based Governance, and Just Transitions

The urgent need for transformative approaches to global climate governance compels us to reassess our frameworks and methodologies. This necessity is particularly pronounced within the context of marginalized and indigenous communities, who have been historically subjected to colonialist structures and capitalistic paradigms. Within these discussions, the concepts of Buen Vivir (Living Well) and Sumak Kawsay serve as essential frameworks for decolonial visions. These concepts, deeply rooted in indigenous traditions, advocate for a pluralistic understanding of well-being that emphasizes harmony with nature, community solidarity, and socio-cultural inclusivity as alternatives to prevailing development paradigms (Acosta, 2015; Barragán, 2020). The incorporation of Buen Vivir into constitutions in countries like Ecuador and Bolivia symbolizes a significant shift toward recognizing the rights of nature and critiquing dominant economic growth narratives (Barragán, 2020).

Buen Vivir represents a socio-ecological ethos that positions individuals within a community and underscores relationships among humans and between humans and nature (Acosta, 2015). This philosophical approach critiques conventional development paradigms that prioritize individualism and consumerism and suggests a recomposition of governance to include diverse epistemologies and ontologies. The application of such concepts reflects a critical engagement with the underlying structures of coloniality and modernity that have historically marginalized many communities, particularly in the Global South (Wagner & Hornidge, 2025). In the face of global climate change (a manifestation of unequal power relationships and unsustainable practices) Buen Vivir can be a powerful tool for generating grassroots movements aimed at ecological restoration and social justice.

In exploring community-based governance, the concept of commons emerges as a central pillar. Community-based governance frameworks, particularly those grounded in the principles of cooperative models and grassroots democracy, emphasize the collective stewardship of shared resources. These frameworks challenge neoliberal privatization and advocate for local management systems that empower communities toward self-determination and sustainable practices (Pérez-Rincón et al., 2019). In the Andean context, communities organize around the notion that resources belong to all, emphasizing a shared responsibility for ecological and social well-being (García-García, 2022). This aligns with the application of Sumak Kawsay within local governance, where indigenous knowledge systems prioritize collaboration and reciprocity over competition and individual gain (de Wit & Mourato, 2022).

Moreover, the concept of "just transition" should be examined within these frameworks. A just transition seeks to address labor rights, ensure renewable energy justice, and include marginalized voices in decision-making processes. Transitioning to renewable energy sources represents both a challenge and an opportunity for labor rights; ensuring that workers, especially those from vulnerable communities, have access to new jobs that are sustainable and just is essential (Izah et al., 2025a, 2025b; Vásquez Bustamante et al., 2021). This necessitates robust policies that not only promote technological advancements but also champion the rights and aspirations of workers affected by systemic shifts in energy consumption and production. The inclusion of marginalized voices, particularly those from indigenous and local communities, ensures that transitions are equitable and reflect the values of Buen Vivir and Sumak Kawsay (Querejazu, 2024).

The notion of solidarity and systemic change is paramount when considering the global nature of climate governance. South-South cooperation and cross-movement alliances can provide robust frameworks for sharing resources, knowledge, and strategies that amplify marginalized voices in climate governance (Castro-Sotomayor & Minoia, 2023). The cross-pollination of strategies among diverse movements can bolster advocacy efforts and facilitate the sharing of decolonial methodologies that respect and integrate local knowledge systems. For instance, indigenous communities participating in transnational alliances can leverage their unique insights into sustainability to directly challenge hegemonic narratives perpetuated by industrialized nations and to promote alternative visions of environmental stewardship (Alkatiri, 2022).

South-South cooperation is essential in fostering resilience and adaptive capacities in countries disproportionately affected by climate change. Economic and political alliances among countries of the Global South can catalyze collective action, pooling knowledge and resources to confront shared vulnerabilities arising from climate impacts (Shade, 2015). These collaborative frameworks can also serve as platforms for discussing and amplifying the rights of communities engaged in just transitions, thereby challenging existing global power dynamics that often sideline their perspectives. Inferences drawn from Buen Vivir and related discourse illuminate pathways toward governance models that foster ecological sustainability and social justice. The pluralistic nature of these approaches encourages a model of governance that doesn't merely impose solutions but co-creates viable pathways forward with affected communities. Such participatory governance frameworks can effectively disrupt patterns of inequality perpetuated by colonial and capitalist legacies, offering alternatives that are firmly rooted in local values and community needs (Barragán, 2020). These discussions must conclude not with mere acknowledgment of indigenous philosophies but with genuine integration into policy and practice that reflects their principles.

Furthermore, decolonial visions, community-based governance, and just transitions are inherently interconnected within climate governance discourse. By challenging modernist conceptions of progress and engaging with indigenous epistemologies, we can foster inclusive governance systems that resonate with the values of Buen Vivir and uphold the rights of nature. Collective action informed by these frameworks can reorient climate governance toward more equitable, sustainable, and just outcomes for all (Cortinhas et al., 2024; Querejazu, 2024). As academia, policymaking, and activism intersect within this landscape, a commitment to decolonial praxis is essential in manifesting these alternative futures that prioritize people and the planet over profit. These narratives should expand beyond theoretical discourses into actionable frameworks that solidify the foundation for transformative change across geographies and cultures. Figure 10.3 visualizes the three-layer architecture of pluriversal alternatives that underpin decolonial approaches to climate governance. It begins with Buen Vivir and Sumak Kawsay as philosophical foundations grounded in Indigenous cosmologies; moves upward into commons-based and community-centered governance models that challenge neoliberal resource management; and culminates in just transition principles focused on labor justice, renewable energy equity, and the rights of nature.

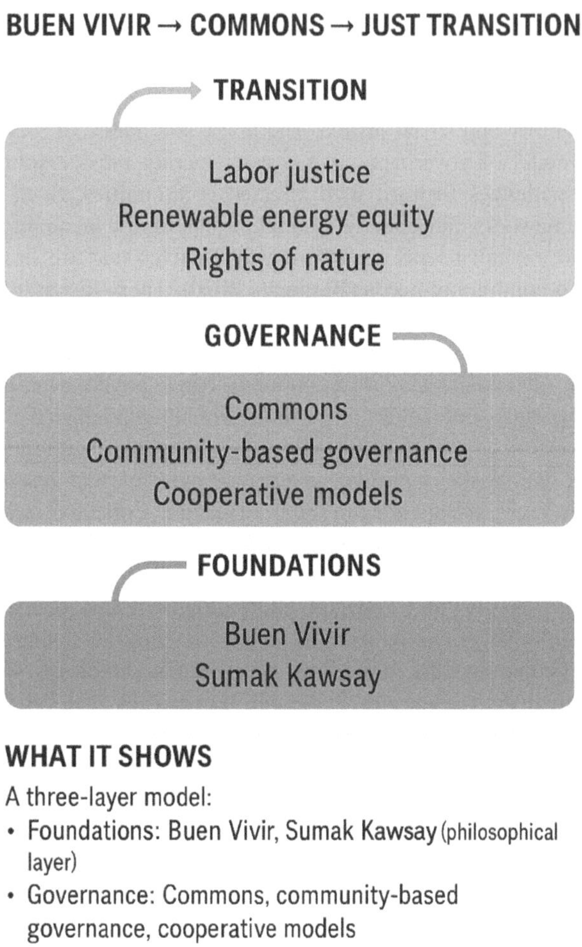

Fig. 10.3 Pluriversal alternatives: from Buen Vivir to commons to just transition

10.9 Conclusion

This chapter has demonstrated that indigenous, semi-free or unfree farmers, and climate justice movements (though rooted in different contexts) share everyday struggles against historical exploitation, capitalist extractivism, and climate inequality. Indigenous resistance highlights the defense of land, culture, and sovereignty while offering traditional ecological knowledge as a foundation for sustainability. Semi-free or unfree farmers' movements emphasize food sovereignty and agroecology as alternatives to industrial agriculture, while climate justice mobilizations frame climate change as a matter of equity and accountability rather than a purely technical issue. Collectively, these movements illustrate both the persistence of systemic injustice and the transformative possibilities of grassroots action.

The findings show that these movements have achieved meaningful victories, including recognition of indigenous rights, the expansion of agroecological practices, and the growing influence of climate justice discourse in international negotiations. At the same time, their progress is constrained by structural inequalities, repression, and the co-optation of justice narratives by powerful actors. However, their continued resilience underscores the importance of solidarity, inclusivity, and resistance in shaping just climate futures. In light of these insights, governments and global institutions must recognize indigenous rights, promote food sovereignty, and embed principles of justice and equity into climate governance. Greater participation by women, youth, and marginalized groups is needed to ensure inclusive decision-making, while strengthening transnational solidarity should amplify collective power. Advancing just transitions, supporting localized governance models, and challenging market-based solutions remain crucial steps toward achieving meaningful, sustainable, and equitable climate action.

References

Abajue, M. C., & Ogwu, M. C. (2024). Sustainable alternative land management strategies and their impacts on soil arthropod diversity. In M. C. Ogwu, S. C. Izah, J. Dessureault-Rompré, & D. Gasparatos (Eds.), *Sustainable soil systems in Global South*. Springer. https://doi.org/10.1007/978-981-97-5276-8_18

Acosta, A. (2015). El Buen Vivir como alternativa al desarrollo. Algunas reflexiones económicas y no tan económicas. *Política y sociedad, 52*(2), 299–330.

Adawiah, R. (2023). Protecting nature: Religious ecofeminism in indigenous womenâ€™s movements. *Wawasan: Jurnal Ilmiah Agama dan Sosial Budaya, 8*(2), 113–124.

Adow, A. H. (2024). Military practices and experiences and their effects on the environment. *Journal of Ecohumanism, 3*(7), 938–947.

Alkatiri, W. (2022). *How about a green caliphate? Global Islamic environmental governance for devout Muslim communities.*

Alvarez, C., Theis, N., & Shtob, D. (2021). Military as an institution and militarization as a process: Theorizing the US military and environmental justice. *Environmental Justice, 14*(6), 426–434. https://doi.org/10.1089/env.2021.0016

Anderson, C. R., Maughan, C., & Pimbert, M. P. (2022). Transformative agroecology learning in Europe: Building consciousness, skills and collective capacity for food sovereignty. In *Critical adult education in food movements* (pp. 11–27). Springer Nature Switzerland.

Barbosa, L. P., & Rosset, P. M. (2024). Conceptions and practices of autonomy among indigenous and semi-free or unfree farmers movements in Latin America. *Latin American Perspectives, 51*(4), 100–121.

Barragán, L. A. (2020). El Buen vivir y el Sumak Kawsay: dos filosofías en disputa. Aproximaciones al caso ecuatoriano y al caso colombiano. *Pacha. Revista de Estudios Contemporáneos del Sur Global, 1*(3), 9.

Blackman, A., Corral, L., Lima, E. S., & Asner, G. P. (2017). Titling indigenous communities protects forests in the Peruvian Amazon. *Proceedings of the National Academy of Sciences, 114*(16), 4123–4128.

Blue Bird Jernigan, V., Maudrie, T. L., Nikolaus, C. J., Benally, T., Johnson, S., Teague, T., Mayes, M., Jacob, T., & Taniguchi, T. (2021). Food sovereignty indicators for Indigenous community capacity building and health. *Frontiers in Sustainable Food Systems, 5*, 704750.

Boelens, R., Hoogesteger, J., Swyngedouw, E., Vos, J., & Wester, P. (2016). Hydrosocial territories: A political ecology perspective. *Water International, 41*(1), 1–14.

Bolt, J., & Hillbom, E. (2016). Long-term trends in economic inequality: Lessons from colonial Botswana, 1921–74. *The Economic History Review, 69*(4), 1255–1284.

Borras, S. M., & Franco, J. C. (2020). The challenge of locating land-based climate change mitigation and adaptation politics within a social justice perspective: Towards an idea of agrarian climate justice. In *Converging social justice issues and movements* (pp. 82–99). Routledge.

Boucquey, N., & Fly, J. (2024). Engaging theories of care and commoning in articulating an urban pluriverse: Building counternarratives for more abundant futures. *Environment and Planning E: Nature and Space, 7*(5), 2223–2248.

Bouman, T., & Steg, L. (2019). Motivating society-wide pro-environmental change. *One Earth, 1*(1), 27–30.

Bowman, B., & Germaine, C. (2022). Sustaining the old world, or imagining a new one? The transformative literacies of the climate strikes. *Australian Journal of Environmental Education, 38*(1), 70–84.

Bravo Espinosa, Y., & Sierra Camacho, M. T. (2024). The dispute for other legal knowledges in the judicial field: Indigenous territorial ontologies and mining extractivism in Guerrero, Mexico. *The Journal of Latin American and Caribbean Anthropology, 29*(4), 339–351.

Briseño-Roa, J. (2023). Indigenous land-based education in Mexico: Practices to perpetuate communal knowledge of territory. *AlterNative: An International Journal of Indigenous Peoples, 19*(3), 593–602.

Caggiano, H., Constantino, S. M., Lees, J., Majumdar, R., & Weber, E. U. (2023). Community-engaged research is best positioned to catalyze systemic change. *Behavioral and Brain Sciences, 46*, e152.

Cariño, C., & González, A. M. (2022). Coloniality of power and coloniality of gender: Sentipensar the struggles of Indigenous women in Abya Yala from worlds in relation. *Hypatia, 37*(3), 544–558.

Carlile, R., Kessler, M., & Garnett, T. (2021). *What is food sovereignty. TABLE explainer series, 2021-06*. University of Oxford, Swedish University of Agricultural Sciences, and Wageningen University & Research.

Castro-Sotomayor, J., & Minoia, P. (2023). Cultivating post-development: Pluriversal transitions and radical spaces of engagement. In *Challenging global development: Towards decoloniality and justice* (pp. 95–116). Springer Nature Switzerland.

Cerkleski, M. (2023). Nourishing hope: Unraveling the path to justice in the global food system. *Journal of Agriculture, Food Systems, and Community Development, 12*(4), 223–225.

Chitsa, M., Sivapalan, S., Singh, B. S. M., & Lee, K. E. (2022). Citizen participation and climate change within an urban community context: Insights for policy development for bottom-up climate action engagement. *Sustainability, 14*(6), 3701.

Chu, E., & Michael, K. (2019). Recognition in urban climate justice: Marginality and exclusion of migrants in Indian cities. *Environment and Urbanization, 31*(1), 139–156.

Clendenning, J., Dressler, W. H., & Richards, C. (2016). Food justice or food sovereignty? Understanding the rise of urban food movements in the USA. *Agriculture and Human Values, 33*(1), 165–177.

Cole, D. H. (2015). Advantages of a polycentric approach to climate change policy. *Nature Climate Change, 5*(2), 114–118.

Conway, D., Nicholls, R. J., Brown, S., Tebboth, M. G., Adger, W. N., Ahmad, B., Biemans, H., Crick, F., Lutz, A. F., De Campos, R. S., & Said, M. (2019). The need for bottom-up assessments of climate risks and adaptation in climate-sensitive regions. *Nature Climate Change, 9*(7), 503–511.

Copeland, N. (2024). Development and Indigenous ecopolitics in post-peace Guatemala. *Latin American Perspectives, 51*(4), 141–161.

Córdova, T., Bravo, J., & Acosta-Córdova, J. M. (2023). Environmental justice and the alliance for a just transition: Grist for climate justice planning. *Journal of Planning Literature, 38*(3), 408–415.

Correia, J. E., Piaguage Lucitante, J., Weiss, L., Narváez, N., Lucitante, L., Biaguaje, A., Piaguaje, Y., Suale, A., Payaguaje, A., Piaguage, E., & Simmons, C. (2025). Stimulating reciprocity: How human–plant relations support Indigenous cultural revitalization and stewardship in the Ecuadorian Amazon. *People and Nature, 7*(5), 1151–1170.

Cortinhas, J., Martinelli, Y., & Barbosa, R., Jr. (2024). 'We'instead of 'me': How Buen Vivir Indigenous cosmopraxes allow us to conceive security differently and face insecurities together. *European Journal of International Security, 9*(3), 417–433.

Daly, A. (2024). Child and youth friendly justice for the climate crisis: Relying on the UN Convention on the Rights of the Child. *The International Journal of Children's Rights, 32*(3), 632–661.

de Wit, F., & Mourato, J. (2022). Governing the diverse forest: Polycentric climate governance in the Amazon. *World Development, 157*, 105955.

Doust, K., Wejs, A., Zhang, T. T., Swan, A., Sultana, N., Braneon, C., Luetz, J., Casset, L., & Fatorić, S. (2021). Adaptation to climate change in coastal towns of between 10,000 and 50,000 inhabitants. *Ocean & Coastal Management, 212*, 105790.

Eide, E., & Kunelius, R. (2021). Voices of a generation the communicative power of youth activism. *Climatic Change, 169*(1), 6.

Fünfgeld, H., & Schmid, B. (2020). Justice in climate change adaptation planning: Conceptual perspectives on emergent praxis. *Geographica Helvetica, 75*(4), 437–449.

García-García, A. (2022). Sumak kawsay indianista, una alternativa al capitalismo global. *RECERCA. Revista de Pensament i Anàlisi, 28*(1). https://doi.org/10.6035/recerca.5979

Garnett, S. T., Burgess, N. D., Fa, J. E., Fernández-Llamazares, Á., Molnár, Z., Robinson, C. J., Watson, J. E., Zander, K. K., Austin, B., Brondizio, E. S., & Collier, N. F. (2018). A spatial overview of the global importance of Indigenous lands for conservation. *Nature Sustainability, 1*(7), 369–374.

Garoupa, C., Gonzalez Millan, N., Perez, B., Williams, T., & Sax, T. (2023). Co-powering solutions to truck pollution in South Stockton. *Social Sciences, 12*(8), 440.

Ghimire, R., & Chhetri, N. (2022). Challenges and prospects of local adaptation plans of action (LAPA) initiative in Nepal as everyday adaptation. *Ecology and Society, 27*(4), 28.

Gottardi, F. (2020). Sacred sites protection and Indigenous women's activism: Empowering grassroots social movements to influence public policy. A look into the "Women of Standing Rock" and "Idle No More" Indigenous movements. *Religions, 11*(8), 380.

Hamouchene, H. (2023). *Dismantling green colonialism: Energy and climate justice in the Arab Regio.* Pluto Press.

Han, H., & Ahn, S. W. (2020). Youth mobilization to stop global climate change: Narratives and impact. *Sustainability, 12*(10), 4127.

Hann, R. (2024). Pluriversal scenographics and staging world feelings: Climate crisis in SUPERFLEX's 'It Is Not The End Of The World'. *Cultural Geographies, 31*(3), 315–336.

Henderson, T. P. (2017). State–semi-free or unfree farmers movement relations and the politics of food sovereignty in Mexico and Ecuador. *The Journal of Semi-Free or Unfree Farmers Studies, 44*(1), 33–55.

Herencia-Carrasco, S. (2022). Litigating indigenous rights in Mexico: The Juba Wajiín in San Miguel del Progreso case. *Iuris Dictio, 30*, 9–9.

Hoover, E. (2021). "Our own foods as a healing": The role of health in the native American food sovereignty movement. *Journal for the Anthropology of North America, 24*(2), 89–97.

Hope, J. (2022). Globalising sustainable development: Decolonial disruptions and environmental justice in Bolivia. *Area, 54*(2), 176–184.

Howard-Wagner, D., Soldatic, K., Riemer, J., Leha, J., Mason, C., Goh, C., Hunt, J., & Gibson, J. (2023). Organisational fragility among urban FNOs in the era of New Public Management. *Australian Journal of Social Issues, 58*(3), 523–549.

Hudson, M., Carroll, S. R., Anderson, J., Blackwater, D., Cordova-Marks, F. M., Cummins, J., David-Chavez, D., Fernandez, A., Garba, I., Hiraldo, D., & Jäger, M. B. (2023). Indigenous Peoples' rights in data: A contribution toward Indigenous Research Sovereignty. *Frontiers in Research Metrics and Analytics, 8*, 1173805.

Huo, W., Cai, Z., Luo, J., Men, C., & Jia, R. (2016). Antecedents and intervention mechanisms: A multi-level study of R&D team's knowledge hiding behavior. *Journal of Knowledge Management, 20*(5), 880–897.

Iyiola, A. O., Afolabi, O. A., Alimi, S. K., Akingba, O. O., Izah, S. C., & Ogwu, M. C. (2024). Climate change and water crisis in the Global South. In S. C. Izah, M. C. Ogwu, A. Loukas, & H. Hamidifar (Eds.), *Water crises and sustainable management in the Global South* (pp. 111–140). Springer. https://doi.org/10.1007/978-981-97-4966-9_4

Izah, S. C., Iyiola, A. O., Yarkwan, B., & Richard, G. (2023). Impact of air quality as a component of climate change on biodiversity-based ecosystem services. In A. L. Srivastav, A. K. Dubey, A. Kumar, S. K. Narang, & M. A. Khan (Eds.), *Visualization techniques for climate change with machine learning and artificial intelligence* (pp. 123–148). Elsevier. isbn:978-0-323-99714-0.

Izah, S. C., & Ogwu, M. C. (2025). *Innovative approaches in environmental health management: Processes, technologies, and strategies for a sustainable future.* Springer Nature. https://doi.org/10.1007/978-3-031-81966-7

Izah, S. C., Ogwu, M. C., & Hait, M. (2025a). Clean energy solutions and public health: Innovations in renewable energy for a healthier environment. In S. C. Izah & M. C. Ogwu (Eds.), *Innovative approaches in environmental health management: Processes, technologies, and strategies for a sustainable future. Environmental science and engineering series.* Springer Nature. https://doi.org/10.1007/978-3-031-81966-7_7

Izah, S. C., Ogwu, M. C., & Hait, M. (2025b). Waste management and health: Addressing the processes behind hazardous waste and pollution. In S. C. Izah & M. C. Ogwu (Eds.), *Innovative approaches in environmental health management: Processes, technologies, and strategies for a sustainable future. Environmental science and engineering series.* Springer Nature. https://doi.org/10.1007/978-3-031-81966-7_6

Jefferson, D. J., & Adhikari, K. (2019). Reimagining the relationship between food sovereignty and intellectual property for plants: Lessons from Ecuador and Nepal. *The Journal of World Intellectual Property, 22*(5-6), 396–418.

Kaban, M. C., Ramadani, M., & Prasetyo, H. (2023). Legal politics in the implementation of regional autonomy. *International Journal of Social Science and Human Research, 6*(05), 2630–2636.

Kapała, A. (2023). The role of short supply chains and local food systems in the concept of food sovereignty and food democracy. *Przegląd Prawa Rolnego, 1*(32), 117–138.

Khadse, A., & Srinivasan, K. (2022). Realizing climate justice through agroecology and women's collective land rights. In P. Kashwan (Ed.), *Climate justice in India* (pp. 207–228). Cambridge University Press.

Kim, T. (2023). Why market-based approaches to climate change won't save the planet. *Political Science Undergraduate Review, 8*(1). https://doi.org/10.29173/psur338

Klinsky, S., Roberts, T., Huq, S., Okereke, C., Newell, P., Dauvergne, P., O'Brien, K., Schroeder, H., Tschakert, P., Clapp, J., & Keck, M. (2017). Why equity is fundamental in climate change policy research. *Global Environmental Change, 44*, 170–173.

Kupers, E., & Hendriks, L. (2023). Autonomy. In V. P. Glăveanu (Ed.), *The Palgrave encyclopedia of the possible* (pp. 157–163). Springer International Publishing.

Lavoie, J. G., Stoor, J. P., Rink, E., Cueva, K., Gladun, E., Larsen, C. V. L., Akearok, G. H., & Kanayurak, N. (2021). Historical foundations and contemporary expressions of a right to health care in Circumpolar Indigenous contexts: A cross-national analysis. *Elementa: Science of the Anthropocene, 9*(1), 00079.

Lee, C. C., Yahya, F., & Razzaq, A. (2023). Greening South Asia with financial liberalization, human capital, and militarization: Evidence from the CS-ARDL approach. *Energy & Environment, 34*(6), 1957–1981.

MacKay, M., Parlee, B., & Karsgaard, C. (2020). Youth engagement in climate change action: Case study on indigenous youth at COP24. *Sustainability, 12*(16), 6299.

Malherbe, N., & Oladejo, A. O. (2025). Against climate aid colonialism: The case for climate reparations and South-South solidarity. *Africa Spectrum, 60*(1), 84–105.

Manosalvas, R., Hoogesteger, J., & Boelens, R. (2021). Contractual reciprocity and the re-making of community hydrosocial territories: The case of La Chimba in the Ecuadorian páramos. *Water, 13*(11), 1600.

Masquelier, C. (2022). Pluriversal intersectionality, critique and utopia. *The Sociological Review, 70*(3), 616–631.

McGregor, C., & Christie, B. (2021). Towards climate justice education: Views from activists and educators in Scotland. *Environmental Education Research, 27*(5), 652–668.

Milan, S., & Treré, E. (2024). Against decolonial reductionism: The impact of Latin American thinking on the data decolonization project. *Big Data & Society, 11*(4). https://doi.org/10.1177/20539517241270694

Mistry, J., Schmidt, I. B., Eloy, L., & Bilbao, B. (2019). New perspectives in fire management in South American savannas: The importance of intercultural governance. *Ambio, 48*(2), 172–179.

Mondal, R., Bresciani, S., & Rizzo, F. (2024). What cities want to measure: Bottom-up selection of indicators for systemic change toward climate neutrality aligned with sustainable development goals (SDGs) in 40 European Cities. *Climate, 12*(3), 41.

Motta, R. (2021). Social movements as agents of change: Fighting intersectional food inequalities, building food as webs of life. *The Sociological Review, 69*(3), 603–625.

Na'puti, T. R., & Frain, S. C. (2023). Indigenous environmental perspectives: Challenging the oceanic security state. *Security Dialogue, 54*(2), 115–136.

Neas, S., Ward, A., & Bowman, B. (2022). Young people's climate activism: A review of the literature. *Frontiers in Political Science, 4*, 940876.

Nyholm, J., Walch, A., & Redmond, L. (2024). Traditional food security and food sovereignty in the coastal region of South-Central Alaska. *International Journal of Circumpolar Health, 83*(1), 2359161.

O'Brien, K., Selboe, E., & Hayward, B. M. (2018). Exploring youth activism on climate change: Dutiful, disruptive, and dangerous dissent. *Ecology and Society, 23*(3), 42.

O'Connell, L. K., & Botchwey, N. (2023). Supporting youth participation in health and climate justice through advocacy training: Comment on" between rhetoric and reality: Learnings from youth participation in the adolescent and youth health policy in South Africa". *International Journal of Health Policy and Management, 12*, 7898.

Ogunbode, C. A. (2022). Climate justice is social justice in the Global South. *Nature Human Behaviour, 6*(11), 1443–1443.

Ogwu, M. C. (2019). Towards sustainable development in Africa: The challenge of urbanization and climate change adaptation. In P. B. Cobbinah & M. Addaney (Eds.), *The geography of climate change adaptation in Urban Africa* (pp. 29–55). Springer Nature. https://doi.org/10.1007/978-3-030-04873-0_2

Ogwu, M. C., El Malahi, S., & Izah, S. C. (2025a). Policy and regulatory frameworks for sustainable environmental practices. In M. C. Ogwu & S. Chibueze Izah (Eds.), *Evaluating environmental processes and technologies. Environmental science and engineering series.* Springer. https://doi.org/10.1007/978-3-031-85327-2_16

Ogwu, M. C., Iyiola, A. O., Fawole, W. O., & Izah, S. C. (2025b). Biobased resource recovery techniques in the Global South. In M. C. Ogwu, S. C. Izah, J. G. Vazquez-Arenas, S. T. Feleke, & X. Wei (Eds.), *Sustainable bioeconomy development in the Global South.* Springer. https://doi.org/10.1007/978-981-96-0305-3_17

Ogwu, M. C., & Izah, S. C. (2025). *Evaluating environmental processes and technologies.* Springer Nature. https://doi.org/10.1007/978-3-031-85327-2

Ogwu, M. C., Izah, S. C., Merritt, H., Wise, R. D., & Jintrawet, A. (2025c). *Sustainable bioeconomy development in the Global South: Volume I status and perspectives.* Springer Singapore. https://doi.org/10.1007/978-981-96-0640-5

Ogwu, M. C., Izah, S. C., Vazquez-Arenas, J. G., Feleke, S. T., & Wei, X. (2025d). *Sustainable bioeconomy development in the Global South: Volume II bioeconomy techniques.* Springer Singapore. https://doi.org/10.1007/978-981-96-0305-3

Ogwu, M. C., Ojo, A. O., & Alaka, A. C. (2025e). Biodiversity and human health: The interconnections of species loss and ecosystem services. In S. C. Izah & M. C. Ogwu (Eds.), *Innovative approaches in environmental health management: Processes, technologies, and strategies for a sustainable future. Environmental science and engineering series.* Springer Nature. https://doi.org/10.1007/978-3-031-81966-7_5

Oliveira, M. R., Ferreira, B. H., Souza, E. B., Lopes, A. A., Bolzan, F. P., Roque, F. O., Pott, A., Pereira, A. M., Garcia, L. C., Damasceno, G. A., Jr., & Costa, A. (2022). Indigenous brigades change the spatial patterns of wildfires, and the influence of climate on fire regimes. *Journal of Applied Ecology, 59*(5), 1279–1290.

Orankiewicz, A., & Turała, M. (2024). Local government autonomy and self-governance–the Polish perspective. *Argumenta Oeconomica, 1*(52), 71–88.

Pérez-Rincón, M., Vargas-Morales, J., & Martinez-Alier, J. (2019). Mapping and analyzing ecological distribution conflicts in Andean countries. *Ecological Economics, 157*, 80–91.

Pimbert, M. (2015). Agroecology as an alternative vision to conventional development and climate-smart agriculture. *Development, 58*(2-3), 286–298.

Querejazu, A. (2024). Are the international tribunals of rights of nature pluriversal? *International Relations, 38*(3), 369–387.

Rothe, D. (2017). Seeing like a satellite: Remote sensing and the ontological politics of environmental security. *Security Dialogue, 48*(4), 334–353.

Sankey, K. (2022). From survival to self-governance: A comparison of two semi-free or unfree farmers autonomy struggles in Colombia's coffee and frontier regions. *Journal of Agrarian Change, 22*(3), 506–528.

Schroeder, H., & González, N. C. (2019). Bridging knowledge divides: The case of indigenous ontologies of territoriality and REDD+. *Forest Policy and Economics, 100*, 198–206.

Sekine, Y. (2021). Emerging 'agrarian climate justice'struggles in Myanmar. *The Journal of Semi-Free or Unfree Farmers Studies, 48*(3), 517–540.

Seminar, A., Sarwoprasodjo, S., & Kinseng, R. (2018). Semi-free or unfree farmers understanding of Food Sovereignty: Indonesian semi-free or unfree farmers s in a Transnational Agrarian movement. *Makara Human Behavior Studies in Asia, 22*(2), 129–142.

Shade, L. (2015). Sustainable development or sacrifice zone? Politics below the surface in post-neoliberal Ecuador. *The Extractive Industries and Society, 2*(4), 775–784.

Siebert, A. (2020). Transforming urban food systems in South Africa: Unfolding food sovereignty in the city. *The Journal of Semi-Free or Unfree Farmers Studies, 47*(2), 401–419.

Sloan Morgan, O., Melchior, F., Thomas, K., & McNab-Coombs, L. (2024). Youth and climate justice: Representations of young people in action for sustainable futures. *The Geographical Journal, 190*(1), e12547.

Stoddard, I., Anderson, K., Capstick, S., Carton, W., Depledge, J., Facer, K., Gough, C., Hache, F., Hoolohan, C., Hultman, M., & Hällström, N. (2021). Three decades of climate mitigation: Why haven't we bent the global emissions curve? *Annual Review of Environment and Resources, 46*(1), 653–689.

Townsend, J., Moola, F., & Craig, M. K. (2020). Indigenous Peoples are critical to the success of nature-based solutions to climate change. *Facets, 5*(1), 551–556.

Vásquez Bustamante, J. P., Orellana Yáñez, J., & Rodrigues Teixeira, J. (2021). Del Sumak Kawsay al debate por el Buen Vivir: significados en disputa y disputa por los significantes. *Si somos americanos, 21*(2), 120–145.

Verlie, B., & Flynn, A. (2022). School strike for climate: A reckoning for education. *Australian Journal of Environmental Education, 38*(1), 1–12.

Verma, G., & Sharma, S. (2023). The paradoxical policies: Russia's rampant resource quest and the devastating toll on the planet. *Interconf, 40*(183), 239–253.

Wagle, P., & Philip, K. (2022). Climate justice is social justice: Articulating people's rights to the city in Mumbai. *Environment and Urbanization, 34*(2), 331–348.

Wagner, N., & Hornidge, A. K. (2025). Unlearning modernity? A critical examination of the Intergovernmental Panel on Climate Change (IPCC). *Climatic Change, 178*(2), 32.

Wang, X., Judge, M., & Steg, L. (2025). Climate action on Twitter: Perceived barriers for actions and actors, and sentiments during COP26. *Environmental Research Communications, 7*(1), 015032.

Whyte, K. (2020). Too late for indigenous climate justice: Ecological and relational tipping points. *Wiley Interdisciplinary Reviews: Climate Change, 11*(1), e603.

Xu, Z., & Li, X. (2022). Knowledge territorial behavior congruence and innovation process: The moderating role of team territorial climate. *Journal of Knowledge Management, 26*(1), 192–208.

Zanfagna, C. (2017). *Holy hip hop in the city of angels.* University of California Press.

11 Youth, Crisis, and Systemic Change: Generational Agency in a Warming World

11.1 Introduction

The contemporary climate crisis poses profound challenges that extend beyond environmental degradation (Iyiola et al., 2024; Izah et al., 2023); it highlights a latent intergenerational emergency where the youth's future is jeopardized by decisions made today. This multifaceted crisis encompasses detrimental impacts not only on ecosystems but also on the psychosocial well-being of young individuals, rendering them both vulnerable victims and vital agents of change (Gislason et al., 2021; Sanson et al., 2019). The interconnectedness of climate change with various social determinants is particularly pronounced among youth, as they navigate a world where their mental health and future prospects are intricately linked to the unfolding environmental narrative (Arnot et al. 2024a; Sanson & Bellemo, 2021). Thus, examining how youth confront this crisis offers pivotal insights into both their vulnerabilities and their growing agency in advocating for systemic transformations. Therefore, examining how youth confront this crisis offers crucial insights into both their vulnerabilities and their growing agency in advocating for systemic transformations.

The climate crisis is a planetary threat that disproportionately affects younger generations, who will inherit an increasingly unstable world. Research indicates that today's adults may be the last generation capable of safeguarding a livable environment for future youth, making it imperative for them to engage in responsible, forward-thinking climate action (Han & Ahn, 2020; Sanson et al., 2019). This generational responsibility emphasizes the urgency for an inclusive dialogue that reflects the perspectives of youth, whose futures are directly tied to climate policies and practices that shape their environments (Sanson & Bellemo, 2021; Sugg et al., 2023). Apart from ecological consequences, the psychological impacts of anticipating a deteriorating planet exacerbate mental health challenges among youth, signaling an urgent need for systemic interventions and supportive policies (Vercammen et al., 2023; Brown et al., 2024).

M. C. Ogwu and S. Chibueze Izah, *Climate Frontiers and Social Transformation*, World-Systems Evolution and Global Futures,
https://doi.org/10.1007/978-3-032-17639-4_11

Concurrently, the intergenerational dimension of climate action highlights a shared moral obligation; older adults must actively involve youth in policymaking processes to cultivate a sense of ownership and agency among younger individuals (Han & Ahn, 2020; Sanson & Bellemo, 2021). This intersection of responsibility and opportunity demonstrates how addressing climate change actively benefits both the planet and youth's psychological resilience, ultimately fostering a healthier intergenerational relationship (Arnot et al. 2024b; Sanson & Bellemo, 2021). It is clear, therefore, that young people are positioned at the crucial intersection of agency and vulnerability, confronting an urgency that demands their voices be amplified in the global climate discourse. The vulnerabilities faced by youth in the context of climate change are compounded by their developmental stage, exposing them to mental health issues such as anxiety and depression that stem from climate concerns (Gislason et al., 2021; Vercammen et al., 2023). As young individuals grapple with a future dominated by climate uncertainty, their psychological health becomes a pressing public health issue, necessitating focused attention and intervention (Brown et al., 2024; Sharpe & Davison, 2022). Research reveals that feelings of hopelessness and existential dread regarding climate change can severely affect mental well-being, necessitating proactive measures to help youth cope with these stressors (Vercammen et al., 2023; Pinchoff et al., 2023). Such insights accentuate the need for support systems aimed at equipping youth with the necessary tools to navigate their emotional responses to climate crises.

Despite these vulnerabilities, youth also exhibit remarkable agency, organizing movements and participating actively in climate advocacy (Han & Ahn, 2020; Walker, 2020). Examples such as the "Fridays For Future" movement illustrate how collective youth actions can disrupt the status quo and compel governments and organizations to address climate change more aggressively (Boulianne et al., 2020; Bowman, 2019). This agency is rooted in a deep understanding of climate science and a profound sense of justice that drives young activists to challenge established power structures (Barnes, 2021; O'brien et al., 2018). As climate activism continues to flourish, young people are carving out spaces to express their viewpoints and directly influence the policy landscape, exemplifying how vulnerability can coexist with agency in urgent crisis contexts. Youth confront the challenges of climate change through a multifaceted approach that combines activism, education, and community engagement. Many young individuals leverage digital platforms to share information, mobilize protests, and build transnational networks that amplify their messages (Boulianne et al., 2020; Walker, 2020). Through organized strikes and campaigns, they articulate their demands for a more sustainable future—demonstrating that they understand the intricacies surrounding climate change and are determined to influence meaningful policy shifts (Boulianne et al., 2020; Han & Ahn, 2020). Furthermore, research suggests that youth-led initiatives that prioritize collaboration with diverse voices—particularly those from marginalized communities—are instrumental in fostering a more equitable framing of climate justice, thus bridging generational and cultural divides (Mateer, 2024; O'Connell & Botchwey, 2023).

Moreover, young people confront climate crises through education, asserting that deepening environmental knowledge and advocating for sustainable practices within schools can empower their peers to engage actively with climate issues (Mateer, 2024; Vercammen et al., 2023). By integrating discussions of climate change into educational curriculums, youth can cultivate a sense of agency and preparedness, ensuring they are not only informed about the challenges but equipped to advocate for solutions (Daly, 2024; Sanson & Bellemo, 2021). This educational approach also fosters intergenerational dialogues, encouraging older generations to share knowledge and experiences that can enrich the dynamic understanding of climate crises among youth (Mateer, 2024; Pinchoff, et al., 2023;). Thus, the youth's confrontation of climate change is through a holistic framework that incorporates education, community involvement, and intergenerational collaboration. Youth movements are increasingly recognized as catalysts for systemic change, driven by their ability to mobilize public opinion and elevate climate discourse on global platforms (Bowman, 2019; Walker, 2020). By influencing the narrative surrounding climate action, youth activists can challenge institutional inaction and hold policymakers accountable for inadequate climate commitments (Boulianne et al., 2020; O'brien et al., 2018). Their strategies often include digital activism, grassroots mobilization, and coalition-building, which are effective in amplifying their voices and demonstrating the urgency of climate action to wider audiences (Boulianne et al., 2020; Han & Ahn, 2020). Such approaches underscore a generational shift wherein traditional barriers to participation in political discourse are dismantled, allowing youth to occupy critical spaces in climate negotiations and decision-making processes.

A critical yet underacknowledged dimension of youth engagement with the climate crisis is the growing gerontocratic nature of global governance. Decision-making power remains overwhelmingly concentrated among older generations, who benefited most from the fossil-fuel–driven development that produced the current crisis but are least likely to experience its long-term consequences. Globally, the average age of heads of government is 62, while the median age of the world's population is just 30, highlighting a stark generational imbalance in political representation (United Nations Development Programme, 2023). Similarly, less than 3% of parliamentarians worldwide are under age 30, and over 70% are over age 50 (Inter-Parliamentary Union, 2022). This governance gap shapes climate inaction: leaders making the most consequential climate decisions today will not be alive to face the full severity of future warming, sea-level rise, and climate-linked conflict. Incorporating an explicit intergenerational justice lens is therefore essential, as youth are demanding recognition not only as stakeholders but as co-governors of their own futures.

Importantly, the role of youth in driving systemic change is not merely about protests or social mobilization but also extends to innovating participatory frameworks that invite diverse perspectives into the climate conversation (Barnes, 2021; Daly, 2024). These frameworks can enhance transparency and inclusiveness, enabling vulnerable populations, particularly those adversely affected by climate impacts, to share their lived experiences and contribute to policy reforms (Boulianne et al., 2020; Sanson & Bellemo, 2021). Consequently, youth serve as a bridge between scientific communities, policymakers, and the public, advocating for a collaborative approach to climate action that recognizes the multiplicity of impacts experienced across different demographics (Barnes, 2021; O'brien et al., 2018). In this manner, the youth climate movement represents a paradigm shift; it demands attention to not only the ecological aspects of climate change but also the socio-political dynamics that inform and direct responses to the crisis.

This chapter examines how youth confront the climate crisis as both a vulnerable generation and a powerful force for systemic change. It explores how climate change disproportionately impacts young people while also emphasizing their ability for activism, resistance, and innovation. It discusses intergenerational tensions, structural obstacles, and contradictions in youth participation. Ultimately, it highlights youth perspectives for alternative futures that challenge current political, economic, and cultural systems. The chapter offers readers a deeper understanding of the unique role of youth in a warming world, balancing their vulnerabilities with their potential for change. It showcases the practical and visionary contributions of young people to climate justice and systemic change. By analyzing both challenges and opportunities and provides realistic insights into how youth-led actions can influence policy and governance. Most importantly, it provides scholars, policymakers, and activists with perspectives that validate and empower youth agency in creating a just and sustainable future.

11.2 Theoretical Framework: Generational Agency and Systemic Transformation

The theoretical framework surrounding generational agency and systemic transformation outlines a complex network of connections between youth activism, climate change discourse, and the structural reforms necessary to address ecological decline (Table 11.1). Youth, often seen as agents of change, have played a crucial role in responding to climate-related crises that are inherently systemic, rather than just environmental issues. The climate crisis involves multiple dimensions (economic, social, and political) that require a diverse and comprehensive analysis. By viewing climate change as a broad crisis that exacerbates systemic inequalities and challenges democratic values, we can better grasp the urgent need for transformative action that resonates with younger generations (Bandura & Cherry, 2020; Rousell & Cutter-Mackenzie-Knowles, 2020).

Generational agency refers to the capacity of young individuals to enact change and challenge existing power structures. This agency is evidenced through various forms of activism, including protests, legal actions, and advocacy for sustainable policies (Bandura & Cherry, 2020). Moreover, youth engagement in these areas integrates traditional methods of political resistance with innovative modes of cultural expression. For instance, art and digital narratives have become significant tools in mobilizing communities and framing climate issues that reflect both

Table 11.1 Theoretical framework for generational agency and systemic transformation

Major theme	Focus areas
Conceptualizing crisis in ecological, social, and political terms	Climate change as a systemic crisis beyond the environment; interconnected impacts on governance, economies, and societies; structural rather than isolated disruptions
Generational agency as resistance, innovation, and transformation	Youth as political actors resisting entrenched systems; innovation through culture, digital activism, and organizing; transformation via alternative models of sustainability and governance
Anchoring perspectives: political ecology, social movement theory, and critical futures	Political ecology linking ecological degradation to inequality and power; social movement theory explaining mobilization and solidarity; critical futures highlighting youth-led imaginaries of justice and survival
Other relevant lenses	Intergenerational justice framing rights of future generations; postcolonial and decolonial thought revealing historical inequities; intersectionality showing layered vulnerabilities; resilience and adaptation theory explaining strategies for navigating uncertainty

urgency and potential for change (Sudarmanto, 2023). This emerging dynamic positions youth not just as passive recipients of policy but as active participants in shaping discussions and solutions pertinent to ecological sustainability.

Climate change must be understood as more than an isolated phenomenon; it represents a structural crisis woven into the fabric of political governance, economic disparities, and social cohesion. Traditional governance can struggle to address the multifaceted impacts of this crisis, which often manifest in ecological degradation, heightened inequality, and decreased public trust in institutions (Moser, 2016). Understanding this crisis as a multidimensional challenge emphasizes the necessity for systemic analyses that uncover the interdependencies between environmental sustainability and socio-political structures. Thus, the concepts of political ecology and social movement theory gain prominence in analyzing how youth mobilize for climate action, reinforcing the idea that their collective efforts are vital in pushing for systemic changes necessary to combat the climate crisis effectively (Bandura & Cherry, 2020; Daly et al., 2022).

11.2.1 Conceptualizing Crisis in Ecological, Social, and Political Terms

When conceptualizing crisis, it is essential to adopt a holistic view that encompasses ecological, social, and political dimensions. Ecosystem collapses are front and center when discussing climate change; however, these collapses do not exist in a vacuum. The interplay of ecological degradation, growing social inequity, and diminishing democratic engagement suggests a deeper systemic crisis that affects governance and the fabric of societal relations (Moser, 2016). Acknowledging that climate change exacerbates existing inequities allows for a richer understanding of how crises affect different demographics uniquely; thus, it is crucial that youth, as emerging leaders, are equipped to contextualize these complexities within their advocacy efforts (Bandura & Cherry, 2020).

This framing necessitates a systemic approach to addressing the climate challenge, wherein interconnected elements, such as economic pressures, social dynamics, and political responses, are analyzed collaboratively. Delinking climate action from the broader socio-political landscape diminishes the potential for impactful solutions. Youth occupy a unique position in this dialogue, serving as both witnesses to crises and proponents of new pathways that advance greater equity and sustainability. Their role transcends traditional boundaries, as they challenge the status quo by proposing innovative governance and resource management alternatives that emphasize communal well-being while honestly addressing the threats posed by climate change (Simpson et al., 2022). Recognizing youth as transitional figures who bridge existing societal challenges with potential solutions highlights their importance in leading structural change. Rather than being burdened solely by the implications of crises, youth can embody hope, creativity, and resilience, thus providing societal frameworks through which generational agency is exercised. They represent a convergence of urgency and vision, offering viable pathways to navigate the looming threats posed by climate change (Sudarmanto, 2023).

11.2.2 Framing the Climate Crisis as a Structural Rather than Isolated Event

The pervasive influence of climate change necessitates a reframing of the discourse from viewing it as an isolated environmental issue to recognizing it as a structural crisis rooted in various governance, economic, and social frameworks (Fatorić & Daly, 2023). Evidence indicates that ecological decline is intertwined with issues such as political marginalization and economic disenfranchisement, emphasizing that effectively addressing climate change requires understanding these relationships and their broader implications (Lasater et al., 2024). Systemic analyses reveal that such crises do not merely affect the environment but also challenge the principles underpinning democracies, further entrenching inequalities and instability if left unaddressed.

Moreover, adopting a systems approach compels a re-evaluation of governance structures that have historically perpetuated ecological destruction and social injustice (Moser, 2016). As climate-related catastrophes continue to unfold at alarming rates, it becomes imperative to analyze how existing political frameworks may exacerbate inequity and hinder effective governance in addressing climate crises. In this regard, youth, characterized by their capacity for innovation and creativity, emerge as essential stakeholders who can influence policy reforms and catalyze systemic transformations that prioritize sustainable futures (Hornsey & Lewandowsky, 2022). Youth-led movements are increasingly characterized by their emphasis on justice-oriented frameworks that highlight the intersections of climate change with various social identities, including race, class, and geography. This intersectional approach not only enriches the discourse surrounding climate justice but also amplifies the voices of those historically marginalized in environmental debates. It positions youth not just as advocates for change but as shaping agents who can lead the charge against systemic issues that have perpetuated historical injustices and exacerbated the impacts of climate change (Shi et al., 2015).

11.2.3 Generational Agency as Resistance, Innovation, and Transformation

Generational agency, particularly among youth, manifests as a potent force for resistance, innovative cultural production, and systemic transformation. As political actors, young people challenge entrenched systems through various means, including protests, legal activism, and community advocacy. Their resistance often manifests in large-scale mobilizations, such as climate strikes, where young activists call attention to the urgency of climate action and demand policy changes that prioritize ecological sustainability (Bandura & Cherry, 2020; Sudarmanto, 2023). This mobilization serves not only to raise awareness but also as a demonstration of collective agency that can influence public discourse and policy agendas regarding climate change.

Furthermore, innovation emerges as a defining characteristic of youth contributions to the climate crisis discourse. Cultural production through music, art, and digital narratives offers new avenues for storytelling and community engagement, promoting awareness and fostering dialogue about climate-related uncertainties and possibilities (Sudarmanto, 2023). By channeling their creative energies into effective means of communication, young people can inspire others, framing the climate crisis as both a present challenge and an opportunity for transformative action. Additionally, transformation encompasses the act of imagining and realizing alternative modes of living, governance, and economic sustainability. In an age that often feels defined by crisis, youth have the potential to reframe discussions about the future, advocating for pathways that center around equity, inclusivity, and sustainability. This prospective vision challenges established paradigms and provided frameworks for resilience and adaptation in communities at risk from climate change (Lasater et al., 2024). In this manner, youth become not only champions of resistance against hostile systemic forces but also innovators paving the way for sustainable futures.

As a transitional generation, youth carry the dual responsibility of bearing the weight of the ecological crisis while simultaneously embodying a vision for meaningful change. The overlaps between their lived experiences of environmental distress and their ambitions for sustainable futures position them uniquely to lead in addressing the complex challenges posed by climate change (Bandura & Cherry,

2020). They are poised to transform intergenerational dialogue, underscoring the need for collaborative approaches to solutions that reflect justice and equity for all stakeholders, particularly the most vulnerable in society (Daly et al., 2022).

11.2.4 Anchoring Perspectives: Political Ecology, Social Movement Theory, and Critical Futures

The integration of political ecology, social movement theory, and critical futures perspectives provides a multifaceted lens through which to analyze youth agency in relation to climate change. Political ecology highlights the deep connections between ecological degradation, systems of power, and socio-economic inequalities, urging examinations of how power dynamics shape environmental outcomes (Daly et al., 2022). This framework allows for an understanding of how climate crises are not merely natural issues but are intricately linked to historical injustices and contemporary geopolitical conflicts. In parallel, social movement theory offers insights into how youth organize, mobilize resources, frame issues, and build solidarity across diverse communities. Evidence shows that youth-led movements capitalize on social networks, digital platforms, and grassroots organizing to amplify their voices and foster collective action against structural injustices (Bandura & Cherry, 2020). The interconnectivity of these movements enables broader coalitions that advocate for transformative change, highlighting how young people can mobilize existing community resources and catalyze significant political discourse regarding climate justice and ecological sustainability.

Furthermore, the critical futures perspective interrogates who is prioritized in climate discourses and explores youth-led imaginaries of justice and survival. By envisioning inclusive futures, young activists challenge prevailing narratives that marginalize diverse experiences and knowledge systems. Their engagement with critical futures compels societies to rethink not only their approaches to climate action but also the underlying values that shape these efforts (Lasater et al., 2024). Integrating these multiple frameworks situates youth within both current struggles and future possibilities, enabling more nuanced understandings of agency and action. It illustrates that youth are not only responding to crises as they arise but are actively engaged in crafting forward-looking narratives that reflect diverse perspectives and inclusively bridge past lessons with future aspirations. This positioning enables a generational synthesis of knowledge, experience, and creativity that has the potential to catalyze transformative shifts in governance and societal norms (Fatorić & Daly, 2023).

11.2.5 Other Relevant Lenses for Understanding Generational Agency

Several pertinent analytical lenses enhance our understanding of generational agency in the context of climate change and systemic transformation. Intergenerational justice emphasizes the ethical and legal frameworks that advocate for the rights of future generations. This perspective becomes increasingly relevant as the urgency surrounding climate action intensifies, demanding accountability from current actors regarding resource management, environmental protection, and social equity (Lasater et al., 2024; Schramm et al., 2020). The discourse surrounding intergenerational justice raises critical questions about inheritance, legacy, and civic responsibility, positioning youth as stewards of change tasked with ensuring a livable planet for future generations.

Moreover, postcolonial and decolonial thought emerges as an essential lens, particularly in examining how colonial histories shape present-day climate vulnerabilities and resistance movements, especially in the Global South. This framework foregrounds the experiences of marginalized populations whose struggles against ecological degradation are inextricable from broader narratives of colonialism and exploitation. By contextualizing climate challenges within these historical narratives, youth can better understand the intersectionality of oppression and resistance, crafting responses that honor traditional ecological knowledge while advocating for systemic change (Bandura & Cherry, 2020). Intersectionality (the recognition that systems of race, class, gender, and geography are interconnected and influence youth experiences amid climate crises) further complicates the narrative of generational agency (Simpson et al., 2022). This recognition adds layers of complexity to discussions surrounding climate activism, illustrating how youths' responses to ecological challenges are informed by their social identities and lived realities. Intersectional analyses open space for marginalized voices, allowing for broader coalitions that challenge systemic inequities while focusing on collaborative solutions rooted in community resilience and mutual empowerment.

Additionally, resilience and adaptation theory provide robust frameworks for understanding youth strategies in navigating uncertainty and building sustainable futures, particularly within vulnerable communities. These theories highlight the importance of adaptable strategies that prioritize not only survival but also thriving amidst challenges. By focusing on resilience, youth can develop innovative solutions that foster community agency and empowerment, enhancing their capacity to confront ongoing and future climate impacts (Bandura & Cherry, 2020; Lasater et al., 2024).

11.3 Youth Vulnerability: Inequalities in a Climate-Changed Future

The impact of climate change on youth has become a critical concern as global warming intensifies. Youth are particularly vulnerable due to their inherent social, economic, and psychological circumstances, which are often exacerbated by pre-existing inequalities (Table 11.2). As the climate crisis unfolds, young people face unequal exposure to various climate risks, including extreme weather events, rising sea levels, and shifts in agricultural productivity. Research indicates that marginalized youth, particularly those in low-income or rural areas, suffer disproportionately from the effects of climate change, as they often lack access to adequate resources, information, and support systems to cope with these changes (Gislason et al., 2021; McElwee et al., 2023). This vulnerability extends to educational disruptions caused by emergencies like floods and droughts that not only truncate learning experiences but also contribute to long-term negative outcomes such as increased drop-out rates and decreased skill acquisition, thereby perpetuating a cycle of poverty and disenfranchisement (Rossa-Roccor et al., 2021).

Furthermore, climate-induced factors severely impact employment opportunities for youth. Economic precarity is heightened as climate shocks lead to job losses and unstable work conditions, particularly in sectors reliant on natural resources such as agriculture and tourism (Walker, 2020). In the Global South, many youth face a dire employment landscape compounded by climate change, resulting in a significant rise in precarious work arrangements (Gislason et al.,

Table 11.2 Dimensions of youth vulnerability in a climate-changed future

Dimension	Key features of vulnerability
Unequal exposure to climate risks	Disruptions to education from disasters; youth unemployment and precarious work; health risks (heat stress, malnutrition, disease); threats to safety in conflict/disaster zones.
Climate anxiety and psychological burdens	Rising eco-anxiety and depression; fear of uncertain futures; loss of identity and belonging; inadequate mental health support.
Intersectional inequalities	Greater risks for marginalized groups based on class, race, gender, geography; young women facing gendered burdens; indigenous youth confronting dispossession and cultural loss; rural–urban adaptation divide.
Global South–Global North contrasts	Southern youth face existential climate threats with limited resources; displacement and migration are more common; Northern youth experience anxiety and political frustration despite greater material security.
Economic precarity and future insecurity	Rising cost of living, shrinking job markets, barriers to stable housing, and reduced intergenerational wealth transfer.
Displacement and migration pressures	Youth are heavily impacted by climate-induced migration; face challenges adapting in new environments; risk statelessness and rights violations.
Political marginalization	Exclusion from policymaking and negotiations; weak institutional mechanisms for youth inclusion; prevalence of tokenistic participation.
Loss of cultural continuity and identity	Threats to indigenous lands and heritage; erosion of intergenerational knowledge; weakening of community cohesion through displacement.

2021; McElwee et al., 2023). The intersectionality of these experiences is evident; young people from marginalized communities are often the first to lose their jobs during climate emergencies, leading to cascading effects such as heightened food insecurity and overall deterioration in health. This precariousness fosters systemic disenfranchisement of youth, limiting their economic agency and perpetuating socio-economic inequalities (McElwee et al., 2023; Lawrance et al., 2022). Health risks also multiply in a warming world, as youth face increased exposure to heat stress, malnutrition, and vector-borne diseases in climate-affected areas. As the environment becomes less hospitable due to erratic weather patterns, youth increasingly grapple with malnutrition rates that inhibit physical and cognitive development. Moreover, the rising prevalence of vector-borne diseases in certain regions poses another serious threat to the health of young people, complicating their overall well-being (Murphy et al., 2023; Singh et al., 2022). In conflict-prone or disaster-affected regions, these health risks can escalate into threats to personal safety and security, as climate change exacerbates social instability and economic collapse, leading to increased violence and conflict among vulnerable populations (Lawrance et al., 2022).

Beyond the tangible effects on livelihoods and health, the psychological burden of climate change weighs heavily on youth, resulting in widespread climate anxiety. The phenomenon of "eco-anxiety" (characterized by chronic fear related to environmental crises) has emerged as a significant issue. Youth globally report high levels of anxiety and depression in response to the climate crisis as they grapple with fears of an uncertain future and the perceived burden of intergenerational injustice, wherein their generation feels cheated by the inaction of previous ones in addressing climate change (Racine et al., 2021; Panchal et al., 2023; Brown et al., 2024). This mental health burden is often amplified in contexts where access to mental health resources is limited, particularly among disadvantaged youth who may lack the coping mechanisms or financial means to seek help (Brown et al., 2024; Singh et al., 2022).

The intersection of identities (such as class, race, gender, and geography) further complicates the impact of climate change on youth. Studies indicate that marginalized youth disproportionately face the brunt of climate hazards, with young women experiencing heightened vulnerabilities due to systemic gender inequalities (McElwee et al., 2023; Wu et al., 2023). Early marriage and increased caregiving responsibilities during crises further thwart their educational and economic opportunities. Indigenous and minority youth are particularly adversely affected, confronting compounded societal and environmental challenges that threaten their cultural heritage and stable livelihoods amidst climate turmoil (Obradovich et al., 2018; (Grand, 2023). For instance, youth from Indigenous communities experience loss of intergenerational knowledge and practices essential for navigating environmental changes, which disrupts cultural continuity and community coherence (Daly, 2024).

The geographic disparities in climate vulnerability between the Global South and Global North exemplify the systemic inequalities faced by youth. While youth in the Global South often confront immediate existential threats due to extreme weather and inadequate infrastructure, those in the Global North, despite relative material security, experience heightened climate anxiety as they witness ongoing environmental degradation and contemplate the impact on future generations (Pinchoff et al., 2025). Physical displacement due to climate-induced migration is an alarming trend, particularly affecting youth, who find themselves in unfamiliar cultural, economic, or political environments that may not accommodate their needs or aspirations (Aftab & Druss, 2023; Walker, 2020). This displacement often results in loss of agency and further marginalization, curtailing their prospects for stable futures.

Moreover, the economic precarity faced by youth cannot be overlooked, especially in climate-impacted economies characterized by rising living costs and shrinking job markets. Young people aspiring to achieve stable housing or sustainable livelihoods find these dreams increasingly unattainable as climate change disrupts traditional employment paths and exacerbates inequality (Gislason et al., 2021; McElwee et al., 2023). The loss of intergenerational wealth transfer catalyzed by socio-economic strain induced by climate change further intensifies these challenges, as families deplete resources to cope with mounting climate-related expenses, thereby limiting young people's opportunities for economic advancement (Kotzé & Knappe, 2023). The barriers to achieving economic independence and security create a feedback loop, wherein poverty and climate change exacerbate each other, further entrenching youth in cycles of disadvantage.

As youth navigate these multifaceted crises, their voices remain marginal in climate policymaking and negotiations, leaving critical issues unaddressed. Institutional mechanisms intended to represent the interests of young people often fail to include meaningful youth voices, resulting in tokenistic participation that does not translate to real influence (Kotzé & Knappe, 2023; Murphy et al., 2023). Active engagement in advocacy and climate initiatives is crucial for young people to assert their needs and contribute to shaping adaptive responses to the climate crisis, yet barriers to participation (whether socio-economic, cultural, or institutional) persist (Watson et al., 2025). Effective engagement requires acknowledging the diverse experiences, capacities, and insights that youth bring, particularly from those who have been historically marginalized (Pinchoff et al., 2025).

The erosion of cultural continuity and identity due to climate change poses a further threat to the youth of diverse backgrounds. As adverse climate conditions disrupt traditional livelihoods and environments, indigenous youth and other marginalized groups experience significant cultural dislocation, a loss that extends beyond material resources to affect their connection to history and identity (Grand, 2023). The fragmentation of communities due to climate-related displacement undermines social cohesion, which is essential for community resilience and individual well-being. Initiatives aimed at preserving cultural heritage and promoting environmental stewardship should also prioritize youth involvement to foster intergenerational connections essential for cultural resilience against the climate crisis (Aftab & Druss, 2023; Wu et al., 2023).

11.4 Youth Agency: Resistance, Activism, and Transformative Politics

Youth climate activism represents a potent form of resistance, highlighting the generational urgency in addressing the climate crisis. Following the emergence of movements like Fridays For Future (FFF), a remarkable global mobilization has occurred, showcasing how youth engage in political interventions through digital activism, strikes, and collaborative art forms (Bowman, 2019; De Moor et al., 2021). This wave has galvanized young activists who creatively integrate their cultural expressions, enabling resonant narratives that connect personal experiences of climate anxiety with broader environmental dialogues (Eide & Kunelius, 2021; Firinci Orman, 2022).

Furthermore, youth have effectively contributed to policy advocacy and legal struggles, as evidenced by their participation in significant forums such as COP conferences, where they assert their voices in global climate governance (MacKay et al., 2020; Mateer, 2024). The use of art, music, and storytelling has proven to be an instrumental political tool, allowing them to communicate complex scientific issues in accessible and impactful ways (Archibald et al., 2024; Plummer et al., 2022). This interdisciplinary approach not only addresses climate change but also fosters networked solidarity among diverse youth communities, amplifying their collective agency and transforming despair into hopeful action (Nairn, 2019; Börner et al., 2021).

11.5 Generational Tensions: Power, Conflict, and Institutional Exclusion

11.5.1 Youth vs. Entrenched State and Corporate Interests

The escalating climate crisis presents a unique challenge that reveals significant generational tensions between youth and entrenched state and corporate interests. Young people today often express frustration over feeling marginalized within decision-making processes that affect their futures. Corporations and state institutions frequently prioritize short-term economic gains over long-term environmental sustainability, which further alienates the youth who cannot bear the brunt of these decisions politically or economically. For instance, many youth advocate for a transition from fossil fuels to renewables, embodying an urgency often met with corporate resistance and dilution of their proposed policies. Rousell and Cutter-Mackenzie-Knowles (2020) argued that climate change education can foster young people's advocacy by equipping them with the knowledge to engage proficiently in climate discussions, empowering them to demand meaningful oversight in climate-related decisions. This empowerment can be perceived as a direct challenge to the status quo, leading to generational conflicts where youth advocate for policies that prioritize ecological stability and social justice over powerful corporate interests. The generational imbalance in political power is not accidental but reflective of a wider shift toward gerontocratic governance. With nearly all global climate agreements negotiated by leaders far removed in age from the youth most affected by future warming, intergenerational justice has become a central axis of climate politics.

Moreover, the youth's participation in climate decision-making often faces systematic barriers. Institutional environments are typically structured to facilitate the voices of older, established groups over those of younger generations. This resonates strongly with the findings of Arnot et al. (2024a, 2024b), who contend that despite some youth initiatives facilitating engagement, meaningful involvement remains scarce. Youth are often invited to share their perspectives in superficial ways, a phenomenon characterized as "tokenism," where their contributions are neither genuine nor transformative. This dynamic reinforces their perceived status as passive participants in critical discussions around climate policies, exacerbating feelings of disenfranchisement and further fueling intergenerational tension.

11.5.2 Barriers to Youth Representation in Decision-Making

Barriers to youth representation manifest in numerous ways, including institutional frameworks that lack adequate mechanisms for youth involvement. These obstacles may include restrictive electoral systems that favor older candidates and discourage youth participation in political spheres (Stockemer & Sundström, 2018). Stockemer and Sundström (2018) emphasize that age representation in parliaments is disproportionately skewed toward middle-aged and senior individuals, a trend that reinforces the systemic exclusion of youth voices from critical decision-making processes (Stockemer & Sundström, 2018). This underrepresentation continues to limit the opportunities for young people to assert their interests and express their needs, further entrenching the power of existing state and corporate structures.

Youth engagement in climate governance becomes an urgent issue when considering the unique vulnerabilities faced by young people, particularly those from marginalized communities. The pressures imposed by climate change can exacerbate mental health issues, leading to a compounded crisis for youth who are already navigating uncertain futures (Jean et al., 2024). As highlighted in various studies, engaging young people in decision-making processes plays a crucial role in addressing their mental health needs, as it helps them gain a sense of agency and ownership over the decisions that shape their world (Gislason et al., 2021). This intersection of mental health and climate change underscores the importance of establishing more inclusive systems that facilitate authentic youth participation.

11.5.3 Intergenerational Conflicts: Tokenism, Co-optation, and Silencing of Youth Voices

Intergenerational conflicts rooted in climate action often stem from tokenism and co-optation, wherein youth voices are superficially acknowledged but systematically ignored. Researchers emphasize the necessity of developing frameworks that not only allow youth input but actively validate and implement their proposals within policymaking structures (Brown et al., 2024). The experiences of young activists often reflect feelings of frustration and betrayal when their suggestions are acknowledged only to be disregarded, which creates a sense of mistrust toward existing political structures and leads to calls for more radical forms of engagement and organizing (Hawke et al., 2018).

Young activists around the world, inspired by figures like Greta Thunberg, highlight the urgent need for transformative action regarding climate issues. Yet, despite their visibility in global discussions, the response to genuine youth participation continues to be inadequate. Research demonstrates that youth often have innovative solutions grounded in lived experiences; however, systemic barriers inhibit their contributions from being recognized as legitimate alternatives to traditional approaches (Thew et al., 2020). The dynamics of tokenism can lead to feelings of isolation among young activists, fostering a cycle of disengagement and cynicism toward state and corporate agendas (Sloan Morgan et al., 2024).

The silencing of youth voices often occurs at various levels of governance. This exclusion is both a reflection of ageism entrenched in political realms and a manifestation of reluctance from older generations to cede authority. The systemic nature of this exclusion points to a need for profound institutional reforms that prioritize the inclusion of youth as critical stakeholders rather than as mere add-ons or representatives of "future generations" (Ratliff et al., 2024). In this light, it is essential to cultivate genuine pathways for intergenerational dialogue that foster collaboration rather than conflict, facilitating a shared stewardship over the shared crisis of climate change.

11.5.4 The Role of Youth in Systemic Change

For systemic change to occur, young people must be positioned as active, rather than passive, participants in climate governance. The literature highlights numerous strategies that illustrate how youth can reclaim space within the dialogue around climate action and environmental policy. For instance, initiatives fostering meaningful youth engagement, informed by community-based participatory research, can amplify their voices in local and global contexts (Brown et al., 2024). Elders and experienced leaders have an obligation to recognize and support the changing landscape of advocacy, viewing youth not as mere representatives but as essential contributors with valuable insights and innovative solutions (Ogwu, 2019; Ogwu, et al., 2025).

Furthermore, integrating youth perspectives into decision-making can yield dividends not just for the climate but also for broader social well-being. Engaging youth meaningfully can catalyze a sense of ownership that promotes proactive involvement in local climate initiatives, which in turn bolsters community resilience (Thew et al., 2020). This participatory model cultivates new leadership paradigms that empower future generations to play active roles in promoting climate justice, thereby reshaping societal structures that have traditionally favored the interests of older generations (Gislason et al., 2021).

11.6 Pathways to Systemic Change: Youth Visions and Alternatives

The intersection of climate change, youth agency, and systemic transformation invites a critical inquiry into the role of youth-led movements in envisioning alternative futures through innovative pathways. As the world grapples with escalating climate crises, youth have emerged as pivotal actors who not only advocate for immediate responses but also articulate long-term visions grounded in principles of justice and sustainability. Scholars have increasingly traced the frameworks that youth mobilizations leverage, illuminating how concepts like degrowth, just transitions, and indigenous resurgence can inform more resilient and equitable societal structures (Akbulut, 2019; Köhler et al., 2019). This multifaceted youth activism highlights a generational agency that is increasingly vocal about the systemic changes necessary to combat the climate emergency while simultaneously fostering intergenerational equity.

Central to these youth-led imaginaries are the principles of degrowth, which challenge the pervasive growth-oriented paradigms characterizing contemporary economies. This movement critiques excessive consumption and production while emphasizing the need for a radical rethinking of societal values toward sustainability and equity (Akbulut, 2019). It advocates for human and ecological well-being over profit, thereby aligning closely with youth demands for a transition away from exploitative systemic practices toward an economy that prioritizes environmental health and social justice (Köhler et al., 2019; Temin, 2024). As documented in various studies, young activists draw on these ideas to envision futures that transcend conventional economic growth, thereby encouraging a cultural shift in how societies perceive progress and success (Köhler et al., 2019; Temin, 2024).

Additionally, in the context of climate action, indigenous resurgence plays a crucial role as it intertwines traditional ecological knowledge with contemporary environmental justice movements. Indigenous communities, recognized for their stewardship of some of the world's most ecologically vital lands, embody frameworks that have long championed sustainable living (Garnett et al., 2018; Mackie et al., 2024). By integrating indigenous perspectives into climate discourse, youth movements not only acknowledge historical injustices but also advocate for policies that restore indigenous rights and knowledge systems essential for holistic approaches to climate adaptation and mitigation (Schneider, 2023; Garnett et al.,

2018). Recognizing these interconnections helps foster alliances between youth advocates and indigenous leaders, amplifying their collective voices and strategies (Mackie et al., 2024; Jessen et al., 2022).

The quest for just transitions, which aim to create equitable pathways to sustainability, further complicates and enriches this discourse. This approach highlights the need for systemic change to address the injustices exacerbated by climate change, particularly in marginalized communities (DeLorey et al., 2022; Nursey-Bray et al., 2022). Youth activism increasingly aligns these movements with feminist, labor, and decolonial struggles, emphasizing that climate justice cannot be achieved in isolation from broader social justice initiatives (Casagrande & Horn, 2023). By fostering cross-movement alliances, youth advocate not only for their generation's needs but also embed their efforts within the context of a diverse array of struggles, emphasizing that systemic transformation requires a holistic understanding of interconnected issues (Idriss et al., 2022).

In addressing the systemic challenges posed by climate change, educational and institutional reforms for intergenerational justice emerge as necessary strategies. Youth demand that education systems evolve to reflect their realities and aspirations, emphasizing the integration of environmental and social justice into curricula (Gabriel, 2023; Msomi & Akhurst, 2023). Institutional reforms must also facilitate greater youth participation in decision-making processes, allowing their perspectives to shape governance structures that directly impact their futures (Carlsson et al., 2023; Correia, 2025). Empowering youth through education and institutional reform enables a more informed, engaged, and impactful generation capable of driving systemic change toward sustainable and equitable futures.

Furthermore, rethinking democracy and governance through youth perspectives is critical in effecting systemic change. Traditional democratic processes often marginalize youth voices, perpetuating a disconnection between policies and the realities faced by younger generations (Akbulut, 2019; Msomi & Akhurst, 2023; Ogwu & Izah, 2025). By promoting inclusive and participatory governance models, youth can reclaim agency over their futures, advocating for adaptations to democracy that genuinely reflect their needs and aspirations (Baer, 2024a, 2024b; Carlsson et al., 2023). This re-imagination necessitates not only a reformulation of democratic engagements but also a commitment to facilitating platforms for collective action and dialogue that transcend demographic divides, allowing diverse voices to inform the democratic discourse.

Researchers have documented a growing trend in youth advocating for the recognition of indigenous perspectives and practices within broader environmental movements, presenting a cohesive challenge to strategies that neglect these traditional knowledges (Jessen et al., 2022; Tennell & Chew, 2024). This recognition is particularly salient in understanding how youth-led initiatives can more effectively engage with indigenous populations, fostering an exchange of knowledge that is both emergent and respectful (Gabriel, 2023; Mackie et al., 2024). As these movements evolve, the synergy between youth activism and indigenous resurgence emerges as an essential component in the development of holistic environmental strategies, one that underscores the importance of social justice and ecological sustainability in tandem.

By fostering cross-cultural dialogues, initiatives developed through these alliances can result in actionable strategies that address multiple forms of oppression simultaneously, showcasing the importance of intersectional approaches within youth-led movements (Casagrande & Horn, 2023; Jessen et al., 2022). As climate change exacerbates existing inequalities, youth movements can serve as vehicles for advocating for solutions that prioritize not only environmental sustainability but also social equity, particularly for marginalized groups (Allard-Tremblay & Coburn, 2023; Ellis-Young, 2022). Engaging multiple stakeholders, including indigenous communities, labor organizations, feminists, and environmental activists, will be essential in forging comprehensive responses that garner broader societal support and participation, ultimately leading to systemic transformative changes.

The stakes have never been higher in terms of youth advocacy for systemic change as climate crises contribute to socio-economic disparities. Youth have begun linking their agendas with global movements that seek to dismantle colonial remnants and challenge capitalist structures that perpetuate exploitation and environmental degradation (DeLorey et al., 2022; Schneider, 2023). These theoretical and practical linkages manifest in myriad forms, from climate strikes and grassroots organizing to participation in global governance platforms where youth can articulate their demands and reshape policy agendas (Baer, 2024a, 2024b). Activism, framed through lenses of justice, aims to cultivate solidarity not only within youth movements but across larger networks of civil society, emphasizing the indispensability of an interconnected approach to tackle the multifaceted crises of our time.

11.7 Challenges and Contradictions in Youth Climate Action

In the realm of climate activism, youth voices command considerable visibility, yet the integration of these voices into real decision-making processes presents significant challenges and contradictions (Table 11.3). Despite the evident dedication and mobilization of young people around climate issues, their symbolic inclusion often fails to translate into substantive power or influence over policy outcomes. The framing of youth as essential stakeholders in climate discourse has grown, yet the opportunities for genuine representation remain limited. Conferences organized under the United Nations Framework Convention on Climate Change (UNFCCC) showcase youth as part of civil society, yet formal mechanisms for integrating their perspectives in significant ways are insufficient, leading to entrenched barriers to meaningful participation (Thew et al., 2020).

The phenomenon of tokenism is particularly significant within governance structures, where youth are invited to participate in dialogues ostensibly aimed at inclusivity, while in practice, their contributions may be overshadowed or ignored. Research highlights that perceived involvement often leans more toward optics than actual engagement, with the substantive demands of youth often sidelined

Table 11.3 Challenges and contradictions in youth climate action

Challenge/contradiction	Narrative
Tokenization and symbolic inclusion	Youth are often invited to global forums for visibility but excluded from meaningful decision-making.
Burnout and precarity	Balancing activism with school, work, and limited resources leads to exhaustion and mental health strain.
Reformist vs. radical divides	Movements face internal tension between incremental reforms and calls for structural transformation.
Unequal access to digital platforms	Activism that relies on digital tools excludes youth in marginalized or resource-poor communities.
Dependence on adult institutions	Reliance on funding and infrastructure from adult organizations risks co-optation of youth agendas.
Generational mistrust	Established elites often dismiss youth perspectives as naïve or unrealistic, limiting collaboration.
Legal and political risks	Youth activists, particularly in authoritarian contexts, face repression, surveillance, and criminalization.
Inclusivity and representation gaps	Some youth movements reproduce inequalities by failing to fully include marginalized voices.
Short-term visibility	Media and public attention often fade quickly, undermining long-term sustainability of movements.
Lifestyle contradictions	Critics highlight inconsistencies between youth critiques of consumerism and their own participation in global systems.

amidst the complexities of policy negotiations (Sithole et al., 2024). This failure to equate visibility with agency underscores a critical disconnect in governance, where youth are celebrated for their activism, yet remain marginalized when it comes to influencing pivotal decisions on climate action.

Additionally, the emotional toll of continuous engagement in activist efforts against a backdrop of slow systemic change is notable. Many young activists find themselves caught in a cycle of burnout, managing the demands of education, work, and unpaid activism, which markedly strains their financial and emotional capacities. This combination of high emotional investment in their causes alongside the slow pace of progress can breed disillusionment, detrimentally affecting mental health outcomes among youth activists (Corvalan et al., 2021; Izah & Ogwu, 2025). The psychological repercussions of advocating for urgent systemic changes in the face of unyielding resistance can lead to feelings of futility, potentially stymying the vibrancy and effectiveness of youth movements.

This exhaustion is often exacerbated by ongoing discourse within activist circles regarding approaches to systemic change, notably the tension between reformist and radical strategies. Some youth advocate for incremental reforms that align with existing systems, while others push for more radical, structural transformations. This divergence in strategic direction can lead to fragmentation within movements, diminishing their collective power and efficacy (Dupuis-Déri, 2021;

Terren & Soler-i-Martí, 2021). In this context, youth organizations often grapple with internal conflicts that complicate their collective messaging and ability to mobilize effectively.

The disparities in access to digital platforms for organizing further compound these challenges. While social media has emerged as a crucial tool for mobilizing and amplifying youth voices in climate action, access is typically skewed toward those with stable internet connections and digital literacy skills, often excluding marginalized young people (particularly in rural or Global South contexts) from mainstream visibility. This uneven digital landscape raises questions about inclusivity and representation within youth-led movements, emphasizing the necessity to address these disparities in order to cultivate a truly representative climate advocacy (Wielk & Standlee, 2021).

Moreover, the reliance of youth movements on adult-led institutions for funding, mentorship, and organizational support introduces dynamics that can compromise their independence. Frequently, these structures prioritize adult perspectives, which can overshadow youth agendas. Dependence on external resources may inadvertently pave the way for the co-optation of youth movements, placing limits on their autonomy and the radical potential of their demands (Terren & Soler-i-Martí, 2021). Adult governance structures may impose constraints that prioritize stability and conventional approaches over the innovative ideas and bold strategies that youth inherently bring.

The generational mistrust between established power structures and young activists further complicates engagement. Older political and corporate elites often dismiss the urgency of youth demands as impractical or naïve, creating barriers to intergenerational collaboration. The resulting ambivalence highlights a critical need for dialogue between generations, as youth possess unique insights into the existential crises of their time, which could inform more effective policymaking (Elliott & Earl, 2018; Thew et al., 2020).

Additionally, legal and political vulnerabilities pose significant risks for youth activists, particularly in repressive political climates where dissent can lead to surveillance or criminalization. Young people in authoritarian or unstable regions may face harsher repercussions for their activism than their counterparts in more liberal democracies, complicating their efforts to mobilize and organize effectively (Kennedy et al., 2023). This reality can deter youth participation in activism, creating an environment of fear that constrains free expression.

Diversity, inclusivity, and representation within youth movements remain crucial areas for development. Movements that fail to adequately incorporate diverse perspectives may inadvertently reinforce the very inequalities they seek to dismantle. Marginalization based on ethnicity, socio-economic status, or additional identities often leads to fragmentation within movement agendas, limiting collective action (Kirkham, 2015; Lee et al., 2022; Ogwu, et al., 2025). Therefore, ensuring that all voices are authentically included within the climate action discourse is vital for fostering solidarity and systemic change.

While media coverage of youth protests and actions tends to spark immediate attention and sympathy, this visibility is often transient. Mainstream media quickly shifts focus, leaving movements struggling to sustain momentum over the long term. The ephemeral nature of media attention can foster a reliance on short-term visibility, which undermines the organic growth of more durable structures for youth-led initiatives. Cultivating a strong foundation for long-term sustainability is critical for the success and impact of youth-led climate initiatives (Bonell, 2022).

Criticism surrounding youth activism often highlights a perceived contradiction between advocating for systemic change while simultaneously engaging in consumer-driven economies. This tension can be leveraged by opponents to delegitimize young activists' claims and strategies, suggesting they are hypocritical or unqualified to demand substantial systemic alterations (Mutalemwa & Walusala, 2024). Addressing these challenges requires a nuanced understanding of how lifestyle choices intersect with broader systemic critiques and collective action and emphasizes the importance of internal reflection within activist movements.

11.8 Future Horizons: Intergenerational Solidarity and Transformative Possibilities

The current discourse surrounding climate change emphasizes not just the ecological impacts but also the social implications and roles that various demographics play in pursuit of sustainability. Among the most vital contributors to climate activism are youth, whose mobilization exemplifies intergenerational solidarity and strategic partnerships aimed at transformative possibilities for a warming world. This phenomenon reflects systemic change agents operating at both global and local governance levels as they articulate their needs and aspirations through organized activism, asserting their significance in climate justice and intergenerational survival (Han & Ahn, 2020; Klinsky et al., 2017).

Youth mobilization has taken various forms, including dutiful, disruptive, and dangerous dissent. O'Brien et al. discuss how young activists assert their leadership when conventional power structures falter, highlighting a powerful narrative of youth resilience and agency in the face of governmental inaction (O'brien et al., 2018). Moreover, this activism challenges entrenched systems and raises vital climate issues often marginalized in adult discourse. This disruptive agency sets the stage for potential structural transformations within climate governance that benefit both current and future generations (O'brien et al., 2018). Their mobilization varies from conventional political processes to more radical forms of engagement, significantly reshaping public perceptions about the role and capacity of youth in environmental governance (O'brien et al., 2018; Ogwu & Izah, 2025; Trott, 2024).

A vital aspect of youth activism is its relationship with generational dynamics, particularly how these young activists relate to older generations and policymakers. Roy and Ayalon (2022) highlight the importance of bridging intergenerational divides by fostering empathy and collaborative efforts toward common climate goals (Roy & Ayalon, 2022). Messages promoting intergenerational solidarity can counteract prevailing narratives that often depict generational tensions as the norm, thus reframing the climate discourse into a shared challenge that transcends age boundaries (Roy & Ayalon, 2022; Sanson et al., 2019). The acknowledgment of such solidarity is crucial, given that climate change disproportionately affects children and young people, intrinsically intertwining their futures with those of their predecessors (Sanson et al., 2019).

Confronting the psychological implications of climate change amplifies the urgency of youth involvement. Research demonstrates that children and youth grapple with climate anxiety, exacerbated by their acute awareness of ecological degradation and its long-term repercussions on their lives (Sanson et al., 2019; Vamvalis, 2023). This emotional terrain necessitates educational frameworks that incorporate climate justice perspectives and empower youth with tools to navigate and mitigate their anxieties while fostering action-oriented responses (Mateer, 2024). Such pedagogical shifts are essential for mobilizing youth potential, allowing them to harness their voices in governance and engage in peer-driven initiatives promoting climate resilience (Sloan Morgan et al., 2024; Trott, 2024).

Intergenerational coalitions can strengthen advocacy for climate justice and systemic change, ensuring a more comprehensive representation of interests within climate negotiations. Engaging youth not only honors the perspectives of future generations but also equips older generations with fresh insights on pressing ecological challenges. Zeal for transformative futures can be energized through initiatives aiming at meaningful interaction between age groups, promoting an exchange of wisdom and innovative approaches to climate action (Trott, 2024; de Paula Sieverding et al., 2024; Börner et al., 2021). The dynamic nature of intergenerational interactions can inspire more equitable climate policies that consider justice across demographics and foster broader participation in the climate movement (Trott, 2024).

Addressing intergenerational conflicts and fostering solidarity can lead to a robust climate action paradigm that respects the voices of youth while addressing the concerns of older generations. Studies elucidate that effective intergenerational communication can mitigate hostility by fostering a shared understanding of the climate crises (Ayalon & Roy, 2023; Roy & Ayalon, 2022). This communal approach propels public narratives toward empathy and interconnectedness, indicating a shift in perception that recognizes the reciprocal benefits of cooperative youth and elder engagement (Chazan & Baldwin, 2019; Hwang et al., 2022).

The potential futures that lie ahead (involving collapse, adaptation, and transformation) are inherently tied to the effectiveness of youth engagement in climate action. By situating youth at the heart of governance discussions, possibilities for adaptive change emerge, countering trajectories that lead to societal disarray. Hence, to foster sustainable futures, youth must be integrated into decision-making processes, thereby driving the pillars of resilience within environmental policy frameworks (Klinsky et al., 2017; Thew et al., 2020).

Moreover, the involvement of youth in climate justice discussions serves as an essential catalyst for cultivating an active societal commitment toward preventive strategies against climate degradation. Initiatives emphasizing the role of youth reflect exponential possibilities for community organizing and soliciting action across various scales, enhancing climate resilience at grassroots levels (Han &

Ahn, 2020; Sloan Morgan et al., 2024). This call for active participation is pivotal in shifting paradigms and co-creating approaches encompassing ethical equity and responsibility across generations (de Paula Sieverding et al., 2024; Klinsky et al., 2017).

Understanding and integrating these perspectives necessitates institutional frameworks that systematically elevate youth voices and experiences within collaborative environmental strategies. Institutionalizing youth engagement within governance systems includes providing platforms for their representation and ensuring they hold meaningful influence over climate-related decisions, fundamental for the sustainability of future socio-ecological systems (Han & Ahn, 2020; Trott, 2024). The establishment of partnerships across generations can be operationalized through education, community engagement, and political deliberation, significantly influencing policies that safeguard the rights and needs of all generations facing the climate crisis (Klinsky et al., 2017; Trott, 2024).

As we chart potential pathways to address climate change, engaging in multifaceted dialogues that examine nuances of intergenerational solidarity vis-a-vis shared struggles against environmental degradation is crucial, echoing critical conversations surrounding social justice (Mateer, 2024; Sloan Morgan et al., 2024). In this context, examining youth narratives and their lived experiences provides a rich source of insight into the manifestations of activism and the quest for justice amidst complexity and uncertainty (Trott, 2024; Vamvalis, 2023). Bridging generational divides is not merely an ethical imperative; it embodies a strategic necessity to combat the climate crisis cohesively and equitably for current and upcoming generations.

Thus, the role of youth within the climate justice framework positions them not only as advocates for their futures but as vital contributors to a broader movement advocating healthy intergenerational relationships and systemic change. By fostering durable alliances across generations and institutionalizing youth agency in governance, societies can shift toward transformative futures characterized by resilience and collective responsibility amid an ever-warming world (de Paula Sieverding et al., 2024; Thew et al., 2020). Such collaborative efforts can serve as a model for existing and emerging activism, revealing opportunities for comprehensive and inclusive climate action that stands the test of time.

11.9 Conclusion

Youth stand at a critical crossroads in the climate crisis, embodying both vulnerability and transformative potential. As the most exposed generation to the long-term consequences of ecological disruption, they experience heightened inequalities, precarity, and psychological strain. Yet this very vulnerability has catalyzed new forms of agency (expressed through activism, litigation, digital mobilization, and creative resistance) that challenge entrenched systems of power and reframe climate change as a justice issue. By articulating visions of just transitions, indigenous resurgence, and sustainable futures, young people demonstrate not only resilience but also the capacity to shape systemic change. Their role highlights the urgent need to reconcile the immediacy of climate threats with a forward-looking vision rooted in equity and hope.

To realize the transformative potential of youth, systemic support and meaningful inclusion are essential. Policymakers and institutions must move beyond symbolic recognition and integrate youth agency into formal decision-making structures at every level. Expanding education for climate literacy, mental health support, and intergenerational learning will empower youth to sustain long-term engagement. Bridging intergenerational divides through collaborative governance and solidarity-based coalitions can mitigate tensions and strengthen collective action. Finally, addressing barriers such as economic precarity, digital exclusion, and tokenism will allow youth movements to thrive as genuine partners in shaping policy and societal transformation. Embedding youth agency at the core of climate governance ensures that responses to the climate crisis are not only urgent but also just, inclusive, and sustainable for present and future generations.

References

Aftab, A., & Druss, B. G. (2023). Addressing the mental health crisis in youth—sick individuals or sick societies? *JAMA Psychiatry, 80*(9), 863–864. https://doi.org/10.1001/jamapsychiatry.2023.1298

Akbulut, B. (2019). The 'state'of degrowth: Economic growth and the making of state hegemony in Turkey. *Environment and Planning E: Nature and Space, 2*(3), 513–527. https://doi.org/10.1177/2514848619835135

Allard-Tremblay, Y., & Coburn, E. (2023). The flying heads of settler colonialism; or the ideological erasures of indigenous peoples in political theorizing. *Political Studies, 71*(2), 359–378.

Archibald, M. M., Makinde, S., & Tongo, N. (2024). Arts-based approaches to priority setting: Current applications and future possibilities. *International Journal of Qualitative Methods, 23*, 16094069231223926. https://doi.org/10.1177/16094069231223926

Arnot, G., Pitt, H., McCarthy, S., Warner, E., & Thomas, S. (2024). 'You can't really separate these risks, our environment, our animals and us': Australian children's perceptions of the risks of the climate crisis. *Health Promotion International, 39*(2), daae023. https://doi.org/10.1093/heapro/daae023

Arnot, G., Thomas, S., Pitt, H., McCarthy, S., & Warner, E. (2024). "Older people will die of old age. I'll die of climate change": Engaging children and young people in climate decision making for public health. *BMC Public Health, 24*(1), 1869. https://doi.org/10.1186/s12889-024-19406-9

Ayalon, L., & Roy, S. (2023). The role of chronological age in climate change attitudes, feelings, and behavioral intentions: The case of null results. *PLoS One, 18*(6), e0286901.

Baer, H. A. (2024a). *Latour's swan songs: Grappling with the ecological crisis By Bruno Latour and Nikolaj Schulz. On the emergence of an ecological class: A memo.* (Julie Rose, Trans. (Vol. 2022, pp. 80–92). Polity Press.

Baer, H. A. (2024b). Latour's swan songs: Grappling with the ecological crisis. *The Australian Journal of Anthropology, 35*(3), 338–343.

Bandura, A., & Cherry, L. (2020). Enlisting the power of youth for climate change. *American Psychologist, 75*(7), 945–951. https://doi.org/10.1037/amp0000512

Barnes, B. R. (2021). Reimagining African women youth climate activism: The case of Vanessa Nakate. *Sustainability, 13*(23), 13214.

Börner, S., Kraftl, P., & Giatti, L. L. (2021). Blurring the '-ism'in youth climate crisis activism: Everyday agency and practices of marginalized youth in the Brazilian urban periphery. *Children's Geographies, 19*(3), 275–283.

Boulianne, S., Lalancette, M., & Ilkiw, D. (2020). "School strike 4 climate": Social media and the international youth protest on climate change. *Media and Communication, 8*(2), 208–218.

Bowman, B. (2019). Imagining future worlds alongside young climate activists: A new framework for research. *Fennia-International Journal of Geography, 197*(2), 295–305.

Brown, M., Mirza, S., Lu, J., & Stewart, S. L. (2024). Walking forward together—The next step: Indigenous youth mental health and the climate crisis. *The Psychiatrist, 6*(1), 210–220.

Bonell, A., Badjie, J., Jammeh, S., Ali, Z., Hydara, M., Davies, A., Faal, M., Ahmed, A. N., Hand, W., Prentice, A. M., Murray, K. A., & Scheelbeek, P. (2022). Grassroots and Youth-Led Climate Solutions From The Gambia. Frontiers in public health, 10, 784915. https://doi.org/10.3389/fpubh.2022.784915

Carlsson, H., Hanell, F., & Engström, L. (2023). Revisiting the notion of the public library as a meeting place: Challenges to the mission of promoting democracy in times of political turmoil. *Journal of Documentation, 79*(7), 178–195. https://doi.org/10.1162/00c13b77.325ea8f8

Casagrande, O., & Horn, P. (2023). Scenes from El Alto: indigenous youth visions for urban Bolivia. *Projections: The Journal of the MIT Department of Urban Studies and Planning, 17*, 1–24.

Chazan, M., & Baldwin, M. (2019). Granny solidarity: Understanding age and generational dynamics in climate justice movements. *Studies in Social Justice, 13*(2), 244–261.

Corvalan, C., Gray, B., Villalobos Prats, E., Sena, A., Hanna, F., & Campbell-Lendrum, D. (2022). Mental health and the global climate crisis. Epidemiology and psychiatric sciences, 31, e86. https://doi.org/10.1017/S2045796022000361

Correia, J. E. (2025). Cartographic imaginaries: Dilemmas of data, erasure, and the threat of latent authoritarianism to indigenous land rights in Paraguay. *Latin American Perspectives, 52*(3), 201–220.

Daly, A. (2024). Child and youth friendly justice for the climate crisis: Relying on the UN convention on the rights of the child. *The International Journal of Children's Rights, 32*(3), 632–661.

Daly, C., Fatorić, S., Carmichael, B., Pittungnapoo, W., Adetunji, O., Hollesen, J., Nakhaei, M., & Diaz, A. H. (2022). Climate change adaptation policy and planning for cultural heritage in low- and middle-income countries. *Antiquity, 96*(390), 1427–1442.

De Moor, J., De Vydt, M., Uba, K., & Wahlström, M. (2021). New kids on the block: Taking stock of the recent cycle of climate activism. *Social Movement Studies, 20*(5), 619–625.

de Paula Sieverding, T., Merten, M., & Kastner, K. (2024). Old for young: Cross-national examination of intergenerational political solidarity. *Group Processes & Intergroup Relations, 27*(6), 1233–1259.

DeLorey, R., Groulx, M., Nooski, L., Ketlo, B., & Ransom, A. (2022). "Unless you get to know us, you're not going to know our passion": From consultation to co-management on Nadleh Whut'en First Nation Territory. *Canadian Planning and Policy/Aménagement et politique au Canada, 2022*, 88–113.

Dupuis-Déri, F. (2021). Youth strike for climate: Resistance of school administrations, conflicts among students, and legitimacy of autonomous civil disobedience—The case of Québec. *Frontiers in Political Science, 3*, 634538.

Eide, E., & Kunelius, R. (2021). Voices of a generation the communicative power of youth activism. *Climatic Change, 169*(1), 6.

Elliott, T., & Earl, J. (2018). Organizing the next generation: Youth engagement with activism inside and outside of organizations. *Social Media+ Society, 4*(1), 2056305117750722.

Ellis-Young, M. (2022). Gentrification as (settler) colonialism? Moving beyond metaphorical linkages. *Geography Compass, 16*(1), e12604.

Fatorić, S., & Daly, C. (2023). Towards a climate-smart cultural heritage management. *Wiley Interdisciplinary Reviews: Climate Change, 14*(6), e855.

Firinci Orman, T. (2022). Youth's everyday environmental citizenship: An analytical framework for studying interpretive agency. *Childhood, 29*(4), 495–511.

Gabriel, M. (2023). Our home is native land: Teachings, perspectives, and experiences of indigenous houselessness. *Canadian Journal of Nonprofit and Social Economy Research, 14*(S1), 17. https://doi.org/10.29173/cjnser574

Garnett, S. T., Burgess, N. D., Fa, J. E., Fernández-Llamazares, Á., Molnár, Z., Robinson, C. J., Watson, J. E., Zander, K. K., Austin, B., Brondizio, E. S., & Collier, N. F. (2018). A spatial overview of the global importance of Indigenous lands for conservation. *Nature Sustainability, 1*(7), 369–374. https://doi.org/10.1038/s41893-018-0100-6

Gislason, M. K., Kennedy, A. M., & Witham, S. M. (2021). The interplay between social and ecological determinants of mental health for children and youth in the climate crisis. *International Journal of Environmental Research and Public Health, 18*(9), 4573. https://doi.org/10.3390/ijerph18094573

Grand, J. (2023). The end of the game: emotional responses of older adults to climate crisis and climate mobilizations in Switzerland. *Frontiers in Climate, 5*, 1264495.

Han, H., & Ahn, S. W. (2020). Youth mobilization to stop global climate change: Narratives and impact. *Sustainability, 12*(10), 4127. https://doi.org/10.3390/su12104127

Hawke, L. D., Relihan, J., Miller, J., McCann, E., Rong, J., Darnay, K., Docherty, S., Chaim, G., & Henderson, J. L. (2018). Engaging youth in research planning, design and execution: Practical recommendations for researchers. *Health Expectations, 21*(6), 944–949. https://doi.org/10.1111/hex.12795

Hornsey, M. J., & Lewandowsky, S. (2022). A toolkit for understanding and addressing climate scepticism. *Nature Human Behaviour, 6*(11), 1454–1464. https://doi.org/10.1038/s41562-022-01463-y

Hwang, W., Kim, J. H., Brown, M. T., & Silverstein, M. (2022). Are filial eldercare norms related to intergenerational solidarity with older parents? A typological developmental approach. *Journal of Family Psychology, 36*(4), 585. https://doi.org/10.1037/fam0000920

Idriss, S., Butler, R., & Harris, A. (2022). Interrogating race, unsettling whiteness: Concepts of transitions, enterprise and mobilities in Australian youth studies. *Journal of Youth Studies, 25*(9), 1251–1267. https://doi.org/10.1080/13676261.2021.1948516

Inter-Parliamentary Union. (2022). *Youth participation in national parliaments 2022*. Inter-Parliamentary Union. https://www.ipu.org/resources/publications. Accessed August 1, 2025.

Iyiola, A. O., Afolabi, O. A., Alimi, S. K., Akingba, O. O., Izah, S. C., & Ogwu, M. C. (2024). Climate change and water crisis in the Global South. In S. C. Izah, M. C. Ogwu, A. Loukas, & H. Hamidifar (Eds.), *Water crises and sustainable management in the Global South* (pp. 111–140). Springer. https://doi.org/10.1007/978-981-97-4966-9_4

Izah, S. C., Iyiola, A. O., Yarkwan, B., & Richard, G. (2023). Impact of air quality as a component of climate change on biodiversity-based ecosystem services. In A. L. Srivastav, A. K. Dubey, A. Kumar, S. K. Narang, & M. A. Khan (Eds.), *Visualization techniques for climate change with machine learning and artificial intelligence* (pp. 123–148). Elsevier.

Izah, S. C., & Ogwu, M. C. (2025). *Innovative approaches in environmental health management: Processes, technologies, and strategies for a sustainable future*. Springer Nature. https://doi.org/10.1007/978-3-031-81966-7

Jean, C., Khan, A. S., Flores, G. M., Donatuto, J., Lopez, P., Perez, C., Martinez, C., & Errett, N. A. (2024). Uncovering effects of climate-sensitive health risks on historically marginalized youth in Washington State: Creating opportunities for maximum involvement. *Environmental Research: Health, 2*(4), 045002.

Jessen, T. D., Ban, N. C., Claxton, N. X., & Darimont, C. T. (2022). Contributions of indigenous knowledge to ecological and evolutionary understanding. *Frontiers in Ecology and the Environment, 20*(2), 93–101.

Kennedy, H., Trujillo, M., Ryan, A., Cooley, D., Martinez, D., McNair, B., & Hunt, C. (2023). Identifying root causes: Evaluation of a program to engage youth in a social justice approach to tobacco control. *Health Promotion Practice, 24*(6), 1151–1162.

Kirkham, S. (2015). Intersectionality and the social meanings of variation: Class, ethnicity, and social practice. *Language in Society, 44*(5), 629–652. https://doi.org/10.1017/s0047404515000585

Klinsky, S., Roberts, T., Huq, S., Okereke, C., Newell, P., Dauvergne, P., O'Brien, K., Schroeder, H., Tschakert, P., Clapp, J., & Keck, M. (2017). Why equity is fundamental in climate change policy research. *Global Environmental Change, 44*, 170–173.

Köhler, J., Geels, F. W., Kern, F., Markard, J., Onsongo, E., Wieczorek, A., Alkemade, F., Avelino, F., Bergek, A., Boons, F., & Fünfschilling, L. (2019). An agenda for sustainability transitions research: State of the art and future directions. *Environmental Innovation and Societal Transitions, 31*, 1–32.

Kotzé, L. J., & Knappe, H. (2023). Youth movements, intergenerational justice, and climate litigation in the deep time context of the Anthropocene. *Environmental Research Communications, 5*(2), 025001.

Lawrance, E. L., Jennings, N., Kioupi, V., Thompson, R., Diffey, J., & Vercammen, A. (2022). Psychological responses, mental health, and sense of agency for the dual challenges of climate change and the COVID-19 pandemic in young people in the UK: An online survey study. *The Lancet Planetary Health, 6*(9), e726–e738.

Lasater, M. E., Prager, G., Choi, Y. A., Groteclaes, T., Rao, D., Kamps, S. P., Altare, C., & Spiegel, P. B. (2025). Understanding relationships among climate change, conflict, migration/displacement and health in humanitarian settings: a scoping review. *Conflict and health, 19*(1), 73. https://doi.org/10.1186/s13031-025-00707-8

Lee, K., O'Neill, S., Blackwood, L., & Barnett, J. (2022). Perspectives of UK adolescents on the youth climate strikes. *Nature Climate Change, 12*(6), 528–531.

MacKay, M., Parlee, B., & Karsgaard, C. (2020). Youth engagement in climate change action: Case study on indigenous youth at COP24. *Sustainability, 12*(16), 6299.

Mackie, A. S., Gray, M., Chappell, A., Dlusskaya, K., Lightning, R., Listener, L., Crier, A., Dumigan-Jackson, B., Thomas, A., Graham, B., & Littlechild, R. (2024). Barriers and supports for Indigenous youth and young adults with childhood-onset chronic health conditions transitioning from pediatric to adult healthcare: a qualitative study. *International Journal for Equity in Health, 23*(1), 268.

Mateer, T. J. (2024). A call for intergenerational solidarity in climate change education: Outlining a strategy to manage climate anxiety and facilitate action. *European Journal of Education, 59*(3), e12659.

McElwee, P., Tuyến, N. P., Huệ, L. T. V., & Hương, V. T. D. (2023). Climate precarity in rural livelihoods: Agrarian transformations and smallholder vulnerability in Vietnam. *Journal of Agrarian Change, 23*(4), 661–686.

Moser, S. C. (2016). Reflections on climate change communication research and practice in the second decade of the 21st century: What more is there to say? *Wiley Interdisciplinary Reviews: Climate Change, 7*(3), 345–369.

Msomi, N. N., & Akhurst, J. (2023). Prominent discourses in South African education from the perspective of community psychology: Challenges and opportunities for youth liberation and well-being. *Journal of Education University of KwaZulu-Natal, 92*, 98–114.

Murphy, N., Rarama, T., Atama, A., Kauyaca, I., Batibasaga, K., Azzopardi, P., Bowen, K. J., & Bohren, M. A. (2023). Changing climates, compounding challenges: A participatory study on how disasters affect the sexual and reproductive health and rights of young people in Fiji. *BMJ Global Health, 8*(Suppl 3), e013299.

Mutalemwa, D., & Walusala, K. (2024). SMEs as agents of peace and security: Insights from the 2023 youth connekt Africa Summit. *Journal of Policy and Development Studies, 16*(2), 1–12.

Nairn, K. (2019). Learning from young people engaged in climate activism: The potential of collectivizing despair and hope. *Young, 27*(5), 435–450.

Nursey-Bray, M., Palmer, R., Chischilly, A. M., Rist, P., & Yin, L. (2022). *Old ways for new days: Indigenous survival and agency in climate changed times* (p. 131). Springer Nature.

Obradovich, N., Migliorini, R., Paulus, M. P., & Rahwan, I. (2018). Empirical evidence of mental health risks posed by climate change. *Proceedings of the National Academy of Sciences, 115*(43), 10953–10958.

O'brien, K., Selboe, E., & Hayward, B. M. (2018). Exploring youth activism on climate change. *Ecology and Society, 23*(3), 230342.

O'Connell, L. K., & Botchwey, N. (2023). Supporting youth participation in health and climate justice through advocacy training: Comment on" between rhetoric and reality: Learnings from youth participation in the adolescent and youth health policy in South Africa". *International Journal of Health Policy and Management, 12*, 7898.

Ogwu, M. C. (2019). Towards sustainable development in Africa: The challenge of urbanization and climate change adaptation. In P. B. Cobbinah & M. Addaney (Eds.), *The geography of climate change adaptation in Urban Africa* (pp. 29–55). Springer Nature. https://doi.org/10.1007/978-3-030-04873-0_2

Ogwu, M. C., El Malahi, S., & Izah, S. C. (2025). Policy and regulatory frameworks for sustainable environmental practices. In M. C. Ogwu & S. Chibueze Izah (Eds.), *Evaluating environmental processes and technologies*. Environmental Science and Engineering Series. Springer. https://doi.org/10.1007/978-3-031-85327-2_16

Ogwu, M. C., Iyiola, A. O., Fawole, W. O., & Izah, S. C. (2025). Biobased resource recovery techniques in the Global South. In M. C. Ogwu, S. C. Izah, J. G. Vazquez-Arenas, S. T. Feleke, & X. Wei (Eds.), *Sustainable bioeconomy development in the Global South*. Springer. https://doi.org/10.1007/978-981-96-0305-3_17

Ogwu, M. C., & Izah, S. C. (2025). *Evaluating environmental processes and technologies*. Springer Nature. https://doi.org/10.1007/978-3-031-85327-2

Ogwu, M. C., Izah, S. C., Merritt, H., Wise, R. D., & Jintrawet, A. (2025). *Sustainable bioeconomy development in the Global South: Volume I Status and Perspectives*. Springer. https://doi.org/10.1007/978-981-96-0640-5

Ogwu, M. C., Izah, S. C., Vazquez-Arenas, J. G., Feleke, S. T., & Wei, X. (2025). *Sustainable bioeconomy development in the Global South: Volume II bioeconomy techniques*. Springer. https://doi.org/10.1007/978-981-96-0305-3

Ogwu, M. C., Ojo, A. O., & Alaka, A. C. (2025). Biodiversity and human health: The interconnections of species loss and ecosystem services. In S. C. Izah & M. C. Ogwu (Eds.), *Innovative approaches in environmental health management: Processes, technologies, and strategies for a sustainable future. Environmental Science and Engineering Series*. Springer Nature. https://doi.org/10.1007/978-3-031-81966-7_5

Panchal, U., Salazar de Pablo, G., Franco, M., Moreno, C., Parellada, M., Arango, C., & Fusar-Poli, P. (2023). The impact of COVID-19 lockdown on child and adolescent mental health: systematic review. *European Child & Adolescent Psychiatry, 32*(7), 1151–1177.

Pinchoff, J., Regules, R., Gómez-Ugarte, A., Abularrage, T., & Bojórquez-Chapela, I. (2023). Coping with climate change: the role of climate related stressors in affecting the mental health of young people in Mexico. *PLOS Global Public Health, 3*(9), e0002219. https://doi.org/10.1371/journal.pgph.0002219

Pinchoff, J., Etetim, E. O., Babatunde, D., Blomstrom, E., Ainul, S., Akomolafe, T. O., Medina Carranza, B., Del Valle, A., & Austrian, K. (2025). How climate change is shaping young people's health: a participatory, youth co-led study from Bangladesh, Guatemala and Nigeria. *BMJ global health, 10*(1), e016788. https://doi.org/10.1136/bmjgh 2024-016788

Plummer, N., Wilson, M., Yaneva-Toraman, I., McKenzie, C., Mitchell, S., Northover, P., Crowley, K., Edwards, T., & Richards, A. (2022). Recipes for resilience: Engaging Caribbean youth in climate action and food heritage through stories and song. *Sustainability, 14*(14), 8717.

Racine, N., McArthur, B. A., Cooke, J. E., Eirich, R., Zhu, J., & Madigan, S. (2021). Global prevalence of depressive and anxiety symptoms in children and adolescents during COVID-19: A meta-analysis. *JAMA Pediatrics, 175*(11), 1142–1150.

Ramirez, L., Levy, S. R., Barbosa, S., Bottin, M., Ocampo-Palacio, J. G., Cadavid, S., Pizarro, A. B., & Quesada, B. (2025). A call for greater collaborative role from psychology in climate change education interventions focused on children/youth. *Canadian Psychology/Psychologie Canadienne, 66*, 296–311.

Ratliff, G. A., Cosgrove, D., Yang, J. O., Sarabia, R., Harvey, T. L., Jeffcoat, N., Lightfoot, M., Adams, S., Lund, I., & Auerswald, C. L. (2024). Keeping young researchers out of harm's way: conducting youth participatory action research with young people experiencing homelessness. *Frontiers in Public Health, 12*, 1386714.

Rossa-Roccor, V., Giang, A., & Kershaw, P. (2021). Framing climate change as a human health issue: Enough to tip the scale in climate policy? *The Lancet Planetary Health, 5*(8), e553–e559.

Rousell, D., & Cutter-Mackenzie-Knowles, A. (2020). A systematic review of climate change education: Giving children and young people a 'voice'and a 'hand'in redressing climate change. *Children's Geographies, 18*(2), 191–208.

Roy, S., & Ayalon, L. (2022). Intergenerational relations in the climate movement: Bridging the gap toward a common goal. *International Journal of Environmental Research and Public Health, 20*(1), 233.

Sanson, A., & Bellemo, M. (2021). Children and youth in the climate crisis. *BJPsych bulletin, 45*(4), 205–209.

Sanson, A. V., Van Hoorn, J., & Burke, S. E. (2019). Responding to the impacts of the climate crisis on children and youth. *Child Development Perspectives, 13*(4), 201–207.

Schneider, L. (2023). Decolonizing conservation? Indigenous resurgence and buffalo restoration in the American West. *Environment and Planning E: Nature and Space, 6*(2), 801–821.

Schramm, P. J., Al Janabi, A. L., Campbell, L. W., Donatuto, J. L., & Gaughen, S. C. (2020). How Indigenous communities are adapting to climate Change: Insights from the climate-ready tribes Initiative: Analysis examines how Indigenous communities are adapting to climate change. *Health Affairs, 39*(12), 2153–2159.

Sharpe, I., & Davison, C. M. (2022). Investigating the role of climate-related disasters in the relationship between food insecurity and mental health for youth aged 15–24 in 142 countries. *PLOS Global Public Health, 2*(9), e0000560.

Shi, J., Visschers, V. H., & Siegrist, M. (2015). Public perception of climate change: The importance of knowledge and cultural worldviews. *Risk Analysis, 35*(12), 2183–2201.

Simpson, N. P., Clarke, J., Orr, S. A., Cundill, G., Orlove, B., Fatorić, S., Sabour, S., Khalaf, N., Rockman, M., Pinho, P., & Maharaj, S. S. (2022). Decolonizing climate change–heritage research. *Nature Climate Change, 12*(3), 210–213.

Singh, G., Xue, S., & Poukhovski-Sheremetyev, F. (2022). Climate emergency, young people and mental health: Time for justice and health professional action. *BMJ paediatrics open, 6*(1), e001375.

Sithole, S. S., Walters, G. M., Mbatha, P., & Matose, F. (2024). Youth engagement in global conservation governance. *Conservation Biology, 38*(6), e14387.

Sloan Morgan, O., Melchior, F., Thomas, K., & McNab-Coombs, L. (2024). Youth and climate justice: Representations of young people in action for sustainable futures. *The Geographical Journal, 190*(1), e12547.

Stockemer, D., & Sundström, A. (2018). Age representation in parliaments: Can institutions pave the way for the young? *European Political Science Review, 10*(3), 467–490.

Sudarmanto, J. A. (2023). The context of murals as youth media in the aesthetic perspective of Marxism. *KnE Social Sciences, 8*, 268–276.

Sugg, M. M., Ryan, S., Spurlock, T., Duncan, S., Hege, A., Ogwu, M. C., Tyson, J., Shay, E., Renwick, K. A., Hendren, C., Roy, M., & Runkle, J. D. (2023). Climate change and health in rural mountain environments: summary of a workshop on knowledge gaps, barriers, and opportunities for action. *GeoJournal, 88*(5), 5397–5409. https://doi.org/10.1007/s10708-023-10916-4

Temin, D. (2024). Wages for earthwork. *American Political Science Review, 119*(1), 179–192. https://doi.org/10.1017/s000305542400

Tennell, C., & Chew, K. A. (2024). Perspectives on relationality in online Indigenous language learning. *AlterNative: An International Journal of Indigenous Peoples, 20*(3), 512–520.

Terren, L., & Soler-i-Martí, R. (2021). "Glocal" and transversal engagement in youth social movements: A Twitter-based case study of Fridays For Future-Barcelona. *Frontiers in Political Science, 3*, 635822.

Thew, H., Middlemiss, L., & Paavola, J. (2020). "Youth is not a political position": Exploring justice claims-making in the UN climate change negotiations. *Global Environmental Change, 61*, 102036.

Trott, C. D. (2024). "It can't just be the younger people": Exploring young activists' perspectives on intergenerational tensions and solidarities for climate justice. *Journal of Community & Applied Social Psychology, 34*(6), e70001.

United Nations Development Programme. (2023). *Democracy and youth political participation: Global trends*. United Nations Development Programme. https://www.undp.org. Accessed June 12, 2024.

Vamvalis, M. (2023). "We're fighting for our lives": Centering affective, collective and systemic approaches to climate justice education as a youth mental health imperative. *Research in Education, 117*(1), 88–112.

Vercammen, A., Yatirajula, S. K., Daniel, M., Maharaj, S., Campbell, M. H., Greaves, N., Guinto, R., Aruta, J. J. B., Peñamante, C. A., Wray, B., & Lawrance, E. L. (2023). Investigating the mental health impacts of climate change in youth: Design and implementation of the international changing worlds study. *Challenges, 14*(3), 34.

Walker, C. (2020). Uneven solidarity: The school strikes for climate in global and intergenerational perspective. *Sustainable Earth, 3*(1), 5.

Watson, D., Hogg, T., Crandon, T., Vercammen, A., Watfern, C., Aruta, J. J. B., Charlson, F., & Lawrance, E. L. (2025). Conceptualizing Young People's Experiences of Climate Change Awareness: A Narrative Review. *Annals of the New York Academy of Sciences, 1554*(1), 45–65. https://doi.org/10.1111/nyas.70114

Wielk, E., & Standlee, A. (2021). Fighting for their future: An exploratory study of online community building in the youth climate change movement. *Qualitative Sociology Review, 17*(2), 22–37.

Wu, J., Long, D., Hafez, N., Maloney, J., Lim, Y., & Samji, H. (2023). Development and validation of a youth climate anxiety scale for the youth development instrument survey. *International Journal of Mental Health Nursing, 32*(6), 1473–1483.

Corporate Climate Strategies and the Future of Green Capitalism

12

12.1 Introduction

Climate change is increasingly recognized as one of the most significant risks to global economic and social stability (Iyiola et al., 2024; Izah et al., 2023). The effects are felt across sectors, leading to stresses such as resource shortages, rising sea levels, and extreme weather events that can disrupt supply chains and economic systems (Wright & Nyberg, 2017; Wright & Nyberg, 2024). Businesses are beginning to understand that climate change transcends environmental issues; it poses an existential threat to long-term profitability and stability (Ogwu, et al., 2025d). As corporations grapple with these realities, there is mounting recognition that proactive engagement in climate strategies is crucial not only for compliance but also for survival. Large organizations, in particular, are under heightened scrutiny from stakeholders, including investors, consumers, and policymakers, demanding transparent climate action (Haque & Islam, 2015; Hormio, 2017).

Moreover, corporations are stepping up to fill gaps where governmental action may falter. As traditional regulatory efforts struggle to keep pace with the scale and speed of climate challenges, firms are increasingly seen as essential players in driving effective climate responses. Proactive execution of climate strategies can significantly enhance corporate reputation and align with market expectations (Backman et al., 2017). This inclination toward corporate responsibility reflects a broader acknowledgment that collaborative efforts between the private sector and public institutions are essential to facilitate meaningful climate action (Almaghrabi, 2023; Dahlmann et al., 2019). In many contexts, businesses are taking the initiative to lead sustainability efforts, investing in innovation and processes that contribute not just to their bottom lines but also to social good (Al-Shaer et al., 2024; Gallagher, 2016). The transition toward "green" practices is no longer merely an ethical consideration but has become a strategic imperative for competitiveness and innovation. Firms that adopt and implement robust environmental

M. C. Ogwu and S. Chibueze Izah, *Climate Frontiers and Social Transformation*, World-Systems Evolution and Global Futures,
https://doi.org/10.1007/978-3-032-17639-4_12

sustainability strategies not only mitigate risks but also position themselves advantageously in an evolving marketplace (Hakovirta et al., 2023; Mohy-ud-Din et al., 2025). Climate-conscious organizations are likely to experience enhanced consumer loyalty, better investor confidence, and improved operational efficiencies, which collectively contribute to long-term viability and success (Principale & Pizzi, 2023; Wright & Nyberg, 2015). This shift motivates companies to see green initiatives as critical components of their business models, redefining their corporate landscapes (Street & Jude, 2019).

As corporations develop and transform their climate strategies, the implications for the future of green business are profound. The integration of climate goals into core business strategies requires comprehensive accountability frameworks, including metrics for performance and sustainability practices (Gallagher, 2016; Vasquez, 2022). Companies are adopting frameworks such as the Task Force on Climate-related Financial Disclosures (TCFD) to strengthen their commitments to transparency and risk management, thereby enhancing their credibility in the eyes of stakeholders (Hormio, 2017). The push for substantive climate action fosters innovation, as firms invest in cutting-edge technology and sustainable practices that not only address climate risks but also harness opportunities within the burgeoning green economy (Ogwu et al., 2025b, d). Therefore, the evolution of corporate climate strategies is likely to propel businesses toward greater adaptability and resilience in the face of ongoing climate challenges (Eleftheriadis & Anagnostopoulou, 2017).

This chapter examines how corporations are designing and implementing climate strategies to address global environmental challenges. It traces the evolution of corporate climate action from basic compliance to sustainability-driven innovation. The discussion highlights the key drivers, strategic approaches, challenges, and governance implications shaping corporate responses. It provides a forward-looking perspective on how businesses can align competitiveness with long-term climate responsibility. The chapter offers a clear understanding of the shifting role of corporations in advancing the green transition. It provides practical insights into strategies and frameworks that can guide both policymakers and business leaders. It will also equip stakeholders with the knowledge to engage critically with the future of green business practices.

12.2 The Evolution of Corporate Climate Strategies

The evolution of corporate climate strategies can be understood as a multi-stage process influenced by changing societal expectations, regulatory frameworks, and the intrinsic motivations of businesses (Table 12.1). In the early stages of corporate climate responses, philanthropic corporate social responsibility (CSR) initiatives dominated. These were often characterized by a focus on charity and community engagement rather than addressing systemic environmental issues. Firms approached climate-related activities largely voluntarily, often viewing CSR as a tool to enhance their reputation rather than a critical component of their operational strategies (Daddi et al., 2018).

The transition from this early stage to a compliance-driven phase was catalyzed mainly by the introduction of environmental regulations targeting waste management and emissions limits. This regulatory pressure forced corporations to adopt reactive compliance approaches, often resulting in minimal adherence to regulatory requirements rather than proactive climate engagement (Ihlen, 2009). During this compliance-driven stage, discussions surrounding environmental sustainability began to gain traction, although corporations typically framed climate issues within

Table 12.1 Evolution of corporate climate strategies

Stage	Major features	Strategic focus	Implications for corporations
Early stage: Philanthropic CSR	Centered on charity, donations, and symbolic environmental projects; minimal climate responsibility	Reputation building and goodwill generation	Limited impact on climate change; treated as a non-core activity
Compliance-Driven Stage	Response to environmental regulations on pollution, waste, and emissions; creation of environmental departments	Reactive compliance and risk avoidance	Climate action is seen as a cost of doing business; minimal innovation
Sustainability stage	Expansion of sustainability reporting (GRI, ESG); adoption of renewable energy and eco-efficiency; triple bottom line approach	Integration of sustainability into business operations and supply chains	Climate strategies linked to competitiveness, investor confidence, and risk management
Climate leadership stage	Net-zero targets, carbon neutrality commitments, renewable sourcing, board-level engagement, and partnerships	Innovation, market positioning, and green leadership	Climate strategies embedded in corporate identity and growth models
Shift to survival-driven innovation	Rising climate risks (supply-chain disruptions, extreme weather, stranded assets); growing investor and regulatory pressure	Business survival through resilience, adaptation, and regenerative practices	Climate action is reframed as essential for long-term viability and competitiveness

the context of risk management rather than as opportunities for innovation (Mohy-ud-Din et al., 2025; Talbot & Boiral, 2015). Moving into the sustainability stage, there was a noticeable shift toward more structured and formalized approaches to climate strategy. This era saw the proliferation of sustainability reporting and the adoption of Environmental, Social, and Governance (ESG) frameworks. These frameworks provided a mechanism for corporations to assess and disclose their climate-related risks and opportunities, and they began integrating these concerns into their core business strategies (Lemma et al., 2021; Wright & Nyberg, 2024). This stage acknowledged the importance of transparency and accountability, with companies publishing detailed reports on their environmental impact in an effort to satisfy stakeholder expectations increasingly (Vasquez, 2022). However, despite these advancements, many corporations faced critiques for adopting a descriptive rather than a prescriptive approach, often criticized for a lack of genuine commitment to transformative practices (Sprengel & Busch, 2011).

The climate leadership stage marks the zenith of corporate climate strategies, characterized by ambitious commitments to net-zero targets, carbon neutrality, and comprehensive climate strategies woven into the fabric of organizational business models. This represents a significant shift from viewing climate action as a peripheral or compliance-related issue to recognizing it as central to corporate survival and operational viability (Dahl & Fløttum, 2019; Linnenluecke & Griffiths, 2010). Companies are now increasingly guided by the principle of climate strategy proactivity (CSP), exemplifying a strategic approach wherein businesses voluntarily engage in practices that exceed legal compliance standards (Kant & Anjali , 2021). This proactive stance is not only a response to regulatory pressures but is also driven by the realization that climate change poses existential threats to business continuity and market competitiveness (Anwar et al., 2024).

Furthermore, the emergent call for corporate collaboration to tackle climate-related challenges reflects a growing recognition of the limitations of isolated approaches. As climate risk escalates, innovative partnerships and multi-stakeholder engagements are increasingly viewed as essential for effective corporate climate action. Firms are now more likely to pursue cooperative strategies that leverage collective expertise and resources, positioning collaboration as a critical element of their climate response (Raupach-Sumiya, 2024; Zhang et al., 2025).

12.3 Key Drivers of Corporate Climate Action

The corporate commitment to climate action is driven by a multitude of interdependent factors, which can be classified into regulatory frameworks, economic incentives, consumer expectations, reputation and brand value, technology and innovation, risk management, and other strategic considerations (Table 12.2). Each of these drivers influences corporate behavior regarding climate change and sustainability.

Regulatory frameworks play a fundamental role in guiding corporate climate strategies. National carbon taxes and levies help internalize the environmental

Table 12.2 Major drivers of corporate climate action

Driver	Expanded clarification
Regulatory frameworks	- National carbon taxes and levies internalize the cost of emissions - Emissions Trading Schemes (ETS) cap and trade corporate emissions - Global commitments like the Paris Agreement push firms toward net-zero - EU's Corporate Sustainability Reporting Directive (CSRD) and other mandatory disclosures enforce transparency - Local laws encourage renewable adoption and penalize high-emission activities
Economic incentives	- Access to green finance instruments (green bonds, climate funds, impact investing) - Carbon markets rewarding emissions reductions and trading credits - Investor demand for ESG compliance and climate risk disclosures - Operational cost savings through energy efficiency and resource optimization - Government subsidies, tax breaks, and credits for clean technology adoption
Consumer expectations	- Strong demand for eco-friendly and ethical products across industries - Calls for transparent supply chains and traceable sourcing - Loyalty shifts toward sustainability-driven brands - Millennials and Gen Z are driving green consumption trends - Boycotts, online activism, and campaigns targeting unsustainable corporations
Reputation & brand value	- Recognition as a green innovator enhances brand competitiveness - Builds trust among consumers, regulators, and investors - Attracts and retains top talent motivated by sustainability values - Avoids reputational risks from climate inaction and greenwashing scandals - Higher rankings in sustainability indices and benchmarks
Technology & innovation	- Expansion of renewable energy (solar, wind, hydrogen) - Electrification of transport and industrial systems - AI, blockchain, and IoT for monitoring carbon footprints and efficiency - Green manufacturing breakthroughs (e.g., sustainable steel, cement, plastics) - Circular economy models promoting recycling, reuse, and waste-to-energy
Risk management	- Addressing physical risks: floods, droughts, wildfires, extreme weather - Tackling transition risks: stranded fossil fuel assets, shifting policies - Mitigating litigation risks from climate lawsuits and false claims - Supply-chain resilience against climate-related disruptions - Safeguarding shareholder value by reducing long-term climate risks

(continued)

Table 12.2 (continued)

Driver	Expanded clarification
Other drivers	- Global competition requires alignment with international standards - Employee pressure for greener corporate practices - Industry benchmarking is pushing more substantial climate commitments - CEO leadership and board-level climate priorities - Collaboration in public–private climate initiatives - Geopolitical influences (trade policies, carbon border taxes) - Shareholder activism demanding corporate accountability

costs of carbon emissions, aligning business operations with broader climate goals (Littlewood et al., 2018). Moreover, emissions trading schemes (ETS) like the European Union ETS impose caps on emissions while allowing trading of allowances, providing financial incentives for firms to reduce their carbon footprints (Adamolekun et al., 2025). Global agreements such as the Paris Agreement amplify these effects by obligating countries to adhere to emissions-reduction commitments, which, by extension, impact corporations (Pattberg & Widerberg, 2017). Stricter reporting standards, such as the European Union's Corporate Sustainability Reporting Directive (CSRD), enhance accountability regarding environmental impacts, compelling firms to develop transparent climate strategies (Lafarre, 2023). Local policies, including renewable energy quotas and pollution penalties, further strengthen this regulatory framework, encouraging businesses to innovate toward greener practices (Pattberg & Widerberg, 2017).

Economic incentives significantly contribute to corporate climate action. The growth of green finance instruments (such as green bonds and climate funds) has provided businesses with pathways to fund sustainable projects, reducing the financial burdens associated with transitioning to greener practices (Eleftheriadis & Anagnostopoulou, 2017). Additionally, participation in carbon markets can financially reward companies for reducing emissions by selling credits, creating direct monetary incentives for proactive environmental behavior (Dietz et al., 2018). Increasing investor scrutiny of environmental, social, and governance (ESG) performance encourages companies to disclose climate-related risks and adopt sustainability practices, often aligning with long-term financial benefits (Wu & Tham, 2023). Improved operational efficiency through energy savings and resource optimization demonstrates the economic viability of green strategies (Lemma et al., 2021).

Consumer expectations have also shifted considerably, with increasing demand for eco-friendly and ethically produced goods influencing corporate behaviors (Haddock-Fraser & Tourelle, 2010). Transparency in supply chains and environmentally responsible sourcing practices has become essential, as consumers favor brands that actively demonstrate sustainability (Margaret et al., 2024). The market's reliance on ethical consumption exerts additional pressure on companies to strengthen their sustainability credentials, with younger consumers (Millennials and Gen Z) advocating for climate action (Lemma et al., 2021; Wu & Tham, 2023). Boycotts and social media activism against unsustainable practices compel companies to make significant operational reforms, highlighting that consumer loyalty increasingly favors responsibly managed firms (Al-Shaer et al., 2024).

Reputation and brand value are closely tied to a corporation's climate strategies. Companies viewed as climate leaders gain competitive advantages through enhanced consumer trust and brand loyalty (Haddock-Fraser & Tourelle, 2010). Sustainability credentials not only improve public perceptions but also attract top talent, reinforcing commitment to sustainable practices (Jansson et al., 2017). Conversely, poor climate performance can lead to reputational damage and backlash from stakeholders and the public (Dietz et al., 2018). Aligning with global sustainability indices and participating in environmental rankings can also elevate a corporation's prestige and market position (Littlewood et al., 2018).

Technology and innovation serve as vital enablers of effective corporate climate strategies. The integration of renewable energy technologies, such as solar and wind, into corporate operations is increasingly becoming standard as firms work to decarbonize their energy sources (Clements, 2021). Innovations in electric vehicles and artificial intelligence for supply-chain optimization further aid emissions tracking and improve operational efficiency (Hakovirta et al., 2023). Additionally, advances in low-carbon manufacturing and circular economy concepts, which emphasize waste reduction and resource reuse, align corporate goals with sustainable development (Hakovirta et al., 2023; Jansson et al., 2017). The management of climate-related risks has become a growing concern for corporations. Companies are preparing for physical risks associated with climate change, such as extreme weather events that can disrupt operations (Hossain et al., 2023). Transition risks, including regulatory changes and potential stranded assets in fossil fuels, require proactive management (Karatayev et al., 2021). Litigation risks linked to inadequate climate action or greenwashing allegations can significantly impact a company's financial standing and operational stability (Hossain, 2025). The need to build resilience into supply chains underscores the importance of strategic risk management in corporate sustainability practices (Eleftheriadis & Anagnostopoulou, 2017).

12.4 Strategic Approaches to Corporate Climate Action

In the context of corporate climate strategies (Table 12.3), effective decarbonization requires a multifaceted approach that involves several key strategic initiatives. One prevalent strategy is transitioning to renewable energy sources, including solar, wind, hydro, and geothermal. The adoption of these energy sources not only aims to reduce greenhouse gas emissions but may also enhance energy security and corporate reputation among stakeholders (Stewart & Niero, 2018; Theeraworawit et al., 2022). Moreover, enhancing energy efficiency across production lines, buildings, and transportation systems is critical for corporate decarbonization strategies, as improved energy efficiency directly correlates with reduced energy consumption and operational costs (Geissdoerfer et al., 2017; Ho & Lin, 2024). The electrification of fleets and industrial processes complements these initiatives by minimizing dependence on fossil fuels and advancing a low-carbon economy (Schöggl et al., 2024).

Low-carbon logistics solutions are essential for minimizing the carbon footprint associated with transportation and distribution. Implementing solutions such as green shipping practices, efficient rail transport, and optimized supply chains enhances logistics sustainability while contributing to overall emissions reductions (Genovese et al., 2017; Theeraworawit et al., 2022). The integration of carbon capture, utilization, and storage (CCUS) technologies is vital in energy-intensive industries, enabling companies to neutralize unavoidable emissions effectively (Mashovic et al., 2022; Wells & Skeete, 2023). Adaptation strategies are equally critical as organizations confront climate risks and vulnerabilities. Strengthening physical infrastructure to withstand severe weather events is crucial in mitigating replacement costs and disruptions (Ogwu, et al., 2025c; Ogwu & Kosoe, 2025a, 2025b). Companies increasingly invest in resilience measures to better withstand floods, hurricanes, and droughts, which climate change exacerbates (Wells & Skeete, 2023). This enhancement extends to supply chains, utilizing climate risk mapping tools and diversifying sourcing strategies to ensure operational continuity amidst climate-related adversities (Mazzucchelli et al., 2022; Zaccone et al., 2022).

The circular economy concept further emphasizes sustainability by advocating for the design of products considering end-of-life recycling and reuse. This approach aims to minimize waste, fostering a transition from linear "take-make-dispose" models to closed-loop production systems. To operationalize these concepts, increasing the utilization of recycled and renewable materials in manufacturing processes is critical (Jones & Comfort, 2018; Urbinati et al., 2017). Encouraging product-service models, such as leasing instead of ownership, extends product lifespans and promotes recycling and repair initiatives among consumers (Genovese et al., 2017; Zaccone et al., 2022). Engagement in such practices can fulfill corporate sustainability goals and drive consumer loyalty and brand enhancement through a greater commitment to environmental stewardship (Zhang et al., 2024). Central to the discourse on climate action is the carbon offsetting mechanism, which involves purchasing verified carbon credits to neutralize emissions that cannot be entirely eliminated. Supporting reforestation and afforestation projects, along with investing in nature-based solutions like mangrove restoration, compensates for unavoidable emissions (Barna et al., 2023; Wondem & Batra, 2019). Addressing carbon offset credibility is essential, requiring stringent transparency and verification measures to maintain stakeholder trust.

Innovation and research are vital in advancing corporate climate strategies, particularly through funding emerging clean energy technologies and developing biodegradable packaging materials. This innovation also entails creating smarter energy grids and utilizing artificial intelligence to enhance efficiency, significantly contributing to sustainability goals. Collaborations across industries, government entities, NGOs, and academia are crucial for knowledge sharing and improving

Table 12.3 Strategic approaches to corporate climate action

Strategic approach	Specific actions	Expected outcomes
Decarbonization	• Transition to renewable energy (solar, wind, hydro, geothermal) • Improve energy efficiency in operations and buildings • Electrify fleets and industrial processes • Adopt low-carbon logistics (green shipping, rail transport, optimized supply chains) • Integrate carbon capture, utilization, and storage (CCUS)	• Reduced greenhouse gas (GHG) emissions • Lower energy costs over time • Compliance with regulations and carbon pricing • Enhanced corporate reputation as a climate leader
Adaptation	• Strengthen infrastructure to withstand floods, droughts, and extreme heat • Diversify and map supply-chain risks • Implement water conservation and sustainable land use practices • Develop insurance and financial risk management tools • Incorporate climate resilience into governance	• Increased resilience to climate shocks • Reduced operational disruptions • Protection of assets and supply chains • Long-term business continuity
Circular Economy	• Design products for recycling, reuse, and minimal waste • Transition from linear to closed-loop systems • Use recycled and renewable materials in manufacturing • Promote product-service models (leasing/sharing) • Encourage consumer participation in recycling initiatives	• Reduced material costs • Lower waste and emissions footprint • Extended product lifecycles • Stronger consumer trust and loyalty
Carbon offsetting	• Purchase verified carbon credits • Support reforestation and afforestation projects • Invest in mangrove restoration and soil carbon sequestration • Explore direct air capture and mineralization • Ensure transparency and third-party verification	• Compensation for unavoidable emissions • Positive biodiversity and ecosystem impacts • Improved credibility with investors and consumers • Contribution to global carbon neutrality

(continued)

Table 12.3 (continued)

Strategic approach	Specific actions	Expected outcomes
Innovation & R&D	• Fund research in hydrogen, biofuels, and emerging clean energy • Develop biodegradable and compostable materials • Innovate smart grids, AI tools, and blockchain tracking • Support sustainable agriculture (vertical farming, precision agri-tech) • Scale pilot projects into market-ready solutions	• Long-term competitiveness through innovation • Breakthrough technologies for low-carbon future • New revenue streams from green products • Stronger intellectual property portfolio
Collaborations	• Join industry alliances for shared best practices • Partner with governments for subsidies and regulatory support • Work with NGOs, academia, and civil society on projects • Engage in multi-stakeholder platforms (UN Global Compact, WBCSD) • Align supply chains globally through partnerships	• Accelerated progress toward climate goals • Access to resources, funding, and knowledge • Stronger legitimacy through multi-stakeholder validation • Harmonized climate action across sectors

accountability toward climate goals, enabling organizations to adhere to climate commitments and innovate commercially viable, scalable solutions (Ogwu, 2019). For corporate climate strategies to be credible, they must integrate **Just Transition principles**, ensuring that workers, communities, and small suppliers are not left behind. This includes providing retraining for workers in fossil-dependent sectors, supporting alternative livelihoods, guaranteeing social protection for vulnerable groups, and co-designing transition plans with affected communities. Without these steps, even ambitious decarbonization strategies risk deepening social and regional inequalities.

12.5 Measuring, Reporting, and Accountability about Corporate Climate Action

The discourse surrounding corporate climate action increasingly emphasizes the importance of measuring, reporting, and ensuring accountability. Central to this discourse are standardized reporting frameworks such as the Global Reporting Initiative (GRI), the Task Force on Climate-Related Financial Disclosures (TCFD), the Sustainability Accounting Standards Board (SASB), and the European Union's Corporate Sustainability Reporting Directive (CSRD). These frameworks guide corporations in disclosing environmental impacts, aligning their sustainability metrics with stakeholder interests and regulatory requirements. Specifically, the CSRD mandates that all large enterprises disclose extensive non-financial information, thereby significantly raising the bar for corporate transparency in sustainability practices across the EU (Makarenko & Makarenko, 2023). This legislative push is further complemented by GRI's role in establishing universal sustainability reporting standards, thereby promoting the level of harmonization essential in an otherwise fragmented landscape (Darnall et al., 2022).

In the context of corporate accountability, Science-Based Targets (SBTs) play a vital role in aligning corporate goals with scientific benchmarks necessary for climate mitigation. These targets ensure that companies commit to emissions reductions grounded in the latest climate science, thereby enhancing the credibility of their sustainability claims. An increasing number of corporations are adopting SBTs as part of their climate strategy, motivated by regulatory frameworks and the growing demand from investors and consumers for accountability in environmental performance (Hahn et al., 2021; Ogwu, El Malahi, & Izah, 2025a). The importance of authentic metric alignment, as facilitated by SBTs, underscores a significant shift in corporate governance toward a more evidence-based approach in addressing climate change (Cortés Arbués et al., 2021).

However, the issue of greenwashing remains prevalent, posing a significant challenge for both corporations and stakeholders seeking genuine sustainability practices (Izah & Ogwu, 2025). Greenwashing involves misleading claims regarding environmental practices, making it essential for stakeholders to differentiate between authentic and deceptive sustainability efforts. Firms often employ various legitimation strategies to present their negative environmental impacts in a favorable light, complicating investors' decision-making processes (Hahn et al., 2021; Ogwu et al., 2023; Ogwu & Izah, 2025). The fragmentation of metrics across different reporting frameworks further exacerbates this issue, creating confusion and reducing the overall credibility of corporate sustainability disclosures. Ongoing scrutiny from regulatory bodies and the public necessitates increased accountability measures to mitigate greenwashing risks (Rim et al., 2019). To address these challenges, there is a growing demand for third-party verification of sustainability claims. Third-party verification is an essential mechanism for enhancing the reliability and transparency of sustainability reports. By ensuring that reported data is independently validated, corporations can reduce perceived risks associated with greenwashing and strengthen investor confidence (Darnall et al., 2022). Furthermore, comprehensive GHG emissions accounting systems are being developed to rigorously assess the impact and effectiveness of corporate climate action initiatives (Cortés Arbués et al., 2021). Such verified disclosures not only meet regulatory requirements but also address stakeholder expectations for transparency and truthfulness in corporate reports (Mijoković et al., 2020). The evolution of these standards, alongside stricter enforcement from regulatory entities, will significantly influence the trajectory of corporate sustainability efforts globally (Stamelos, 2022).

12.6 Barriers and Challenges to Corporate Climate Action

Corporate climate action is an urgent agenda in the current socio-economic landscape, yet various barriers and challenges impede its advancement (Table 12.4). One of the foremost barriers is greenwashing, a deceptive practice where companies exaggerate or misrepresent their environmental performance to attract consumers and investors. The deceptive nature of greenwashing undermines public trust, which is crucial for fostering genuine sustainability initiatives in corporate sectors (Österblom et al., 2022). When corporations engage in superficial branding, they create disillusionment among consumers and dilute the impact of authentic sustainability efforts by those truly investing in climate strategies. This trend leads to unfair competition, disadvantaging responsible firms in the marketplace (Bulińska-Stangrecka & Bagieńska, 2021). Moreover, regulatory bodies and watchdogs increasingly scrutinize sustainability claims made by corporations, heightening the potential for reputational and legal repercussions for companies involved in greenwashing (Hsueh, 2019).

Transitioning to renewable energy sources, sustainable manufacturing, and green logistics entails high upfront capital costs, which act as significant barriers to corporate climate action. Smaller firms, tiny and medium-sized enterprises (SMEs), often face formidable challenges securing the capital needed for such investments, placing them at a severe disadvantage compared to larger counterparts (Caputo et al., 2024). Established corporations also encounter difficulties due to competing budget priorities, with pressures surrounding immediate profitability often delaying investments in long-term climate initiatives (Akomea-Frimpong et al., 2022). A critical hesitation lies in the uncertainties surrounding returns from green investments, further dissuading corporations from committing to ambitious climate goals (Hakovirta et al., 2023). This financial discourse underscores a crucial challenge: the need for innovative financing solutions to support the transition to sustainable practices across diverse enterprise scales.

The disparity in resources between large corporations and SMEs exacerbates inequalities in corporate climate action capabilities. Multinational enterprises are typically better equipped, with access to substantial financial resources, specialized expertise, and expansive global networks to implement ambitious climate strategies effectively (Caputo et al., 2024). Conversely, smaller firms often grapple with restrictions in accessing green financing, coupled with limited expertise and bargaining power in their supply chains (Bulińska-Stangrecka & Bagieńska, 2021). This divide creates a "two-speed transition" phenomenon, whereby only large corporations advance toward sustainability while SMEs remain tethered to conventional practices (Raupach-Sumiya, 2024). Without policy frameworks that foster inclusivity and equitable support for SME initiatives, the risk of leaving these smaller entities behind intensifies, aggravating existing socio-economic inequalities (Akomea-Frimpong et al., 2022).

Additionally, the demand for short-term profits places immense pressure on corporations, often pitting shareholder expectations against the requirements for sustainable, long-term climate investments. Many firms operate under stringent

Table 12.4 Barriers and challenges concerning corporate climate action

Barrier	Narrative	Possible solutions / recommendations
Greenwashing: Cosmetic branding without substantive action	Companies sometimes exaggerate or misrepresent environmental performance to gain market appeal, undermining trust and exposing themselves to reputational and legal risks	Enforce stricter sustainability reporting standards; introduce third-party verification; strengthen penalties for false claims; encourage transparent communication with stakeholders
High costs of transition: Capital investment in renewable infrastructure and sustainable technologies	Moving to renewable energy, low-carbon manufacturing, and sustainable logistics requires significant upfront costs. SMEs especially lack capital, while even large firms face competing priorities and uncertain returns	Provide access to green finance and subsidies; establish climate funds to support SMEs; encourage public–private partnerships to share costs; promote tax incentives for sustainable investment
Unequal capacity: Large corporations have resources; SMEs often struggle to keep up	Multinationals have capital, expertise, and networks to act on climate, while SMEs face financing gaps, limited knowledge, and weak bargaining power—leading to a two-speed transition	Build capacity through training and technical support; improve access to climate financing for SMEs; integrate SMEs into corporate sustainability supply chains; create collaborative green hubs
Short-term profit pressures: Shareholder expectations sometimes conflict with long-term climate investments	Corporations prioritize quarterly profits, making it difficult to invest in long-term climate goals. Executives may avoid ambitious strategies if short-term returns are uncertain	Shift investor culture toward long-term value; expand ESG-based investment models; reward companies for sustainable performance; align executive compensation with climate targets
Global inequities: Developing nations and supply-chain partners may bear disproportionate burdens	Corporations often transfer costs of climate compliance to suppliers in the Global South, where resources are limited and climate vulnerabilities are greater	Adopt fair trade and equitable sourcing policies; support suppliers with technology transfer and capacity building; include climate justice principles in corporate strategy; engage in cross-border collaboration
Policy uncertainty: Fluctuating regulations and political resistance can discourage long-term planning	Shifting policies (carbon taxes, emissions caps) and weak global harmonization discourage corporations from investing in green technologies	Develop stable, long-term regulatory frameworks; harmonize international climate standards; strengthen public–private dialogue on policy; implement carbon border adjustment mechanisms to ensure fairness

quarterly reporting timelines that prioritize immediate financial returns, complicating the alignment of corporate governance strategies with sustainability initiatives (Wahab et al., 2024). The focus on short-term performance undermines the feasibility of pursuing climate strategies that require extended timelines and delayed financial payoffs, prompting executives to shy away from ambitious climate goals for fear of inadequate short-term financial performance (Zhang et al., 2025). This conflict represents a broader governance challenge: how to align investor priorities with sustainable practices at a corporate level while fulfilling financial obligations.

The context of global inequities further complicates corporate climate action, as companies frequently transfer the costs of climate-related decisions onto vulnerable supply chains in developing nations, where profit margins are already marginal. The disproportionate burden placed on suppliers in the Global South sheds light on complex questions of climate justice and corporate accountability, particularly in environments where regulatory frameworks are less robust and enforcement of green standards may be weak (Fankhauser & Stern, 2016). Climate vulnerabilities arising from extreme weather and resource scarcity exacerbate the risks faced by suppliers and laborers in low-income regions, raising the stakes for corporate responsibility within global value chains (Yang et al., 2020). Consequently, corporations are compelled to reconcile their operating practices with ethical imperatives that address the broader implications of their actions on marginalized communities.

Despite the urgent need for corporate climate action, policy uncertainty remains a significant challenge, driven by fluctuating regulations and political resistance. Sudden policy shifts, such as changes in carbon taxes or emissions regulations, generate a climate of unpredictability that hampers long-term corporate planning and investment in sustainable practices (Jabbour et al., 2020). Political dynamics, including lobbying by fossil fuel interests, can further erode the integrity of climate policies, leading to inconsistent regulatory frameworks across regions (Averchenkova et al., 2021). This scenario complicates operations, especially for corporations with international supply chains, where adherence to disparate regulations can incur substantial compliance costs (Schnyder, 2021). Stable and cohesive policy signals are essential for fostering a predictable environment conducive to investment in green technologies. The scale of corporate responsibility becomes clearer when viewed against the concentration of emissions in the global energy system. Recent Carbon Majors analyses show that a relatively small set of producers accounts for a large share of historical and current emissions, with state-owned and investor-owned fossil fuel companies together accounting for a substantial share of global CO_2 output since the industrial era. This concentration reinforces two linked implications for green capitalism: first, that meaningful progress requires targeting high-emitting sectors and firms with enforceable decarbonization timelines; and second, that voluntary corporate pledges are insufficient without state mechanisms that curb fossil expansion, regulate lobbying, and redirect capital toward low-carbon infrastructure.

Beyond these barriers, the role of corporate governance and culture in shaping climate strategies cannot be overstated. Indeed, aligning organizational culture with sustainable practices is a vital aspect of overcoming resistance to implementing green initiatives (Bulińska-Stangrecka & Bagieńska, 2021). A corporate governance structure that encapsulates sustainability at its core can catalyze action across all levels, fostering a culture of accountability and innovation (Akomea-Frimpong et al., 2022). This perspective underscores the relevance of establishing frameworks to encourage collaborative innovation as a vital mechanism for overcoming challenges posed by climate risk (Zhang et al., 2025). When organizations prioritize sustainability in their governance, they enhance their reputation and contribute meaningfully to the broader mission of climate resilience in their communities.

Crucially, the dynamic interplay between stakeholders influences the effectiveness of corporate climate strategies. Engaging diverse stakeholders (including employees, supply-chain partners, regulators, and communities) can facilitate a cooperative approach to overcoming barriers and mobilizing resources for collective climate action (Raupach-Sumiya, 2024). Studies suggest that meaningful engagement and dialogue among stakeholders enhance corporate climate performance and foster a supportive environment for driving systematic changes across industries (Jabbour et al., 2020). In this highly interconnected landscape, corporations that effectively harness stakeholder networks for climate action may find opportunities for growth and innovation, demonstrating that collective efforts can yield significant progress toward shared sustainability goals.

12.7 The Future of Green: Emerging Trends

Despite persistent obstacles, many corporations are demonstrating meaningful progress. As of 2024, more than 4000 companies across 70 countries have adopted science-based emissions targets, and global investments in clean technologies surpassed $1.8 trillion, outpacing fossil investments for the first time (IEA [International Energy Agency], 2024). These developments indicate that corporate actors are increasingly positioning climate action as a core business imperative rather than a reputational exercise. The transition toward more innovative corporate practices to address climate change has led to the emergence of several trends that collectively represent a transformative vision for sustainability in the business sector. Among these, regenerative business models have gained traction as companies move beyond merely minimizing their ecological footprints to actively restoring ecosystems and enriching biodiversity (Ogwu, Ojo, & Alaka, 2025f). Research indicates that successful regenerative business models redefine value creation beyond profit maximization, advocating for a holistic approach that recognizes businesses as integral components of social and environmental systems (Hahn & Tampe, 2021; Ryan et al., 2023). This shift necessitates a deeper engagement with local ecosystems and an understanding of how corporate activities can contribute to ecosystem resilience (Drupsteen & Wakkee, 2024; Ryan et al., 2023).

Digital transformation is another notable trend, in which businesses increasingly leverage advanced technologies such as artificial intelligence (AI), blockchain, and the Internet of Things (IoT) to enhance supply-chain transparency and accurately track emissions. The integration of these digital tools facilitates more accountable and efficient reporting on climate impact, ensuring that companies can substantiate their commitment to sustainability (Drupsteen & Wakkee, 2024; Ryan et al., 2023). For instance, blockchain can offer immutability and transparency in transactions, thus enabling better tracking of sustainability claims throughout supply chains, while IoT devices can provide real-time data on emissions, facilitating immediate corrective actions when necessary (Caldera et al., 2022; Neisig, 2025).

The phenomenon of climate litigation serves as a significant pressure point for corporations, with an increase in lawsuits targeting organizations that fail to meet climate action expectations. This rise indicates growing public awareness and legal advocacy demanding corporate accountability in addressing climate change hazards. Legal frameworks are evolving, holding companies liable for perceived insufficient action against climate change, prompting them to enhance their sustainability agendas (Drupsteen & Wakkee, 2024; Hahn & Tampe, 2021).

In the financial landscape, a green finance revolution is underway, marked by significant expansion in sustainable bonds, climate funds, and investments aligned with Environmental, Social, and Governance (ESG) criteria. This revolution reflects an increasing recognition among investors that the future profitability of businesses will hinge on their ability to navigate climate-related risks and contribute to a sustainable future (Dhaliwal et al., 2019; Jorge et al., 2022). Green finance is establishing frameworks that not only drive capital toward environmentally beneficial endeavors but also create accountability for the promised sustainability outcomes.

Concurrently, a cultural shift is emerging in corporate environments, especially among younger employees who demand meaningful commitments to sustainability from their employers. This generational change is propelling organizations to embed sustainability into their core business strategies and corporate cultures, recognizing that employee engagement in sustainability can lead to enhanced morale, retention, and productivity (Drupsteen & Wakkee, 2024; Hahn & Tampe, 2021; Neisig, 2025).

As governments and regulatory bodies gradually shift from voluntary pledges to mandatory climate regulations, organizations should now adapt to increasingly stringent laws regarding emissions reporting and reduction targets. This transformation reflects a global trend toward standardized climate accountability, where compliance becomes essential rather than optional, prompting companies to reassess their sustainability strategies to align with evolving legal standards (Dhaliwal et al., 2019; Jorge et al., 2022). Additionally, global convergence in climate accounting practices is gaining momentum as calls for harmonized standards increase to address discrepancies that can afford corporations loopholes. Internationally, collaborative efforts are underway to streamline reporting and compliance standards to enable comparability and transparency among businesses (Neisig, 2025). These harmonization efforts are crucial for creating a level playing field in which corporate sustainability claims can be reliably evaluated and standardized across jurisdictions.

12.8 Governance and Policy Implications

Corporate climate action is increasingly deemed essential in addressing the urgent challenges posed by climate change. Governance and policy implications in this context emphasize not only the critical role of corporations but also the significant influence of governments and global institutions in facilitating or hindering the implementation of effective climate strategies.

12.8.1 Governance and Policy Implications

Corporate governance should adapt to incorporate climate action into core business strategies. This involves a shift from voluntary Environmental, Social, and Governance (ESG) practices to mandatory sustainability frameworks, which can enforce compliance among companies to meet short- and long-term climate goals. Such regulatory frameworks can help align corporate strategies with global climate targets, particularly as international expectations evolve. As indicated in recent analyses, the historical reliance on voluntary compliance has been inadequate, prompting calls for more stringent regulations to ensure that corporate activities mitigate, rather than exacerbate, climate harm (Branger & Quirion, 2014; Tagliapietra & Wolff, 2021).

12.8.2 Role of Governments: Creating Enabling Environments

Governments play a crucial role in creating an enabling environment by imposing policies, subsidies, and penalties that promote climate readiness among companies. They can implement border carbon adjustment mechanisms (BCAs) to discourage carbon leakage—situations where businesses transfer emissions-intensive operations to countries with less stringent regulations, thereby undermining domestic efforts to reduce greenhouse gas emissions (Majumder et al., 2025; Arvanitopoulos et al., 2021). Such adjustments help maintain competitive parity while incentivizing corporations to invest in cleaner technologies, aligning with larger national and international climate strategies. In recent discussions, the effectiveness of BCAs has been varied; while they may protect local industries, critics warn that they could instigate protectionist measures that distort international trade (Kemajou et al., 2019; Van den Bergh et al., 2020). This duality complicates the implementation of BCAs and underscores the importance of nuanced policy design.

12.8.3 Public–Private Collaboration

Collaboration among corporations, governments, and NGOs can drive innovative climate initiatives that foster greater environmental responsibility. Multi-stakeholder partnerships have been shown to expand the scope and impact of these initiatives by leveraging diverse resources and expertise. For instance, initiatives may include joint investments in renewable energy projects or coordinated efforts in carbon management (Isaksson & Çelik, 2013; Oberthür & Dupont, 2021). Such partnerships can also facilitate information sharing, particularly in regions where corporate sustainability capacities may be lacking. Eventually, strong alliances between the public and private sectors can catalyze more significant climate action, as collective strides toward decarbonization become increasingly necessary to comply with both local and global mandates.

12.8.4 Global Institutions and Trade Competitiveness

Global institutions such as the United Nations (UN) and World Trade Organization (WTO) are key players in climate policy, working to standardize climate governance across nations. Their efforts seek to harmonize laws and regulations that can facilitate international trade while tackling climate issues effectively (Weitzel et al., 2012; Winter, 2023). The introduction of carbon border adjustment mechanisms presents an opportunity for these institutions to shape trade regulations in a way that considers both environmental sustainability and competitive equity. However, the implementation of such measures should adhere to WTO rules to prevent potential conflicts that could undermine progress (Buttner & Lowe, 2017; Jakob, 2023). As various nations grapple with the implications of global decarbonization, BCAs could serve as tools not only to curb carbon leakage but also to enhance the competitive positioning of compliant firms globally (Bao et al., 2013; Smith et al., 2024). This balance is crucial for maintaining economic stability while advancing environmental objectives, as the burdens of climate regulation should not disproportionately affect developing economies (Arvanitopoulos et al., 2021; Burniaux et al., 2013).

12.8.5 Equity Dimensions

Recent global assessments show that 91% of climate-related deaths occur in low- and middle-income countries, even though these regions contribute the least to cumulative emissions (UNDRR, 2022). A significant dimension of corporate climate action involves addressing equity and ensuring that corporate strategies do not reinforce global inequalities in resource distribution. The narrative surrounding climate responsibility often intersects with social equity, highlighting disparities in who bears the brunt of climate impacts and policy responses (Keen & Kotsogiannis, 2014; Latinović & Obradović, 2013). Ensuring that policy measures include protective factors for vulnerable populations is essential to achieving just climate transitions. The socio-economic implications of inequitable climate policies may exacerbate existing global inequalities, necessitating a more holistic and inclusive approach to corporate governance in the climate arena (Balistreri & Rutherford, 2012; Park et al., 2023). A further governance priority is aligning corporate climate action with a justice-centered reform agenda at the state level. While many firms are moving toward net-zero targets, circular models, and resilience investments, their progress is constrained or distorted by policy environments that still privilege fossil systems. Globally, fossil fuel subsidies remain enormous, estimated by the IMF at about $7 trillion in 2022 (≈7% of global GDP), most of it implicit under-pricing of environmental and health damages (IMF [International Monetary Fund], 2023). Subsidy reform therefore, represents one of the clearest governmental levers for accelerating genuine green transitions, since it would reduce perverse incentives, free large public revenues, and level the playing field for low-carbon innovation.

In parallel, the democratization of climate governance is essential to prevent "green capitalism" from reproducing old inequalities in new forms. Expanding democratic participation—through stronger transparency rules, labor and community representation in transition planning, and protection for civic and Indigenous land rights—helps ensure that corporate decarbonization does not externalize costs onto vulnerable workers, suppliers, or frontline communities. This matters because climate harms are patterned by power: those least responsible for emissions are most exposed to losses, and without explicit justice safeguards, corporate strategies can deepen extraction, displacement, or labor precarity in the Global South. Consequently, equity-oriented public policy (including social protection, just transition financing, and enforceable human rights due diligence in supply chains) should be treated not as an "add-on" to corporate strategy but as a core condition for credible and durable climate action.

12.8.6 Future Regulation

Looking ahead, the transition from voluntary to mandatory sustainability frameworks is imperative for driving substantial corporate climate action. This evolution in policy can help standardize expectations and establish benchmarks for corporate sustainability reports, thereby enhancing transparency and accountability across sectors (Oberthür & Dupont, 2021; Winter, 2023). Countries need to harmonize their approaches to carbon pricing, tax mechanisms, and regulatory practices to minimize competitive disadvantages while concurrently meeting their climate commitments (Böhringer et al., 2010; Van den Bergh et al., 2020). Future regulations should thus focus not only on compliance but also on fostering innovation, enabling businesses to pivot toward sustainability as a core tenet of their operational models.

12.8.7 Climate Justice and Corporate Responsibility

A justice-centered approach to corporate climate action is essential because climate impacts fall disproportionately on communities with the least responsibility for global emissions. As of 2023, the lowest-income 50% of the world's population produces just 12% of global CO_2 emissions, yet experiences the majority of climate-related deaths, displacement, and livelihood losses (Oxfam, 2023). Corporate strategies, if not designed with distributive justice in mind, risk shifting climate burdens onto already disadvantaged workers, Indigenous communities, and supply-chain partners in the Global South. Embedding equity safeguards—including living-wage guarantees, Indigenous consultation protocols, fair trade procurement, and climate-loss compensation mechanisms—ensures that the transition to a green economy does not reproduce historical inequalities.

12.9 Conclusion

Corporate climate strategies are no longer optional but fundamental to business survival in a rapidly changing environmental and economic landscape. At the same time, the chapter's evidence indicates that the corporate world is not uniformly resistant to change. Across sectors, many companies are adjusting to climate risk, investing in renewables, redesigning products for circularity, and treating decarbonization as a competitiveness strategy rather than mere compliance. These shifts reveal an important, if uneven, opening for transformation: firms can become accelerators of transition when policy, finance, and public accountability reward genuine emissions cuts and penalize delay or deception. Yet corporate momentum will remain partial and fragile unless paired with structural public reforms—especially subsidy removal, mandatory reporting, and expanded democratic oversight—so that green capitalism evolves from symbolic commitments into a system capable of delivering real mitigation and fair outcomes. The evolution from philanthropy-driven efforts to sophisticated, integrative climate leadership reflects a growing recognition that corporations are critical agents in shaping a sustainable future. Transitioning from harm reduction to systemic regeneration will define the "future of green," demanding that climate considerations be embedded in every layer of decision-making—operations, governance, supply chains, and innovation. An interplay of regulatory frameworks, economic incentives, consumer expectations, reputational pressures, and risk management imperatives drives this trajectory. Yet, challenges such as financial burdens, regulatory uncertainties, inequities between corporations, and persistent greenwashing remain barriers that should be confronted with resilience and creativity.

To overcome these challenges, corporations should adopt comprehensive climate strategies that combine decarbonization, adaptation, circular economy practices, innovation, and collaborative partnerships. Science-based targets and transparent reporting are essential for credibility, while digital tools can enhance monitoring and accountability. Policymakers should strengthen regulatory harmonization, create incentives for sustainable investment, and extend tailored support to SMEs, enabling them to participate meaningfully in the green transition. Equally, stakeholders across sectors (businesses, governments, civil society, and global institutions) should collaborate to ensure equity, policy coherence, and systemic transformation. Companies that proactively embrace regenerative business models, leverage digital transformation, engage in climate finance, and align with emerging mandatory regulations will not only remain competitive but also position themselves as leaders in sustainability. Eventually, the future of green is collaborative, technologically enabled, and accountability-driven, shaping not just corporations but the entire global economy.

References

Adamolekun, G., Ahmed, A., Kwansa, N. A., Lawal, R., & Sakariyahu, R. (2025). Corporate climate risk and membership of emission trading schemes. *Business Strategy and the Environment, 34*(5), 6054–6077.

Akomea-Frimpong, I. (2022). Systematic review of sustainable corporate governance of SMEs: Conceptualisation and propositions. *Corporate Ownership & Control, 19*(3), 74–91.

Akomea-Frimpong, I., Asogwa, I. E., & Tenakwah, E. J. (2022). Systematic review of sustainable corporate governance of SMEs: Conceptualisation and propositions. *Corporate Ownership & Control, 19*(3), 74–91.

Almaghrabi, K. S. (2023). Climate change exposure and firm performance: Does managerial ability matter? *Sustainability, 15*(17), 12878.

Al-Shaer, H., Liu, Y. S., & Albitar, K. (2024). Driving businesses towards a better climate: Macro and micro mechanisms to protect the planet. *Business Strategy and the Environment, 33*(3), 1810–1833.

Anwar, S., Maulidiyah, D. R., Rusanti, E., & Mochlasin, M. (2024). Book review: Sustainable finance and ESG; risk, management, regulations and implications for financial institutions. *Rajagiri Management Journal, 18*(1), 99–103.

Arvanitopoulos, T., Garsous, G., & Agnolucci, P. (2021). *Carbon leakage and agriculture: A literature review on emissions mitigation policies*. OECD.

Averchenkova, A., Fankhauser, S., & Finnegan, J. J. (2021). The influence of climate change advisory bodies on political debates: Evidence from the UK committee on climate change. *Climate Policy, 21*(9), 1218–1233.

Backman, C. A., Verbeke, A., & Schulz, R. A. (2017). The drivers of corporate climate change strategies and public policy: A new resource-based view perspective. *Business & Society, 56*(4), 545–575.

Balistreri, E. J., & Rutherford, T. F. (2012). Subglobal carbon policy and the competitive selection of heterogeneous firms. *Energy Economics, 34*, S190–S197.

Bao, Q., Tang, L., Zhang, Z., & Wang, S. (2013). Impacts of border carbon adjustments on China's sectoral emissions: Simulations with a dynamic computable general equilibrium model. *China Economic Review, 24*, 77–94.

Barna, C., Zbuchea, A., & Stănescu, S. M. (2023). Social economy enterprises contributing to the circular economy and the green transition in Romania. *CIRIEC-España, revista de economía pública, social y cooperativa, 107*, 47–69.

Böhringer, C., Fischer, C., & Rosendahl, K. E. (2010). The global effects of subglobal climate policies. *The BE Journal of Economic Analysis & Policy, 10*(2), 2583.

Branger, F., & Quirion, P. (2014). Would border carbon adjustments prevent carbon leakage and heavy industry competitiveness losses? Insights from a meta-analysis of recent economic studies. *Ecological Economics, 99*, 29–39.

Bulińska-Stangrecka, H., & Bagieńska, A. (2021). Culture-based green workplace practices as a means of conserving energy and other natural resources in the manufacturing sector. *Energies, 14*(19), 6305.

Burniaux, J. M., Chateau, J., & Duval, R. (2013). Is there a case for carbon-based border tax adjustment? An applied general equilibrium analysis. *Applied Economics, 45*(16), 2231–2240.

Buttner, E. H., & Lowe, K. B. (2017). Addressing internal stakeholders' concerns: The interactive effect of perceived pay equity and diversity climate on turnover intentions. *Journal of Business Ethics, 143*(3), 621–633.

Caldera, S., Hayes, S., Dawes, L., & Desha, C. (2022). Moving Beyond Business as Usual Toward Regenerative Business Practice in Small and Medium-Sized Enterprises. *Frontiers in Sustainability, 3*, 799359. /https://doi.org/10.3389/frsus.2022.799359

Caputo, A., Pizzi, S., & Santini, E. (2024). Sustainability and SMEs: Opening the black box. *Journal of Management & Organization, 30*(3), 413–420.

Chiappetta Jabbour, C. J., Fiorini, P. D. C., Ndubisi, N. O., Queiroz, M. M., & Piato, É. L. (2020). Digitally-enabled sustainable supply chains in the 21st century: A review and a research agenda. *Science of The Total Environment*, 725, 138177. https://doi.org/10.1016/j.scitotenv.2020.138177

Clements, P. (2021). *How should a global carbon budget be estimated?*. https://doi.org/10.21203/rs.3.rs-336107/v1

Cortés Arbués, I., Chatzivasileiadis, T., Storm, S., Ivanova, O., & Filatova, T. (2025). Private investments in climate change adaptation are increasing in Europe, although sectoral differences remain. Communications earth & environment, 6(1), 470. https://doi.org/10.1038/s43247-025-02454-3

Daddi, T., Todaro, N. M., De Giacomo, M. R., & Frey, M. (2018). A systematic review of the use of organization and management theories in climate change studies. *Business Strategy and the Environment, 27*(4), 456–474.

Dahl, T., & Fløttum, K. (2019). Climate change as a corporate strategy issue: A discourse analysis of three climate reports from the energy sector. *Corporate Communications: An International Journal, 24*(3), 499–514.

Dahlmann, F., Branicki, L., & Brammer, S. (2019). Managing carbon aspirations: The influence of corporate climate change targets on environmental performance. *Journal of Business Ethics, 158*(1), 1–24.

Darnall, N., Ji, H., Iwata, K., & Arimura, T. H. (2022). Do ESG reporting guidelines and verifications enhance firms' information disclosure? *Corporate Social Responsibility and Environmental Management, 29*(5), 1214–1230.

Dhaliwal, D., He, W., Li, Y., & Pereira, R. (2019). Accounting standards harmonization and financial integration. *Contemporary Accounting Research, 36*(4), 2437–2466.

Dietz, S., Fruitiere, C., Garcia-Manas, C., Irwin, W., Rauis, B., & Sullivan, R. (2018). An assessment of climate action by high-carbon global corporations. *Nature Climate Change, 8*(12), 1072–1075.

Drupsteen, L., & Wakkee, I. (2024). Exploring characteristics of regenerative business models through a Delphi-inspired approach. *Sustainability, 16*(7), 3062.

Eleftheriadis, I., & Anagnostopoulou, E. (2017). Measuring the level of corporate commitment regarding climate change strategies. *International Journal of Climate Change Strategies and Management, 9*(5), 626–644.

Fankhauser, S., & Stern, N. (2016). Climate change, development, poverty and economics. In K. Basu, D. Rosenblatt, & C. Sepúlveda (Eds.), *The state of economics, the state of the world*. MIT Press.

Gallagher, D. R. (2016). Climate change leadership as sustainability leadership: From the C-suite to the conference of the parties. *Journal of Leadership Studies, 9*(4), 60–64.

Geissdoerfer, M., Savaget, P., Bocken, N. M., & Hultink, E. J. (2017). The circular economy–A new sustainability paradigm? *Journal of Cleaner Production, 143*, 757–768.

Genovese, A., Acquaye, A. A., Figueroa, A., & Koh, S. L. (2017). Sustainable supply chain management and the transition towards a circular economy: Evidence and some applications. *Omega, 66*, 344–357.

Haddock-Fraser, J. E., & Tourelle, M. (2010). Corporate motivations for environmental sustainable development: Exploring the role of consumers in stakeholder engagement. *Business Strategy and the Environment, 19*(8), 527–542.

Hahn, R., Reimsbach, D., Kotzian, P., Feder, M., & Weißenberger, B. E. (2021). Legitimation strategies as valuable signals in nonfinancial reporting? Effects on investor decision-making. *Business & Society, 60*(4), 943–978.

Hahn, T., & Tampe, M. (2021). Strategies for regenerative business. *Strategic Organization, 19*(3), 456–477.

Hakovirta, M., Kovanen, K., Martikainen, S., Manninen, J., & Harlin, A. (2023). Corporate net zero strategy—Opportunities in start-up driven climate innovation. *Business Strategy and the Environment, 32*(6), 3139–3150.

Haque, S., & Islam, M. A. (2015). Stakeholder pressures on corporate climate change-related accountability and disclosures: Australian evidence. *Business and Politics, 17*(2), 355–390.

Ho, Y. H., & Lin, C. Y. (2024). Circular economy implementation and sustainability performance of the textile industry in an emerging economy. *International Journal of Clothing Science and Technology, 36*(5), 887–906.

Hormio, S. (2017). Can corporations have (moral) responsibility regarding climate change mitigation? *Ethics, Policy & Environment, 20*(3), 314–332.

Hossain, M. Z. (2025). *A quantitative analysis of greenwashing practices in corporate carbon accounting and its impact on stakeholder trust and regulatory compliance*. Available at SSRN 5255546 or https://doi.org/10.2139/ssrn.5255525

Hossain, A., Masum, A. A., Saadi, S., Benkraiem, R., & Das, N. (2023). Firm-level climate change risk and CEO equity incentives. *British Journal of Management, 34*(3), 1387–1419.

Hsueh, L. (2019). Opening up the firm: What explains participation and effort in voluntary carbon disclosure by global businesses? An analysis of internal firm factors and dynamics. *Business Strategy and the Environment, 28*(7), 1302–1322.

IEA [International Energy Agency]. (2024). *World energy investment report*. https://www.iea.org/reports/world-energy-investment-2023/overview-and-key-findings. Accessed on June 2, 2025.

Ihlen, Ø. (2009). Business and climate change: the climate response of the world's 30 largest corporations. *Environmental Communication, 3*(2), 244–262.

IMF [International Monetary Fund]. (2023). *Fiscal monitor: The price of inaction—Fiscal costs of climate change*. International Monetary Fund. https://www.imf.org/en/Publications/FM/Issues/2023/10/04/fiscal-monitor-october-2023. Accessed on April 30 2025.

Isaksson, M., & Çelik, S. (2013). Equity markets, corporate governance and value creation. *OECD Journal: Financial Market Trends, 2013*(1), 53–84.

Iyiola, A. O., Afolabi, O. A., Alimi, S. K., Akingba, O. O., Izah, S. C., & Ogwu, M. C. (2024). Climate change and water crisis in the Global South. In S. C. Izah, M. C. Ogwu, A. Loukas, & H. Hamidifar (Eds.), *Water crises and sustainable management in the Global South* (pp. 111–140). Springer. https://doi.org/10.1007/978-981-97-4966-9_4

Izah, S. C., Iyiola, A. O., Yarkwan, B., & Richard, G. (2023). Impact of air quality as a component of climate change on biodiversity-based ecosystem services. In A. L. Srivastav, A. K. Dubey, A. Kumar, S. K. Narang, & M. A. Khan (Eds.), *Visualization techniques for climate change with machine learning and artificial intelligence* (pp. 123–148). Elsevier.

Izah, S. C., & Ogwu, M. C. (2025). *Innovative approaches in environmental health management: Processes, technologies, and strategies for a sustainable future*. Springer Nature. https://doi.org/10.1007/978-3-031-81966-7

Jakob, M. (2023). The political economy of carbon border adjustment in the EU. *Oxford Review of Economic Policy, 39*(1), 134–146.

Jansson, J., Nilsson, J., Modig, F., & Hed Vall, G. (2017). Commitment to sustainability in small and medium-sized enterprises: The influence of strategic orientations and management values. *Business Strategy and the Environment, 26*(1), 69–83.

Jones, P., & Comfort, D. (2018). Winning hearts and minds: A commentary on circular cities. *Journal of Public Affairs, 18*(4), e1726.

Jorge, S., Dabbicco, G., Aggestam-Pontoppidan, C., & Vaz de Lima, D. (2022). New development: The development of standardized charts of accounts in public sector accounting. *Public Money & Management, 42*(7), 530–533.

Kant, N. & Anjali, K. (2021). Climate Strategy Proactivity (CSP). *Advances in Business Information Systems and Analytics*, 103-119. https://doi.org/10.4018/978-1-7998-6788-3.ch006

Karatayev, V. A., Vasconcelos, V. V., Lafuite, A. S., Levin, S. A., Bauch, C. T., & Anand, M. (2021). A well-timed shift from local to global agreements accelerates climate change mitigation. *Nature Communications, 12*(1), 2908.

Keen, M., & Kotsogiannis, C. (2014). Coordinating climate and trade policies: Pareto efficiency and the role of border tax adjustments. *Journal of International Economics, 94*(1), 119–128.

Kemajou, A., Jaligot, R., Bosch, M., & Chenal, J. (2019). Assessing motorcycle taxi activity in Cameroon using GPS devices. Journal of Transport Geography, 79, 102472. https://doi.org/10.1016/j.jtrangeo.2019.102472

Lafarre, A. (2023). The proposed Corporate Sustainability due diligence directive: Corporate liability design for social harms. *European Business Law Review, 34*(2), 213–237.

Latinovic, M., & Obradovic, T. (2013). The performance of socially responsible investments. *Entrepreneurial Business and Economics Review, 1*(2), 29–39. https://doi.org/10.15678/eber.2013.010203

Lemma, T. T., Lulseged, A., & Tavakolifar, M. (2021). Corporate commitment to climate change action, carbon risk exposure, and a firm's debt financing policy. *Business Strategy and the Environment, 30*(8), 3919–3936.

Littlewood, D., Decelis, R., Hillenbrand, C., & Holt, D. (2018). Examining the drivers and outcomes of corporate commitment to climate change action in European high emitting industry. *Business Strategy and the Environment, 27*(8), 1437–1449.

Linnenluecke, M. K., & Griffiths, A. (2010). Corporate sustainability and organizational culture. *Journal of World Business, 45*(4), 357-366. https://doi.org/10.1016/j.jwb.2009.08.006

Majumder, P., Mathur, S., & Pohit, S. (2025). Impact of the European union's carbon border adjustment mechanism: evidence from India and other selected trading partners of EU. *Green and Low-Carbon Economy, 3*(3), 220–229.

Makarenko, I., & Makarenko, S. (2023). Multi-level benchmark system for sustainability reporting: EU experience for Ukraine. *Accounting and Financial Control, 4*(1), 41–48.

Margaret, I., Schoubben, F., & Verwaal, E. (2024). When do investors see value in international environmental management certification of multinational corporations? A study of ISO 14001 certification after the Paris agreement. *Global Strategy Journal, 14*(1), 25–55.

Mashovic, A., Ignjatović, J., & Kisin, J. (2022). Circular economy as an imperative of sustainable development in North Macedonia and Serbia. *Ecologica, 29*(106), 169–177.

Mazzucchelli, A., Chierici, R., Del Giudice, M., & Bua, I. (2022). Do circular economy practices affect corporate performance? Evidence from Italian large-sized manufacturing firms. *Corporate Social Responsibility and Environmental Management, 29*(6), 2016–2029.

Mijoković, M., Knežević, G., & Mizdraković, V. (2020). Analysing the link between CSR reporting and financial performance variables of Belgrade stock exchange companies. *TEME: Casopis za Društvene Nauke, 44*(4), 1369–1389.

Mohy-ud-Din, K., Shahbaz, M., & Du, A. M. (2025). Corporate social responsibility and climate change mitigation: Discovering the interaction role of green audit and sustainability committee. *Corporate Social Responsibility and Environmental Management, 32*(1), 1198–1212.

Neisig, M. (2025). The regenerative semantics and structural change: Social systems using nature as a regenerative medium. *Systems Research and Behavioral Science, 42*(2), 531–544.

Oberthür, S., & Dupont, C. (2021). The European Union's international climate leadership: towards a grand climate strategy? *Journal of European Public Policy, 28*(7), 1095–1114.

Ogwu, M. C. (2019). Towards sustainable development in Africa: The challenge of urbanization and climate change adaptation. In P. B. Cobbinah & M. Addaney (Eds.), *The geography of climate change adaptation in Urban Africa* (pp. 29–55). Springer Nature. https://doi.org/10.1007/978-3-030-04873-0_2

Ogwu, M. C., El Malahi, S., & Izah, S. C. (2025a). Policy and regulatory frameworks for sustainable environmental practices. In M. C. Ogwu & S. Chibueze Izah (Eds.), *Evaluating environmental processes and technologies*. Environmental Science and Engineering Series. Springer. https://doi.org/10.1007/978-3-031-85327-2_16

Ogwu, M. C., Iyiola, A. O., Fawole, W. O., & Izah, S. C. (2025b). Biobased resource recovery techniques in the global south. In M. C. Ogwu, S. C. Izah, J. G. Vazquez-Arenas, S. T. Feleke, & X. Wei (Eds.), *Sustainable bioeconomy development in the global south*. Springer. https://doi.org/10.1007/978-981-96-0305-3_17

Ogwu, M. C., & Izah, S. C. (2025). *Evaluating environmental processes and technologies*. Springer Nature. https://doi.org/10.1007/978-3-031-85327-2

Ogwu, M. C., Izah, S. C., Merritt, H., Wise, R. D., & Jintrawet, A. (2025c). *Sustainable bioeconomy development in the global south: Volume I status and perspectives*. Springer. https://doi.org/10.1007/978-981-96-0640-5

Ogwu, M. C., Izah, S. C., Vazquez-Arenas, J. G., Feleke, S. T., & Wei, X. (2025d). *Sustainable bioeconomy development in the global south: Volume II bioeconomy techniques*. Springer. https://doi.org/10.1007/978-981-96-0305-3

Ogwu, M. C., & Kosoe, E. A. (2025a). Integrating green infrastructure into sustainable agriculture to enhance soil health, biodiversity, and microclimate resilience. *Sustainability, 17*(9), 3838. https://doi.org/10.3390/su17093838

Ogwu, M. C., & Kosoe, E. A. (2025b). Environmental economics: Valuing ecosystem services and green technologies. In M. C. Ogwu & S. Chibueze Izah (Eds.), *Evaluating environmental processes and technologies*. Environmental Science and Engineering Series. Springer. https://doi.org/10.1007/978-3-031-85327-2_15

Ogwu, M. C., Kosoe, E. A., Alaka, A. C., & Xu, J. (2025e). Green infrastructure and environmental health security: The role of nature-based solutions in improving urban health. In S. C. Izah & M. C. Ogwu (Eds.), *Innovative approaches in environmental health management: Processes, technologies, and strategies for a sustainable future*. Environmental science and engineering series. Springer Nature. https://doi.org/10.1007/978-3-031-81966-7_9

Ogwu, M. C., Malikia, C. N., Stansfield, A., Gonzales-Torres, A. D., & Izah, S. C. (2023). Sustainable food processing waste management for environmental protection. In A. L. Srivastav, A. S. Grewal, M. Tiwari, & T. D. Pham (Eds.), *Role of green chemistry in ecosystem restoration to achieve environmental sustainability* (pp. 291–299). Elsevier. https://doi.org/10.1016/B978-0-443-15291-7.00022-5

Ogwu, M. C., Ojo, A. O., & Alaka, A. C. (2025f). Biodiversity and human health: The interconnections of species loss and ecosystem services. In S. C. Izah & M. C. Ogwu (Eds.), *Innovative approaches in environmental health management: Processes, technologies, and strategies for a sustainable future*. Environmental science and engineering series. Springer Nature. https://doi.org/10.1007/978-3-031-81966-7_5

Österblom, H., Bebbington, J., Blasiak, R., Sobkowiak, M., & Folke, C. (2022). Transnational corporations, biosphere stewardship, and sustainable futures. *Annual Review of Environment and Resources, 47*(1), 609–635.

Oxfam. (2023). *Climate equality: A planet for the 99%*. Oxfam International.

Park, C.-Y., Yamamoto, Y. & Doong, M. A. L. 2023. *EU carbon border adjustment mechanism: Economic impact on Asia*, Asian Development Bank. Philippines. Retrieved Aug 25, 2025, from https://coilink.org/20.500.12592/v6wwsdk

Pattberg, P. H., & Widerberg, O. E. (2017). The climate change regime. In *Oxford research encyclopedia of climate science* (pp. 1–35). The Oxford University Press. https://doi.org/10.1093/acrefore/9780190228620.013.46

Principale, S., & Pizzi, S. (2023). The determinants of TCFD reporting: A focus on the Italian context. *Administrative Sciences, 13*(2), 61.

Raupach-Sumiya, J. (2024). *Relevance, potentials and success factors of regional multi-stakeholder cooperation for corporate decarbonization–evidence from German business*. or https://doi.org/10.21203/rs.3.rs-5364626/v1

Rim, H., Kim, J., & Dong, C. (2019). A cross-national comparison of transparency signaling in corporate social responsibility reporting: The United States, South Korea, and China cases. *Corporate Social Responsibility and Environmental Management, 26*(6), 1517–1529.

Ryan, N., Beesemyer, L., Caulliez, S., Waiyaki, J., Nayak, M., Chakrabarty, R., Kim, S. K., & Vladimirova, D. (2023). Introducing a novel framework for regenerative business. In *New business models conference proceedings 2023*. Maastricht University Press. https://doi.org/10.26481/mup.2302.40

Schnyder, G. (2021). Making cg research fit for the 21st century: Time for courage a note. *Corporate Governance Insight, 3*(1), 1.

Schöggl, J. P., Stumpf, L., & Baumgartner, R. J. (2024). The role of interorganizational collaboration and digital technologies in the implementation of circular economy practices—Empirical evidence from manufacturing firms. *Business Strategy and the Environment, 33*(3), 2225–2249.

Smith, I. D., Overland, I., & Szulecki, K. (2024). The EU's CBAM and its 'significant others': Three perspectives on the political fallout from Europe's unilateral climate policy initiative. *JCMS: Journal of Common Market Studies, 62*(2), 603–618.

Sprengel, D. C., & Busch, T. (2011). Stakeholder engagement and environmental strategy–the case of climate change. *Business Strategy and the Environment, 20*(6), 351–364.

Stamelos, C. (2022). Corporate sustainability and ESG factors in Greece and Cyprus: Compliance, laws and business practices, towards a holistic approach. Inter EU Law East. *Journal for the international and european law, economics and market integrations, 9*(2), 289–313.

Stewart, R., & Niero, M. (2018). Circular economy in corporate sustainability strategies: A review of corporate sustainability reports in the fast-moving consumer goods sector. *Business Strategy and the Environment, 27*(7), 1005–1022.

Street, R. B., & Jude, S. (2019). Enhancing the value of adaptation reporting as a driver for action: Lessons from the UK. *Climate Policy, 19*(10), 1340–1350.

Tagliapietra, S., & Wolff, G. B. (2021). Form a climate club: United States, European Union and China. *Nature, 591*(7851), 526–528.

Talbot, D., & Boiral, O. (2015). Strategies for climate change and impression management: A case study among Canada's large industrial emitters. *Journal of Business Ethics, 132*(2), 329–346.

Theeraworawit, M., Suriyankietkaew, S., & Hallinger, P. (2022). Sustainable supply chain management in a circular economy: A bibliometric review. *Sustainability, 14*(15), 9304.

United Nations Office for Disaster Risk Reduction. (2022). *Global assessment report on disaster risk reduction*. Accessed on April 15,2025. or https://digitallibrary.un.org/record/3969846?v=pdf

Urbinati, A., Chiaroni, D., & Chiesa, V. (2017). Towards a new taxonomy of circular economy business models. *Journal of Cleaner Production, 168*, 487–498.

Van den Bergh, J. C., Angelsen, A., Baranzini, A., Botzen, W. W., Carattini, S., Drews, S., Dunlop, T., Galbraith, E., Gsottbauer, E., Howarth, R. B., & Padilla, E. (2020). A dual-track transition to global carbon pricing. *Climate Policy, 20*(9), 1057–1069.

Vasquez, R. (2022). CSR, CSA, or CPA? Examining corporate climate change communication strategies, motives, and effects on consumer outcomes. *Sustainability, 14*(6), 3604.

Wahab, S. N. A. A., Azman, N. H. N., & Ahmad, Z. (2024). Environmental, social, and governance (ESG) and corporate governance in Asean: A threshold effects of corporate strategy. *International Journal of Business and Society, 25*(2), 660–695.

Weitzel, M., Hübler, M., & Peterson, S. (2012). Fair, optimal or detrimental? Environmental vs. strategic use of border carbon adjustment. *Energy Economics, 34*, S198–S207.

Wells, P., & Skeete, J. P. (2023). Circular business models as instruments of corporate power. In *Proceedings of the 8th international conference on new business models*. Maastricht University Press.https://doi.org/10.26481/mup.2302.36

Winter, J. (2023). The challenge of border carbon adjustments as a mechanism for climate clubs. *PLOS Climate, 2*(2), e0000135. https://doi.org/10.1371/journal.pclm.0000135

Wondem, B. A., & Batra, G. S. (2019). The impact of corporate governance practices on corporate financial performance in Ethiopia. *International Journal of Accounting Research, 7*(1), 1–10.
Wright, C., & Nyberg, D. (2015). *Climate change, capitalism, and corporations*. Cambridge University Press.
Wright, C., & Nyberg, D. (2017). An inconvenient truth: How organizations translate climate change into business as usual. *Academy of Management Journal, 60*(5), 1633–1661.
Wright, C., & Nyberg, D. (2024). Corporations and climate change: An overview. *Wiley Interdisciplinary Reviews: Climate Change, 15*(6), e919.
Wu, Y., & Tham, J. (2023). The impact of executive green incentives and top management team characteristics on corporate value in China: The mediating role of environment, social and government performance. *Sustainability, 15*(16), 12518.
Yang, A., Liu, W., & Wang, R. (2020). Cross-sector alliances in the global refugee crisis: An institutional theory approach. *Business Ethics: A European Review, 29*(3), 646–660.
Zaccone, M. C., Santhià, C., & Bosone, M. (2022). How hybrid organizations adopt circular economy models to foster sustainable development. *Sustainability, 14*(5), 2679.
Zhang, J., Wang, M., & Li, M. (2024). Does environmental management system certification affect green innovation performance?-Based on a moderated mediating effects model. *Frontiers in psychology*, 14, 1264207. https://doi.org/10.3389/fpsyg.2023.1264207
Zhang, Q., Lin, Y., Wang, Y., & Cao, Y. (2025). Band together or go it alone? Climate risk and corporate collaborative innovation. *Humanities and Social Sciences Communications, 12*(1), 1–20.

Part IV

Futures and Transformations

Modeling Future Climate and World-System Interactions: Risk, Collapse, and Adaptation

13

13.1 Introduction

The issue of climate change has emerged as one of the most pressing stressors on global systems, impacting not just the environment but also the socio-ecological fabric of human societies. Recent assessments underscore the escalating risks facing the world system. The IPCC (2022) reports that 3.3–3.6 billion people live in highly climate-vulnerable contexts. Economic analyses project that climate change could reduce global GDP by 11%–14% by 2050 under high-emission scenarios (Swiss Re Institute, 2021). Human mobility is also expected to accelerate, with as many as 216 million climate migrants by 2050 across Sub-Saharan Africa, South Asia, and Latin America alone (World Bank, 2021). These figures highlight the urgency of adopting modeling approaches capable of capturing cross-scale risks, cascading failures, and socio-political consequences. As we delve into the intricacies of climate–world-system interactions, it is crucial to explore the co-evolution of Earth systems (particularly climatic and ecological factors) with human systems, including economies, governance, and societal structures. The current literature emphasizes that recognizing the interconnectedness of these systems is essential to addressing the multifaceted challenges posed by climate change. For instance, Steffen et al. (2015) argue for understanding planetary boundaries and their implications for sustainability, as breaches of these thresholds can lead to drastic shifts in Earth systems and human societies.

Moreover, the resilience of human systems in the face of emerging environmental challenges relies heavily on robust modeling frameworks that enable anticipatory governance and proactive adaptation. Such frameworks are fundamental in identifying potential collapse pathways, revealing complex interdependencies, and guiding necessary adaptations across various scales of governance, social organization, and environmental management. Climate change has historically been framed primarily as an environmental challenge; however, it is increasingly recognized as a socio-ecological destabilizer. The impacts of climate change extend

M. C. Ogwu and S. Chibueze Izah, *Climate Frontiers and Social Transformation*, World-Systems Evolution and Global Futures,
https://doi.org/10.1007/978-3-032-17639-4_13

beyond environmental degradation, affecting societal stability, economic viability, and governance frameworks. In this context, authors such as Anderies et al. (2023) emphasize the paradox of climate change: both an instigator of socio-ecological vulnerabilities and a catalyst for the evolution of adaptive responses within societies. For instance, significant financial investments in renewable energy, resource conservation, and sustainable practices have emerged as key strategic adaptations to mitigate climate risks. Research indicates that systemic responses to climate change need to incorporate socio-political dynamics to address inequality exacerbated by environmental degradation and climate-induced migration. For example, Anderies et al. (2023) explore the implications of social inequality on adaptation strategies and emphasize the importance of development and policy frameworks that integrate equity considerations in socio-ecological resilience building. They argue that maintaining resilience in this context requires a nuanced understanding of social dynamics and their interplay with ecological shifts.

Furthermore, understanding the multiple stressors that climate change imposes on human systems is essential. As demonstrated by Bury et al. (2019), a deeper appreciation for socio-climate modeling reveals how individual perceptions, behaviors, and social learning dynamics play crucial roles in shaping collective responses to environmental challenges. These factors are deeply intertwined with broader socio-economic contexts that shape public awareness and engagement with climate issues. To fully comprehend the interactions between climate and human systems, it is imperative to define and frame "climate–world-system interactions" clearly. This concept reflects the dynamic relationships between biophysical components of the Earth, such as climate and ecosystems, and the social, economic, and governance systems shaped by humans. Donges et al. (2018) argue for approaching these interactions through a multidisciplinary lens encompassing planetary-scale systems, emphasizing that climate change requires holistic analytical frameworks that incorporate both natural and social factors. As highlighted by Little et al. (2023), the evolution of Earth systems into what is now termed the "Anthropocene" has been marked by significant feedback loops between human actions and planetary conditions. The recognition of these interdependencies points toward the need for integrated models that account for both societal and ecological dimensions as co-evolving processes. For example, integrating human behavioral traits, economic imperatives, and ecological insights can illuminate pathways through which climate change impacts trigger socio-economic destabilization.

Moreover, modeling these interactions is critical for stress-testing responses to climate challenges. By exploring scenarios of resilience and adaptability, methodologies informed by systems science—such as those proposed by Calvin and Bond-Lamberty (2018)—form the backbone of advancing our understanding of the complexities inherent in climate–world-system interactions. The rationale for developing robust modeling frameworks extends far beyond mere predictive capacity; it encompasses the essential need to prepare societies to encounter and respond to climate-induced risks. Modeling allows researchers and policymakers to visualize potential collapse pathways and the tipping points beyond which sustaining human and ecological systems may become increasingly complex. Meehl et al. (2020) explore how advanced climate models enable scientists to iterate scenarios of climate sensitivity and systemic responses, providing vital insights into future trajectories under various emission scenarios.

The emergence of computational approaches in Earth system science offers substantial advantages in anticipating climate impacts through rigorous simulations of socio-ecological dynamics. As demonstrated by Müller et al. (2021), such anticipatory frameworks not only enhance our understanding of the potential cascading effects of climate change on human systems but also foster necessary dialogues regarding adaptive governance structures. Effective modeling can guide policymakers toward preemptive actions that mitigate the most catastrophic consequences of climate change while emphasizing flexible governance mechanisms that can evolve with emerging challenges. Additionally, integrating the social sciences into climate modeling is paramount for creating a comprehensive picture of societal responses to climate risk. For instance, studies of risk perception, such as those conducted by Lee et al. (2015), confirm the importance of understanding public awareness and collective risks in shaping adaptive behaviors and policy support. Improved comprehension of these dynamics enables tailored interventions, potentially increasing community resilience in the face of natural disasters exacerbated by climate change. Furthermore, recognizing the interactive roles that various stakeholders play amid changing climatic conditions is crucial. The intersection of climate interventions with community dynamics—analyzed by scholars like Heinze et al. (2024)—provides valuable perspectives on how "tipping points" in one domain of the Earth system can cascade into others, thereby altering human-economic interactions. This understanding can act as a catalyst for broader, interconnected strategies designed to uphold the integrity of both environmental and socio-economic systems.

The focus of this chapter is to examine how future climate change interacts with the global world system, generating risks that may lead to systemic stress, partial or total collapse, and the need for adaptation. By integrating perspectives from Earth system science, world-systems theory, and complex systems modeling, the chapter highlights how climate shocks propagate through economic, social, ecological, and geopolitical networks. The benefit of this chapter lies in providing researchers, policymakers, and practitioners with a structured understanding of modeling approaches that can anticipate cascading risks and identify strategies for building resilience and adaptation. In doing so, it bridges theoretical insights with practical tools, offering guidance on preparing for uncertain futures in a climate-constrained world.

13.2 Theoretical and Conceptual Foundations

13.2.1 World-Systems Theory

World-systems theory provides an essential framework for understanding the systemic interactions between societal structures and environmental changes on a global scale. Central to this theory is the categorization of countries into core, semi-periphery, and periphery regions, which influences their vulnerability to climate shocks. Core nations often dominate global wealth and technological advancements, enabling them to mitigate climate impacts more effectively than semi-peripheral and peripheral nations, where limited resources exacerbate fragility in the face of climate variability (Dalby, 2025; Iyiola et al. 2023, 2025a, 2025b; Steffen et al., 2018). The discourse surrounding hierarchical interdependencies reveals that peripheral nations bear the brunt of climate-induced disruptions while being the least equipped to adapt or recover. This phenomenon is evident in regions such as East Africa, where climate change disproportionately affects agricultural outputs critical for survival due to historic economic inequities and lack of adaptive infrastructure (Gebrechorkos et al., 2023a, 2023b). The interplay of socio-economic status and environmental stressors indicates a widening gap in vulnerability based on national categorization, suggesting that climate resilience strategies should account not only for environmental determinants but also for socio-political dynamics underlying these disparities (Gebrechorkos et al., 2023a, 2023b; Steel et al., 2022).

Moreover, the historical record reveals the cyclical nature of climate and societal outcomes, underscoring the need to examine past civilizations for lessons on vulnerability (Samnuzulsari et al., 2023). Civilizations such as the Maya experienced systemic decline amid climate fluctuations, as evidenced by archeological studies associating periods of drought with socio-political unrest and eventual collapse (Kennett et al., 2022; Vora & Singh, 2022). Understanding these patterns through a world-systems lens can inform contemporary assessments of resilience and adaptation strategies globally.

13.2.2 Earth System Science

Earth system science advances our understanding of the intricate interdependencies among climate variables, ecosystems, and human activities, emphasizing feedback mechanisms that significantly shape global systems. It posits that biosphere-climate interactions play a crucial role in shaping environmental outcomes, and that disturbances in one domain can have cascading effects throughout the system (Steffen et al., 2018; Yang et al., 2023). Research indicates that self-reinforcing feedback loops, such as the release of carbon from thawing permafrost, can exacerbate climate change beyond human control (Steffen et al., 2018). This complexity necessitates comprehensive modeling approaches that incorporate biological, physical, and chemical processes to project future climate scenarios accurately (Ogwu & Izah, 2025).

The Coupled Model Intercomparison Project Phase 6 (CMIP6) exemplifies such approaches, providing frameworks to assess diverse climatic conditions and their implications on global ecosystems (Schewe et al., 2019; Yin et al., 2016). By utilizing high-resolution climate data derived from extensive simulations, researchers can identify nuanced patterns of climate extremes and subsequent ecological impacts (Eyring et al., 2016; Vora & Singh, 2022). These advanced modeling frameworks not only help in illustrating potential future states of the Earth system but also highlight vulnerabilities inherent to various biogeographical regions.

Additionally, the interplay between climate systems and human socio-economic factors should be continually scrutinized. Assessing these interactions elucidates how societal adaptations (or lack thereof) can change landscape resilience. Areas particularly vulnerable to climatic extremes, such as coastal regions impacted by sea-level rise, further illustrate the precarious balance between natural systems and human exploitation, necessitating policies that foster sustainable development and ecological conservation (Gebrechorkos et al., 2023a, 2023b).

13.2.3 Complex Systems Thinking

Complex systems thinking enriches our understanding of climate dynamics by emphasizing nonlinear relationships, emergent risks, and system thresholds. Adaptation to climate change is often hampered by this inherent complexity, where diverse ecological and social factors interact unpredictably, leading to unforeseen consequences (Fitzpatrick & Dunn, 2019; Yang et al., 2023). Climate-related crises in agriculture, infrastructure, and public health reveal the fragility of systems thought to be stable under historical conditions. For instance, the increased frequency of extreme weather events such as droughts under future climate scenarios necessitates robust integration of systems thinking into resilience planning (Mazarrasa et al., 2022; Yang et al., 2023).

An essential aspect of understanding complex systems is recognizing that thresholds may exist where systems could drastically shift to new equilibria under stress, such as the transition to a "Hothouse Earth" condition, where warming continues despite emission reductions (Steffen et al., 2018). This perspective aligns with historical analyses that demonstrate how past civilizations have surpassed critical thresholds, leading to societal upheaval and resource scarcity (Kennett et al., 2022; Vora & Singh, 2022). Thus, the integration of complex systems thinking can enhance predictive capabilities and preparedness efforts for future environmental shocks.

Furthermore, recognizing cascading effects (where initial impacts trigger a series of secondary consequences) underscores the importance of comprehensive risk assessment in climate action. Events such as widespread crop failures or community displacements due to climate extremes exemplify how single stressors can unravel complex socio-ecological networks, ultimately leading to severe socio-economic implications (Shin et al., 2022). Stakeholders should thus engage in interdisciplinary dialogues that incorporate ecological, economic, and social dimensions to formulate holistic climate adaptation strategies.

13.2.4 Historical Analogues: Lessons from Past Civilizations

The analysis of historical civilizations facing climate stress offers critical insights into modern vulnerability and resilience frameworks. Evidence from archeological studies indicates that ancient societies such as the Maya experienced significant societal transformations aligned with climate fluctuations, often leading to state collapse under prolonged environmental stress (Kennett et al., 2022; Samnuzulsari et al., 2023; Vora & Singh, 2022). These historical analogues serve as cautionary frameworks highlighting the potential consequences of failing to adapt to or mitigate climate risks.

For instance, the collapse of the Maya civilization provides compelling evidence of the intricate connections between climate variability, resource management, and social stability. Studies reveal that significant drought events were closely associated with agricultural failures and subsequent societal fragmentation (Vora & Singh, 2022). Contemporary societies can learn from these dynamics, particularly in how resource scarcity and governance failures can compound the effects of climate change on vulnerable populations.

Moreover, the historical perspective underscores the role of cultural resilience and adaptive strategies in shaping civilizational stability amidst environmental changes. The varying responses of ancient societies to climate stress (ranging from migratory patterns to agricultural innovations) demonstrate that adaptive capacity is crucial in navigating climate hazards (Dalby, 2025; Xu et al., 2019). Implementing localized adaptations informed by historical experiences can significantly enhance contemporary resilience strategies to climate change, especially in areas with similar socio-ecological contexts. The significance of comprehensive historical analysis is further underscored by the urgent need to address current climate vulnerabilities. The ongoing examination of past climate-related transitions can illuminate the potential pathways modern civilizations might take under similar stressors, guiding policymakers and stakeholders in preemptive action to avert crises (Dalby, 2025; Tahir & Nuri, 2024). Ultimately, a historical lens is indispensable in recognizing the intricate connections between climate change and human societal evolution, driving the need for proactive rather than reactive measures.

13.3 Modeling Approaches for Climate–World-System Dynamics

Despite rapid advancements in climate and world-system modeling, empirical studies show that researchers often rely heavily on a single modeling tradition, even when system complexity demands multiple perspectives. A meta-analysis of 1242 climate–risk modeling studies found that 62% used only one modeling framework, while fewer than 14% incorporated two or more complementary models (Beckage et al., 2018). This over-reliance limits the ability to capture nonlinear feedbacks, cross-scale interactions, and emergent behavior that characterize climate–world-system dynamics. Modelers can easily become addicted to a single model. Increasingly, scholars argue that model pluralism, or the intentional integration of multiple frameworks, improves robustness and reduces uncertainty in climate futures assessment (Raihani & Aplin, 2022). Therefore, this chapter foregrounds the value of combining system dynamics, agent-based models, network models, and integrated assessment models as a strategic approach for capturing interconnected risk pathways.

13.3.1 Integrated Assessment Models

Integrated Assessment Models (IAMs) serve as a crucial interface for examining the multifaceted relationships and interdependencies between climate dynamics, economic trends, and societal responses. They synthesize knowledge from various disciplines, enabling researchers and policymakers to assess the implications of climate change across diverse sectors. IAMs often incorporate feedback loops, capturing the cascading effects of climate impacts on economic output, resource use, and social well-being. Advances in modeling aerosol processes have refined the capacity of IAMs to simulate interactions with the Atlantic Meridional Overturning Circulation (AMOC) and other oceanic currents, highlighting the complex relationships between anthropogenic emissions, climate systems, and economic repercussions (Menary et al., 2020; Mulcahy et al., 2018). Table 13.1 shows overview of modeling approaches for climate–world-system dynamics.

Research indicates that IAMs can gain improved accuracy by integrating new climate dynamics, such as updated aerosol-radiation interactions that were not adequately represented in earlier models. The efficacy of IAMs in predicting social responses, such as migration patterns influenced by climate-induced scarcity, hinges on the accuracy of climate projections utilized within these models. Empirical evidence suggests that deeper integration of climate projections into economic decision-making can yield resilience strategies that mitigate risks associated with abrupt climate changes (Bettis et al., 2017; van Westen et al., 2025). However, the inherent uncertainty in projecting future climate impacts poses significant challenges for IAMs. Various models have reported differing outcomes based on assumptions about human behavior, technological advancements, and ecological feedback mechanisms. Strengthening the links between IAMs and real-world data can significantly enhance the predictive validity of these models (Steel et al., 2022, 2024). The findings underscore the essential nature of ongoing refinement and validation efforts in IAMs for effective policymaking in the context of climate change.

Why Model Integration Matters

- IAMs underestimate risks from climate extremes by up to 40%, compared with Earth system models (Schewe et al., 2019).
- Network models show that a shock affecting only 10% of nodes in a global supply chain can depress global GDP by 3%–6% (Inoue & Todo, 2019).

Table 13.1 Modeling approaches for climate–world-system dynamics

Modeling approach	Focus area	Strengths	Limitations	Relevance to climate–world-system interactions
Integrated assessment models	Climate–economy–policy linkages	Widely used, policy relevance (IPCC)	Oversimplification of social complexity	Projects climate–economic risks and policy outcomes
System dynamics models (e.g., World3)	Long-term feedback and limits	Captures nonlinear dynamics	Static assumptions, outdated data	Useful for modeling collapse pathways
Agent-based models	Micro-level actors (households, firms, migrants)	Captures heterogeneity, emergent behavior	Data-intensive, computationally heavy	Simulates adaptation behaviors under stress
Network & complexity models	Interconnected systems (trade, finance, supply chains)	Highlights cascading failures	Requires complex datasets	Helpful in studying systemic collapse risks
AI & machine learning models	Pattern recognition in large datasets	Adaptive, predictive capacity	Lack of transparency ("black box")	Enhances forecasting of climate–social interactions

- Agent-based models reveal that climate-induced migration can rise by 200% by 2050 under high-emission scenarios (Kumari et al., 2021).
- System dynamics models predict that failing to integrate economic–ecological feedbacks can lead to 50%–70% underestimation of collapse likelihood (Motesharrei et al., 2014).

13.3.2 System Dynamics Models

System dynamics models (Table 13.1), particularly the influential World3 model employed in the "Limits to Growth," provide a foundational framework for understanding the potential risks of societal collapse in the face of environmental stressors. These models simulate the interplay between population growth, resource depletion, pollution, and industrial output, revealing feedback loops that can lead to unsustainable progress and eventual societal downturn (Steel et al., 2022, 2024). The fundamental lesson derived from such models is the critical importance of recognizing limits to growth imposed by ecological constraints. Recent studies have revisited the model's insights in light of current climate science, demonstrating parallels between predicted collapse scenarios and contemporary evidence, such as rising temperatures and diminishing natural resources (Steel et al., 2022). The hysteresis effect observed in AMOC dynamics exemplifies how systems can abruptly shift from one state to another when certain thresholds are crossed—observations that influence contemporary discussions about climate resilience and adaptive capacity in human societies (Ditlevsen & Ditlevsen, 2023; Liu et al., 2017).

Moreover, incorporating new climate data and system parameters into dynamic models can help elucidate potential scenarios under which civilization may become vulnerable to abrupt changes. Understanding these patterns aids in forecasting cascading failures across interconnected systems, such as food security, water resources, and public health initiatives amid climate-induced stresses. By grasping the cyclical patterns revealed in these models, researchers and policymakers can initiate preemptive measures which could mitigate the adverse effects of impending systemic collapse (Pérez-Quezada et al., 2023; Steel et al., 2022, 2024).

13.3.3 Agent-Based Models

Agent-Based Models (ABMs) have emerged as a vital tool for exploring micro-level responses to climate change, particularly concerning adaptation strategies and migration patterns (Table 13.1). These models simulate the behavior of individual agents (human beings, households, etc.) who can make decisions based on environmental cues, socio-economic factors, and personal experiences of climate impacts (Carstens & Wing, 2022; Pérez-Quezada et al., 2023). Within the context of climate stress, ABMs provide insights into how localized communities adapt, revealing a spectrum of responses ranging from immediate behavioral changes to long-term migration in search of better environmental conditions. Research using

ABMs has shown that adaptive behavior is heavily influenced by social networks and individual perceptions of risk (Noguerales et al., 2024; Rooney-Varga et al., 2021). Communities informed by climate change simulation exercises have shown increased consensus and proactive adaptation strategies. Such findings underscore the importance of integrating community perspectives into ABMs to forecast how climate change might impact not only physical landscapes but also social structures and cultural identities (Flood et al., 2018; Pérez-Quezada et al., 2023). By modeling society at the individual agent level, researchers can better understand how varying socio-economic statuses can lead to disparate adaptation capabilities and, subsequently, different migration trends.

Moreover, ABMs can be particularly useful in understanding emergent phenomena, allowing researchers to explore complex interactions between climate impacts and human behavior at unprecedented granularity. Studies have showcased how an increase in extreme weather events may lead to heightened outmigration from affected areas, contributing to demographic shifts and socio-economic challenges in regions receiving migrants (Rooney-Varga et al., 2021; van Westen & Dijkstra, 2023). These models thus serve as an essential component in the toolkit for climate change analysis, equipping policymakers with better foresight regarding migration policies and social protections for vulnerable populations.

13.3.4 Network and Complexity Models

Network and complexity models have significantly expanded our ability to investigate the interdependencies among socio-economic systems affected by climate change (Table 13.1). These models highlight how international trade, financial markets, and resource flows are intertwined with ecological systems, demonstrating that disruptions in one area can have cascading effects worldwide. The complexities surrounding trade in food and energy resources underscore the vulnerability of nations to climate shocks, which can lead to significant economic disruptions (Bettis et al., 2017; Noguerales et al., 2024). Recent developments in network analysis approaches illustrate how climate events can impact global supply chains, potentially disrupting food security—a direct link to how climate change can lead to social instability (Rooney-Varga et al., 2021; Steel et al., 2024). Such insights have been critical for analyzing systemic risks, in which localized failures can propagate through global networks, causing widespread crises. Furthermore, the use of complexity models can elucidate how financial systems adapt (or fail to adapt) to climate risks, offering critical decision-making inputs for policymakers navigating a sustainable economic transition (Noguerales et al., 2024; Steel et al., 2022).

The incorporation of feedback mechanisms and nonlinear interactions into these models allows for a more nuanced understanding of climate risks and adaptation pathways. Studies have shown that introducing climate risk assessments into financial planning can lead to transformational adaptations in investment strategies, significantly reducing vulnerability to fossil fuel dependence and enhancing resilience against climate impacts. However, the ongoing challenge remains effectively quantifying the complex relationships and feedback loops inherent in these systems to derive actionable insights (Bettis et al., 2017; Lima et al., 2023; Pérez-Quezada et al., 2023).

13.3.5 AI and Machine Learning

Artificial intelligence (AI) and machine learning are revolutionizing predictive models in climate-world-system analysis by facilitating the processing of vast datasets involving climate variables, socio-economic information, and historical trends (Table 13.1). These advanced computational techniques allow for the identification of patterns and relationships within data that traditional models may overlook. The potential for real-time analysis and forecasting can enhance responses to climate dynamics, especially in understanding complex, interrelated systems and emergent behaviors. Machine learning applications have been particularly fruitful in climate impact assessments, yielding enhanced predictive accuracy for phenomena such as extreme weather events and their socio-economic implications (Bettis et al., 2017; Noguerales et al., 2024). AI algorithms can be harnessed to optimize agricultural productivity in response to changing climate conditions, identifying the most effective adaptation strategies within specific regional contexts. This helps ensure food security amidst variability, which remains vital in an era of climate uncertainty (Carstens & Wing, 2022; Pérez-Quezada et al., 2023).

Nonetheless, the implementation of AI and machine learning in climate modeling is not without limitations. Reliance on large datasets can introduce biases, and models' opacity often undermines users' ability to understand how conclusions are reached. Furthermore, the risk of generating misleading forecasts due to overfitting or underestimating uncertainties is a significant challenge that necessitates scrutiny of model outputs and assumptions. Therefore, fostering collaboration between AI researchers and climate scientists is vital for addressing these challenges and leveraging AI's full potential for climate-world-system analysis (Noguerales et al., 2024; Pérez-Quezada et al., 2023).

13.3.6 Model Addiction and Confirmation Bias in Climate Modeling

A growing body of research shows that climate modelers often remain confined to a single modeling paradigm—whether IAMs, system dynamics, GCMs, or agent-based models—despite the complexity of climate–world-system interactions. Surveys indicate that over 70% of modeling teams specialize in only one model type, and fewer than 12% routinely integrate multiple approaches (Edmonds et al., 2020; Lamontagne et al., 2018). Institutional incentives, training patterns, and disciplinary silos reinforce this tendency.

This narrow focus can lead to confirmation bias, in which researchers select models that align with their assumptions. A meta-analysis of 1242 climate risk studies found that 62% used a single model type even when multi-model frameworks would have reduced uncertainty (Beckage et al., 2018). Economic–climate studies relying solely on IAMs were shown to underestimate climate damages by 20%–40%, compared to analyses incorporating networks or agent-based models (Pindyck, 2017; Raihani & Aplin, 2022). Such methodological monocultures create blind spots—particularly for cascading migration, food-system instability, governance fragility, and social tipping points. To address this, scholars emphasize model pluralism, the intentional combination of multiple modeling traditions to capture nonlinearity, emergent risks, and cross-scale feedbacks (Beckage et al., 2018; Raihani & Aplin, 2022). Interdisciplinary training and hybrid frameworks are increasingly recognized as essential for overcoming "model addiction" and improving the robustness and policy relevance of climate–world-system assessments.

13.3.7 Strengths, Limitations, and Uncertainties in Each Approach

Across the discussed modeling approaches, each methodology presents distinct strengths and limitations, contributing uniquely to our understanding of climate-world-system interactions. IAMs excel at synthesizing interdisciplinary knowledge but often struggle to project future scenarios accurately due to the complexity of human responses to climate change (Table 13.2; Bettis et al., 2017; Steel et al., 2022). System dynamics models adeptly reveal feedback loops and potential tipping points but can oversimplify socio-political dimensions, which are crucial for understanding societal collapses (Steel et al., 2024; van Westen & Dijkstra, 2023). Agent-Based Models (ABMs) excel at elucidating micro-level adaptations and migration decisions, though their configurational complexity may complicate generalizing results across contexts (Carstens & Wing, 2022; Pérez-Quezada et al., 2023). Complexity models successfully portray interdependencies among global systems but struggle to quantify the nonlinear relationships inherent in climate interactions, leading to ambiguity in model conclusions (Lima et al., 2023; Noguerales et al., 2024). AI and machine learning offer unprecedented capabilities in analyzing massive datasets but hinge on the quality of input data and face scrutiny concerning the transparency and interpretability of their outputs (Carstens & Wing, 2022; Noguerales et al., 2024). Effectively navigating these strengths, limitations, and uncertainties is paramount for researchers and policymakers aiming to devise effective responses to the multifaceted challenges posed by climate change.

Table 13.2 Cross-model comparison for climate–world-system analysis

Model type	Best use cases	Worst use cases	Data requirements	Computational demands	Policy relevance
Integrated assessment models	Long-term economy–policy interactions; carbon pricing; mitigation pathways	Short-term behavioral responses; local-scale conflict or migration	Medium–high (economic, energy, emissions, climate projections)	Moderate	High (IPCC central models)
General circulation models	Physical climate dynamics; temperature/ precipitation projections; extreme events	Social dynamics, governance, migration behavior	Very high (global atmospheric & oceanic datasets)	Very high	High for physical risk assessment
System dynamics models	Feedback loops; overshoot-collapse dynamics; resource depletion; long-run transitions	Micro-scale decisions; heterogeneous agent behavior	Low–medium	Low–moderate	Medium–high (narrative scenarios)
Agent-based models	Migration; household adaptation; conflict; collective action; risk perception	Long-term global projections; macroeconomic outcomes	Medium (micro-level behavioral data)	Moderate–high	Medium (behavior-sensitive)
Network & complexity models	Systemic risk; cascading failures; food/energy/financial networks	Isolated systems; uniform populations	High (trade, supply chains, financial flows, mobility networks)	Moderate–high	High for fragility & resilience planning
AI/machine learning models	Pattern detection; early warning for extremes; real-time forecasts	Theory-building; transparency-required decisions	Very high (big data, historical & sensor datasets)	Very high	Medium (concerns about interpretability)

13.3.8 Choosing the Right Model for the Right Question

Selecting the appropriate modeling framework requires aligning the model with the specific research question, since no single paradigm can capture the full complexity of climate–world-system dynamics (Table 13.2). Model choice should reflect the scale of analysis, the structure of feedback, the degree of uncertainty, and the importance of human behavior. IAMs are best suited for evaluating economy–policy linkages, long-term emissions trajectories, and mitigation trade-offs, making them central to macro-level climate–economic assessments (Steel et al., 2024). GCMs are essential when physical climate dynamics—such as temperature change, precipitation variability, or ocean–atmosphere interactions—are the primary focus (Menary et al., 2020). ABMs are most appropriate when heterogeneity, local decision-making, and emergent behavior shape outcomes, such as migration responses or household adaptation strategies (Pérez-Quezada et al., 2023). Network and complexity models are preferred for analyzing cascading failures—such as disruptions in food systems, global supply chains, or financial networks—because they capture interdependencies that linear models overlook (Inoue & Todo, 2019). SDMs excel when the research question centers on long-term feedback loops, tipping dynamics, and global nonlinearity—making them especially useful for examining collapse pathways or limits-to-growth scenarios (Steel et al., 2022). In essence, IAMs answer economy–policy questions; GCMs address physical climate processes; ABMs reveal micro-behavioral adaptations; network models detect interdependent risks; and SDMs capture global feedback complexity. Recent scholarship argues that hybrid or pluralistic modeling frameworks provide the most robust insights by reducing structural blind spots and integrating multiple perspectives (Beckage et al., 2018; Raihani & Aplin, 2022).

13.3.9 Communicating Model Results Responsibly

Communicating complex model outcomes requires balancing urgency with accuracy. Overstating risks can generate alarmism that undermines public trust, while minimizing threats may produce complacency and maladaptation. Because different modeling approaches yield different levels of uncertainty, transparency about assumptions, limitations, and confidence intervals is essential. Poor communication—whether through oversimplified graphics, misleading certainty, or lack of context—can distort risk perception, influence political decision-making, and produce maladaptive policy responses. Effective communication, therefore, forms a crucial bridge between modeling and real-world governance.

Five Principles for Next-Generation Climate–World-System Modeling

1. Multimodel integration—Combine IAMs, SDMs, ABMs, networks, and GCMs to capture cross-scale feedbacks.
2. Behavioral realism—Incorporate social learning, adaptation strategies, risk perception, and heterogeneity.

3. Socio-political equity—Embed justice, representation, and global power asymmetries in model structure.
4. Early-warning thresholds—Detect tipping points, nonlinearities, and abrupt transitions using integrated signals.
5. Participatory modeling—Engage policymakers, communities, and vulnerable populations in scenario co-design.

13.4 Climate–World-System Interactions: Risk Pathways

The dynamic relationship between climate change and various socio-economic systems highlights the complexity and interdependence inherent in global interactions. Climate change serves as a "risk multiplier," exacerbating vulnerabilities in economic, social, ecological, and geopolitical spheres (Table 13.3). This phenomenon has been well documented, illustrating how interconnected systems amplify risks across domains, compelling nations and communities to reassess their vulnerabilities and preparedness for climate impacts. For instance, rising global temperatures, altered precipitation patterns, and increased frequency of extreme weather events significantly disrupt agricultural productivity, leading to food insecurity and heightened economic risks for communities reliant on consistent crop yields (Perez et al., 2015; Pérez-Quezada et al., 2023). Economic risks manifest in numerous ways, often as trade disruptions and resource scarcity resulting from climate-induced shocks. These trade disruptions can be exacerbated by failed harvests due to severe droughts or floods, leading to skyrocketing commodity prices and exacerbating economic inequality (Falkner, 2016; Guzman et al., 2020). There exists a robust correlation between climate vulnerability and economic disadvantage, as poorer communities lack the capacity to adapt effectively to these challenges, further entrenching cyclical poverty and inequality (Deléglise et al., 2022). In southern Africa, for example, numerous studies indicate a direct relationship between climatic stress and heightened socio-economic fragility, illustrating how climate change magnifies existing economic disparities (Knutsen & NyquistPedersen, 2024).

Social risks linked to climate change comprise forced migration, public health crises, and the potential for inequality-driven unrest. As environmental conditions degrade, populations are often forced to migrate, creating social tensions in both sending and receiving regions due to competition for resources (Rooney-Varga et al., 2021). Moreover, the resultant climate-related health issues—such as the rise of vector-borne diseases like dengue—are becoming increasingly evident, putting additional pressure on public health systems that are often already under strain (Ide, 2023; Ivanescu et al., 2023). Public health crises can also stem from malnutrition, respiratory diseases exacerbated by pollution, and heat-related illnesses, all of which are linked to climate variations (Atanasiu, 2023). These demographic shifts not only challenge existing social structures but can also trigger wide-scale civil unrest. In regions where governance is weak or where social inequities are prevalent, climate-induced migration and resource competition have been linked to

Table 13.3 Climate–world-system risk pathways

Risk domain	Climate drivers	Systemic impacts	Feedback effects
Economic	Droughts, storms, sea-level rise	Reduced productivity, disrupted trade, and inflation	Economic inequality intensifies social instability
Social	Extreme heat, disasters	Health crises, migration, unrest	Weak institutions worsen vulnerabilities
Ecological	Ocean acidification, habitat loss	Biodiversity decline, food-system stress	Loss of ecosystem services undermines human security
Geopolitical	Resource scarcity, transboundary risks	Territorial disputes, conflict, and governance breakdown	Conflict further undermines cooperation on climate action

an upsurge in conflict (Gonçalves et al., 2024; Knutsen & NyquistPedersen, 2024). Research indicates that nations experiencing environmental degradation are often more prone to instability, creating a feedback loop where climate change exacerbates governance failures, leading to conflict escalation (Falkner, 2016; Skains et al., 2022).

13.4.1 Interdependence of Risks → Cascading Failures in the Global System

Environmental interdependence and the cascading nature of risks emphasize the severity and urgency of addressing climate change comprehensively (Iyiola et al., 2024; Izah et al., 2023). Extremes in weather patterns do not solely impact individual regions but can lead to global supply-chain disruptions, effectively linking localized climate events to broader economic ramifications (Butler et al., 2016; Klose et al., 2020). The concept of "tipping points" in climate systems illustrates the precarious balance of ecosystems, in which specific thresholds can trigger dramatic shifts across ecological and human systems, leading to cascading failures (Jiranek et al., 2023).

When ecosystems, such as coral reefs or rainforests, reach these thresholds, the ramifications are profound, impacting biodiversity and the ecosystem services they provide, such as carbon sequestration and water purification (Bony et al., 2015). Biodiversity collapse can, in turn, affect agricultural systems and food security, leading to increased economic instability and further complicating humanitarian responses (Cherp et al., 2018; Mbah et al., 2021). Such interconnections highlight the necessity for an integrated approach to climate governance, recognizing the shared vulnerabilities across socio-ecological systems and prioritizing risk assessments that reflect these complexities (Deléglise et al., 2022; Flood et al., 2018).

Critical evaluations reflect a growing recognition that mitigating climate risks involves a multifaceted strategy that incorporates techno-economic, socio-technical, and political perspectives (Sibiya et al., 2022). Moving toward sustainable solutions requires a concerted effort at all levels, urging stakeholders across sectors to cooperate and prioritize climate change mitigation (Guzman et al., 2020). Furthermore, participatory governance that empowers marginalized groups is essential to fostering inclusive climate action and ensuring that all voices are considered in shaping resilient communities.

13.5 Pathways to Systemic Risk and Collapse

13.5.1 Defining "Collapse" in Climate–World-System Terms

The definition of "collapse" within the climate–world system can take two forms: gradual decay and abrupt breakdown. Gradual decay typically refers to the slow degradation of ecological and socio-political systems, resulting from compounding environmental stresses such as climate change, resource depletion, and biodiversity loss. Conversely, abrupt breakdown characterizes a sudden, irreversible shift in system stability, often precipitated by crossing critical thresholds, also known as tipping points. These definitions are crucial in understanding the various trajectories of climate impacts and societal adaptations. Recent research emphasizes that, under anthropogenic pressures, systems may transition from stability to chaos without clear warning signs, making them particularly vulnerable to collapse (Kemp et al., 2022; Steffen et al., 2018)

The notion of tipping points is central to discussions of collapse. For example, Steffen et al. illustrate that self-reinforcing feedback loops, particularly as they relate to climate change, can usher the Earth System into a Hothouse state beyond current stabilization efforts (Steffen et al., 2018). This framework elucidates where thresholds lie and the consequences of crossing them—events that could lead to significant and possibly unmanageable shifts in climate stability. Kemp et al. further reinforce this by highlighting the differential impacts of climate change, which may lead to catastrophic outcomes if early warning systems fail, particularly for vulnerable ecosystems and societies (Kemp et al., 2022). The interaction of these concepts underscores the profound significance of monitoring and managing thresholds to avert irreversible breakdowns in ecological integrity and social systems.

13.5.2 Potential Trajectories: Resource Exhaustion and Food Insecurity

Resource exhaustion and food insecurity are critical dimensions of potential collapse trajectories exacerbated by climate change. As climatic conditions shift, particularly with rising temperatures and altered precipitation patterns, agricultural systems face unprecedented challenges. Historical precedents demonstrate that failure in agricultural systems has precipitated the collapse of civilizations, such as the Maya and the Roman Empire (Durach et al., 2025). With current trajectories indicating warming exceeding 1.5 °C, many regions could experience drastic reductions in crop yields, elevating the risk of famine and communal strife (Armstrong McKay et al., 2022). The interdependencies within the global food supply chain mean that resource exhaustion in one region can resonate across the globe, highlighting the interconnectedness of food systems and climate resilience.

Recent research indicates that extreme weather phenomena—intensified by climate change—pose immediate threats to food security, particularly in susceptible regions (Lenton et al., 2019). For instance, droughts, floods, and heatwaves can devastate local crops and disrupt food supply networks. The nexus between climatic variability, agricultural productivity, and socio-economic stability thus constitutes a multifaceted risk landscape. Furthermore, new models predict that as ecosystems shift or collapse, biodiversity loss can further impair agricultural resilience and food security (Riddell et al., 2019). This trajectory paves the way for geopolitical tensions and governance issues as states grapple with resource scarcity, demanding urgent attention to adaptive strategies within agricultural and environmental policy frameworks.

13.5.3 Political Destabilization and Governance Fragility

The linkage between climate impacts and political destabilization is increasingly evident as resource scarcity heightens competition, leading to governance fragility. Historical assessments have highlighted how eco-political systems become vulnerable under environmental stressors, often igniting conflicts over resources (Durach et al., 2025). As climate change exacerbates environmental degradation, the likelihood of state fragility increases, particularly in regions already beset by socio-economic injustices. Systems experiencing incremental environmental challenges may face sudden political unrest, thereby redefining national security discussions in the context of climate change.

Research emphasizes that governance structures should evolve to address the dual challenges of environmental degradation and socio-political instability. Insights drawn from geohistorical analogs, such as the Pliocene epoch, provide crucial data on how past climates influenced societal resilience (Burke et al., 2018; Iyiola & Ogwu, 2025). A proactive governance approach that integrates climate science and anticipatory risk management may mitigate trajectories leading toward collapse. Moreover, Cooper et al. (2020) elucidate how challenging ecosystems may shift disproportionately, necessitating governance that is adaptable and resilient to socio-environmental shifts. This presents a fundamental challenge: political adaptability should parallel ecological and climatic transformations to avert systemic collapse.

13.5.4 Climate Tipping Points

Climate tipping points are critical thresholds beyond which significant, often irreversible changes occur across various Earth systems. The melting of polar ice, dieback in the Amazon rainforest, and slowdowns in ocean circulation (especially the Atlantic Meridional Overturning Circulation, AMOC) exemplify powerful tipping elements whose behaviors can dramatically alter global climates (Ben-Yami et al., 2024; Rahmstorf, 2024). Recent studies indicate that these events can trigger cascading effects in weather patterns, ocean currents, and ultimately, global temperatures, leading to what researchers term "Hothouse Earth" scenarios (Steffen et al., 2018). Each tipping point is interlinked with others, suggesting that the failure of one can trigger a domino effect. Critical thresholds around polar ice caps and key ecosystems, such as the Amazon basin, have been evaluated to determine when these tipping points might be triggered. Armstrong McKay et al. (2022) argue that crossing crucial global warming thresholds (such as the 1.5 °C mark) could trigger multiple tipping points simultaneously, leading to catastrophic shifts in weather and climate patterns. Such scenarios indicate that monitoring these focal tipping elements is vital for forecasting potential trajectories of the climate system. Consequently, the interconnected nature of these tipping points necessitates a holistic understanding of the climate system to inform policies aimed at reducing emissions and safeguarding critical ecosystems before thresholds are irrevocably crossed.

13.5.5 Cascading Collapse Scenarios

The concept of cascading collapse sheds light on the interdependencies among ecological and human systems, illustrating how failure in one area can propagate through multiple systems, exacerbating risks of widespread societal and environmental breakdown. Research on the AMOC highlights how its potential collapse could disrupt weather patterns not only in the North Atlantic but globally, affecting agricultural systems, water supplies, and regional governance structures (Ditlevsen & Ditlevsen, 2023; Liu et al., 2017). In these scenarios, ecological tipping points can amplify vulnerabilities in human systems, creating a complex, interlinked fabric of risks.

Cascading failures are often triggered by local shocks that reverberate across interconnected systems, amplifying systemic risks and compounding existing uncertainties. Chemison et al. (2022) indicate that the cascading effects from an AMOC collapse involve changes to monsoon systems and rainfall distribution, further endangering agricultural output and freshwater resources. Thus, the study of cascading collapse scenarios is imperative for understanding potential futures wherein isolated incidents lead to extensive ramifications across multiple sectors. Building resilience within these interlinked systems requires comprehensive strategies that account for the significant potential for secondary effects arising from initial disruptions.

13.5.6 Comparing Partial Collapse (Regional, Sectoral) vs. Systemic Collapse (Global Breakdown)

The differentiation between partial and systemic collapse is crucial for understanding the spectrum of climate-induced risks. Partial collapse may manifest as localized, regional, or sectoral breakdowns, while systemic collapse refers to widespread, global failures affecting numerous interconnected systems. Historical instances of partial collapses (ranging from regional droughts leading to local food shortages to sector-specific crises) exemplify how crises can emerge within certain spheres without necessarily resulting in global failure (Newton et al., 2024). However, systemic collapse remains a profound concern, where interplay between regional failures can culminate in a comprehensive breakdown of global systems. Research elucidates that systemic failures are characterized by a lack of resilience in critical infrastructure, governance, and ecological frameworks, suggesting profound implications for human societies in an interconnected global economy (Ogwu et al., 2025b, 2025c). Studies, including those by Lenton et al. (2019), illustrate the precarious nature of our climate systems and emphasize the need to rethink resilience strategies to mitigate risks associated with potential systemic breakdowns. It becomes increasingly clear that understanding how individual collapses interrelate is key to preparing for and preventing broader systemic failings.

13.6 Modeling Adaptation and Resilience

13.6.1 Adaptation Modeling

The intersection of climate change and technological innovation has become a focal point of adaptation modeling, necessitating exploration of renewable energy transitions and the integration of social protection systems (Table 13.4). Technological advancements play a crucial role in shaping energy systems that are resilient to climate impacts, as highlighted in studies assessing energy transitions in light of climate resilience requirements (Change, 2023; Iyiola et al., 2025a). These adaptive methodologies not only incorporate renewable technologies but also underscore the importance of social equity in energy access, ensuring that vulnerable populations are included in the transition process (Cong et al., 2022). Energy equity can be an important metric for evaluating energy poverty, revealing how shifts in temperature affect energy consumption among different income groups (Cong et al., 2022; Iyiola et al., 2025b).

Moreover, the need for social protection systems is underscored by correlations between climate impacts, economic vulnerability, and adaptive capacity (Change, 2023). The call for renewable energy solutions as part of climate-resilient development stresses that without robust social frameworks, the benefits of technological adaptation are unlikely to be equitably distributed across different strata of society. When adaptive capacities are constrained by socio-economic factors, renewable

Table 13.4 Adaptation and resilience strategies in modeling

Adaptation strategy	Scale of application	Modeling approach used	Potential benefits	Challenges
Technological innovation (renewables, carbon capture)	Global/ national	IAMs, scenario planning	Reduces emissions, fosters energy security	Cost, unequal access
Social protection and migration management	National/ local	ABMs, system dynamics	Enhances equity, reduces unrest	Political resistance, funding
Ecosystem-based adaptation (reforestation, water mgmt)	Regional/ local	Coupled climate–ecosystem models	Improves resilience, supports biodiversity	Slow returns, land competition
Transformational adaptation (new governance/ economic models)	Global/ systemic	Scenario modeling, complexity approaches	Long-term systemic resilience	Uncertainty, political feasibility

energy projects may serve only certain communities, thereby exacerbating existing inequalities. This highlights the importance of integrating socio-economic considerations into adaptation modeling, particularly in regions disproportionately affected by climate change (Change, 2023; Cong et al., 2022; Howard et al., 2021).

13.6.2 Resilience Frameworks

Resilience frameworks serve as essential tools for assessing the adaptive capacities of societies facing climate change, enabling the analysis of feedback loops and redundancy within systems. The multifaceted dimensions of resilience are reflected in the ability of systems and societies to absorb shocks, employ adaptive governance, and integrate feedback mechanisms that mitigate negative impacts of climate variations (Change, 2023; Sultana et al., 2020). For instance, frameworks that consider local governance and community-level actions have shown that enhancing feedback mechanisms can significantly bolster resilience against climate-induced disruptions (Ogwu & Kosoe, 2025a, 2025b; Ogwu et al., 2025d; Sultana et al., 2020). This approach insists that resilience is not solely an infrastructure issue but is intricately tied to social dynamics and communal adaptability.

Redundancy in systems has emerged as another crucial element of resilience, providing a buffer against climate stressors by ensuring that alternative options are available when primary systems fail. The concept is bolstered by comparative studies in water supply systems, which advocate diversifying water sources to mitigate risks posed by climate variability (Howard et al., 2021; Sweya & Wilkinson,

2021). Moreover, resilience is closely related to the health of ecosystems; systems that retain redundancy and adaptability serve not only human needs but also preserve biodiversity, which is critical for long-term sustainability (Change, 2023; Ogwu et al., 2025e; Sweya & Wilkinson, 2021). The imperative for transformational adaptation has taken the forefront, shifting from incremental fixes toward comprehensive changes in values, governance, and economic structures. This transition is vital, as societies should develop systems that are not just resilient but also regenerative, creating a paradigm in which climate adaptation is intrinsically linked to sustainable development goals (Change, 2023; Khatibi et al., 2024). Transformational adaptation challenges conventional paradigms by advocating for systems that can evolve beyond existing limitations, ensuring future resilience in the face of escalating climatic threats (Howard et al., 2021).

13.6.3 Modeling Under Uncertainty

The complexity of climate change and its varied impacts necessitate modeling under uncertainty, which incorporates scenario planning, stress testing, and adaptive policy pathways. The inherent uncertainty in climate models dictates that policymakers should embrace flexible frameworks that can adapt to shifting realities. This highlights the significance of scenario planning for uncovering potential futures shaped by different climate parameters, enabling stakeholders to anticipate and prepare for diverse outcomes (Change, 2023; Khatibi et al., 2024). Stress-testing existing infrastructure against various climate scenarios can reveal vulnerabilities and inform necessary adjustments within urban and regional planning to enhance resilience (Mosleh et al., 2023; Ogwu et al. 2025a).

Furthermore, adaptive policy pathways offer a dynamic approach to governance, in which policies evolve based on new information and experience (Change, 2023; Khatibi et al., 2024). This iterative process fosters environments where stakeholders can learn from climate impacts and successes, adjusting strategies to bolster long-term resilience in communities (Izah & Ogwu, 2025). The application of these models emphasizes that, in a world characterized by uncertainty, the ability to pivot and adapt becomes a paramount trait for local and global governance structures (Cong et al., 2022; Sweya & Wilkinson, 2021).

13.6.4 Role of Equity and Justice in Ensuring Global Resilience

The role of equity and social justice is vital in ensuring global resilience against climate change. It is necessary to explore how social inequalities shape vulnerabilities and access to resources, particularly in adaptation strategies (Cong et al., 2022; Jenkins et al., 2016). Climate justice frameworks advocate that the benefits of resource allocation and adaptation measures should be equitably shared, especially as marginalized populations often face the brunt of climate impacts (Nijhawan et al., 2022). This equity lens highlights how failure to incorporate justice considerations in adaptation strategies not only perpetuates existing inequalities but can also exacerbate them in the face of climate-related challenges.

The recognition of the energy equity gap presents an opportunity to address disparities in energy consumption and access, which is essential in crafting effective adaptation policies (Cong et al., 2022; Howard et al., 2021). Advocating for energy justice involves ensuring that low-income communities receive adequate resources and support to adapt to climate change without incurring undue burdens. This equitable approach not only strengthens communal resilience but also fosters social solidarity, which is crucial for managing collective climate challenges (Change, 2023; Jenkins et al., 2016).

13.7 Ethical, Political, and Methodological Challenges

13.7.1 Modeling Uncertainty

In addressing the multifaceted relationships between climate change and global systems, modeling uncertainty is a significant challenge that impacts the predictability of outcomes (Table 13.5). Due to the inherent complexity of climate systems, predictions often carry substantial uncertainty. While climate models have improved in their ability to simulate global temperatures and predict trends, they still demonstrate limitations in spatial and temporal resolution, complicating efforts to accurately forecast specific regional impacts of climate change (Busby, 2024; Koubi, 2019). The interactions between climatic factors and human activities (such as land use changes and socio-economic dynamics) further complicate predictive accuracy (Busby, 2024; Koubi, 2019). There is a critical need to continuously refine these models to incorporate a broader range of variables, including socio-political factors that significantly influence climate adaptation and resilience (Busby, 2024; Koubi, 2019). Consequently, reliance on past data to forecast future trends can mislead policymakers, as emergent phenomena and unprecedented interactions challenge historical norms and predictive frameworks (Busby, 2024).

13.7.2 Normative Tensions

The discourse surrounding climate change and its potential societal impacts often oscillates between alarmist and optimistic perspectives (Table 13.5). Alarmism is characterized by highlighting the severe potential consequences of climate change, which can heighten the sense of urgency for action (Mozgovoy et al., 2022). Conversely, the optimistic viewpoint emphasizes resilience and human adaptation, focusing on innovative solutions and sustainable practices (Mozgovoy et al., 2022). The polarized narratives within these discourses can shape public perception and policy decisions, influencing resource allocation for climate adaptation strategies (Ruiu, 2021). Alarmist rhetoric can, paradoxically, create a sense of despair among

Table 13.5 Ethical, political, and methodological challenges in modeling

Challenge	Key issues	Implications for modeling
Uncertainty	Nonlinear dynamics, unknown tipping points	Limits predictability; requires scenario-based approaches
Normative framing	Alarmism vs. optimism	Shapes policy uptake; risk of politicization
Political dimensions	Power asymmetries, representation	Models may privilege Global North perspectives
Inclusivity	Marginalized voices are often absent	Limits realism and justice in adaptation scenarios
Methodological	Data scarcity, reliance on economic proxies	Misses cultural/social resilience factors

affected populations, undermining collective action, while overly optimistic narratives may foster complacency and hinder necessary intervention (Mozgovoy et al., 2022; Ruiu, 2021). Thus, reconciling these conflicting perspectives is paramount for formulating effective climate policy that addresses potential collapse while fostering adaptive resilience.

13.7.3 Political Implications

The political ramifications of climate collapse modeling are profound, as these models significantly influence global governance structures and power dynamics (Table 13.5). Different nations and regions often prioritize resources based on their perceived vulnerability and predicted impacts of climate collapse (Barreiro et al., 2020; Busby, 2024; Koubi, 2019). Consequently, collapse narratives can shape diplomatic agendas, fuel international conflicts over resources, and dictate funding allocations for adaptation efforts (Busby, 2024; Ruiu, 2021). As climate change exacerbates inequalities, nations in the Global South (despite facing some of the most severe consequences) often receive inadequate support relative to their needs due to these power structures (Barreiro et al., 2020; Koubi, 2019). This inequitable resource distribution underscores the need for inclusive governance frameworks that account for diverse socio-economic contexts and promote justice in climate-related decision-making (Bisare Bitire, 2023; Busby, 2024; Koubi, 2019). Effective adaptation requires not only financial resources but also the empowerment of marginalized voices in the global discourse surrounding climate change impacts and responses.

13.7.4 Inclusivity

The inclusivity of climate models often reflects a bias toward the priorities of nations in the Global North, marginalizing the voices and needs of those in the

Global South (Table 13.5). Climate modeling frameworks tend to prioritize data and inputs based on experience from developed nations, neglecting the localized adaptations required in less affluent regions (Barreiro et al., 2020; Koubi, 2019). This oversight creates a disconnect between model outputs and the realities faced by vulnerable populations, who may bear the brunt of climate change (Barreiro et al., 2020; Busby, 2024). Furthermore, incorporating Indigenous knowledge and local practices into modeling efforts can enhance prediction accuracy and resilience strategies, underscoring the critical need for diverse representation in climate research (Bisare Bitire, 2023). The complexity of climate interactions requires a holistic approach to inclusivity that ensures a wide range of social, economic, and geographical perspectives are integrated into predictive modeling and decision-making processes (Barreiro et al., 2020; Bisare Bitire, 2023; Ogwu et al., 2023).

13.7.5 Methodological Challenges

A significant methodological challenge in climate change research is the reliance on quantitative economic indicators, which may inadequately address the nuanced realities faced by various populations (Bisare Bitire, 2023; Busby, 2024; Koubi, 2019). This emphasis on economic measures often overshadows other critical factors, such as cultural and social resilience, which play vital roles in community adaptation and recovery in the face of climate stressors (Bisare Bitire, 2023; Koubi, 2019). Additionally, data limitations stemming from inadequate monitoring systems and a lack of accessible repositories further compound the issue, impeding a comprehensive understanding of local vulnerabilities and potential adaptive capacities (Barreiro et al., 2020; Bisare Bitire, 2023). Exploring alternative methodologies that integrate qualitative assessments and local knowledge could enhance the depth and applicability of climate modeling, enabling more accurate portrayals of community resilience and adaptation strategies (Bisare Bitire, 2023; Koubi, 2019). The connection of these ethical, political, and methodological challenges highlights the complex landscape within which climate change operates, necessitating a concerted effort to cultivate more balanced and informed approaches to climate research and policy (Table 13.5).

13.8 Future Directions for Climate–World-System Modeling

Climate change presents one of the most pressing global challenges, demanding innovative approaches to understanding its interactions with the world system (Table 13.6). Future directions for climate modeling should focus on integrating climate data, socio-economic factors, and advanced analytics to produce actionable insights for adaptation and mitigation strategies. This interdisciplinary approach is vital to ensuring that climate responses align with societal needs and capacities.

Table 13.6 Future directions in climate–world-system modeling

Future direction	Narrative	Potential contributions
Data integration	Merging climate, socio-economic, and big data sources	Improves accuracy and real-time analysis
AI-driven adaptive modeling	Machine learning for dynamic scenario adjustment	Enhances forecasting under uncertainty
Multi-scale integration	Linking global models with local/community knowledge	Ensures contextual relevance of models
Anticipatory adaptation planning	Moving beyond collapse prediction to resilience pathways	Supports proactive governance
Global resilience frameworks	Coordinated modeling platforms for shared risks	Promotes equitable global adaptation strategies

Ethical challenges arise not only from interpretation of model outputs but also from how models are constructed and whose perspectives they privilege. Intergenerational justice is central: long-term projections profoundly affect populations who have no voice today. Digital inequity also shapes global modeling capacities, as AI-based tools depend on computational power disproportionately concentrated in wealthy nations. Furthermore, concerns about data colonialism highlight how Global South data are frequently extracted, modeled, and published by Global North institutions without equitable co-authorship, interpretation rights, or benefit-sharing. Ethical modeling thus requires transparency, inclusivity, and equitable collaboration across communities and regions.

13.8.1 Integration of Climate Data, Socio-economic Data, and Big Data Analytics

The integration of diverse data sources is crucial for developing robust climate models that reflect the complex interactions between environmental and human systems. Coupling climate and socio-economic data enables a comprehensive assessment of vulnerability and resilience across regions. For instance, the relationship between climate change impacts on health systems and socio-economic vulnerabilities has been emphasized through the establishment of evidence-based policies and strategic governance (Mosadeghrad et al., 2023). Moreover, the use of big data analytics can enhance predictive capabilities, enabling stakeholders to visualize potential climatic outcomes and mitigate associated risks before they materialize. The development of metrics, such as the Vegetation Sensitivity Index (VSI), highlighted by Seddon et al. (2016), merges ecological data with socio-economic contexts to reveal critical thresholds that ecosystems face under changing climates. This systemic view fosters a deeper understanding of how different components of the world system interact in response to climate variations.

Additionally, leveraging artificial intelligence (AI) in this integration process can facilitate real-time scenario adjustments as new data emerges. AI-driven models can dynamically evaluate climate projections while accounting for economic trends, providing a clearer picture of potential future scenarios in a rapidly changing world (Mehryar et al., 2024). The incorporation of community-level knowledge systems into these frameworks can further enrich model robustness, recognizing that local contexts significantly influence the efficacy of climate responses (Kamakaula, 2024). By combining high-tech advancements with traditional knowledge, models can better support localized climate adaptation strategies, leading to more equitable and sustainable outcomes.

13.8.2 AI-Driven Adaptive Modeling for Real-Time Scenario Adjustment

AI's role in climate modeling extends beyond data processing capabilities. It also provides powerful tools for adaptive modeling that incorporate feedback loops and data from diverse sources, enabling real-time scenario adjustments. The utilization of AI can simulate numerous climate scenarios, helping stakeholders envision varied futures and explore the potential implications of different policy choices (Mehryar et al., 2024). This capability is particularly valuable in regions facing rapid climate changes, where traditional modeling methods may struggle to keep pace. Furthermore, AI can enhance learning outcomes through continuous feedback from environmental monitoring systems, thereby improving resilience strategies over time (Davies & Arrieta, 2024). AI-driven models that simulate complex interactions across ecosystems, economies, and social structures can yield insights that traditional models may overlook. For example, incorporating local agricultural practices as adaptive strategies demonstrates how communities adjust to climate stressors, underscoring the importance of cultural perspectives in resilience building (Das & Mishra, 2023; Kamakaula, 2024). This integrated modeling approach aligns with calls for systems mapping and collective learning frameworks that emphasize knowledge integration across sectors and scales (Morais et al., 2025).

13.8.3 Bridging Global-Scale Models with Local, Community-Level Knowledge Systems

Bridging global-scale models with local, community-level knowledge systems enhances the relevance and applicability of climate solutions. Local knowledge is often a rich source of insights into adaptive practices and the effects of climate change on community livelihoods (Kamakaula, 2024). Community-based insights can inform global models to ensure that they accurately reflect local realities, enhancing the legitimacy and acceptance of climate action. This symbiotic relationship between scales also enables nuanced adaptation strategies tailored to specific contexts, which is vital given the diverse environmental, social, and economic landscapes across the globe (Ernawati et al., 2024; Li et al., 2018).

Moreover, integrating local knowledge into climate models can help identify specific vulnerabilities within communities, utilizing participatory approaches that engage stakeholders at all levels (Lori et al., 2025). Such engagement is crucial to establishing localized climate resilience initiatives that consider economic, cultural, and environmental dimensions. For instance, incorporating traditional agricultural practices, as illustrated in studies of adaptation in coastal Bangladesh, can enhance food security and build resilience (Ashik-Ur-Rahman et al., 2024). By valuing and integrating local knowledge alongside scientific data and models, policy and governance frameworks can produce more nuanced insights that are essential for effective policy and governance.

13.8.4 Moving from Collapse Prediction to Anticipatory Adaptation Planning

Transitioning from collapse prediction to anticipatory adaptation planning represents a significant shift in how climate change is addressed. Anticipatory adaptation emphasizes preparing for change rather than merely responding to crises as they unfold. This proactive approach requires robust modeling that can simulate various possible futures, incorporating diverse scenarios of climate impacts and societal responses (Guan et al., 2025). Current research indicates that predicting potential collapse scenarios can lead to paralysis rather than action, suggesting a need for models that foster empowerment through anticipatory strategies (Farooq et al., 2025; Mosadeghrad et al., 2023). Effective anticipatory adaptation planning relies on understanding both local and global drivers of change. For instance, examining the interlinkages among climate change, economic stability, and governance can elucidate the pathways to adaptation (Ernawati et al., 2024; Li et al., 2018). By fostering adaptability within systems, societies can build resilience against potential shocks rather than relying on reactive measures that may prove insufficient in the face of unprecedented environmental changes (Rölfer et al., 2022). Models should therefore focus on strengthening adaptive capacities as a buffer against collapse, utilizing integrated frameworks that involve economic, environmental, and social factors.

13.8.5 Toward a Global Resilience Modeling Framework for Policy, Governance, and Sustainable Futures

The development of a global resilience modeling framework is essential for guiding policy and governance toward sustainable futures. This framework should synthesize diverse evidence sources, from community-level data to scientific research, to formulate comprehensive resilience strategies that are actionable at various scales. The multifaceted nature of climate impacts necessitates cross-sectoral collaboration, as evidenced by initiatives such as the Sustainable Development Goals (SDGs), which encourage countries to incorporate resilience into national planning and policy (Rölfer et al., 2022). Furthermore, a successful resilience framework should also navigate the intricacies of global governance, acknowledging the diverse political and economic landscapes in which climate actions are implemented (Davies & Arrieta, 2024). It is crucial to recognize the role of international cooperation in enhancing resilience strategies, particularly in regions most affected by climate impacts. Building partnerships across governments, civil society, and the private sector can facilitate the sharing of best practices and resources, fostering a culture of collaboration essential to tackling climate change (Ernawati et al., 2024).

13.9 Conclusion

The modeling of future climate–world-system interactions highlights the deep interconnections between climate change and the economic, social, ecological, and political dimensions of human society. Climate change acts as a risk multiplier, amplifying vulnerabilities and creating cascading threats that can destabilize global systems and even trigger systemic collapse. Addressing this complexity requires models that move beyond linear forecasts of warming or sea-level rise toward frameworks capable of representing feedback loops, tipping points, and co-evolving human–environment dynamics. Such approaches not only illuminate high-risk trajectories but also identify leverage points where targeted interventions can build resilience and prevent catastrophic breakdowns. This underscores the need for interdisciplinary integration—combining Earth system science, world-systems theory, social sciences, and computational innovations such as network modeling, agent-based simulations, and machine learning—to more effectively capture uncertainty and guide responses.

At the same time, models should serve as forward-looking tools for adaptation and transformation, not just as warnings of collapse. By simulating pathways for renewable energy transitions, food and water security, strengthened social safety nets, and cooperative international governance, adaptation modeling demonstrates how societies can navigate instability while fostering long-term resilience. Ethical and political challenges (particularly issues of representation, equity, and inclusivity) remain central to the design and application of these models. Eventually, modeling climate–world-system interactions is both a scientific and ethical endeavor: it equips policymakers with anticipatory governance tools, empowers communities to adapt, and fosters global cooperation in building a just and sustainable future. Preparing for uncertainty, therefore, demands more than predicting collapse; it requires actively designing resilient pathways that enable societies to thrive in an era of profound climate disruption.

References

Anderies, J. M., Barfuss, W., Donges, J. F., Fetzer, I., Heitzig, J., & Rockström, J. (2023). A modeling framework for World-Earth system resilience: Exploring social inequality and Earth system tipping points. *Environmental Research Letters, 18*(9), 095001.

Armstrong McKay, D. I., Staal, A., Abrams, J. F., Winkelmann, R., Sakschewski, B., Loriani, S., Fetzer, I., Cornell, S. E., Rockström, J., & Lenton, T. M. (2022). Exceeding 1.5 °C global warming could trigger multiple climate tipping points. *Science, 377*(6611), eabn7950.

Ashik-Ur-Rahman, M., Swarnokar, S. C. C., Ahasan, S. N., Mohibbullah, M., Mou, S. I., & Gain, A. K. (2024). Farmers' adaptation practices in climate-stressed coastal Bangladesh: A systematic review. *Environmental Research Communications, 7*(1), 012001.

Atanasiu, M. (2023). Vulnerabilities transposed by climate change effects in ecological risks and threats to the national Security. *Strategic Impact, 88*(3), 72–88.

Barreiro, J., Lopes, R., Ferreira, F., Brito, R., Telhado, M. J., Matos, J. S., & Matos, R. S. (2020). Assessing urban resilience in complex and dynamic systems: The RESCCUE project approach in Lisbon research site. *Sustainability, 12*(21), 8931.

Beckage, B., Lacasse, K., Winter, J. M., Horton, R. M., Baskett, M. L., Gross, L. J., et al. (2018). Systemic modeling of climate change adaptation and mitigation in coupled human–natural systems. *Nature Climate Change, 8*(9), 757–763.

Ben-Yami, M., Good, P., Jackson, L. C., Crucifix, M., Hu, A., Saenko, O., Swingedouw, D., & Boers, N. (2024). Impacts of AMOC collapse on monsoon rainfall: A multi-model comparison. *Earth's Future, 12*(9), e2023EF003959.

Bettis, O. D., Dietz, S., & Silver, N. G. (2017). The risk of climate ruin. *Climatic Change, 140*(2), 109–118.

Bisare Bitire, B. (2023). Appraisal of climate change mitigation and adaptation regulatory frameworks in Ethiopia and their congruency with the UN climate change convention. *International Journal of Climate Change Strategies and Management, 15*(5), 638–651.

Bony, S., Stevens, B., Frierson, D. M., Jakob, C., Kageyama, M., Pincus, R., Shepherd, T. G., Sherwood, S. C., Siebesma, A. P., Sobel, A. H., & Watanabe, M. (2015). Clouds, circulation and climate sensitivity. *Nature Geoscience, 8*(4), 261–268.

Burke, K. D., Williams, J. W., Chandler, M. A., Haywood, A. M., Lunt, D. J., & Otto-Bliesner, B. L. (2018). Pliocene and Eocene provide best analogs for near-future climates. *Proceedings of the National Academy of Sciences, 115*(52), 13288–13293.

Bury, T. M., Bauch, C. T., & Anand, M. (2019). Charting pathways to climate change mitigation in a coupled socio-climate model. *PLoS Computational Biology, 15*(6), e1007000.

Busby, J. W. (2024). Climate security: How to write about the future without lapsing into prophecy. *PS: Political Science & Politics, 57*(1), 45–49.

Butler, A. H., Arribas, A., Athanassiadou, M., Baehr, J., Calvo, N., Charlton-Perez, A., Déqué, M., Domeisen, D. I., Fröhlich, K., Hendon, H., & Imada, Y. (2016). The climate-system historical forecast project: Do stratosphere-resolving models make better seasonal climate predictions in boreal winter? *Quarterly Journal of the Royal Meteorological Society, 142*(696), 1413–1427.

Calvin, K., & Bond-Lamberty, B. (2018). Integrated human-earth system modeling—state of the science and future directions. *Environmental Research Letters, 13*(6), 063006.

Carstens, J. D., & Wing, A. A. (2022). A spectrum of convective self-aggregation based on background rotation. *Journal of Advances in Modeling Earth Systems, 14*(5), e2021MS002860.

Chemison, A., Defrance, D., Ramstein, G., & Caminade, C. (2022). Impact of an acceleration of ice sheet melting on monsoon systems. *Earth System Dynamics, 13*(3), 1259–1287.

Cherp, A., Vinichenko, V., Jewell, J., Brutschin, E., & Sovacool, B. (2018). Integrating techno-economic, socio-technical and political perspectives on national energy transitions: A meta-theoretical framework. *Energy Research & Social Science, 37*, 175–190.

Cong, S., Nock, D., Qiu, Y. L., & Xing, B. (2022). Unveiling hidden energy poverty using the energy equity gap. *Nature Communications, 13*(1), 2456.

Cooley, S. R., Schoeman, D. S., Bopp, L., Boyd, P., Donner, S., Ito, S.-I., Kiessling, W., Martinetto, P., Ojea, E., Racault, M.-F., Rost, B., Skern-Mauritzen, M., & Yemane Ghebrehiwet, D. (2023). Oceans and coastal ecosystems and their services. In *Climate change 2022: Impacts, adaptation and vulnerability: Contribution of working group II to the sixth assessment report of the intergovernmental panel on climate change.* Cambridge University Press.

Cooper, G. S., Willcock, S., & Dearing, J. A. (2020). Regime shifts occur disproportionately faster in larger ecosystems. *Nature Communications, 11*(1), 1175.

Dalby, S. (2025). History, geopolitics and climate security. *Environment and Security, 3*(1), 108–114.

Das, S., & Mishra, A. J. (2023). Climate change and the Western Himalayan community: Exploring the local perspective through food choices. *Ambio, 52*(3), 534–545.

Davies, J. S., & Arrieta, T. (2024). The limits of "resilience": Relationalities, contradictions, and re-appropriations. *Wiley Interdisciplinary Reviews: Climate Change, 15*(6), e911.

Deléglise, C., François, H., Dodier, H., Crouzat, E., Samacoïts, R., Morin, S., Bray, F., & Nettier, B. (2022). Agro-climatic profiles of summer mountain pastures in the French Alps: Towards a monitoring tool to contribute to climate risk assessment. *Agronomy for Sustainable Development, 42*(3), 40.

Ditlevsen, P., & Ditlevsen, S. (2023). Warning of a forthcoming collapse of the Atlantic meridional overturning circulation. *Nature Communications, 14*(1), 1–12.

Donges, J. F., Lucht, W., Heitzig, J., Barfuss, W., Cornell, S. E., Lade, S. J., & Schlüter, M. (2018). Taxonomies for structuring models for World-Earth system analysis of the Anthropocene: Subsystems, their interactions and social-ecological feedback loops. *Earth System Dynamics Discussions, 2018*, 1–30.

Durach, C. F., Simpson, D., Wiengarten, F., & Wu, Z. (2025). Beyond the yield: Enhancing agricultural sustainability through operations management. *Journal of Operations Management.*

Edmonds, J. A., McJeon, H., Clarke, L., Kyle, P., & Calvin, K. (2020). Limits of integrated assessment modeling for climate policy. *Energy Policy, 139*, 111350.

Ernawati, E., Madi, R. A., & Asri, M. (2024). Correlation macroeconomic, government efficiency, infrastructure, and climate change vulnerability: A cross-country analysis. In *IOP conference series: Earth and environmental science* (Vol. 1302(1), p. 012074). IOP Publishing.

Eyring, V., Bony, S., Meehl, G. A., Senior, C. A., Stevens, B., Stouffer, R. J., & Taylor, K. E. (2016). Overview of the Coupled Model Intercomparison Project Phase 6 (CMIP6) experimental design and organization. *Geoscientific Model Development, 9*(5), 1937–1958.

Falkner, R. (2016). The Paris Agreement and the new logic of international climate politics. *International Affairs, 92*(5), 1107–1125.

Fitzpatrick, M. C., & Dunn, R. R. (2019). Contemporary climatic analogs for 540 North American urban areas in the late 21st century. *Nature Communications, 10*(1), 1–7.

Flood, S., Cradock-Henry, N. A., Blackett, P., & Edwards, P. (2018). Adaptive and interactive climate futures: Systematic review of 'serious games' for engagement and decision-making. *Environmental Research Letters, 13*(6), 063005.

Gebrechorkos, S., Leyland, J., Slater, L., Wortmann, M., Ashworth, P. J., Bennett, G. L., Boothroyd, R., Cloke, H., Delorme, P., Griffith, H., & Hardy, R. (2023a). A high-resolution daily global dataset of statistically downscaled CMIP6 models for climate impact analyses. *Scientific Data, 10*(1), 611.

Gebrechorkos, S. H., Taye, M. T., Birhanu, B., Solomon, D., & Demissie, T. (2023b). Future changes in climate and hydroclimate extremes in East Africa. *Earth's Future, 11*(2), e2022EF003011.

Gonçalves, V. K., Ribeiro, T. L., Inoue, C. Y. A., & Lins, J. (2024). Indigenous climate finance and the worlding of International Relations: Climate justice in motion. *International Relations, 38*(3), 388–406.

Guan, C., Meng, J., & Xu, Q. (2025). Structural resilience of climate regime in the global conflicts. *Environmental Research Communications, 7*(3), 035024.

Guzman, P., Fatorić, S., & Ishizawa, M. (2020). Monitoring climate change in world heritage properties.

Heinze, C., Michel, C., Torsvik, T., Schwinger, J., & Tjiputra, J. F. (2024). More frequent abrupt marine environmental changes expected. *Geophysical Research Letters, 51*(2), e2023GL106192.
Howard, G., Nijhawan, A., Flint, A., Baidya, M., Pregnolato, M., Ghimire, A., Poudel, M., Lo, E., Sharma, S., Mengustu, B., & Ayele, D. M. (2021). The how tough is WASH framework for assessing the climate resilience of water and sanitation. *npj Clean Water, 4*(1), 39.
Ide, T. (2023). Rise or recede? How climate disasters affect armed conflict intensity. *International Security, 47*(4), 50–78.
Inoue, H., & Todo, Y. (2019). Firm-level propagation of shocks through supply-chain networks. *Nature Sustainability, 2*(9), 841–847.
Ivanescu, L. M., Bodale, I., Grigore-Hristodorescu, S., Martinescu, G., Andronic, B., Matiut, S., Azoicai, D., & Miron, L. (2023). The risk of emerging of dengue fever in Romania, in the context of global warming. *Tropical Medicine and Infectious Disease, 8*(1), 65.
Iyiola, A. O., Afolabi, O. A., Alimi, S. K., Akingba, O. O., Izah, S. C., & Ogwu, M. C. (2024). Climate change and water crisis in the Global South. In S. C. Izah, M. C. Ogwu, A. Loukas, & H. Hamidifar (Eds.), *Water crises and sustainable management in the Global South* (pp. 111–140). Springer. https://doi.org/10.1007/978-981-97-4966-9_4
Iyiola, A.O., Akingba, O.O., Alimi, S.K., Akinsorotan, A.M., Osei, S.A., Idowu, C.F., Izah, S.C., and Ogwu, M.C. (2025a). Impact assessment of biobased products in the Global South. In M.C. Ogwu, S.C. Izah, J.G. Vazquez-Arenas, S.T. Feleke, and X. Wei (Eds.), *Sustainable bioeconomy development in the Global South: Volume II bioeconomy techniques*. Springer. doi:https://doi.org/10.1007/978-981-96-0305-3_16.
Iyiola, A. O., Akingba, O. O., Asiedu, B., Osei, S. A., Ogwu, M. C., & Morya, S. (2025b). Biobased value-added products and services from biological resources in the Global South. In M. C. Ogwu, S. C. Izah, H. Merritt, R. D. Wise, & A. Jintrawet (Eds.), *Sustainable bioeconomy development in the Global South: Volume I status and perspectives*. Springer. https://doi.org/10.1007/978-981-96-0640-5_9
Iyiola, A. O., Kolawole, A. S., Ajayi, F. O., Ogidi, O. I., & Ogwu, M. C. (2023). Sustainable water use and management for agricultural transformation in Africa. In S. Madhav, A. L. Srivastav, S. C. Izah, & E. van Hullebusch (Eds.), *Water resources management for rural development* (pp. 287–299). Elsevier. https://doi.org/10.1016/B978-0-443-18778-0.00014-3
Iyiola, A. O., & Ogwu, M. C. (2025). Environmental management using renewable biological resources in the Global South. In M. C. Ogwu, S. C. Izah, J. G. Vazquez-Arenas, S. T. Feleke, & X. Wei (Eds.), *Sustainable bioeconomy development in the Global South: Volume II bioeconomy techniques*. Springer. https://doi.org/10.1007/978-981-96-0305-3_11
Izah, S. C., Iyiola, A. O., Yarkwan, B., & Richard, G. (2023). Impact of air quality as a component of climate change on biodiversity-based ecosystem services. In A. L. Srivastav, A. K. Dubey, A. Kumar, S. K. Narang, & M. A. Khan (Eds.), *Visualization techniques for climate change with machine learning and artificial intelligence* (pp. 123–148). Elsevier. isbn:978-0-323-99714-0.
Izah, S. C., & Ogwu, M. C. (2025). *Innovative approaches in environmental health management: Processes, technologies, and strategies for a sustainable future*. Springer Nature. https://doi.org/10.1007/978-3-031-81966-7
Jenkins, K., McCauley, D., Heffron, R., Stephan, H., & Rehner, R. (2016). Energy justice: A conceptual review. *Energy Research & Social Science, 11*, 174–182.
Jiranek, J., Miller, I. F., An, R., Bruns, E., & Metcalf, C. J. E. (2023). Mechanistic models to meet the challenge of climate change in plant–pathogen systems. *Philosophical Transactions of the Royal Society B, 378*(1873), 20220017.
Kamakaula, Y. (2024). Cultural perspectives on climate change adaptation in agriculture an ethnoecological study. *West Science Interdisciplinary Studies, 2*, 743–748.
Kemp, L., Xu, C., Depledge, J., Ebi, K. L., Gibbins, G., Kohler, T. A., Rockström, J., Scheffer, M., Schellnhuber, H. J., Steffen, W., & Lenton, T. M. (2022). Climate endgame: Exploring catastrophic climate change scenarios. *Proceedings of the National Academy of Sciences, 119*(34), e2108146119.

Kennett, D. J., Masson, M., Lope, C. P., Serafin, S., George, R. J., Spencer, T. C., Hoggarth, J. A., Culleton, B. J., Harper, T. K., Prufer, K. M., & Milbrath, S. (2022). Drought-induced civil conflict among the ancient Maya. *Nature Communications, 13*(1), 3911.

Khatibi, H., Wilkinson, S., Sweya, L. N., Baghersad, M., & Dianat, H. (2024). Navigating climate change challenges through smart resilient cities: A comprehensive assessment framework. *Land, 13*(3), 266.

Klose, A. K., Karle, V., Winkelmann, R., & Donges, J. F. (2020). Emergence of cascading dynamics in interacting tipping elements of ecology and climate. *Royal Society Open Science, 7*(6), 200599.

Knutsen, B. O., & NyquistPedersen, M. (2024). How to understand climate change as a threat multiplier in the arctic. *Arctic Review on Law and Politics, 15*(1), 153–176.

Koubi, V. (2019). Climate change and conflict. *Annual Review of Political Science, 22*(1), 343–360.

Lamontagne, J. R., Reed, P. M., Marangoni, G., Garner, G. G., Keller, K., & Keller, E. (2018). Robust abatement pathways to tolerable climate futures require immediate global action. *Nature Climate Change, 8*(8), 592–597.

Lee, T. M., Markowitz, E. M., Howe, P. D., Ko, C. Y., & Leiserowitz, A. A. (2015). Predictors of public climate change awareness and risk perception around the world. *Nature Climate Change, 5*(11), 1014–1020.

Lenton, T. M., Rockström, J., Gaffney, O., Rahmstorf, S., Richardson, K., Steffen, W., & Schellnhuber, H. J. (2019). Climate tipping points—too risky to bet against. *Nature, 575*(7784), 592–595.

Li, D., Wu, S., Liu, L., Zhang, Y., & Li, S. (2018). Vulnerability of the global terrestrial ecosystems to climate change. *Global Change Biology, 24*(9), 4095–4106.

Lima, M., Gayó, E. M., Gurruchaga, A., Estay, S. A., & Santoro, C. M. (2023). 1000 years of population, warfare, and climate change in pre-Columbian societies of the Central Andes. *PLoS One, 18*(11), e0278730.

Little, J. C., Kaaronen, R. O., Hukkinen, J. I., Xiao, S., Sharpee, T., Farid, A. M., Nilchiani, R., & Barton, C. M. (2023). Earth systems to anthropocene systems: An evolutionary, system-of-systems, convergence paradigm for interdependent societal challenges. *Environmental Science & Technology, 57*(14), 5504–5520.

Liu, W., Xie, S. P., Liu, Z., & Zhu, J. (2017). Overlooked possibility of a collapsed Atlantic Meridional Overturning Circulation in warming climate. *Science Advances, 3*(1), e1601666.

Lori, T., Osei-Kwarteng, M., Ojija, F., Ojo, A. O., & Ogwu, M. C. (2025). Participatory approaches to environmental health management. In S. C. Izah & M. C. Ogwu (Eds.), *Innovative approaches in environmental health management: Processes, technologies, and strategies for a sustainable future. Environmental science and engineering series.* Springer Nature. https://doi.org/10.1007/978-3-031-81966-7_11

Mazarrasa, I., Garcia-Orellana, J., Puente, A., & Juanes, J. A. (2022). Coastal engineering infrastructure impacts Blue Carbon habitats distribution and ecosystem functions. *Scientific Reports, 12*(1), 19352.

Mbah, M., Ajaps, S., & Molthan-Hill, P. (2021). A systematic review of the deployment of indigenous knowledge systems towards climate change adaptation in developing world contexts: Implications for climate change education. *Sustainability, 13*(9), 4811.

Meehl, G. A., Senior, C. A., Eyring, V., Flato, G., Lamarque, J. F., Stouffer, R. J., Taylor, K. E., & Schlund, M. (2020). Context for interpreting equilibrium climate sensitivity and transient climate response from the CMIP6 Earth system models. *Science Advances, 6*(26), eaba1981.

Mehryar, S., Yazdanpanah, V., & Tong, J. (2024). AI and climate resilience governance. *Iscience, 27*(6), 109812.

Menary, M. B., Robson, J., Allan, R. P., Booth, B. B., Cassou, C., Gastineau, G., Gregory, J., Hodson, D., Jones, C., Mignot, J., & Ringer, M. (2020). Aerosol-forced AMOC changes in CMIP6 historical simulations. *Geophysical Research Letters, 47*(14), e2020GL088166.

Morais, L. M. D. O., Borde, E., Tatah, L., Thondoo, M., Wayas, F. A., Gordon-Strachan, G., Obonyo, C., Assah, F., Oni, T., Caiaffa, W. T., & Garcia, L. (2025). Collective learning for

resilience in Global South cities: A community-based systems mapping approach to integrated climate and health action. *Frontiers in Public Health, 13*, 1582550.
Mosadeghrad, A. M., Isfahani, P., Eslambolchi, L., Zahmatkesh, M., & Afshari, M. (2023). Strategies to strengthen a climate-resilient health system: A scoping review. *Globalization and Health, 19*(1), 62.
Mosleh, L., Negahban-Azar, M., & Pavao-Zuckerman, M. (2023). Stormwater green infrastructure resilience assessment: A social-ecological framework for urban stormwater management. *Water, 15*(9), 1786.
Mozgovoy, L., Dubinina, V., Malikova, Y., Steshenko, V., Butko, Y., & Slabouz, V. (2022). Alarmism vs eschatology: Conceptual analysis (on material of Ukrainian discourse). *Wisdom, 2*(22), 68–80.
Mulcahy, J. P., Jones, C., Sellar, A., Johnson, B., Boutle, I. A., Jones, A., Andrews, T., Rumbold, S. T., Mollard, J., Bellouin, N., & Johnson, C. E. (2018). Improved aerosol processes and effective radiative forcing in HadGEM3 and UKESM1. *Journal of Advances in Modeling Earth Systems, 10*(11), 2786–2805.
Müller, P. M., Heitzig, J., Kurths, J., Lüdge, K., & Wiedermann, M. (2021). Anticipation-induced social tipping: Can the environment be stabilised by social dynamics? *The European Physical Journal Special Topics, 230*(16), 3189–3199.
Newton, A. C., Coward, F., Elliott, S., Jenkins, E., Linden, M. V., Riris, P., & Silva, F. (2024). Understanding long-term human ecodynamics through the lens of ecosystem collapse. *The Holocene, 34*(10), 1439–1453.
Nijhawan, A., Howard, G., Poudel, M., Pregnolato, M., Eunice Lo, Y. T., Ghimire, A., Baidya, M., Geremew, A., Flint, A., & Mulugeta, Y. (2022). Assessing the climate resilience of community-managed water supplies in Ethiopia and Nepal. *Water, 14*(8), 1293.
Noguerales, V., Arjona, Y., García-Olivares, V., Machado, A., López, H., Patiño, J., & Emerson, B. C. (2024). Genetic legacies of mega-landslides: Cycles of isolation and contact across flank collapses in an oceanic island. *Molecular Ecology, 33*(9), e17341.
Ogwu, M. C. (2019). Towards sustainable development in Africa: The challenge of urbanization and climate change adaptation. In P. B. Cobbinah & M. Addaney (Eds.), *The geography of climate change adaptation in Urban Africa* (pp. 29–55). Springer Nature. https://doi.org/10.1007/978-3-030-04873-0_2
Ogwu, M. C., El Malahi, S., & Izah, S. C. (2025a). Policy and regulatory frameworks for sustainable environmental practices. In M. C. Ogwu & S. Chibueze Izah (Eds.), *Evaluating environmental processes and technologies. Environmental science and engineering series.* Springer. https://doi.org/10.1007/978-3-031-85327-2_16
Ogwu, M. C., & Izah, S. C. (2025). *Evaluating environmental processes and technologies.* Springer Nature. https://doi.org/10.1007/978-3-031-85327-2
Ogwu, M. C., Izah, S. C., Merritt, H., Wise, R. D., & Jintrawet, A. (2025b). *Sustainable bioeconomy development in the Global South: Volume I status and perspectives.* Springer Singapore. https://doi.org/10.1007/978-981-96-0640-5
Ogwu, M. C., Izah, S. C., Vazquez-Arenas, J. G., Feleke, S. T., & Wei, X. (2025c). *Sustainable bioeconomy development in the Global South: Volume II bioeconomy techniques.* Springer Singapore. https://doi.org/10.1007/978-981-96-0305-3
Ogwu, M. C., & Kosoe, E. A. (2025a). Integrating green infrastructure into sustainable agriculture to enhance soil health, biodiversity, and microclimate resilience. *Sustainability, 17*(9), 3838. https://doi.org/10.3390/su17093838
Ogwu, M. C., & Kosoe, E. A. (2025b). Environmental economics: Valuing ecosystem services and green technologies. In M. C. Ogwu & S. Chibueze Izah (Eds.), *Evaluating environmental processes and technologies. Environmental science and engineering series.* Springer. https://doi.org/10.1007/978-3-031-85327-2_15
Ogwu, M. C., Kosoe, E. A., Alaka, A. C., & Xu, J. (2025d). Green infrastructure and environmental health security: The role of nature-based solutions in improving urban health. In S. C. Izah & M. C. Ogwu (Eds.), *Innovative approaches in environmental health management: Processes,*

technologies, and strategies for a sustainable future. Environmental science and engineering series. Springer Nature. https://doi.org/10.1007/978-3-031-81966-7_9

Ogwu, M. C., Malikia, C. N., Stansfield, A., Gonzales-Torres, A. D., & Izah, S. C. (2023). Sustainable food processing waste management for environmental protection. In A. L. Srivastav, A. S. Grewal, M. Tiwari, & T. D. Pham (Eds.), *Role of green chemistry in ecosystem restoration to achieve environmental sustainability* (pp. 291–299). Elsevier. https://doi.org/10.1016/B978-0-443-15291-7.00022-5. isbn:978-0-443-15291-7.

Ogwu, M. C., Ojo, A. O., & Alaka, A. C. (2025e). Biodiversity and human health: The interconnections of species loss and ecosystem services. In S. C. Izah & M. C. Ogwu (Eds.), *Innovative approaches in environmental health management: Processes, technologies, and strategies for a sustainable future. Environmental science and engineering series*. Springer Nature. https://doi.org/10.1007/978-3-031-81966-7_5

Perez, C., Jones, E. M., Kristjanson, P., Cramer, L., Thornton, P. K., Förch, W., & Barahona, C. A. (2015). How resilient are farming households and communities to a changing climate in Africa? A gender-based perspective. *Global Environmental Change, 34*, 95–107.

Pérez-Quezada, J. F., Barichivich, J., Urrutia-Jalabert, R., Carrasco, E., Aguilera, D., Bacour, C., & Lara, A. (2023). Warming and drought weaken the carbon sink capacity of an endangered paleoendemic temperate rainforest in South America. *Journal of Geophysical Research: Biogeosciences, 128*(4), e2022JG007258.

Pindyck, R. S. (2017). The use and misuse of models for climate policy. *Review of Environmental Economics and Policy, 11*(1), 100–114.

Rahmstorf, S. (2024). Is the Atlantic overturning circulation approaching a tipping point? *Oceanography, 37*(3), 16–29.

Raihani, N. J., & Aplin, L. M. (2022). Model pluralism in complex systems research. *Trends in Ecology & Evolution, 37*(10), 885–895.

Riddell, E. A., Iknayan, K. J., Wolf, B. O., Sinervo, B., & Beissinger, S. R. (2019). Cooling requirements fueled the collapse of a desert bird community from climate change. *Proceedings of the National Academy of Sciences, 116*(43), 21609–21615.

Rölfer, L., Celliers, L., & Abson, D. (2022). Resilience and coastal governance: Knowledge and navigation between stability and transformation. *Ecology and Society, 27*(2).

Rooney-Varga, J. N., Hensel, M., McCarthy, C., McNeal, K., Norfles, N., Rath, K., Schnell, A. H., & Sterman, J. D. (2021). Building consensus for ambitious climate action through the world climate simulation. *Earth's Future, 9*(12), e2021EF002283.

Ruiu, M. L. (2021). Representation of climate change consequences in British newspapers. *European Journal of Communication, 36*(5), 478–493.

Sabil, F. M., Nazia, F., Shun, C. T., & Riaz, A. (2025). Foreign direct investment and climate change risks: A panel data study of China's OBOR initiative. *Environment and Urbanization ASIA, 16*(1). https://doi.org/10.1177/09754253251339557

Samnuzulsari, T., Elsera, M., Artini, Y. D., Susanto, H., Yudiatmaja, W. E., & Utari, D. S. (2023). Undertaking climate change risk reduction: A case study of Green Siak in Indonesia. In *IOP conference series: Earth and environmental science* (Vol. 1148(1), p. 012043). IOP Publishing.

Schewe, J., Gosling, S. N., Reyer, C., Zhao, F., Ciais, P., Elliott, J., Francois, L., Huber, V., Lotze, H. K., Seneviratne, S. I., & Van Vliet, M. T. (2019). State-of-the-art global models underestimate impacts from climate extremes. *Nature Communications, 10*(1), 1005.

Seddon, A. W., Macias-Fauria, M., Long, P. R., Benz, D., & Willis, K. J. (2016). Sensitivity of global terrestrial ecosystems to climate variability. *Nature, 531*(7593), 229–232.

Shin, N., Bowers, J., Roderick, S., McIntyre, C., Stephens, A. L., Eidin, E., Krajcik, J., & Damelin, D. (2022). A framework for supporting systems thinking and computational thinking through constructing models. *Instructional Science, 50*(6), 933–960.

Sibiya, N., Sithole, M., Mudau, L., & Simatele, M. D. (2022). Empowering the voiceless: Securing the participation of marginalised groups in climate change governance in South Africa. *Sustainability, 14*(12), 7111.

Skains, R. L., Rudd, J. A., Horry, R., & Ross, H. (2022). Playing for change: Teens' attitudes towards climate change action as expressed through interactive digital narrative play. *Frontiers in Communication, 6*, 789824.

Steel, D., Belotti, G., Mittiga, R., & Mintz-Woo, K. (2024). A dynamic collapse concept for climate change. *Environmental Values, 33*(6), 606–625.

Steel, D., DesRoches, C. T., & Mintz-Woo, K. (2022). Climate change and the threat to civilization. *Proceedings of the National Academy of Sciences, 119*(42), e2210525119.

Steffen, W., Richardson, K., Rockström, J., Cornell, S. E., Fetzer, I., Bennett, E. M., Biggs, R., Carpenter, S. R., De Vries, W., De Wit, C. A., & Folke, C. (2015). Planetary boundaries: Guiding human development on a changing planet. *Science, 347*(6223), 1259855.

Steffen, W., Rockström, J., Richardson, K., Lenton, T. M., Folke, C., Liverman, D., Summerhayes, C. P., Barnosky, A. D., Cornell, S. E., Crucifix, M., & Donges, J. F. (2018). Trajectories of the earth system in the anthropocene. *Proceedings of the National Academy of Sciences, 115*(33), 8252–8259.

Sultana, S., Das, P. K., Saikia, D., & Barman, I. (2020). Farmer's extent of adoption of climate resilient agro-technologies. *International Journal of Environment and Climate Change, 10*(5), 53–56.

Sweya, L. N., & Wilkinson, S. (2021). Tool development to measure the resilience of water supply systems in Tanzania: Economic dimension. *Jàmbá: Journal of Disaster Risk Studies, 13*(1), a860.

Tahir, Z., & Nuri, A. J. (2024). Examining the impact of religion on civilization: Insights from Ibn Khaldun and Arnold Toynbee's Theories. *DIROSAT: Journal of Education, Social Sciences & Humanities, 2*(2), 106–123.

van Westen, R. M., & Dijkstra, H. A. (2023). Asymmetry of AMOC hysteresis in a state-of-the-art global climate model. *Geophysical Research Letters, 50*(22), e2023GL106088.

van Westen, R. M., Kliphuis, M., & Dijkstra, H. A. (2025). Collapse of the Atlantic meridional overturning circulation in a strongly eddying ocean-only model. *Geophysical Research Letters, 52*(6), e2024GL114532.

Vora, A., & Singh, R. (2022). Improving rainfall-runoff model reliability under nonstationarity of model parameters: A hypothesis testing based framework. *Water Resources Research, 58*(11), e2022WR032273.

Xu, D., Lu, H., Chu, G., Liu, L., Shen, C., Li, F., Wang, C., & Wu, N. (2019). Synchronous 500-year oscillations of monsoon climate and human activity in Northeast Asia. *Nature Communications, 10*(1), 4105.

Yang, R., Zhou, A., Zhang, H., Chen, L., Cao, K., Huang, Y., Lu, Y., & Dong, W. (2023). Mid and late Holocene climate changes recorded by biomarkers in the sediments of Lake Gouchi and their relationship with the cultural evolution of northern Shaanxi. *Progress in Physical Geography: Earth and Environment, 47*(5), 721–740.

Yin, J., Fang, X., & Su, Y. (2016). Correlation between climate and grain harvest fluctuations and the dynastic transitions and prosperity in China over the past two millennia. *The Holocene, 26*(12), 1914–1923.

Toward a Just Energy Transition: Policy, Power, and Post-Carbon Futures

14

14.1 Introduction

The twenty-first century is marked by unprecedented challenges posed by climate change, revealing the unsustainability of reliance on fossil fuels (Iyiola et al., 2024; Izah et al., 2023). This crisis highlights the urgent need for a global energy transition from fossil fuels to renewable energy resources such as solar, wind, hydro, geothermal, and bioenergy. The increase in global temperatures, coupled with environmental degradation and socio-economic disparities, necessitates a comprehensive shift to energy systems that can foster sustainable and equitable growth. Emissions from fossil fuels have reached alarmingly high levels, increasing greenhouse gas concentrations in the atmosphere and leading to catastrophic climate consequences (McCauley & Heffron, 2018). This context places immense pressure on policymakers, stakeholders, and societies to rethink energy paradigms toward renewable alternatives, emphasizing that the transition is more than a mere swap of technologies; it is a comprehensive metamorphosis of political, economic, and social structures (Sun et al., 2023a, 2023b).

The transition entails profound implications for power dynamics, equity, and justice in the energy sector. It raises critical inquiries regarding who benefits from the new energy systems and who bears the burdens, especially in regions historically marginalized in energy discourse (Gyamfi et al., 2015). A crucial concept emerging within this context is that of a "just energy transition," which aligns decarbonization efforts with social justice principles, ensuring that the benefits and burdens associated with energy transitions are equitably distributed (McCauley & Heffron, 2018). The uneven benefits of renewable technologies, which may lead to land dispossession or inequitable infrastructure development, highlight that success requires not only transitioning energy resources but also addressing injustices produced by previous energy regimes (Huang & Liu, 2021). Efforts to realize a just energy transition necessitate a framework that integrates climate, energy, and environmental justice. Such integration ensures that clean technology promotion

M. C. Ogwu and S. Chibueze Izah, *Climate Frontiers and Social Transformation*, World-Systems Evolution and Global Futures,
https://doi.org/10.1007/978-3-032-17639-4_14

is linked with job creation and community empowerment, significantly enhancing inclusivity and effectiveness of the transition (McCauley & Heffron, 2018). The Paris Agreement has catalyzed global acknowledgment of the urgency of this transformation, prompting countries to develop strategies that harmonize climate goals with socio-economic considerations (McCauley et al., 2019). These intertwined considerations challenge existing power relations and call for new governance models that embrace public participation and accountability (Sun et al., 2023a, 2023b).

As nations strive for substantial reductions in greenhouse gas emissions, it becomes critical to systematically analyze and address the social impacts of these energy transitions. Scholars have noted the limited yet growing body of research that delineates how social assessments could enhance understanding of the multifaceted impacts of renewable energy deployment (Healy & Barry, 2017). Advancing empirical research on these dimensions is fundamental for ensuring that the energy transition does not replicate past injustices but rather serves as a catalyst for social and economic advancement (Sun et al., 2023a, 2023b). Approximately 3.3–3.6 billion people live in settings highly vulnerable to climate change, with vulnerability concentrated in the Global South (IPCC, 2022). Macroeconomic risk is also large: under high-emissions pathways, climate change could reduce global GDP by about 11%–14% by 2050, deepening inequality unless transitions are equity-centered (Swiss Re Institute, 2021). Displacement pressures will intensify, with projections of up to 216 million internal climate migrants by 2050, reinforcing the need for energy policies that reduce risk while expanding livelihoods and access (World Bank, 2021).

Transitioning to renewable energy is not merely an environmental imperative but also an economic opportunity. The renewable energy sector has demonstrated potential for job creation and economic development, particularly in the Global South, where investment in clean technologies can provide pathways for sustainable development (Gyamfi et al., 2015). Countries such as Ghana, Pakistan, and Indonesia have begun adopting policies tailored to their specific circumstances. However, the varying levels of success highlight the importance of contextualized approaches to policy development (Ali et al., 2022). Observing these efforts brings to light the role of governance, risk management, and political dynamics in shaping energy policies (Huang & Liu, 2021; Saadaoui & Chtourou, 2023).

Given the considerations of global equity and justice within energy strategies, the implications for developing countries are particularly pronounced. Renewable energy development can enhance access to reliable electricity and create opportunities for economic growth; yet without adequate policy support, these initiatives risk failing to meet their potential. Strategies should therefore be formulated to prioritize local needs while fostering technical and financial investments in renewable energy frameworks (Eze et al., 2023; Kettner & Kletzan-Slamanig, 2020). Moreover, engaging local communities and recognizing indigenous rights can serve as critical components of policy development to ensure that the transition serves all segments of society (Gyamfi et al., 2015; Zahedi et al., 2022).

This chapter focuses on advancing the understanding of a just energy transition by examining how policy, power, and equity shape the pathways toward post-carbon futures. It emphasizes that energy transition is not simply a technological shift from fossil fuels to renewable energy resources such as solar, wind, hydro, geothermal, and bioenergy, but a more profound socio-political transformation that determines who benefits, who is excluded, and how justice is embedded in decision-making. By analyzing theoretical foundations, governance frameworks, and the political economy of renewable energy, the chapter highlights both the opportunities and the risks inherent in ongoing transitions. The benefits of this analysis lie in its ability to inform inclusive policy design, promote equitable access to renewable energy, support vulnerable communities and workers affected by the decline of fossil fuels, and anticipate unintended consequences, such as energy colonialism or ecological harm from large-scale renewable projects. Eventually, the chapter equips policymakers, scholars, and communities with insights to ensure that the global move toward post-carbon futures is democratic, participatory, and just.

14.2 Theoretical Foundations of the Energy Transition

At its core, the energy transition represents a substantial shift from fossil-fuel-dominated systems to those reliant on low-carbon and renewable energy sources. This transformation is not merely technical; it encapsulates social and political phenomena, deeply intertwined with economic and governance structures. Energy systems have far-reaching implications for societal organization; thus, changes in energy production, distribution, and consumption are sure to provoke significant sociocultural shifts and alterations in power dynamics (Herranz-Surralles, 2024; Sovacool & Brisbois, 2019). The transition influences everyday life, instigating changes in consumption patterns, community dynamics, and economic structures, while simultaneously redefining power relationships and governance mechanisms (Bogdanov et al., 2019).

A juxtaposition of the just energy transition framework elucidates critical dimensions of justice integral to this shift. Distributive justice addresses the equitable allocation of costs and benefits arising from the transition, focusing on whether new renewable energy jobs and affordable electricity are distributed fairly across diverse segments of the population. Procedural justice highlights the need for inclusive decision-making processes, ensuring that marginalized communities, including Indigenous peoples and women, have a substantive voice in shaping the energy landscape (Zhang et al., 2025). This dimension is especially salient given historical injustices, in which energy projects often intersect with Indigenous lands or resource-dependent economies (Greim et al., 2020). Additionally, recognition justice calls for a meaningful acknowledgment of diverse cultural perspectives, while intergenerational justice emphasizes safeguarding future generations' rights against the backdrop of today's choices in the unfolding energy landscape (Jacobson et al., 2015; Kühne et al., 2022).

Furthermore, the political economy underpinning the energy transition reveals how power relations significantly influence its outcomes. For instance, established fossil fuel incumbents often resist transformative change, seeking to maintain their dominant position in energy markets, while emerging renewable energy sectors increasingly vie for attention and influence. This dynamic is further complicated by stark disparities between the Global North and Global South regarding access to financing, technological advancements, and the establishment of renewable energy infrastructure (Basit et al., 2020; Herranz-Surralles, 2024; Ogwu & Kosoe, 2025a, 2025b). Critical theories, such as decolonial, feminist, and Indigenous perspectives, challenge prevailing narratives of energy transitions, highlighting the potential for current models to replicate extractive, inequitable logics masquerading as principles of sustainability (Mohammadi & Khabbazan, 2022).

14.2.1 Implications for Policy and Power Dynamics

The implications of energy transition processes extend into policy-making domains, as effective governance becomes essential for successful transformations. Policymakers face the intricate task of designing systems that efficiently allocate resources while upholding social equity and environmental sustainability. This involves crafting policy frameworks that promote technological advancement in renewable energy, thereby reducing costs and enhancing access to cleaner energy (Lin et al., 2022; Liu et al., 2024). As energy systems evolve, the convergence of geopolitical interests (especially in resource-rich regions) complicates the formulation and implementation of coherent policy strategies (Herranz-Surralles, 2024). For instance, Indonesia's energy transition to renewable sources illustrates the tension with existing fossil-fuel infrastructure, highlighting both economic imperatives and international commitments, such as the Paris Agreement. The integration of renewable energy efforts into legal frameworks has prompted campaigns to stimulate governmental and corporate constituencies to adopt a proactive stance on sustainability (Sandria et al., 2023). Similarly, the nuances of renewable energy market design across contexts, such as China, highlight the interplay between policy and market dynamics that can either facilitate or hinder the transition (Liu et al., 2021).

The roles of elite actors in shaping energy policy further complicate the transition landscape. As elites often wield significant influence over governance and decision-making, their interests can either bolster or undermine efforts to achieve decarbonization and social equity (Mohammadi & Khabbazan, 2022; Sovacool & Brisbois, 2019). Managing energy transitions requires not only technological advancements but also strategic political maneuvering to navigate the entrenched power structures that can obstruct progressive energy policies or manipulate outcomes in favor of established incumbents.

14.2.2 Navigating Justice and Equity in Transition Processes

The theme of justice permeates the discourse surrounding energy transitions, demanding a multifaceted approach that involves distributive, procedural, recognition, and intergenerational justice. Each dimension highlights the expectation that a transition will yield equitable benefits across societal strata rather than perpetuate existing inequalities. Distributive justice ensures that the advantages of renewable energy proliferation (such as job creation and lower costs) are accessible to all societal segments, including marginalized groups often sidelined in decision-making processes (Liu et al., 2024). Procedural justice seeks to institutionalize the inclusion of diverse voices and perspectives in the design of energy frameworks, a process necessitating genuine engagement with communities that are traditionally underrepresented in energy discussions (Simpao, 2025).

Indeed, the intersections with Indigenous rights and local community participation further enrich the conversation about recognition justice. Energy policies must include mechanisms to acknowledge and address the historical injustices faced by Indigenous peoples, a sentiment echoed in numerous scholarly accounts advocating greater recognition of cultural perspectives in energy strategies (Greim et al., 2020). Furthermore, intergenerational justice posits that policies should account for the rights and well-being of future generations, protecting them from the adverse repercussions of today's energy choices, such as climate change and ecological destruction (Liao et al., 2022; Rodina, 2019). As practitioners and policymakers embrace the complex realities of transitioning energy landscapes, they should be vigilant in structuring equitable policies that explicitly address power dynamics and prioritize justice throughout the process. The importance of stakeholders' active participation cannot be overstated, as fostering a collaborative environment could provide the necessary framework for grassroots innovations that truly reflect diverse societal needs (Rodina, 2019).

A just energy transition requires a holistic understanding of how social, political, cultural, and temporal factors shape the distribution of benefits and burdens within global energy systems. To guide this analysis, Fig. 14.1 presents a conceptual framework outlining four core dimensions of justice—distributive, procedural, recognition, and intergenerational. These pillars offer a coherent lens for evaluating equity concerns in energy transitions and support the theoretical foundation developed in this section of the book.

14.3 Policy Frameworks Guiding the Energy Transition

The complex and multifaceted challenge of transitioning from fossil fuels to renewable energy requires a robust, just policy framework. This transition is not merely about technological shifts; it is fundamentally about societal transformation, involving the redistribution of resources, power, and opportunities across communities (Siciliano et al., 2021). Therefore, the equitable design of policy frameworks is critical, as these frameworks not only dictate the pace of fossil fuel

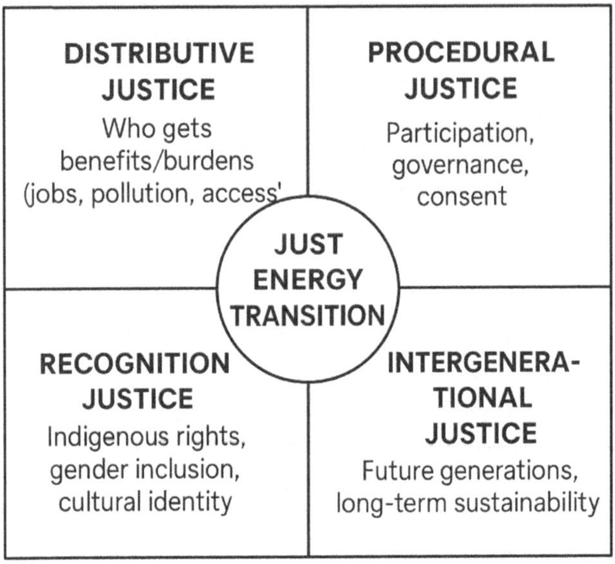

Fig. 14.1 Four pillars of a just energy transition

phase-out and renewable energy integration but also influence who gains or loses in the process (Müller et al., 2021). The potential for cleaner air, affordable electricity, and green jobs can only be fully realized if the policy landscape is intentionally crafted to promote inclusivity and address the needs of all stakeholders, especially the marginalized (Jenkins et al., 2016).

14.3.1 International Agreements and Their Implications

International agreements such as the Paris Agreement play a vital role in shaping national policies and commitments toward decarbonization. These agreements highlight the principles of equity and "common but differentiated responsibilities," recognizing the historical context of emissions and the need for developed nations to take the lead in facilitating transitions (Brennecke et al., 2024). Furthermore, the UN Sustainable Development Goals (SDGs), particularly Goal 7, highlight the global ambition for affordable and clean energy; however, operationalizing these goals often faces significant barriers, particularly in financing and technology transfer to developing regions (Noorman et al., 2023). Although the framework set by international agreements can provide a motivational baseline, the actual implementation remains a challenge due to a lack of accountability and structured support for lower-income nations (Brennecke et al., 2024; Müller et al., 2021).

14.3.2 National Policy Frameworks and Renewable Energy

National governments wield considerable influence in shaping their countries' energy transition pathways through various policies and strategies, including carbon pricing mechanisms, renewable energy incentives, and coal phase-out timelines (Huang & Liu, 2021). For instance, feed-in tariffs in Germany and substantial subsidies for solar installation across the European Union have accelerated the deployment of renewable technologies. Such initiatives illustrate how deliberate policy choices can catalyze market transformations and technology adoption, thereby contributing to substantial increases in renewable energy capacity (Huang & Liu, 2021; Siciliano et al., 2021). However, an analysis of such national policies reveals that they can also perpetuate inequalities when projects are located solely in economically affluent regions or when dichotomous subsidies simultaneously support both fossil fuels and renewables, leading to confused market signals and a lack of coherent strategy (Si et al., 2025).

14.3.3 Regional Cooperation in Energy Transition Policies

Regional initiatives offer invaluable opportunities for collective action in the energy transition. For instance, the European Union's Green Deal aims to achieve a carbon-neutral continent by 2050 and to integrate ambitious renewable energy targets with mechanisms for economic justice (Jenkins et al., 2016). This proactive stance not only accelerates the transition but also incorporates protections for workers and communities vulnerable to economic displacement as a result of the shift. In the African context, the Agenda 2063 Roadmap emphasizes expanding renewable energy capacity while tackling persistent energy poverty, demonstrating a coherent, regionally tailored vision for sustainable development (Awolesi et al., 2024; Goforth & Nock, 2022). Similar regional frameworks in ASEAN are beginning to harmonize energy markets through cross-border collaborations that integrate diverse renewable energy sources (Kovacic et al., 2021). Such cooperative measures highlight the importance of meaningful partnerships and collaboration in advancing energy transition efforts more efficiently (Goforth & Nock, 2022).

14.3.4 Achieving Accountability and Inclusivity

One of the most persistent challenges in the realm of energy transition policy is ensuring accountability and inclusivity during policy implementation (McCauley et al., 2019). Many energy policies continue to prioritize industrial and economic growth over social equity, resulting in the exclusion of marginalized communities from crucial decision-making processes. As highlighted by recent literature, financing mechanisms such as green bonds and climate finance can facilitate the scaling of renewable energy projects; however, they should also be structured in ways that empower disadvantaged groups rather than reinforcing existing power dynamics (Jones Jr & Reyes, 2023; Telli et al., 2025). These justice dimensions are critical to overcoming the inherent inequities in current energy systems, which often leave the voices of the most affected communities unheard (Awolesi et al., 2024; Sovacool et al., 2019).

14.3.5 Opportunities and Challenges of Energy Transition Policies

The potentialities and pitfalls inherent in energy transition policies necessitate careful design and implementation to maximize benefits and minimize adverse outcomes (Mrozowska et al., 2021). Policies can provide enabling environments conducive to renewable energy growth and social protection, effectively balancing climate goals with social responsibilities. However, without stringent measures to ensure equity and justice are prioritized, there is a risk that such policies might enable a rapidly inequitable transition that further marginalizes vulnerable groups (Jones Jr & Reyes, 2023; Sovacool et al., 2017). Therefore, establishing a policy landscape that upholds justice is paramount not only for the success of renewable energy projects but also for sustaining public trust and engagement throughout the transition process (Jones Jr & Reyes, 2023; McCauley et al., 2019; Sovacool et al., 2017). Through this comprehensive approach to designing and implementing energy transition policies, stakeholders can nurture an inclusive ecosystem that equitably shares both the burdens and benefits of transitioning to a decarbonized future. Such frameworks should continuously evolve, aligning with community needs and values while addressing the transitioning challenges inherent in shifting energy architectures worldwide (McCauley et al., 2019; Siciliano et al., 2021).

14.4 Power and Politics in the Energy Transition

The energy transition is not merely a technical process of substituting fossil fuels with renewable energy resources; it is a deeply political transformation that reshapes power relations within and across societies. Decisions about who controls renewable energy infrastructure, how resources are distributed, and which communities benefit are determined by political struggles that extend from local councils to global governance arenas (Köhler et al., 2019; Ogwu et al., 2025d). Understanding these dynamics is crucial to ensuring that the transition is not only rapid but also just. Table 14.1 provides an overview of power and politics in the energy transition. Recent research on sustainability transitions emphasizes the importance of power, agency, and politics within these transformations, suggesting that without addressing these dynamics, the desired outcomes may not materialize (Haas, 2021; Köhler et al., 2019).

Figure 14.2 illustrates the complex configuration of political and economic forces that shape the direction, pace, and equity of the global energy transition. It

Table 14.1 Power and politics in the energy transition

Dimension	Fossil fuel era	Renewable energy transition	Justice implications
Corporate power	Oil, coal, and gas companies dominate policy and lobbying	Renewable firms and tech corporations are gaining influence	Need for regulation to prevent new monopolies
Geopolitics	Control over oil and gas reserves shapes global politics	Critical minerals (lithium, cobalt, rare earths) and renewable tech control	Risk of "energy colonialism" in resource extraction
Labor and employment	Coal miners and oil workers are concentrated in extractive regions	Transition to renewable jobs (solar, wind, hydrogen, etc.)	Requires retraining and just transition funds
North–south divide	Developed countries dominate fossil fuel profits	Unequal financing and technology transfer for renewable projects	Justice demands equitable finance and technology sharing

highlights four major actor groups—incumbent fossil fuel power, emerging renewable energy interests, community and civil society movements, and government and policy institutions—and maps the dynamic ways in which each group influences transition pathways. Power asymmetries, lobbying, regulatory frameworks, resource control, and justice-oriented advocacy interact continuously, producing tensions that either delay transformative change or push the transition toward more democratic, inclusive, and sustainable outcomes.

At the heart of this contestation lies the incumbent power of fossil fuel industries, which continue to exert significant influence over policy and investment decisions. Corporate lobbying, political alliances, and entrenched financial interests often slow the pace of coal, oil, and gas phase-outs (Brown & Spiegel, 2019; Christophers, 2022a, 2022b). For instance, in many countries, subsidies for fossil fuels still far outweigh incentives for renewable energy, reflecting the political clout of established energy actors. This resistance complicates efforts to redirect subsidies and public financing toward solar, wind, and other renewable technologies, as these powerful actors leverage their influence to maintain the status quo (Babayomi et al., 2022; Brown & Spiegel, 2019). The persistence of these subsidies underscores the challenges of creating a truly equitable energy transition and highlights the need to scrutinize the political and economic frameworks that facilitate fossil fuel dependency (Babayomi et al., 2022; Christophers, 2021).

The transition also generates new centers of power. Companies that control solar panel manufacturing, wind turbine technologies, or battery storage systems are emerging as key players in the global energy economy (Goldthau et al., 2020). Control over critical minerals—such as lithium, cobalt, and rare earths, which are essential for renewable technologies—has introduced new geopolitical dynamics (Tornel, 2023). As countries in the Global South, particularly in Africa and Latin America, become sites of extraction for these minerals, there are rising concerns

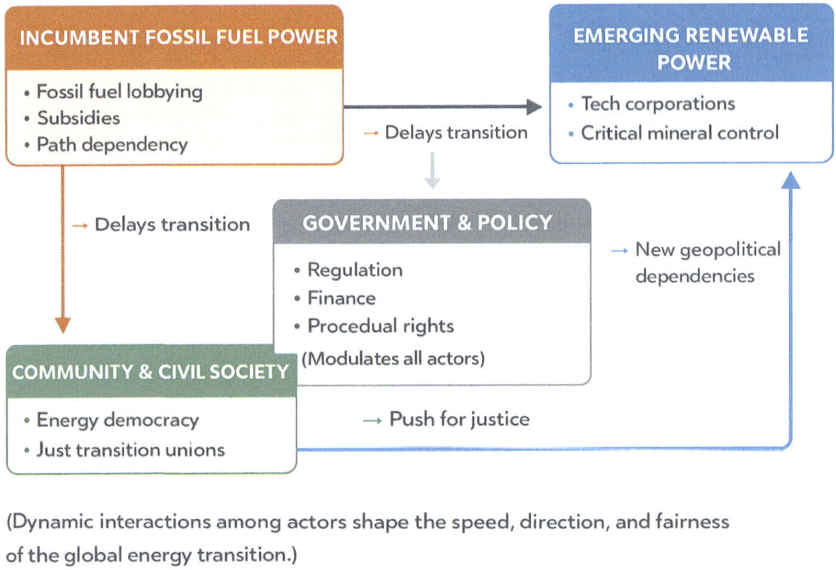

Fig. 14.2 Political–economic forces shaping energy transition outcomes

over "green colonialism." The argument here is that resource wealth may serve global demand without delivering equitable benefits to local populations, echoing historical patterns of exploitation (Goldthau et al., 2020) and highlighting the need to reevaluate how energy justice is conceptualized and realized (Tornel, 2023).

On a broader scale, the Global North–South divide shapes how the benefits and burdens of the energy transition are distributed. Wealthier nations often have greater access to financing, advanced technologies, and renewable infrastructure, while poorer nations face barriers to entry despite being among the most vulnerable to climate change (Lees, 2021; Mulia et al., 2023). The lack of adequate climate finance and fair technology transfer perpetuates dependency, making it difficult for many countries in the Global South to chart independent, just transition pathways (Mulia et al., 2023; Ogwu, 2019). The imbalances in resources not only create disparities in energy access but also in the capacity to adapt to changing climate conditions, highlighting a systemic issue within the global energy landscape that should be addressed through collective political action (Healy & Barry, 2017).

Power dynamics are equally visible within societies. Workers in coal and oil-dependent regions often resist transition policies when they perceive them as threats to livelihoods (Allan et al., 2021). Trade unions play a vital role in advocating for retraining programs and social safety nets to protect workers as industries shift (Haas, 2021). At the community level, questions of land rights, resource ownership, and participation in decision-making highlight the political stakes of renewable energy development. Large-scale solar farms or wind parks can empower communities through shared ownership models, but if imposed without consent, they risk marginalizing local populations (Allan et al., 2021; Healy & Barry, 2017). The necessity of including diverse voices in the energy transition discourse is crucial to achieving equitable outcomes.

Eventually, the politics of the energy transition reveal that justice cannot be assumed simply by expanding renewables. Without addressing entrenched power structures, the transition risks reproducing old inequalities under a new green economy (Healy & Barry, 2017; Ogwu et al., 2025b, 2025c). As researchers have noted, the energy transition requires a radical rethinking of how power operates within the sector (Babayomi et al., 2022; Christophers, 2021). Ensuring that the energy transition is democratic, participatory, and inclusive requires not only technological innovation but also political courage to challenge the dominance of influential actors and prioritize the needs of vulnerable populations (Brown & Spiegel, 2019; Chipato et al., 2020). Future policies should be crafted with an awareness of historical injustices and the objective of dismantling inequitable power dynamics that have persisted over extended periods.

14.5 Social and Economic Dimensions of the Energy Transition

The energy transition represents a significant shift in both the environmental and socio-economic landscapes of societies worldwide (Table 14.2). As countries strive to reduce carbon emissions and advance toward renewable energy sources, the implications for employment, community empowerment, and poverty alleviation become increasingly significant. However, there are inherent risks associated with this transition, particularly if the changes are not managed equitably. The social and economic dimensions of the energy transition fundamentally determine whether it promotes inclusivity and sustainability or perpetuates and exacerbates existing inequalities.

One of the most immediate and observable effects of the shift toward renewable energy is its impact on employment and labor markets. The renewable energy sector is rapidly emerging as one of the most promising areas for job creation globally, with industries such as solar, wind, hydro, and bioenergy witnessing remarkable growth. For instance, jobs for solar photovoltaic (PV) installers and wind turbine technicians are among the fastest-growing professions, highlighting a significant shift in labor demand from fossil fuels to renewable energy technologies (Taşkın et al., 2020). Nevertheless, while new job opportunities arise, the decline of traditional fossil fuel jobs, notably in coal and oil sectors, raises concerns about worker displacement and the necessity of effective transition strategies to support affected individuals (Okkonen & Lehtonen, 2016). A just transition is therefore critical, which requires policies that facilitate retraining, income support, and alternative employment for those displaced by the energy transformation. Labor unions and worker associations play a crucial role in advocating for these policies, ensuring that affected communities are not marginalized in the process.

Table 14.2 Social and economic dimensions of the energy transition

Aspect	Challenges	Opportunities with renewables	Justice considerations
Employment	Job losses in the fossil fuel industries	Renewable energy jobs: solar installers, wind engineers, etc.	Skills development, labor protection
Energy access	Persistent energy poverty in many developing countries	Decentralized renewables (solar microgrids, community wind)	Ensuring universal and affordable clean energy
Community participation	Centralized decision-making in energy projects	Community-owned renewable cooperatives, local involvement	Procedural justice, empowerment of communities
Gender equity	Women are marginalized in fossil fuel and energy policy	Women as leaders in clean cooking, solar entrepreneurship	Recognition, justice, and gender inclusion
Intergenerational justice	Future generations are burdened with climate change costs	Investing in sustainable and long-term renewable systems	Protecting future rights to energy and the environment

Moreover, the energy transition presents a unique opportunity to tackle energy poverty and enhance access to affordable clean electricity. Decentralized renewable energy systems, such as community-based solar projects and micro-hydropower, can significantly influence areas where traditional energy grids have failed to reach, particularly in rural and underserved communities (Idoko et al., 2023). These systems not only provide reliable energy access but also help reduce reliance on expensive, environmentally damaging energy sources like diesel. Successful initiatives can lead to improved living standards and the nurturing of local economies; however, their benefits hinge on prioritizing accessibility and affordability in policy frameworks rather than framing renewable energy purely as a market commodity (Allan et al., 2020). Comprehensive policies are needed to harness these potentials, ensuring that the focus remains on equitable energy distribution and not solely on economic profit. Community participation and stakeholder engagement are also crucial for fostering social justice within the energy transition. Projects designed with input from local communities tend to be more successful in delivering socio-economic benefits, including stable income streams and reduced energy costs (Alnaqbi & Alami, 2023). Conversely, energy projects that overlook community involvement can lead to displacement, disruption, and the violation of cultural rights. Inclusion of Indigenous peoples and marginalized voices is paramount for developing trust and legitimizing transition policies; failure to do so may heighten social tensions and resentment toward renewable energy initiatives.

The gendered impacts of the energy transition further highlight the importance of equity within the renewable energy ecosystem. Women in many developing countries disproportionately face challenges related to energy poverty, primarily through their reliance on biomass and limited access to electricity. Renewable energy initiatives that focus on empowering women—such as training programs for female engineers or supporting women-led energy cooperatives—demonstrate transformative potential at both household and community levels (Bennear, 2022). Such gender-sensitive approaches are essential to ensure that the energy transition does not entrench existing disparities but instead dismantles them, contributing to more equitable socio-economic outcomes. Addressing intergenerational justice is another critical dimension of the energy transition. The decisions made and investments undertaken today will reverberate through future generations, affecting their security, health, and overall quality of life (Izah & Ogwu, 2025). In prioritizing long-term sustainability, it is necessary to integrate educational and skills-development pathways that equip young people for careers in a renewable energy economy. This framework aligns the transition toward a sustainable energy future with broader goals of equity and fairness, acknowledging the responsibilities of current generations to future ones. Such a vision promises a comprehensive approach to the energy transition, enabling not only reductions in carbon emissions but also the fostering of social equity and economic resilience.

14.6 Post-Carbon Futures and Energy Transition Pathways

The global energy transition is focused not only on reducing carbon emissions but also on establishing post-carbon futures characterized by social justice, technological innovation, and environmental sustainability (Table 14.3). A post-carbon future envisions a systematic reorganization of energy frameworks that pivots toward renewable resources such as solar, wind, hydropower, geothermal energy, and bioenergy. This shift necessitates a reconfiguration of both economies and societies, aiming to curtail dependencies on extractive and polluting industries that have historically contributed to environmental degradation and social inequities (Sahoo et al., 2023; Sohag et al., 2015). The energy pathways selected today will have significant ramifications on the resilience and equity of future societies, potentially entrenching existing inequalities or fostering a more just energy landscape (Zhang et al., 2020).

Figure 14.3 synthesizes major post-carbon transition pathways discussed in the literature. It shows how alternative models—100% renewable scenarios, technological innovation routes, circular/doughnut economies, degrowth and well-being paradigms, and energy-democracy movements—branch from the present energy system toward distinct but overlapping post-carbon futures. Together, these pathways highlight that decarbonization trajectories vary in their technological, political, and justice implications, reinforcing the need for equity-centered governance in shaping sustainable outcomes.

Technological innovation and system integration are critical components of this transition (Izah et al., 2024). The rapid expansion of solar and wind energy has catalyzed significant transformations in global electricity markets. However, the

Table 14.3 Post-carbon futures and energy transition pathways

Pathway/model	Key features	Role of renewable energy	Justice potential
100% renewable scenarios	World powered entirely by solar, wind, hydro, geothermal, and bioenergy	Eliminates fossil dependence, decentralized systems	Reduces emissions; should ensure equitable access
Technological innovation	Smart grids, AI, green hydrogen, storage solutions	Integrates and balances renewable generation	Accessibility and affordability are critical
Circular/doughnut economies	Focus on sustainability, limits to growth, and planetary boundaries	Renewable energy as the foundation of a low-impact economy	Prioritizes ecological balance and social equity
Degrowth and well-being models	Reducing material/energy demand; prioritizing quality of life	Emphasizes efficiency and sufficiency in energy use	Benefits marginalized groups, lessens overconsumption
Energy-democracy movements	Grassroots-driven, community control of energy systems	Community-owned solar, wind, micro-hydro	Enhances participation, distributes benefits fairly

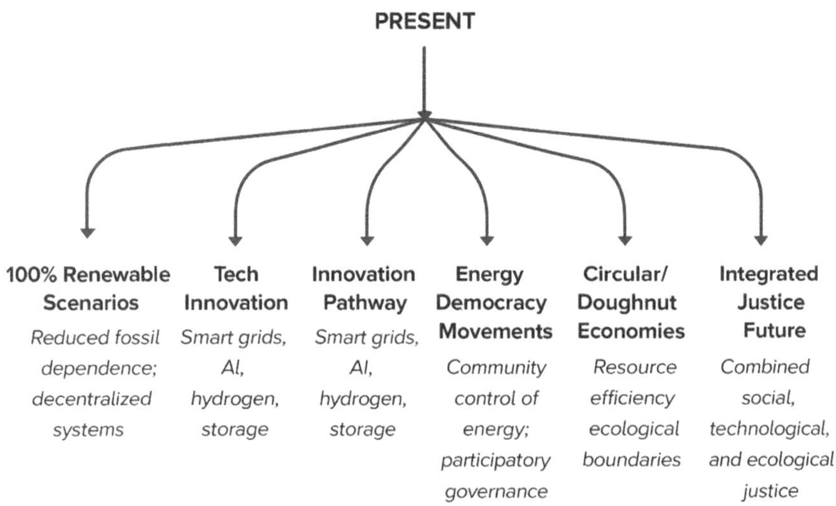

Fig. 14.3 Multiple pathways toward post-carbon futures

sustained advancement of these technologies depends on the development of complementary frameworks, including large-scale energy storage solutions, smart grid technology, and demand response management systems. Innovative structures like offshore wind farms, floating solar installations, and enhanced geothermal systems exemplify how renewable energy can augment supply diversification and bolster energy security (Fu et al., 2023; Li & Liao, 2020). Moreover, emerging technologies such as green hydrogen represent a multifaceted energy carrier that facilitates the storage of excess renewable energy and aids in decarbonizing traditionally challenging industries like steel, cement, and aviation (Adkin, 2019; Wang & Chen, 2024). Collectively, these advancements signal a future where renewable resources dominate not only the electricity sector but also transportation, industrial processes, and heating systems (Xie et al., 2024).

Another critical pathway emphasizes the importance of decentralized, community-focused energy systems. Within this model, small-scale renewable installations (including solar rooftops and microgrids) enhance local communities' capabilities to generate and manage their energy supply independently. This decentralization mitigates vulnerabilities associated with centralized power structures and stimulates local economic development while fostering participatory governance in energy matters (Kunskaja et al., 2023; Zhao & Tang, 2024). For marginalized and rural communities, decentralized renewable systems provide more than just environmental solutions; they represent pathways to social equity and political autonomy (Guinot et al., 2022; Wang et al., 2024).

The evolving discourse surrounding post-carbon futures invites a reevaluation of existing development models. Conventional frameworks often categorize energy transition as a driver of economic growth, mainly through the promotion of green industries and new market opportunities. However, alternative perspectives advocate for development models that transcend mere growth-centric paradigms. Degrowth movements emphasize the need to reduce consumption in affluent societies to diminish ecological footprints, thereby fostering sustainability (Kang et al., 2024). The circular economy model promotes concepts of reuse, recycling, and resource efficiency, aligning economic activities with ecological imperatives. Frameworks such as doughnut economics and well-being economies shift the focus from GDP-centric growth toward holistic human flourishing and ecological integrity (Koilo et al., 2022; Zhao et al., 2023). It is also paramount to acknowledge the roles of social movements and civil society in shaping these energy transition pathways. Grassroots mobilizations advocating for "climate justice" and "energy democracy" are crucial in pressuring both governments and corporations to expedite the phase-out of fossil fuels, thus expanding renewable energy initiatives that prioritize human over financial interests (Cao & Qian, 2023). The landscape of post-carbon futures is likely to be a contested arena, profoundly influenced by both technological advancements and active political engagement (Cheng et al., 2022).

In evaluating the relationship between energy use and technological progress, it is evident that improvements in energy efficiency and reductions in emissions from fossil fuels are crucial for addressing the interconnected challenges of economic growth, energy security, and environmental sustainability (Guo et al., 2022). For instance, substantial advancements in green technologies have been shown to improve corporate environmental performance while meeting escalating regulatory mandates for sustainable practices (Sun et al., 2023a, 2023b; Zhao & Zhao, 2024). The facilitation of these processes through financial development and innovation is essential in countries striving to achieve elevated standards of green productivity (Long et al., 2023). Addressing the dynamics of the energy transition requires examining regional differences and their impacts on industrial strategies and carbon emission profiles (Wang et al., 2021). Technology diffusion, particularly in new energy sectors, serves the dual purpose of fostering economic integration while enabling the societal benefits derived from widespread technological advances (Borowski, 2021). The interactions among technological innovation, energy efficiency, and economic growth further underscore the urgency of coherent policy frameworks that support sustainable development.

14.7 Equity Challenges in the Global Energy Transition

While the energy transition offers unprecedented opportunities for decarbonization and sustainable development, it also raises profound equity challenges. The shift toward renewable energy is unfolding unevenly across regions, classes, and communities, creating winners and losers. Addressing these disparities is central to ensuring that the global energy transition is not only rapid but also just. The concept of energy justice emphasizes the fair distribution of energy benefits and burdens, which is crucial in the current global energy landscape (Sovacool et al., 2017). Table 14.4 provides an overview of equity challenges in the global energy transition.

One of the most pressing equity challenges concerns the unequal distribution of renewable energy resources and infrastructure. Wealthier countries in the Global North have made significant progress in scaling solar, wind, and storage technologies, while many nations in the Global South continue to struggle with limited access to financing, skilled labor, and technology transfer (Zhou et al., 2023). This asymmetry reflects historical injustices: those least responsible for greenhouse gas emissions are often the ones least equipped to invest in renewable energy, even though they are the most vulnerable to climate impacts (Durá-Alemañ et al., 2023). For instance, many small island developing states depend heavily on imported fossil fuels, making them particularly vulnerable to global market fluctuations and the effects of climate change (Van de Ven et al., 2021).

A second challenge is the risk of land-use conflicts and displacement associated with large-scale renewable energy projects. Solar farms, wind parks, and hydropower dams often require significant land, which can lead to the dispossession of local communities, particularly Indigenous peoples and small-scale farmers (Yenneti et al., 2016). In some instances, renewable energy projects can replicate patterns of exclusion and exploitation historically linked to fossil fuel

Table 14.4 Equity challenges in the global energy transition

Challenge	Example/ manifestation	Impact	Justice lens
Unequal resource distribution	Solar and wind are concentrated in specific geographies	Some regions benefit more, others lag behind	Calls for regional balancing and fair financing
Land-use conflicts	Displacement for solar farms, hydropower dams, and wind parks	Loss of livelihoods, Indigenous land rights	Requires consent, compensation, recognition
Green extractivism	Mining lithium, cobalt, and rare earths in the Global South	Exploitation, environmental harm	Need ethical sourcing, fair trade
Vulnerable regions	SIDS are dependent on renewable imports and financing	High costs, energy insecurity	International support and adaptation funding
Technology gaps	North dominates renewable technology patents and expertise	Exclusion of Global South innovators	Policies for tech transfer and local capacity

development, thereby raising concerns about "green colonialism" (Yenneti et al., 2016). This highlights the need to ensure meaningful community participation and the recognition of land rights to avoid deepening social injustices under the guise of sustainability. The concept of spatial justice is pertinent here, as it highlights the need for equitable land distribution in the context of renewable energy developments (Kiesecker et al., 2023).

Another major concern is the extraction of resources for renewable technologies. The demand for critical minerals such as lithium, cobalt, and rare earth elements (needed for batteries, wind turbines, and solar panels) has surged (Fu et al., 2024). Much of this extraction occurs in resource-rich but economically marginalized regions of Africa, Latin America, and Asia, raising significant equity concerns regarding labor practices and environmental sustainability (Fu et al., 2024). Without robust governance, mining activities risk replicating the exploitative structures familiar in fossil-fuel economies, thereby creating a "green" form of exploitation. For example, lithium extraction in South America often results in severe social and environmental repercussions for local communities, underscoring the need for ethical resource extraction practices in the transition to renewable energy (Fu et al., 2024). Small Island Developing States (SIDS) and climate-vulnerable regions face unique equity challenges in the global energy transition. Although their contribution to global emissions is minimal, they bear disproportionate risks from rising sea levels, extreme weather, and energy insecurity (Van de Ven et al., 2021). These states often depend on imported fossil fuels, which raises their susceptibility to market volatility. Transitioning to renewables, such as solar and wind, could enhance resilience, but financial and technological barriers continue to impede their progress. This situation illustrates the importance of international support and climate finance to ensure that vulnerable nations are not left behind in the global energy transition paradigm (Durá-Alemañ et al., 2023).

Equity challenges are not only global but also manifest within societies. Within countries, marginalized groups (such as low-income households, rural populations, and women) often have the least access to renewable energy and the least influence over transition policies (Ayllón & Jenkins, 2023). Without deliberate measures, subsidies, and inclusive decision-making processes, the transition may inadvertently reinforce existing social divides rather than bridge them. In particular, localized policy frameworks that emphasize community engagement and representation can help rectify disparities within nations by aligning energy policies with equity principles outlined in energy justice discussions (Zhou et al., 2023). The path toward a just and inclusive energy transition requires policies that not only prioritize sustainability but also actively address past injustices. Principles of energy justice and the active involvement of marginalized communities should guide the design and implementation of renewable energy projects to avoid exacerbating existing inequalities (Sovacool et al., 2017). Policymakers should ensure that all voices are heard, particularly those of vulnerable populations, to create a balanced and equitable transition to a post-carbon future that equitably distributes both the benefits and burdens of energy resources (Ayllón & Jenkins, 2023). Ethical questions sit at the core of post-carbon futures. Intergenerational justice demands that today's transition choices do not transfer ecological debt, stranded risks, or unaffordable energy systems onto future generations; this requires long-horizon planning, youth inclusion, and safeguards against short-term political capture. Digital inequity is becoming a new fault line as AI-enabled grid management, forecasting, and finance tools expand; without equitable access to data, skills, and digital infrastructure, poorer regions may be locked out of the benefits of "smart" transitions and left with higher costs or unreliable systems. Data colonialism is a rising concern in which Global North institutions extract energy, mineral, and community data from the Global South without fair reciprocity, governance authority, or shared benefits. Ethical transitions, therefore, require data sovereignty, transparent partnerships, and benefit-sharing agreements that respect local ownership of knowledge and resources.

Figure 14.4 visualizes the structural inequities that shape the global energy transition. While the Global North possesses the financial, technological, and infrastructural advantages that enable rapid decarbonization, the Global South continues to face systemic barriers, including limited access to climate finance, challenges in technology transfer, extractive pressures, and heightened climate vulnerability. Presenting these contrasts side by side underscores the asymmetrical foundations of the transition and reinforces the need for justice-centered frameworks that prioritize equity, capacity building, and inclusive governance.

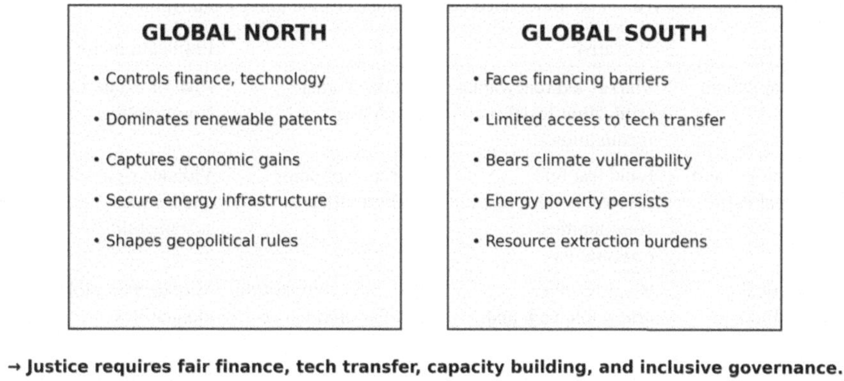

Fig. 14.4 Structural inequities in the global energy transition

14.8 Barriers, Risks, and Unintended Consequences of the Energy Transition

Although the global energy transition, aimed at reducing carbon emissions and enhancing sustainability, embodies significant potential, it is fraught with numerous barriers and risks that can hinder its effective implementation. The interaction of social, economic, and geopolitical factors complicates the transition from fossil fuels to renewable energy systems. These complexities necessitate an analysis of potential unintended consequences, policy frameworks, and market dynamics. A key barrier to the transition is the persistent tendency of governments to subsidize fossil fuels, which has not only fostered dependency on traditional energy sources but has also perpetuated an uneven playing field that favors non-renewable energy investments over renewables. According to Jewell et al. (2018), removing fossil fuel subsidies may improve the prospects for renewable energy adoption, but such measures yield limited emission reductions unless they are implemented in energy-exporting regions, where reliance on fossil fuel revenues is especially pronounced. This situation reflects the entrenched structures that impede progress toward a sustainable energy landscape. Table 14.5 provides the barriers, risks, and unintended consequences of energy transition.

The ongoing political and economic dominance of fossil fuel companies shapes energy policy and investment decisions. Breetz et al. (2018) contend that the political logics driving clean energy transitions highlight global disparities in energy cost curves, allowing countries to participate in the transition differently based on their political advantages. Additionally, systemic inequalities, often perpetuated by fossil fuel subsidies and corporate lobbying, can inhibit equitable participation in clean energy development and result in new forms of economic exploitation. As highlighted by Manowska et al. (2024), the transition to renewable energy can indeed create economic opportunities and foster innovation; however, without

Table 14.5 Barriers, risks, and unintended consequences of the energy transition

Barrier/risk	Narrative	Example	Implications for justice
Green capitalism	Market-led renewables replicating fossil fuel inequalities	Privatized solar monopolies	Risk of excluding poor communities
Displacement and ecological harm	Land use for renewables displaces communities/ecosystems	Hydropower dams flooding villages	Violates rights, undermines sustainability
Geopolitical competition	Rivalries over renewable tech and critical minerals	US–China competition for solar dominance	May deepen global inequalities
Technological lock-ins	Over-reliance on specific renewable technologies	Overdependence on lithium for storage	Vulnerable to supply shocks, injustice in resource chains
Rebound effects	Efficiency gains drive more consumption	Cheaper renewables lead to higher use	Risks negating emissions reductions

deliberate interventions, new energy systems may replicate the social injustices witnessed in the fossil fuel era.

Moreover, ecological risks associated with renewable energy technologies warrant significant attention. While they are designed to minimize greenhouse gas emissions, considerable resource extraction and land-use change can have detrimental environmental impacts. Capellán-Pérez et al. identify specific biophysical constraints associated with land availability for solar energy, underscoring the need for thorough environmental assessments during project planning (Capellán-Pérez et al., 2017). For instance, while the construction of major hydropower facilities may yield energy benefits, it can also disrupt local ecosystems and communities. Thus, renewable technologies can become new ecological burdens unless mitigative measures are implemented diligently. Resource dependencies on critical minerals essential to the energy transition also pose substantial risks. The spike in demand for minerals like lithium, cobalt, and rare earth elements used in battery technologies raises concerns about environmental degradation, labor rights, and geopolitical tensions, especially in the Global South, where mining operations often exploit vulnerable communities. Scholars like Kern and Rogge emphasize that governance frameworks should evolve to address these dependencies effectively or risk reproducing the "resource curse" seen in fossil fuel markets (Kern & Rogge, 2016). The potential for geopolitical factions to emerge over control of clean energy supply chains may further complicate the international landscape. In their analysis, Yang et al. note that competition for dominance in renewable technology markets could narrow opportunities for equitable energy transitions, particularly affecting nations lacking the requisite technological infrastructure or economic power (Yang et al., 2023).

Another notable concern is rebound effects, in which improvements in energy efficiency inadvertently lead to increased overall energy consumption. In light of this, Rosenow et al. assert that without addressing systemic consumption patterns alongside technological advancements, the ecological gains from renewable sources may be overshadowed by rising demands (Rosenow et al., 2017). This dynamic emphasizes the essential need for a cultural shift toward sustainable lifestyles, which can help mitigate the negative implications of increased energy efficiency by fostering demand for genuinely sustainable consumption practices.

14.9 Governance and Policy Recommendations for a Just Energy Transition

Achieving a just energy transition requires more than technological innovation; it demands governance structures and policies that ensure inclusivity, accountability, and sustainability. The transition to renewable energy will succeed only if it actively addresses the social, economic, and political dimensions of justice while delivering on decarbonization goals. It is essential to recognize the multifaceted implications of energy policies, where the journey toward decarbonization should also embrace equity, solidarity, and comprehensive planning to ensure no group is left behind in this critical transformation.

A primary policy priority is redirecting subsidies and investments. Governments should phase-out fossil fuel subsidies, which continue to lock economies into carbon-intensive pathways, and redirect these funds toward solar, wind, hydro, geothermal, and emerging energy storage technologies. Numerous studies highlight the efficacy of such measures, highlighting that phasing out fossil fuel subsidies can free up significant capital for investment in sustainable energy solutions, eventually aligning with the goals set by international agreements like the Paris Agreement (Green, 2021). Strategic public investments should prioritize decentralized, community-owned renewable projects to ensure the benefits of the transition are shared more equitably, particularly in marginalized and vulnerable communities (Muttitt & Kartha, 2020). For instance, Lai points out that innovative funding mechanisms can harness green finance for local-level applications, fostering community resilience and empowerment in steering their energy futures toward sustainability (Lai, 2025).

Second, strong regulatory frameworks and participatory governance are essential to prevent exploitation and exclusion during the energy transition. The deployment of large-scale renewable projects should be subject to thorough environmental and social impact assessments, accompanied by meaningful community engagement at every stage (Earsom, 2024). As documented by Diezmartínez and Gianotti, cities are increasingly recognizing the significance of justice in climate action plans, thereby defining a path toward inclusive energy governance that respects local context and community dynamics (Diezmartínez & Short Gianotti, 2022). Mechanisms such as participatory budgeting and energy cooperatives can empower local actors while reducing the risk of "green colonialism," which refers to scenarios where communities are displaced or marginalized in the name of renewable energy projects (Lori et al., 2025; Ogwu & Izah, 2025; Zografos & Robbins, 2020). Maintaining transparency in governance structures can help mitigate the potential for corporate power concentration, further enhancing accountability in energy planning (Gifford & Sauls, 2024).

International cooperation and climate finance should play a central role in addressing global inequities that characterize the energy transition landscape. Wealthy nations bear a historical responsibility to support developing countries in their energy transition—not only through financial transfers but also by facilitating technology transfer, capacity building, and debt relief measures (Barbier & Burgess, 2020; Skovgaard et al., 2023). As articulated by Mohan (2023), larger, poorer island nations often attract more climate finance, leaving smaller but equally vulnerable nations at a disadvantage. Thus, it is imperative to create a global just transition framework that guarantees that nations vulnerable to climate impacts—including Small Island Developing States and resource-dependent economies—are not neglected (Nacke et al., 2024). The establishment of a Loss and Damage Fund under the UNFCCC could serve as a transformative mechanism to bolster climate finance flows, ensuring that those disproportionately affected by climate events receive the necessary support (Uri et al., 2024).

Additionally, governments must recognize the labor dimensions of the energy transition. Workers in the fossil fuel industry face the risk of job losses as demand for coal and oil declines. Just transition policies should include comprehensive retraining programs, income support for displaced workers, and pathways into renewable energy jobs to ensure that no group is disproportionately burdened by the structural changes toward clean energy (Muttitt & Kartha, 2020). As noted by Gifford and Sauls, climate justice scholarship is increasingly acknowledging the systemic inequalities present in energy transitions and advocating for policies that promote equity across various sectors (Gifford & Sauls, 2024). Trade unions, civil society groups, and employers should be key partners in the design of these retraining programs to facilitate a smoother transition into new, sustainable job markets (Jenkins et al., 2020). Another critical dimension of governance for a just energy transition is sustainable resource management, particularly the surge in demand for critical minerals used in renewable technologies (Iyiola & Ogwu, 2025; Ogwu et al., 2025a). Robust international standards for responsible mining, supply chain transparency, and the protection of labor rights are paramount to prevent exploitative practices and ecological degradation associated with mineral extraction (Iyiola et al., 2025b; Joshipura et al., 2025). Investment should also be directed toward recycling initiatives, circular economy practices, and material substitution to minimize reliance on new resource frontiers, as emphasized by El Khoury et al. (2024). This approach would not only mitigate environmental impact but also alleviate the socio-economic consequences that typically accompany resource extraction (Iyiola et al., 2025a).

Furthermore, integrated planning across sectors is vital for effective energy governance, as energy intersects significantly with housing, transport, food systems, and water use (Iyiola et al., 2023). Policies that connect renewable energy expansion with improvements in public transportation and sustainable agriculture can create co-benefits and mitigate unintended consequences (Harrington et al., 2016). This holistic approach to governance advocates for viewing energy policy within a broader socio-economic and environmental context, enabling systems transformations that go beyond merely aiming for decarbonization (Tyson & Weiss, 2025). Such cross-sectoral strategies should be employed to foster systemic resilience, ensuring the energy transition is grounded in the realities of community needs and environmental imperatives (Harrington et al., 2016). Communicating evidence from energy transition models and scenario assessments requires balancing urgency with accuracy. Over-alarmist framing can erode trust and provoke political backlash, while overly optimistic or technocratic communication can minimize risk and delay necessary action. Because energy transitions affect livelihoods, land, and sovereignty, results should be presented with explicit assumptions, uncertainty ranges, and distributional implications. Effective communication also means translating findings into locally meaningful narratives, co-interpreting results with communities, and avoiding one-size-fits-all claims that can misrepresent realities in the Global South. In justice-centered transitions, communication is not only about clarity but also about procedural fairness—who gets to interpret results, whose risks count, and whose futures are prioritized.

Five Principles for Next-Generation Just Energy Transitions

1. **Multilevel integration:** Link national decarbonization goals with local energy access, land rights, and livelihood priorities.
2. **Behavioral and social realism:** Treat communities, workers, and households as active transition agents, not passive recipients.
3. **Equity by design:** Build distributive, procedural, recognition, and intergenerational justice into policy metrics and project approval.
4. **Early-warning governance:** Monitor mineral, land use, price, and displacement thresholds to prevent cascading harms.
5. **Participatory accountability:** Co-design, co-monitor, and co-benefit transitions with marginalized groups, ensuring democratic control over outcomes.

Achieving a just energy transition requires a coordinated and multi-level policy architecture that aligns global commitments with national strategies and locally grounded governance. Figure 14.5 outlines this hierarchical yet interconnected structure, illustrating how international frameworks shape national policy instruments, which, in turn, influence community-based decision-making processes.

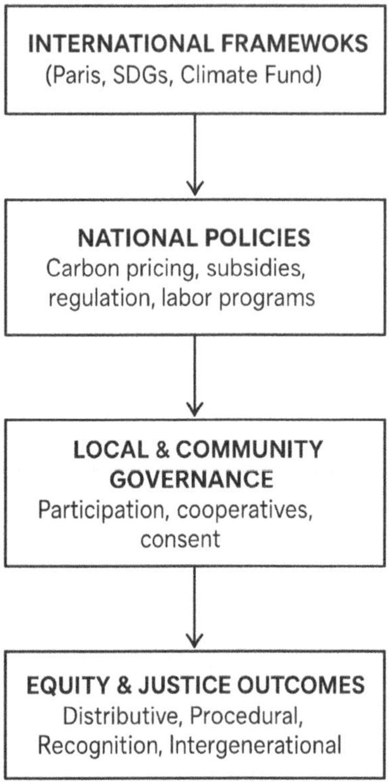

Fig. 14.5 Policy architecture for a just energy transition

14.10 Conclusion

The energy transition is far more than a technological shift; thus, it represents a societal transformation that will define the twenty-first century. Moving from fossil fuels to renewable energy sources such as solar, wind, hydro, geothermal, and bioenergy offers enormous opportunities for climate mitigation, sustainable growth, and energy access. However, without deliberate attention to justice, equity, and inclusivity, this transition risks replicating historical injustices under the guise of new technologies. Embedding democratic governance, dismantling entrenched fossil fuel interests, and ensuring meaningful participation of workers, communities, and marginalized groups are essential for making renewable energy a tool for social and environmental justice rather than another form of exclusion.

At the same time, the complexities of the energy transition (ranging from ecological risks and resource dependencies to global inequalities and power struggles) demand justice-centered frameworks that go beyond technology deployment. By integrating innovation with comprehensive policies on labor rights, gender equality, environmental stewardship, and intergenerational equity, societies can ensure that the benefits of renewable energy are broadly shared. Eventually, the success of the energy transition will not be measured solely by the scale of clean technology adoption but by the transformation of the social relations and power structures that underpin energy systems. If equity and fairness are aggressively pursued, the transition can serve not only as a response to climate change but as the foundation for reimagined post-carbon futures built on justice, sustainability, and shared prosperity.

This book has shown that climate change is not only a scientific or policy challenge but a profound test of global justice and democratic participation. Addressing the climate crisis requires amplifying the voices of communities that have historically been excluded from decision-making—Indigenous peoples, youth, women, frontline workers, and nations disproportionately affected by environmental degradation. By integrating world-systems perspectives with analyses of corporate power, uneven development, and geopolitical inequality, this book reveals how global warming magnifies pre-existing structural injustices while also creating possibilities for transformative change. Building a just and sustainable future depends on collective action, equitable governance, and an unwavering commitment to listening to those most affected.

References

Adkin, L. E. (2019). Technology innovation as a response to climate change: The case of the climate change emissions management corporation of alberta. *Review of Policy Research, 36*(5), 603–634.

Ali, S., Yan, Q., Irfan, M., Ameer, W., Atchike, D. W., & Acevedo-Duque, A. (2022). RETRACTED: Green investment for sustainable business development: The influence of policy instruments on solar technology adoption. *Frontiers in Energy Research, 10*, 874824.

Allan, G., Comerford, D., Connolly, K., McGregor, P., & Ross, A. G. (2020). The economic and environmental impacts of UK offshore wind development: The importance of local content. *Energy, 199*, 117436.

Allan, J., Lemaadel, M., & Lakhal, H. (2022). Oppressive energopolitics in Africa's last colony: Energy, subjectivities, and resistance. *Antipode, 54*(1), 44–63.

Alnaqbi, S. A., & Alami, A. H. (2023). Sustainability and renewable energy in the UAE: A case study of Sharjah. *Energies, 16*(20), 7034.

Awolesi, O., Salter, C. A., & Reams, M. (2024). A systematic review on the path to inclusive and sustainable energy transitions. *Energies, 17*(14), 3512.

Ayllón, L. M. S., & Jenkins, K. E. (2023). Energy justice, Just Transitions and Scottish energy policy: A re-grounding of theory in policy practice. *Energy Research and Social Science, 96*, 102922.

Babayomi, O. O., Dahoro, D. A., & Zhang, Z. (2022). Affordable clean energy transition in developing countries: Pathways and technologies. *iScience, 25*(5), 104178.

Barbier, E. B., & Burgess, J. C. (2020). Sustainability and development after COVID-19. *World Development, 135*, 105082.

Basit, M. A., Dilshad, S., Badar, R., & Sami ur Rehman, S. M. (2020). Limitations, challenges, and solution approaches in grid-connected renewable energy systems. *International Journal of Energy Research, 44*(6), 4132–4162.

Bennear, L. S. (2022). Energy justice, decarbonization, and the clean energy transformation. *Annual Review of Resource Economics, 14*, 647–668.

Bogdanov, D., Farfan, J., Sadovskaia, K., Aghahosseini, A., Child, M., Gulagi, A., Oyewo, A. S., de Souza Noel Simas Barbosa, L., & Breyer, C. (2019). Radical transformation pathway towards sustainable electricity via evolutionary steps. *Nature Communications, 10*(1), 1–16.
Borowski, P. F. (2021). Innovation strategy on the example of companies using bamboo. *Journal of Innovation and Entrepreneurship, 10*(1), 3.
Breetz, H., Mildenberger, M., & Stokes, L. (2018). The political logics of clean energy transitions. *Business and Politics, 20*(4), 492–522.
Brennecke, M., Jurek, D., Rieger, A., & Roth, T. (2024). Towards social justice in energy transitions: An information systems perspective. In *Proceedings of the 57th Hawaii International Conference on System Sciences (HICSS),* Honolulu, USA.
Brown, B., & Spiegel, S. J. (2019). Coal, climate justice, and the cultural politics of energy transition. *Global Environmental Politics, 19*(2), 149–168.
Cao, Z., & Qian, L. (2023). Crowding effect of institutional openness based on the big data algorithm on the efficiency of new energy technology innovation. *Frontiers in Bioengineering and Biotechnology, 11*, 1178737.
Capellán-Pérez, I., De Castro, C., & Arto, I. (2017). Assessing vulnerabilities and limits in the transition to renewable energies: Land requirements under 100% solar energy scenarios. *Renewable and Sustainable Energy Reviews, 77*, 760–782.
Cheng, S., Meng, L., & Xing, L. (2022). Energy technological innovation and carbon emissions mitigation: Evidence from China. *Kybernetes, 51*(3), 982–1008.
Chipato, F., Wang, L., Zuo, T., & Mudimu, G. T. (2020). The politics of youth struggles for land in post-land reform Zimbabwe. *Review of African Political Economy, 47*(163), 59–77.
Christophers, B. (2022a). Fossilised capital: Price and profit in the energy transition. *New Political Economy, 27*(1), 146–159.
Christophers, B. (2022b). Taking renewables to market: Prospects for the after-subsidy energy transition: The 2021 Antipode RGS-IBG Lecture. *Antipode, 54*(5), 1519–1544.
Diezmartínez, C. V., & Short Gianotti, A. G. (2022). US cities increasingly integrate justice into climate planning and create policy tools for climate justice. *Nature Communications, 13*(1), 5763.
Durá-Alemañ, C. J., Moleón, M., Pérez-García, J. M., Serrano, D., & Sánchez-Zapata, J. A. (2023). Climate change and energy crisis drive an unprecedented EU environmental law regression. *Conservation Letters, 16*, e12958.
Earsom, J. (2024). Fit for purpose? Just Energy Transition Partnerships and accountability in international climate governance. *Global Policy, 15*(1), 135–141.
El Khoury, R., Nasrallah, N., Marashdeh, H., Atayah, O. F., & Mathiyazhagan, K. (2024). The nexus of Green finance and renewable energy on CO_2 emissions. *Business Strategy and the Environment, 33*(8), 8911–8928.
Eze, V. H. U., Edozie, E., Wisdom, O., & Uche, C. K. A. (2023). A comparative analysis of renewable energy policies and its impact on economic growth: A review. *International Journal of Education, Science, Technology, and Engineering (IJESTE), 6*(2), 41–46.
Fu, C., Luo, C., & Liu, Y. (2024). Is the economy, environment and energy (3E) system sustainable?—An analysis of the coordination degree of carbon decoupling and energy equity in the Yangtze River Economic Belt. *Sustainability, 16*(13), 5817.
Fu, J., Xue, H., Wang, F., & Wang, L. (2023). The impact of high-quality energy development and technological innovation on the real economy of the Yangtze River Economic Belt in China: A spatial economic and threshold effect analysis. *Sustainability, 15*(2), 1453.
Gifford, L., & Sauls, L. A. (2024). Defining climate finance justice: Critical geographies of justice amid financialized climate action. *Geography Compass, 18*(11), e70008.
Goforth, T., & Nock, D. (2022). Air pollution disparities and equality assessments of US national decarbonization strategies. *Nature Communications, 13*(1), 7488.
Goldthau, A., Eicke, L., & Weko, S. (2020). The global energy transition and the global south. In *The geopolitics of the global energy transition* (pp. 319–339). Springer International Publishing.
Green, J. F. (2021). Does carbon pricing reduce emissions? A review of ex-post analyses. *Environmental Research Letters, 16*(4), 043004.

Greim, P., Solomon, A. A., & Breyer, C. (2020). Assessment of lithium criticality in the global energy transition and addressing policy gaps in transportation. *Nature Communications, 11*(1), 4570.

Guinot, J., Barghouti, Z., & Chiva, R. (2022). Understanding green innovation: A conceptual framework. *Sustainability, 14*(10), 5787.

Guo, Q., Ding, C., Guo, T., & Liu, S. (2022). Dynamic effects and regional differences of industrialization and urbanization on China's energy intensity under the background of "dual carbon". *Sustainability, 14*(16), 9948.

Gyamfi, S., Modjinou, M., & Djordjevic, S. (2015). Improving electricity supply security in Ghana—The potential of renewable energy. *Renewable and Sustainable Energy Reviews, 43*, 1035–1045.

Haas, T. (2021). From green energy to the green car state? The political economy of ecological modernisation in Germany. *New Political Economy, 26*(4), 660–673.

Harrington, L. J., Frame, D. J., Fischer, E. M., Hawkins, E., Joshi, M., & Jones, C. D. (2016). Poorest countries experience earlier anthropogenic emergence of daily temperature extremes. *Environmental Research Letters, 11*(5), 055007.

Healy, N., & Barry, J. (2017). Politicizing energy justice and energy system transitions: Fossil fuel divestment and a "just transition". *Energy Policy, 108*, 451–459.

Herranz-Surralles, A. (2024). The EU energy transition in a geopoliticizing world. *Geopolitics, 29*(5), 1882–1912.

Huang, P., & Liu, Y. (2021). Toward just energy transitions in authoritarian regimes: Indirect participation and adaptive governance. *Journal of Environmental Planning and Management, 64*(1), 1–21.

Idoko, I.̀ P., Ayodele, T. R., Abolarin, S. M., & Ewim, D. R. E. (2023). Maximizing the cost effectiveness of electric power generation through the integration of distributed generators: Wind, hydro and solar power. *Bulletin of the National Research Centre, 47*(1), 166.

Iyiola, A. O., Afolabi, O. A., Alimi, S. K., Akingba, O. O., Izah, S. C., & Ogwu, M. C. (2024). Climate change and water crisis in the Global South. In S. C. Izah, M. C. Ogwu, A. Loukas, & H. Hamidifar (Eds.), *Water crises and sustainable management in the Global South* (pp. 111–140). Springer. https://doi.org/10.1007/978-981-97-4966-9_4

Iyiola, A.O., Akingba, O.O., Alimi, S.K., Akinsorotan, A.M., Osei, S.A., Idowu, C.F., Izah, S.C., and Ogwu, M.C. (2025a). Impact assessment of biobased products in the Global South. In M.C. Ogwu, S.C. Izah, J.G. Vazquez-Arenas, S.T. Feleke, and X. Wei (Eds.), *Sustainable bioeconomy development in the Global South: Volume II bioeconomy techniques*. Springer. doi:https://doi.org/10.1007/978-981-96-0305-3_16.

Iyiola, A. O., Akingba, O. O., Asiedu, B., Osei, S. A., Ogwu, M. C., & Morya, S. (2025b). Biobased value-added products and services from biological resources in the Global South. In M. C. Ogwu, S. C. Izah, H. Merritt, R. D. Wise, & A. Jintrawet (Eds.), *Sustainable bioeconomy development in the Global South: Volume I status and perspectives*. Springer. https://doi.org/10.1007/978-981-96-0640-5_9

Iyiola, A. O., Kolawole, A. S., Ajayi, F. O., Ogidi, O. I., & Ogwu, M. C. (2023). Sustainable water use and management for agricultural transformation in Africa. In S. Madhav, A. L. Srivastav, S. C. Izah, & E. van Hullebusch (Eds.), *Water resources management for rural development* (pp. 287–299). Elsevier. https://doi.org/10.1016/B978-0-443-18778-0.00014-3

Iyiola, A. O., & Ogwu, M. C. (2025). Environmental management using renewable biological resources in the Global South. In M. C. Ogwu, S. C. Izah, J. G. Vazquez-Arenas, S. T. Feleke, & X. Wei (Eds.), *Sustainable bioeconomy development in the Global South: Volume II bioeconomy techniques*. Springer. https://doi.org/10.1007/978-981-96-0305-3_11

Izah, S. C., Amabie, T., & Aigberua, A. O. (2024). Technological advancements for sustainable clean air. In M. C. Ogwu & S. C. Izah (Eds.), *Sustainable strategies for air pollution mitigation: Development, economics and technologies. Part of the book series: The handbook in environmental chemistry* (pp. 279–310). Springer. https://link.springer.com/chapter/10.1007/698_2024_1120

Izah, S. C., Iyiola, A. O., Yarkwan, B., & Richard, G. (2023). Impact of air quality as a component of climate change on biodiversity-based ecosystem services. In A. L. Srivastav, A. K. Dubey, A. Kumar, S. K. Narang, & M. A. Khan (Eds.), *Visualization techniques for climate change with machine learning and artificial intelligence* (pp. 123–148). Elsevier. isbn:978-0-323-99714-0.

Izah, S. C., & Ogwu, M. C. (2025). *Innovative approaches in environmental health management: Processes, technologies, and strategies for a sustainable future*. Springer Nature. https://doi.org/10.1007/978-3-031-81966-7

Izah, S. C., & Ohimain, E. I. (2013). The challenge of biodiesel production from oil palm feedstock in Nigeria. *Greener Journal of Biological Sciences, 3*(1), 1–12.

Jacobson, M. Z., Delucchi, M. A., Cameron, M. A., & Frew, B. A. (2015). Low-cost solution to the grid reliability problem with 100% penetration of intermittent wind, water, and solar for all purposes. *Proceedings of the National Academy of Sciences, 112*(49), 15060–15065.

Jenkins, K., McCauley, D., Heffron, R., Stephan, H., & Rehner, R. (2016). Energy justice: A conceptual review. *Energy Research and Social Science, 11*, 174–182.

Jenkins, K. E., Sovacool, B. K., Błachowicz, A., & Lauer, A. (2020). Politicising the just transition: Linking global climate policy, nationally determined contributions and targeted research agendas. *Geoforum, 115*, 138–142.

Jewell, J., McCollum, D., Emmerling, J., Bertram, C., Gernaat, D. E., Krey, V., Paroussos, L., Berger, L., Fragkiadakis, K., Keppo, I., & Saadi, N. (2018). Limited emission reductions from fuel subsidy removal except in energy-exporting regions. *Nature, 554*(7691), 229–233.

Jones, E. C., Jr., & Reyes, A. (2023). Identifying themes in energy poverty research: Energy justice implications for policy, programs, and the clean energy transition. *Energies, 16*(18), 6698.

Joshipura, M., Nasrallah, N., & Kedia, N. (2025). Green and climate finance for sustainability: Hybrid review and framework development. *Business Strategy and the Environment, 34*(3), 3618–3634.

Kang, L., Lv, J., & Zhang, H. (2024). Can the water resource fee-to-tax reform promote the "three-wheel drive" of corporate green energy-saving innovations? Quasi-natural experimental evidence from China. *Energies, 17*(12), 2866.

Kern, F., & Rogge, K. S. (2016). The pace of governed energy transitions: Agency, international dynamics and the global Paris agreement accelerating decarbonisation processes? *Energy Research and Social Science, 22*, 13–17.

Kettner, C., & Kletzan-Slamanig, D. (2020). Is there climate policy integration in European Union energy efficiency and renewable energy policies? Yes, no, maybe. *Environmental Policy and Governance, 30*(3), 141–150.

Kiesecker, J. M., Nagaraju, S. K., Oakleaf, J. R., Ortiz, A., Lavista Ferres, J., Robinson, C., Krishnaswamy, S., Mehta, R., Dodhia, R., Evans, J. S., & Heiner, M. (2023). The road to India's renewable energy transition must pass through crowded lands. *Land, 12*(11), 2049.

Köhler, J., Geels, F. W., Kern, F., Markard, J., Onsongo, E., Wieczorek, A., Alkemade, F., Avelino, F., Bergek, A., Boons, F., & Fünfschilling, L. (2019). An agenda for sustainability transitions research: State of the art and future directions. *Environmental Innovation and Societal Transitions, 31*, 1–32.

Koilo, V. (2022). The interplay between technological innovation, energy efficiency, and economic growth: Evidence from 30 European countries. *Problems and Perspectives in Managemen, 20*(3), 448–464.

Kovacic, Z., Musango, J. K., Buyana, K., Ambole, A., Smit, S., Mwau, B., Ogot, M., Lwasa, S., & Brent, A. (2021). Building capacity towards what? Proposing a framework for the analysis of energy transition governance in the context of urban informality in Sub-Saharan Africa. *Local Environment, 26*(3), 364–378.

Kühne, O., Parush, D., Shmueli, D., & Jenal, C. (2022). Conflicted energy transition—Conception of a theoretical framework for its investigation. *Land, 11*(1), 116.

Kunskaja, S., Bauer, J. F., Budzyński, A., & Jitea, I. C. (2023). A research analysis: The implementation of innovative energy technologies and their alignment with SDG 12. *Eastern-European Journal of Enterprise Technologies, 5*(13), 6–25.

Lai, K. P. (2025). Financial geography II–green finance and climate transition. *Progress in Human Geography, 49*(2), 215–226.

Lees, N. (2021). The Brandt Line after forty years: The more North–South relations change, the more they stay the same? *Review of International Studies, 47*(1), 85–106.

Li, T., & Liao, G. (2020). The heterogeneous impact of financial development on green total factor productivity. *Frontiers in Energy Research, 8*, 29.

Liao, H., Chen, Y. T., Chen, L., & Chen, J. F. (2022). Development of a bidirectional DC–DC converter with rapid energy bidirectional transition technology. *Energies, 15*(13), 4583.

Lin, J., Abhyankar, N., He, G., Liu, X., & Yin, S. (2022). Large balancing areas and dispersed renewable investment enhance grid flexibility in a renewable-dominant power system in China. *iScience, 25*(2), 103749.

Liu, M., Tang, L., Zeng, J., Huang, G., Liu, X., Yao, S., He, G., Shang, N., Tao, H., Ren, S., & Wang, P. (2024). Promoting decarbonization in China: Revealing the impact of various energy policies on the power sector based on a coupled model. *Energies, 17*(13), 3234.

Long, F., Xiao, Y., Wan, J., & Zhao, Q. (2023). "Renewable energy + energy storage" business model innovation — An emerging sustainable business model innovation. In *SHS web of conferences* (Vol. 163, p. 02007). EDP Sciences.

Lori, T., Osei-Kwarteng, M., Ojija, F., Ojo, A. O., & Ogwu, M. C. (2025). Participatory approaches to environmental health management. In S. C. Izah & M. C. Ogwu (Eds.), *Innovative approaches in environmental health management: Processes, technologies, and strategies for a sustainable future. Environmental science and engineering series*. Springer Nature. https://doi.org/10.1007/978-3-031-81966-7_11

Manowska, A., Bluszcz, A., Chomiak-Orsa, I., & Wowra, R. (2024). Towards energy transformation: A case study of EU countries. *Energies, 17*(7), 1778.

McCauley, D., & Heffron, R. (2018). Just transition: Integrating climate, energy and environmental justice. *Energy Policy, 119*, 1–7.

McCauley, D., Ramasar, V., Heffron, R. J., Sovacool, B. K., Mebratu, D., & Mundaca, L. (2019). Energy justice in the transition to low carbon energy systems: Exploring key themes in interdisciplinary research. *Applied Energy, 233*, 916–921.

Mohammadi, N., & Khabbazan, M. M. (2022). The influential mechanisms of power actor groups on policy mix adoption: Lessons learned from feed-in tariffs in the renewable energy transition in Iran and Germany. *Sustainability, 14*(7), 3973.

Mohan, P. S. (2023). Financing needs to achieve nationally determined contributions under the Paris agreement in Caribbean Small island developing states. *Mitigation and Adaptation Strategies for Global Change, 28*(5), 26.

Mrozowska, S., Wendt, J. A., & Tomaszewski, K. (2021). The challenges of Poland's energy transition. *Energies, 14*(23), 8165.

Mulia, A. H., Wukirasih, S., & Suryadinata, W. H. (2023). Whither just transition? A case study of Energy Transition Mechanism (ETM) country platform in Indonesia. *Global South Review, 5*(1), 31–46.

Müller, F., Neumann, M., Elsner, C., & Claar, S. (2021). Assessing African energy transitions: Renewable energy policies, energy justice, and SDG 7. *Politics and Governance, 9*(1), 119–130.

Muttitt, G., & Kartha, S. (2020). Equity, climate justice and fossil fuel extraction: Principles for a managed phase out. *Climate Policy, 20*(8), 1024–1042.

Nacke, L., Vinichenko, V., Cherp, A., Jakhmola, A., & Jewell, J. (2024). Compensating affected parties necessary for rapid coal phase-out but expensive if extended to major emitters. *Nature Communications, 15*(1), 3742.

Noorman, M., Espinosa Apráez, B., & Lavrijssen, S. (2023). AI and energy justice. *Energies, 16*(5), 2110.

Ogwu, M. C. (2019). Towards sustainable development in Africa: The challenge of urbanization and climate change adaptation. In P. B. Cobbinah & M. Addaney (Eds.), *The geography of climate change adaptation in Urban Africa* (pp. 29–55). Springer Nature. https://doi.org/10.1007/978-3-030-04873-0_2

Ogwu, M. C., El Malahi, S., & Izah, S. C. (2025a). Policy and regulatory frameworks for sustainable environmental practices. In M. C. Ogwu & S. Chibueze Izah (Eds.), *Evaluating environmental processes and technologies. Environmental science and engineering series*. Springer. https://doi.org/10.1007/978-3-031-85327-2_16

Ogwu, M. C., & Izah, S. C. (2025). *Evaluating environmental processes and technologies*. Springer Nature. https://doi.org/10.1007/978-3-031-85327-2

Ogwu, M. C., Izah, S. C., Merritt, H., Wise, R. D., & Jintrawet, A. (2025b). *Sustainable bioeconomy development in the Global South: Volume I status and perspectives*. Springer Singapore. https://doi.org/10.1007/978-981-96-0640-5

Ogwu, M. C., Izah, S. C., Vazquez-Arenas, J. G., Feleke, S. T., & Wei, X. (2025c). *Sustainable bioeconomy development in the Global South: Volume II bioeconomy techniques*. Springer Singapore. https://doi.org/10.1007/978-981-96-0305-3

Ogwu, M. C., & Kosoe, E. A. (2025a). Integrating green infrastructure into sustainable agriculture to enhance soil health, biodiversity, and microclimate resilience. *Sustainability, 17*(9), 3838. https://doi.org/10.3390/su17093838

Ogwu, M. C., & Kosoe, E. A. (2025b). Environmental economics: Valuing ecosystem services and green technologies. In M. C. Ogwu & S. Chibueze Izah (Eds.), *Evaluating environmental processes and technologies. Environmental science and engineering series*. Springer. https://doi.org/10.1007/978-3-031-85327-2_15

Ogwu, M. C., Kosoe, E. A., Alaka, A. C., & Xu, J. (2025d). Green infrastructure and environmental health security: The role of nature-based solutions in improving urban health. In S. C. Izah & M. C. Ogwu (Eds.), *Innovative approaches in environmental health management: Processes, technologies, and strategies for a sustainable future. Environmental science and engineering series*. Springer Nature. https://doi.org/10.1007/978-3-031-81966-7_9

Ohimain, E. I., & Izah, S. C. (2014). Potential of biogas production from palm oil mills' effluent in Nigeria. *Sky Journal of Soil Sciences and Environmental Management, 3*(5), 50–58.

Okkonen, L., & Lehtonen, O. (2016). Socio-economic impacts of community wind power projects in Northern Scotland. *Renewable Energy, 85*, 826–833.

Pingkuo, L., Pengbo, G., & Chen, Z. (2021). How to promote energy transition with market design: A review on China's electric power sector. *Frontiers in Energy Research, 9*, 709272.

Rodina, L. (2019). Risk management of the alternative power engineering. In *E3S web of conferences* (Vol. 124, p. 04024). EDP Sciences.

Rosenow, J., Kern, F., & Rogge, K. (2017). The need for comprehensive and well targeted instrument mixes to stimulate energy transitions: The case of energy efficiency policy. *Energy Research and Social Science, 33*, 95–104.

Saadaoui, H., & Chtourou, N. (2023). Does improvement in capital intensity facilitate the transition to renewable energies? Evidence from Tunisia. *Environmental Science and Pollution Research, 30*(18), 54059–54072.

Sahoo, S., Kumar, A., & Upadhyay, A. (2023). How do green knowledge management and green technology innovation impact corporate environmental performance? Understanding the role of green knowledge acquisition. *Business Strategy and the Environment, 32*(1), 551–569.

Sandria, A., Adji, K. D., & Prabowo, J. J. (2023). Exceptions for stop building coal-fired power plants in Indonesia. In *E3S web of conferences* (Vol. 467, p. 05007). EDP Sciences.

Si, Y., Kinol, A., Yi, H., & Hernández, D. (2025). Space for justice in sustainable development goal 7 (SDG7) discourse: Examining hashtag co-occurrence networks and energy justice messaging on social media. *Sustainable Development, 33*(3), 3941–3957.

Siciliano, G., Wallbott, L., Urban, F., Dang, A. N., & Lederer, M. (2021). Low-carbon energy, sustainable development, and justice: Towards a just energy transition for the society and the environment. *Sustainable Development, 29*(6), 1049–1061.

Simpao, F. T., & Espina, R. U. (2025). Thermal power plants in Mindanao: Trends, challenges, and opportunities. In *IOP conference series: Earth and environmental science* (Vol. 1513(1), p. 012016). IOP Publishing.

Skovgaard, J., Adams, K. M., Dupuy, K., Dzebo, A., Funder, M., Fejerskov, A. M., & Shawoo, Z. (2023). Multilateral climate finance coordination: Politics and depoliticization in practice. *Global Environmental Politics, 23*(2), 125–147.

Sohag, K., Begum, R. A., Abdullah, S. M. S., & Jaafar, M. (2015). Dynamics of energy use, technological innovation, economic growth and trade openness in Malaysia. *Energy, 90*, 1497–1507.

Sovacool, B. K., & Brisbois, M. C. (2019). Elite power in low-carbon transitions: A critical and interdisciplinary review. *Energy Research and Social Science, 57*, 101242.

Sovacool, B. K., Burke, M., Baker, L., Kotikalapudi, C. K., & Wlokas, H. (2017). New frontiers and conceptual frameworks for energy justice. *Energy Policy, 105*, 677–691.

Sovacool, B. K., Martiskainen, M., Hook, A., & Baker, L. (2019). Decarbonization and its discontents: A critical energy justice perspective on four low-carbon transitions. *Climatic Change, 155*(4), 581–619.

Sun, W., Li, R., Cai, R., Ji, Z., & Cheng, M. (2023a). The impact of solar energy investment in multilateral development banks on technological innovation: Evidence from a multi-period DID method. *Frontiers in Energy Research, 10*, 1085012.

Sun, Z., Zhang, F., Wang, Y., & Shao, Z. (2023b). Literature review and analysis of the social impact of a just energy transition. *Frontiers in Sustainable Food Systems, 7*, 1119877.

Taşkın, D., Vardar, G., & Okan, B. (2020). Does renewable energy promote green economic growth in OECD countries? *Sustainability Accounting, Management and Policy Journal, 11*(4), 771–798.

Telli, I., Kırısık, A., Quaranta, E., Kuriqi, A., Kasiulis, E., & Muntean, S. (2025). Hydropower's role in enhancing energy justice: Preliminary insights from the EU and Turkey. In *IOP conference series: Earth and environmental science* (Vol. 1442(1), p. 012009). IOP Publishing.

Tornel, C. (2023). Decolonizing energy justice from the ground up: Political ecology, ontology, and energy landscapes. *Progress in Human Geography, 47*(1), 43–65.

Tyson, L. D., & Weiss, D. (2025). Climate finance: Progress, challenges, and opportunities on the path to a sustainable planet. *California Management Review, 67*(2), 5–39.

Uri, I., Robinson, S. A., Roberts, J. T., Ciplet, D., Weikmans, R., & Khan, M. (2024). Equity and Justice in loss and damage finance: A narrative review of catalysts and obstacles. *Current Climate Change Reports, 10*(3), 33–45.

Van de Ven, D. J., Capellan-Peréz, I., Arto, I., Cazcarro, I., De Castro, C., Patel, P., & Gonzalez-Eguino, M. (2021). The potential land requirements and related land use change emissions of solar energy. *Scientific Reports, 11*(1), 2907.

Wang, L., He, Y., & Wu, R. (2024). Digitization meets energy transition: Shaping the future of environmental sustainability. *Energies, 17*(4), 767.

Wang, M., Li, Y., & Liao, G. (2021). Research on the impact of green technology innovation on energy total factor productivity, based on provincial data of China. *Frontiers in Environmental Science, 9*, 710931.

Wang, Y., & Chen, X. (2024). Impact mechanism of renewable energy technology innovation on carbon productivity based on spatial Durbin model. *Sustainability, 16*(5), 2100.

Xie, M., Zhao, S., & Lv, K. (2024). The impact of green finance and financial technology on regional green energy technological innovation based on the dual machine learning and spatial econometric models. *Energies, 17*(11), 2521.

Yang, Y., Xia, S., & Qian, X. (2023). Geopolitics of the energy transition. *Journal of Geographical Sciences, 33*(4), 683–704.

Yenneti, K., Day, R., & Golubchikov, O. (2016). Spatial justice and the land politics of renewables: Dispossessing vulnerable communities through solar energy mega-projects. *Geoforum, 76*, 90–99.

Zahedi, R., Zahedi, A., & Ahmadi, A. (2022). Strategic study for renewable energy policy, optimizations and sustainability in Iran. *Sustainability, 14*(4), 2418.

Zhang, M., Li, B., & Yin, S. (2020). Is technological innovation effective for energy saving and carbon emissions reduction? Evidence from China. *IEEE Access, 8*, 83524–83537.

Zhang, Z., Chen, G., Wu, J., Fang, P., Gu, X., Xiong, L., Tang, J., & Huang, P. (2025). Coal power and new energy collaborative low-carbon transformation model and path. In *Journal of physics: Conference series* (Vol. 3015(1), p. 012014). IOP Publishing.

Zhao, Q., & Tang, C. (2024). The impact of economic policy uncertainty on green technology innovation of new energy vehicle enterprises in China. *Sustainability, 16*(10), 4206.

Zhao, X., Xu, H., Yin, S., & Zhou, Y. (2023). Threshold effect of technological innovation on carbon emission intensity based on multi-source heterogeneous data. *Scientific Reports, 13*(1), 19054.

Zhao, X., & Zhao, J. (2024). Digital finance and inequality in renewable energy technology innovation. *Energy and Environment, 35*(7), 3673–3702.

Zhou, S., Gao, X., Wellstead, A. M., & Kim, D. M. (2023). Operationalizing social equity in public policy design: A comparative analysis of solar equity policies in the United States. *Policy Studies Journal, 51*(4), 741–772.

Zografos, C., & Robbins, P. (2020). Green sacrifice zones, or why a green new deal cannot ignore the cost shifts of just transitions. *One Earth, 3*(5), 543–546.